Stochastic Methods for Flow in Porous Media
Coping with Uncertainties

Stochastic Methods for Flow in Porous Media

Coping with Uncertainties

Dongxiao Zhang

Earth and Environmental Sciences Division
Los Alamos National Laboratory
Los Alamos, NM 87545, USA

ACADEMIC PRESS

A Harcourt Science and Technology Company

San Diego   San Francisco   New York
Boston   London   Sydney   Tokyo

Copyright © 2002 by ACADEMIC PRESS

Academic Press
A division of Harcourt Inc.
525 B Street, Suite 1900, San Diego, California 92101-4495, USA
http://www.academicpress.com

Academic Press
A division of Harcourt Inc.
Harcourt Place, 32 Jamestown Road, London NW1 7BY, UK
http://www.academicpress.com

ISBN 0-12-779621-5

Library of Congress Catalog Number: 2001089406

A catalogue record of this book is available from the British Library

Printed and bound by CPI Group (UK) Ltd, Croydon, CR0 4YY
Transferred to Digital Print 2011

Typeset by Newgen Imaging Systems (P) Ltd., Chennai, India

CONTENTS

Colour plate sections between pages 211 and 212 and 256 and 257.

FOREWORD

This book deals with issues of fluid flow in complex geologic environments under uncertainty. The resolution of such issues is important for the rational management of water resources, the preservation of subsurface water quality, the optimization of irrigation and drainage efficiency, the safe and economic extraction of subsurface mineral and energy resources, and the subsurface storage of energy and wastes.

Hydrogeologic parameters such as permeability and porosity have been traditionally viewed as well-defined local quantities that can be assigned unique values at each point in space. Yet subsurface fluid flow takes place in a complex geologic environment whose structural, lithologic and petrophysical characteristics vary in ways that cannot be predicted deterministically in all of their relevant details. These characteristics tend to exhibit discrete and continuous variations on a multiplicity of scales, causing flow parameters to do likewise. In practice, such parameters can at best be measured at selected well locations and depth intervals, where their values depend on the scale (support volume) and mode (instrumentation and procedure) of measurement. Estimating the parameters at points where measurements are not available entails a random error. Quite often, the support of measurement is uncertain and the data are corrupted by experimental and interpretive errors. These errors and uncertainties render the parameters random and the corresponding flow equations stochastic.

The recognition that geology is complex and uncertain has prompted the development of geostatistical methods to help reconstruct it on the basis of limited data. The most common approach is to view parameter values, determined at various points within a more-or-less distinct hydrogeologic unit, as a sample from a spatially correlated random field defined over a continuum. This random field is characterized by a joint (multivariate) probability density function or, equivalently, its joint ensemble moments. The field fluctuates randomly from point to point in the hydrogeologic unit and from one realization to another in probability space. Its spatial statistics are obtained by sampling the field in real space across the unit, and its ensemble statistics are defined in terms of samples collected in probability space across multiple random realizations. Geostatistical analysis consists of inferring such statistics (most commonly the two leading ensemble moments, mean and variance-covariance) from a discrete set of measurements at various locations within the hydrogeologic unit.

Once the statistical properties of relevant random parameters have been inferred from data, the next step is to solve the corresponding stochastic flow equations. This is the subject of the present book. Following a lucid introduction to the theory of correlated random fields, the book details a number of methods for the solution of stochastic flow problems under steady state and transient, single- and two-phase conditions in porous and fractured media. The most common approach is to solve such stochastic flow equations numerically by Monte Carlo simulation. This entails generating numerous equally likely random realizations of the parameter fields, solving a deterministic flow equation for each realization by standard numerical methods, and averaging the results to obtain sample moments of the solution. The approach is conceptually straightforward and has the advantage of applying to a very broad range of both linear and nonlinear flow problems. It however has a number of conceptual and computational drawbacks. The book therefore focuses more heavily on direct methods of solution, which allow one to compute leading statistical moments of hydrogeologic variables, such as fluid pressure and flux, without having to generate multiple realizations of these variables.

One direct approach is to write a system of partial differential equations satisfied approximately by leading ensemble moments and to solve them numerically. Though the approach has been known for some time, the book emphasizes its recent application to statistically nonhomogeneous media in which the moments of hydrogeologic parameters, most notably permeability, vary across the field. Such nonhomogeneity may arise from systematic spatial variability of the parameters, proximity to sources and boundaries, and conditioning on measured parameter values. The corresponding partial differential moment equations are derived in a straightforward manner and lend themselves to solution by standard finite difference methods. They form the basis for most applications and computational examples described in the book.

Another direct approach is to write exact or approximate integro-differential equations for moments of interest. Exact integro-differential moment equations have been developed in recent years for steady state and transient flows in saturated porous media and for steady state flow in unsaturated soils in which hydraulic conductivity varies exponentially with capillary pressure head (as well as for advective-dispersive solute transport in random velocity fields). In addition to being mathematically rigorous and elegant due to their exact and compact nature, they are extremely useful in revealing the nonlocal nature of stochastic moment solutions, the effect of information content (scale, quantity and quality of data) on these solutions, the conditions under which nonlocal integro-differential formulations can be localized to yield approximate partial differential moment equations, the nature and properties of corresponding local effective parameters, the relationship between localized moment equations and standard deterministic partial-differential equations of flow (and transport), and the implications of this relationship *vis-a-vis* the application of standard deterministic models to randomly heterogeneous media under uncertainty. The integro-differential approach relies on Green's functions, which are independent of internal sources and the magnitudes of boundary terms. Once these functions have been computed for a given boundary configuration, they can be used repeatedly to obtain solutions for a wide range of internal sources and boundary terms. The book focuses on the mechanics of how exact integro-differential moment equations are derived, approximated and solved numerically by finite elements. It points out that numerical solutions based on partial-differential and integro-differential moment formulations must ultimately be similar. Computational examples demonstrating the accuracy of the integro-differential approach when applied to complex flow problems in strongly heterogeneous media may be found in the cited literature.

The hydrogeologic properties of natural rocks and soils exhibit spatial variations on a multiplicity of scales. Incorporating such scaling in geostatistical and stochastic analyses of hydrogeologic phenomena has become a major challenge. The book provides a brief but useful introduction to this fascinating subject together with some key references, which the reader is encouraged to explore.

Students, teachers, researchers and practitioners concerned with hydrogeologic uncertainty analysis will find much in this book that is instructive, useful and timely.

Shlomo P. Neuman   May 2, 2001
Regents' Professor
Department of Hydrology and Water Resources
University of Arizona, Tucson

PREFACE

It has now been well recognized that flow in porous media is strongly influenced by medium spatial variabilities and is subject to uncertainties. Since such a situation cannot be accurately modeled deterministically without considering the uncertainties, it has become quite common to approach the subsurface flow problem stochastically. This book aims to systematically introduce a number of stochastic methods for describing flow in complex porous media under uncertainties. The fundamentals of the various stochastic methods are given in a tutorial way so that no prior in-depth knowledge of stochastic processes is needed to comprehend these methods. Among the methods discussed are perturbative expansion, Green's function method, spectral method, adjoint state method, Adomian decomposition, closure approximations, and Monte Carlo simulations. Some emerging techniques, such as renormalization, renormalization group, and Feynman diagrams, are briefly introduced. The potential and limitations are discussed for each method, and solution techniques and illustrative examples are presented for some selected ones. The types of flow discussed range from steady-state to transient flow, from saturated, unsaturated to two-phase flow, and from nonfractured to fractured porous flow. It is hoped that after studying the various stochastic methods the reader will be able to determine appropriate methods and use them for the problems of his/her interest, which may not necessarily be those of flow in porous media. For the reader who is interested in having hands-on experience with the methods and applying them to real problems, a number of computer codes developed on the basis of the algorithms discussed in the book are available from the author upon request.

This book differs from others in this area in a number of ways. First, it surveys a broad range of approaches used by researchers in the community, not restricted to just some particular ones. Second, unlike others that mainly deal with stationary flow, this book presents methodologies for both stationary and nonstationary flows. The flow nonstationarity may stem from finite domain boundaries, complex flow configurations (e.g., fluid pumping and injecting), and medium multiscale, nonstationary features (e.g., distinct geological layers, zones, and facies), all of which are important factors to consider in real-world applications. Third, most of the material in the book is presented in a tutorial way so that it is accessible to those without prior in-depth knowledge of stochastic processes. Some stochastic methods are compared in terms of their characteristics and illustrated with examples, a number of well-known results and useful formulae are compiled, and a few future research topics are discussed.

Although this book is mainly a scientific monograph, it can also be used as a textbook for graduate students and perhaps upper-level undergraduates as well. For this purpose, a few exercises are included at the end of each chapter. These exercises are designed to either help understand some techniques introduced in the text or supplement the material conveyed in that chapter.

I hope that this book is useful to graduate students, scientists, and professionals in the fields of hydrology, petroleum engineering, soil physics, geological engineering, agriculture engineering, civil and environmental engineering, and applied mathematics.

ACKNOWLEDGMENTS

I would like to thank Professor Shlomo P. Neuman at the University of Arizona for introducing me to the fascinating field of stochastic hydrogeology in 1991. Since then Shlomo has truly been my "father-in-school" (a translation of the combination of "mentor" and "teacher" in Chinese). Without his constant encouragement, guidance, and friendship, my learning and research experience in the field would never have been the same.

I also want to thank the following colleagues who have either collaborated with me on the topics of the book or reviewed portions of the book and helped to improve it: Kathy Campbell, Shiyi Chen, Thomas Harter, Kuo-Chin Hsu, Bill Hu, Liyong Li, Guoping Lu, Zhiming Lu, Shlomo Neuman, Shlomo Orr, Rajesh Pawar, Alex Sun, Hamdi Tchelepi, John Wilson, Larry Winter, Jichun Wu, and You-Kuan Zhang.

Last but not least, the arduous task of writing this book would not be possible without the kind support and encouragement of my family. I wish to dedicate this monograph to them – my wife Liheng, my children Benjamin and Grace, and my mother Jieqiu Gong.

1

INTRODUCTION

1.1 STOCHASTIC PARTIAL DIFFERENTIAL EQUATIONS

Although geological formations are intrinsically deterministic, we usually have incomplete knowledge of their properties. Formation material properties, including fundamental parameters such as permeability and porosity, are ordinarily observed at only a few locations despite the fact that they exhibit a high degree of *spatial variability* at all length scales. This combination of significant spatial heterogeneity with a relatively small number of observations leads to *uncertainty* about the values of the formation properties and thus, to uncertainty in estimating or predicting flow in such formations.

While uncertainty in the values of properties can be reduced by improved geophysical techniques, it can never be entirely eliminated. When computational models of flow are used to manage and predict groundwater resources or to assess the benefits of petroleum reservoir development, the degree of uncertainty in the prediction must be quantified in terms of uncertainty in formation parameters.

The theory of *stochastic processes* provides a natural method for evaluating uncertainties. In the stochastic formalism, uncertainty is represented by *probability* or by related quantities like *statistical moments*. Since material parameters such as permeability and porosity are not purely random, they are treated as *random space functions* (RSFs) whose variabilities exhibit some *spatial correlation structures*. The spatial correlations may be quantified by *joint* (multivariate and/or multipoint) *probability distributions* or *joint statistical moments* such as *cross-* and *auto-covariances*. Many terms (e.g., RSF, probability distribution, and covariance) are given either with loose definitions or without definitions in this chapter but will be defined more rigorously in Chapter 2.

Let us use steady-state, single-phase fluid flow as an example. The flow satisfies the following continuity equation and Darcy's law [e.g., Bear, 1972; Freeze and Cherry, 1979; de Marsily, 1986],

$$\nabla \cdot \mathbf{q}(\mathbf{x}) = g(\mathbf{x}) \tag{1.1}$$

$$q_i(\mathbf{x}) = -K_S(\mathbf{x}) \frac{\partial h(\mathbf{x})}{\partial x_i} \tag{1.2}$$

1

subject to boundary conditions

$$h(\mathbf{x}) = H_B(\mathbf{x}), \quad \mathbf{x} \subset \Gamma_D$$
$$\mathbf{q}(\mathbf{x}) \cdot \mathbf{n}(\mathbf{x}) = Q(\mathbf{x}), \quad \mathbf{x} \in \Gamma_N$$

$$(1.3)$$

where $\mathbf{x} = (x_1, \ldots, x_d)^T$ is the vector of space coordinates (where d is the number of space dimensions and T indicates transpose), $\mathbf{q}(x) = (q_1, \ldots, q_d)^T$ is the specific discharge (flux) vector at point $\mathbf{x}$, ∇ is the grad operator with respect to $\mathbf{x}$, $h(\mathbf{x})$ is the hydraulic head, $K_S(\mathbf{x})$ is the saturated hydraulic conductivity, $g(\mathbf{x})$ is the source/sink term (due to recharge, pumping, or injecting; positive for source and negative for sink), $H_B(\mathbf{x})$ is the prescribed head on Dirichlet boundary segments Γ_D, $Q(\mathbf{x})$ is the prescribed flux across Neumann boundary segments Γ_N, and $\mathbf{n}(\mathbf{x}) = (n_1, \ldots, n_d)^T$ is an outward unit vector normal to the boundary.

Equations (1.1)–(1.3) may be solved to yield a deterministic set of values for the dependent variables $h(\mathbf{x})$ and $\mathbf{q}(\mathbf{x})$ if the hydraulic conductivity $K_S(\mathbf{x})$ is specified at each point in the domain (or, at each node on a numerical grid) and if the boundary conditions are given. The quality of the modeling results will strongly depend on how well the natural spatial variability of the K_S field is accounted for. Illustrated in Fig. 1.1 are such variabilities of permeability and porosity, measured based on core samples from a borehole in a sandstone aquifer [Bakr, 1976]. Other examples of medium spatial variabilities are reported by Hoeksema and Kitanidis [1985a] and Sudicky [1986]. However, as the rule rather than the exception, only limited measurements are available at a few locations (boreholes) because of the high cost associated with subsurface measurements. Hence, there is uncertainty about the conductivity (permeability) values at points between (sparse) measurements. In addition, another type of uncertainty may arise from measurement and/or interpretation errors.

When the hydraulic conductivity field $K_S(\mathbf{x})$ is treated as a random space function (or, a *spatial stochastic process*), the dependent variables $h(\mathbf{x})$ and $\mathbf{q}(\mathbf{x})$ also become random space functions. In turn, Eqs. (1.1) and (1.2) become *stochastic differential equations* whose solutions are no longer deterministic numbers but probability distributions of the dependent variables.

Like the hydraulic conductivity field, the boundary conditions and the forcing term may also be treated as random variables or random space functions because of spatial variabilities, measurement errors, or incomplete information. For example, for the situation of a *confined aquifer* bounded by two streams, the boundary conditions have to be treated as random variables either if the stream stages are influenced by some external force (e.g., precipitation) in a random fashion, or if the record of the stages is not available for the particular year of interest. Recharge $g(\mathbf{x})$ into an *unconfined aquifer* may be modeled as a random space function when there is not enough information to deterministically describe its spatial variability. Another example is that when one attempts to forecast the pumping or injecting rate $g(\mathbf{x}_o)$ at a particular well, it may be more appropriate to express $g(\mathbf{x}_o)$ as a random variable than to assign a certain value to it.

In summary, the governing flow equations become stochastic partial differential equations (PDEs) in any or some combinations of the following cases: (1) When the material properties such as hydraulic conductivity and porosity are treated as spatial stochastic processes; (2) when boundary and/or initial conditions are prescribed as random variables

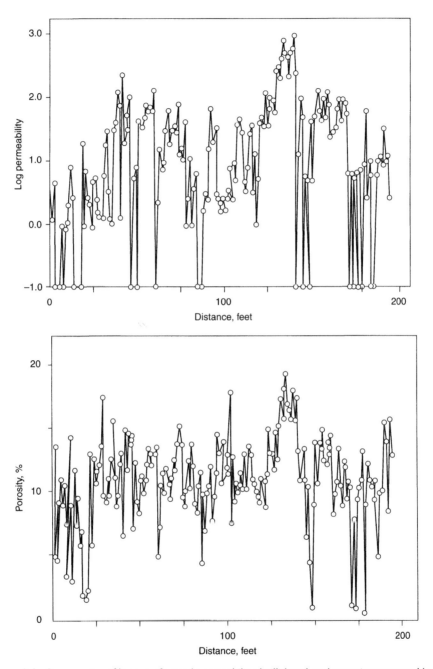

Figure 1.1. Space series of log transformed permeability (millidarcy) and porosity measured based on core samples from a borehole in Mt. Simon sandstone aquifer in Illinois. [Adapted from Bakr, 1976.]

or random spatial functions; and (3) when the forcing term is best considered as a stochastic process. Because the solutions of stochastic PDEs are no longer deterministic values but are statistical quantities, the techniques for solving these equations may be significantly different than those for solving deterministic PDEs. In the rest of this chapter, we illustrate some techniques for solving stochastic differential equations based on one-dimensional examples of saturated and unsaturated flows in porous media.

1.2 SATURATED FLOW WITH RANDOM FORCING TERMS

In this section, we consider the case of groundwater flow subject to recharge, as illustrated in Fig. 1.2, where the hydraulic conductivity $K_S(\mathbf{x})$ is assumed to be uniform and known with certainty and the boundary conditions are also specified with certainty. Hence, for the situation of flow in a one-dimensional domain with only prescribed heads the governing equation can be rewritten from Eqs. (1.1)–(1.3) as

$$\frac{d^2 h(x)}{dx^2} = w(x) \tag{1.4}$$

subject to

$$
\begin{aligned}
h(x) &= H_o, \quad x = 0 \\
h(x) &= H_L, \quad x = L
\end{aligned}
\tag{1.5}
$$

In the above, $w(x) = -g(x)/K_S$ is a forcing term accounting for recharge, fluid pumping, or injecting. In many situations, the forcing term could not be described deterministically because of incomplete information, the inaccuracy of measurements, or the attempt of making forecast. Hence, one may treat $w(x)$ statistically. In turn, $w(x)$ is described by its *probability density function* (PDF) or the corresponding statistical moments. For w at point

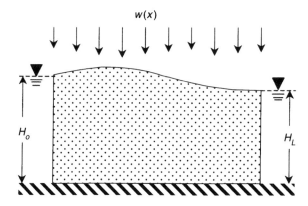

Figure 1.2. Sketch of groundwater flow subject to random recharge.

x_1, the *expected value* $\langle w(x_1) \rangle$, the *mean square value* $\langle w^2(x_1) \rangle$ and other *higher moments* can be given as

$$\langle w^m(x_1) \rangle = \int [w(x_1)]^m p_1[w(x_1)] \, dw(x_1) \tag{1.6}$$

where m is a nonnegative integer, $\langle \ \rangle$ indicates *expectation*, and $p_1[w(x_1)] \, dw(x_1)$ is the probability that $w(x_1)$ lies between $w(x_1)$ and $w(x_1) + dw(x_1)$. However, the joint PDF $p_N[w(x_1), w(x_2), \ldots, w(x_N)]$ as $N \to \infty$ is required to completely characterize the forcing term $w(x)$ at all points in space. With this PDF and through appropriate integrations, one obtains $\langle w^m(x_i) \rangle$, $\langle w^m(x_i)w^l(x_j) \rangle$, and higher moments. In particular, the *two-point moments* $\langle w^m(x_i)w^l(x_j) \rangle$ are given as

$$\langle w^m(x_i)w^l(x_j) \rangle = \iint [w(x_i)]^m [w(x_j)]^l p_2[w(x_i), w(x_j)] \, dw(x_i) \, dw(x_j) \tag{1.7}$$

where $p_2[w(x_i), w(x_j)]$ is the *two-point joint* probability density function, and $p_2[w(x_i), w(x_j)] \, dw(x_i) \, dw(x_j)$ is the probability that $w(x_i)$ lies between $w(x_i)$ and $w(x_i) + dw(x_i)$ and that $w(x_j)$ lies between $w(x_j)$ and $w(x_j) + dw(x_j)$.

For a *continuous random field* such as $w(x)$ which requires an infinite number of points to completely describe its statistical structure, it is mathematically more concise to define the PDF in a *functional form* [e.g., Beran, 1968]. In that form, $p[w(x)]$ is used to denote $p_N[w(x_1), w(x_2), \ldots, w(x_N)]$ for $N \to \infty$. In such a notation, p is a function of the function $w(x)$ and x is a continuous variable.

Since $w(x)$ is a stochastic process, so is the hydraulic head $h(x)$ in Eq. (1.4). Thus, $h(x)$ can only be described statistically. The complete solution of $h(x)$ is the PDF $p[h(x)]$, which stands for $p_M[h(x_1), h(x_2), \ldots, h(x_M)]$ with $M \to \infty$, as a function of the input PDF $p[w(x)]$. That is, the solution is either the conditional PDF $p[h(x)|w(x)]$, or the joint PDF $p[h(x), w(x)]$.

In principle, it is possible to derive an equation governing the joint PDF $p[h(x), w(x)]$ or the conditional PDF $p[h(x)|w(x)]$ from the original stochastic differential equations (1.4) and (1.5) by the so-called *PDF methods*. PDF methods include but are not limited to the *characteristic functional method* [e.g., Beran, 1968] and the *PDF transport equation method* [e.g., Pope, 1985, 1994]. The $p[h(x), w(x)]$ equation would contain all information about the statistical nature of the head field if it could be solved in terms of the input PDF $p[w(x)]$. However, it is usually difficult to solve *PDF equations* directly. In the field of turbulent flows, much progress has recently been made in solving PDF equations with the help of *Monte Carlo simulations* [e.g., Pope, 1985, 1994] and through *some closure approximations* such as the *mapping closure* of Chen et al. [1989]. To the writer's knowledge, the PDF methods have not been used, in any significant way, for studying flow in porous media. Furthermore, it is almost always impossible to obtain complete information about the functional PDF of the input random fields, such as $w(x)$ in this case and hydraulic conductivity $K(x)$ in other cases, from limited measurements. Instead, only the first few moments of these variables are usually available. Therefore, in this book we will not attempt to derive the PDF equations or directly obtain the PDFs of dependent variables, except for some very special cases. As an alternative, we aim to derive equations governing the statistical moments of the dependent variables and solve the *moment equations* in terms of the first few moments of the input random variables.

1.2.1 Moment Differential Equations

In this section, we show how to directly derive equations governing the statistical moments of $h(x)$. For Eqs. (1.4) and (1.5), the procedure for deriving moment equations in terms of the moments of $w(x)$ is rather straightforward. Taking ensemble expectation on both sides of Eqs. (1.4) and (1.5) yields

$$\frac{d^2 \langle h(x) \rangle}{dx^2} = \langle w(x) \rangle \tag{1.8}$$

subject to

$$\begin{aligned} \langle h(x) \rangle &= H_o, \quad x = 0 \\ \langle h(x) \rangle &= H_L, \quad x = L \end{aligned} \tag{1.9}$$

In the above, we have utilized the property that differentiation (with respect to x) and expectation are *commutative*, namely,

$$\left\langle \frac{dh(x)}{dx} \right\rangle = \frac{d \langle h(x) \rangle}{dx} \tag{1.10}$$

This property follows from the definition of a derivative,

$$\begin{aligned} \left\langle \frac{dh(x)}{dx} \right\rangle &= \left\langle \lim_{\Delta x \to 0} \frac{h(x + \Delta x) - h(x)}{\Delta x} \right\rangle \\ &= \lim_{\Delta x \to 0} \frac{\langle h(x + \Delta x) \rangle - \langle h(x) \rangle}{\Delta x} = \frac{d \langle h(x) \rangle}{dx} \end{aligned} \tag{1.11}$$

if $h(x)$ is *mean square differentiable* (see Section 2.4.3 for details). It is clear that the same applies to a partial derivative.

The two-point moments $\langle h(x)h(y) \rangle$ can be obtained by performing the self-product of Eqs. (1.4) and (1.5) applied at two different locations x and y, and taking expectation. The resulting equation reads as

$$\frac{\partial^4 \langle h(x)h(y) \rangle}{\partial x^2 \, \partial y^2} = \langle w(x)w(y) \rangle \tag{1.12}$$

subject to

$$\begin{aligned} \langle h(x)h(y) \rangle &= H_o^2, \quad && x = 0, \;\; y = 0 \\ \langle h(x)h(y) \rangle &= H_o H_L, \quad && x = 0, \;\; y = L \\ \langle h(x)h(y) \rangle &= H_o H_L, \quad && x = L, \;\; y = 0 \\ \langle h(x)h(y) \rangle &= H_L^2, \quad && x = L, \;\; y = L \end{aligned} \tag{1.13}$$

Equations governing higher moments of $h(x)$ can be given in a similar manner. It is of interest to note that for the case under investigation, the equations governing the first two moments (and higher moments) are derived without invoking any approximation. However, this may not be true for most of the cases presented in the book. It is also seen that for this special case, $\langle h(x) \rangle$ is only a function of $\langle w(x) \rangle$, and $\langle h(x)h(y) \rangle$ does not depend on any other moments but is only a function of $\langle w(x)w(y) \rangle$. One may prove that this pattern is also true for higher moments in this special case. Again, this nice property may or may not hold for other cases.

The first moment $\langle h(x) \rangle$ can be solved from Eqs. (1.8) and (1.9) in terms of $\langle w(x) \rangle$, and the second moment $\langle h(x)h(y) \rangle$ from Eqs. (1.12) and (1.13) in terms of $\langle w(x)w(y) \rangle$. The first moment $\langle h(x) \rangle$ estimates the random head field $h(x)$. The second moment $\langle h(x)h(y) \rangle$ is related to the covariance function of $h(x)$, defined as

$$C_h(x, y) \equiv \langle h'(x)h'(y) \rangle = \langle h(x)h(y) \rangle - \langle h(x) \rangle \langle h(y) \rangle \tag{1.14}$$

where $h'(x) = h(x) - \langle h(x) \rangle$ is the *zero-mean fluctuation* of $h(x)$. The covariance function measures the spatial correlation structure of the head field. When $x = y$, the covariance reduces to the *variance* $\sigma_h^2(x)$, which is a measure of the *magnitude of variability* of h at x. Similarly, the covariance function $C_w(x, y) = \langle w(x)w(y) \rangle - \langle w(x) \rangle \langle w(y) \rangle$ and the variance $\sigma_w^2(x) = \langle w^2(x) \rangle - \langle w(x) \rangle^2$ measure the spatial structure and the magnitude of variability of the random forcing term $w(x)$, respectively.

1.2.2 Moment Integral Equations

The moment equations (1.8), (1.9) and (1.12), (1.13) are differential equations. Statistical moments can also be derived from *integral equations*. This is generally done with the aid of *Green's function*. The solution for Eqs. (1.4) and (1.5) reads as [Cheng and Lafe, 1991]

$$h(x) = \int_\Omega G(x, \chi)w(\chi) \, d\chi + T_{bc} \tag{1.15}$$

where Ω is the interior of the domain, $G(x, \chi)$ is Green's function for *Laplace equation* of $d^2h/dx^2 = 0$ subject to *homogeneous boundary conditions* (i.e., $h(x) = 0$ at $x = 0$ and L), and T_{bc} is an integral accounting for the boundaries. Note that $G(x, \chi)$ is a deterministic function and that the form of T_{bc} will be given in Section 1.4.3. For this special case, the solution of $h(x)$ can be given explicitly by the following stochastic integral expression [Cheng and Lafe, 1991]:

$$h(x) = H_o + \int_0^x \int_0^\chi w(\chi') \, d\chi' \, d\chi$$
$$+ \frac{x}{L} \left[H_L - H_o - \int_0^L \int_0^\chi w(\chi') \, d\chi' \, d\chi \right] \tag{1.16}$$

Since $w(x)$ is random, so is $h(x)$.

The expected value $\langle h(x) \rangle$ can be obtained by directly taking expectation of Eq. (1.16),

$$\langle h(x) \rangle = \bar{H}_o + \int_0^x \int_0^\chi \langle w(\chi') \rangle \, d\chi' \, d\chi$$

$$+ \frac{x}{L} \left[H_L - H_o - \int_0^L \int_0^\chi \langle w(\chi') \rangle \, d\chi' \, d\chi \right] \qquad (1.17)$$

where we have used the property that integration and expectation are commutative. One may prove this commutative property in a way similar to Eq. (1.11). It should be easy to verify that Eq. (1.17) is indeed the solution of Eqs. (1.8) and (1.9). With Eq. (1.17), one obtains the head fluctuation $h'(x)$ by subtracting it from Eq. (1.16),

$$h'(x) = \int_0^x \int_0^\chi w'(\chi') \, d\chi' \, d\chi - \frac{x}{L} \int_0^L \int_0^\chi w'(\chi') \, d\chi' \, d\chi \qquad (1.18)$$

Hence, one obtains the covariance equation by multiplying $h'(x)$ with $h'(y)$ rewritten from Eq. (1.18) at a different location y and taking expectation,

$$C_h(x, y) = \int_0^x \int_0^y \int_0^\chi \int_0^\eta C_w(\chi', \eta') \, d\eta' \, d\chi' \, d\eta \, d\chi$$

$$- \frac{x}{L} \int_0^L \int_0^y \int_0^\chi \int_0^\eta C_w(\chi', \eta') \, d\eta' \, d\chi' \, d\eta \, d\chi$$

$$- \frac{y}{L} \int_0^x \int_0^L \int_0^\chi \int_0^\eta C_w(\chi', \eta') \, d\eta' \, d\chi' \, d\eta \, d\chi$$

$$+ \frac{xy}{L^2} \int_0^L \int_0^L \int_0^\chi \int_0^\eta C_w(\chi', \eta') \, d\eta' \, d\chi' \, d\eta \, d\chi \qquad (1.19)$$

Higher moments can be derived similarly. Like the moment differential equations, the integral equations derived for the head moments are *exact* in this case.

So far, we have not made any assumption about the statistical properties of w. Thus, w may take any distributional form. However, many or even an infinite number of moments are required to characterize some forms of the joint PDF $p[w(x)]$. In turn, it is likely that one needs to compute other moments beyond the mean $\langle h(x) \rangle$ and the covariance $C_h(x, y)$ in order to better characterize the random field $h(x)$. If $w(x)$ is a *Gaussian* random field, the first two moments $\langle w(x) \rangle$ and $C_w(x, y)$ completely characterize the joint PDF of $w(x)$. Then, the salient question is whether $h(x)$ is Gaussian. This question may be partially answered by evaluating higher moments or can be fully addressed by seeking the PDF of $h(x)$ with a PDF approach mentioned earlier. However, either has seldom been done in the literature of flow in porous media. The common practice is to concentrate on the first two moments of dependent variables without knowing the distributional forms. This practice is justified because the first two moments often suffice to approximate *confidence intervals* for a random field if it is not far from being Gaussian.

The problem is usually further simplified by invoking the assumption of *second-order* (or *weak*) *stationarity* (or *statistical homogeneity*). This assumption requires the mean and the variance to be constant in space and the covariance to be only dependent on the *separation vector*. With this requirement, the information needed to characterize the random forcing term $w(x)$ reduces to $\langle w(x) \rangle \equiv \langle w \rangle$ and $C_w(r)$ where $r = x - y$. This is a desirable property in that the input statistics are less demanding. However, one should be careful in invoking this assumption. If a random field is strongly *nonstationary* (location dependent), a *forced stationarity* will significantly exaggerate the variability of the input variable and thus result in excessive variability and uncertainty in the prediction. We will elaborate on this in Section 2.3.3.

1.2.3 Solutions of Statistical Moments

Under the condition of second-order stationary $w(x)$, the expected value of $h(x)$ is given from Eq. (1.17) as

$$\langle h(x) \rangle = H_o + \frac{x}{L}(H_L - H_o) - \frac{\langle w \rangle}{2}(L - x)x \tag{1.20}$$

The covariance $C_h(x, y)$ can be evaluated from Eq. (1.19) after specifying the forcing term covariance $C_w(r)$. A common type of covariance function is the *exponential model*:

$$C_w(r) = \sigma_w^2 \exp\left(-\frac{|r|}{\lambda_w}\right) \tag{1.21}$$

where λ_w is a parameter characterizing the *correlation length* of w. Other covariance functions will be introduced in Section 2.3.2. Though tedious it should be relatively straightforward to obtain an explicit expression of $C_h(x, y)$ for the exponential or any given $C_w(r)$.

For the sake of illustration, we discuss a special case, in which recharge is fully correlated in space. This case follows from Eq. (1.21) by setting $\lambda_w \to \infty$ so that $C_w(r) = \sigma_w^2$ [Cheng and Lafe, 1991]. In this simple case, integrating Eq. (1.19) yields

$$C_h(x, y) = \frac{\sigma_w^2}{4}xy(L - x)(L - y)$$
$$\tag{1.22}$$
$$\sigma_h^2(x) = \frac{\sigma_w^2}{4}x^2(L - x)^2$$

Figure 1.3 shows the mean head $\langle h \rangle$ and the normalized head variance $\sigma_h^2/(\sigma_w^2 L^4)$ as a function of x/L, for $\langle w \rangle = -0.1\,[1/L]$, 0 and $0.1\,[1/L]$, where L in $[\ ]$ denotes an arbitrary length unit. By recalling that $w = -g/K_S$ where K_S is the hydraulic conductivity, we see that a negative value of w stands for recharge and a positive one for discharge. We set $L = 10\,[L]$, $\sigma_w = 0.1\,[1/L]$, $H_o = 7\,[L]$, and $H_L = 5\,[L]$. The recharge/discharge is assumed to be fully correlated in space. Compared to the case that the mean recharge is zero, a recharge (discharge) increases (decreases) the hydraulic head in the domain. However, the

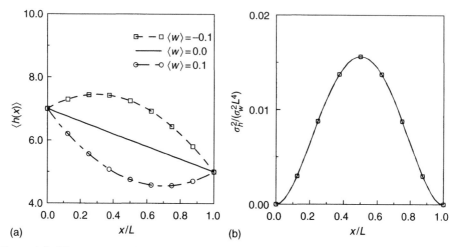

Figure 1.3. The expected value and variance of head in one-dimensional domain subject to random recharge.

magnitude of the mean recharge has no effect on the head variance in the case of uniform hydraulic conductivity and deterministic boundary heads. It is seen from Fig. 1.3b that the variance is zero at the two boundary ends and is the largest at the center of the domain. It is seen from Eq. (1.22) that the magnitude of the head variance is proportional to $\sigma_w^2 L^4$. Thus, the head variance is unbounded for an infinite domain (i.e., $L \to \infty$). In practical terms, this means that it is increasingly more difficult to predict or estimate the head field under a random recharge as the domain size increases. It is obvious that the prediction uncertainty increases with the magnitude of σ_w^2.

Figure 1.4 illustrates how to interpret and utilize the results of the first two moments of head. Figure 1.4a shows the mean head field in a domain of $L = 10$ [L] subject to a fully correlated random recharge with the given statistics $\langle w \rangle = -0.1$ [1/L] and $\sigma_w = 0.1$ [1/L]. Also presented are the curves obtained by adding and subtracting one head standard deviation from the mean quantity. These curves correspond to the 68% confidence intervals for the head, meaning that the probability for $h(x)$ to fall within the intervals specified at x is about 0.68. This is in an approximate sense because the head may not be normally distributed. Nevertheless, these intervals provide a way to quantify uncertainties due to incomplete information about the recharge distribution. It is seen that the boundary conditions have a strong impact on the confidence intervals.

1.2.4 Monte Carlo Simulation Method

Figure 1.4b shows 20 different head profiles corresponding to 20 possible values of the recharge w. The 20 values are drawn from a normal distribution with the above specified statistics of w, and hence these 20 values constitute 20 realizations of the fully correlated random recharge field. At each realization, the head profile is obtained from Eq. (1.16)

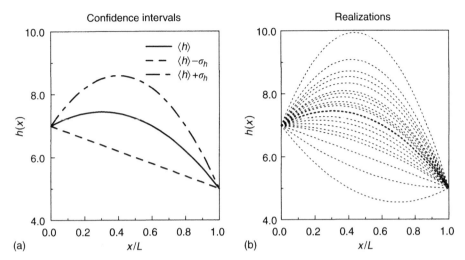

Figure 1.4. The confidence intervals of a head field under random recharge with $\langle w \rangle = 0.1$ and $\sigma_w = -0.1$, and 20 realizations of such a head field.

by using a generated value of w. It is seen that although some head profiles lie outside of the 68% confidence intervals, most of them fall within the intervals. At each point, one may compute the sample mean and sample variance from these 20 head values. When the number of realizations increases, the sample statistics should approach those from the moment equation method, which is exact in this case. This intuitive, multiple-realization procedure is the so-called *Monte Carlo simulation method*. It is a complementary method to the moment equation and the PDF methods. A detailed discussion on the Monte Carlo method will be given in Section 3.11.

1.3 SATURATED FLOW WITH RANDOM BOUNDARY CONDITIONS

As mentioned in Section 1.1, the boundary conditions (BCs) may have to be treated as stochastic processes. This could be the case for a confined aquifer connected to two rivers with random stages. For the sake of easy discussion, we again assume hydraulic conductivity K_S to be a known constant. Flow in such a situation is described by Eqs. (1.4)–(1.6) but with random boundary conditions H_o and H_L. We further let $w \equiv 0$ to single out the effect of random boundary conditions. One may write a set of differential equations for the moments of $h(x)$. But a more straightforward way for solving this simple case is to directly work on the random solution of $h(x)$, given as

$$h(x) = H_o + \frac{x}{L}(H_L - H_o) \tag{1.23}$$

Equation (1.23) is called a *random solution* in that although it is the *formal* solution of the flow problem, it is still in terms of random variables H_o and H_L. In spite of its exactness, the solution itself does not reveal any practical information on the head value if we do not know the values of H_o and H_L. If the statistical properties of H_o and H_L are given, this random solution provides a great vehicle for deriving the statistical moments or even the probability density function of the dependent variable $h(x)$.

1.3.1 Direct Moment Method

The first few head moments may be easily obtained. By decomposing the random boundary terms into their means and fluctuations in the following manner: $H_o = \langle H_o \rangle + H'_o$ and $H_L = \langle H_L \rangle + H'_L$, one obtains the head mean and fluctuation, respectively, as

$$\langle h(x) \rangle = \langle H_o \rangle + \frac{x}{L}(\langle H_L \rangle - \langle H_o \rangle) \tag{1.24}$$

$$h'(x) = H'_o + \frac{x}{L}(H'_L - H'_o) \tag{1.25}$$

With Eq. (1.25), the head covariance and variance can be given as

$$C_h(x, y) = \sigma^2_{H_o} + \frac{x + y}{L}(\sigma_{H_o H_L} - \sigma^2_{H_o}) + \frac{xy}{L^2}(\sigma^2_{H_o} - 2\sigma_{H_o H_L} + \sigma^2_{H_L})$$

$$\sigma^2_h(x) = \sigma^2_{H_o} + \frac{2x}{L}(\sigma_{H_o H_L} - \sigma^2_{H_o}) + \frac{x^2}{L^2}(\sigma^2_{H_o} - 2\sigma_{H_o H_L} + \sigma^2_{H_L})$$

(1.26)

where $\sigma^2_{H_o}$ and $\sigma^2_{H_L}$ are the respective variance of the random boundary head at $x = 0$ and $x = L$, and $\sigma_{H_o H_L}$ is the covariance between the two boundary heads. The two boundary heads may or may not be correlated depending on the processes influencing them. In any case, it is seen that the head variance strongly depends on the location in the domain and that the heads at two points x and y are correlated even in the case that the two boundary heads are uncorrelated (i.e., $\sigma_{H_o H_L} = 0$).

Note that higher head moments can be derived with the help of Eq. (1.25). Also, note that differential equations governing the statistical moments of head can be easily derived with the same procedure outlined in Section 1.2.1. However, it is more straightforward to obtain these moments on the basis of a random solution whenever such a solution is explicitly available.

1.3.2 Direct PDF Method

We may also derive some of the $h(x)$ PDFs from Eq. (1.23), which can be rewritten as

$$h(x) = \frac{L - x}{L} H_o + \frac{x}{L} H_L \tag{1.27}$$

Let us assume that the PDFs of H_o and H_L are known. Define

$$Z_1(x) = \frac{L - x}{L} H_o \quad \text{and} \quad Z_2(x) = \frac{x}{L} H_L$$

The *univariate* (*marginal*) PDFs of these two new random functions can be expressed as functions of $p_{H_o}(H_o)$ and $p_{H_L}(H_L)$, the respective marginal PDFs of H_o and H_L (see Section 2.1.3),

$$p_{Z_1}[Z_1(x)] = \frac{L}{L-x} p_{H_o}\left[\frac{L}{L-x}Z_1\right]$$

$$p_{Z_2}[Z_2(x)] = \frac{L}{x} p_{H_L}\left[\frac{L}{x}Z_2\right] \tag{1.28}$$

The joint PDF of Z_1 and Z_2 can be given as a function of the joint PDF of H_o and H_L, $p_{H_o H_L}(H_o, H_L)$, as a result of the transformation of variables (see Section 2.2.3). The joint PDF reads as

$$p_{Z_1 Z_2}[Z_1(x), Z_2(x)] = \frac{L^2}{(L-x)x} p_{H_o H_L}\left[\frac{L}{L-x}Z_1, \frac{L}{x}Z_2\right] \tag{1.29}$$

With this PDF, one may express the (one-point) PDF for $h(x) = Z_1(x) + Z_2(x)$ as [e.g., Ross 1997, p. 55]

$$p_1[h(x)] = \int_{-\infty}^{\infty} p_{Z_1 Z_2}[h(x) - Z_2, Z_2]\, dZ_2$$

$$= \frac{L^2}{x(L-x)} \int_{-\infty}^{\infty} p_{H_o H_L}\left\{\frac{L}{L-x}[h(x) - Z_2], \frac{L}{x}Z_2\right\} dZ_2 \tag{1.30}$$

Therefore, the one-point PDF of $h(x)$ can be evaluated from Eq. (1.30) with the given PDFs $p_{H_o H_L}$. Equation (1.30) is exact no matter what distributional forms H_o and H_L take. However, the one-point PDF does not provide a complete description of the random head field. It provides information at each point separately, but no joint information at two or more points. For example, the mean $\langle h(x) \rangle$ and the variance $\sigma_h^2(x)$ of the head at each point can be evaluated with Eq. (1.30) on the basis of the definition (1.6), but the covariance $C_h(x, y)$ cannot be evaluated yet. Because the covariance measures how hydraulic heads at two points are correlated, it is often an important piece of information to obtain. In order (for the PDF method) to evaluate this quantity, one must first obtain the two-point PDF of $h(x)$ and $h(y)$.

Recognizing that

$$h(x) = \frac{L-x}{L}H_o + \frac{x}{L}H_L \quad \text{and} \quad h(y) = \frac{L-y}{L}H_o + \frac{y}{L}H_L$$

the joint PDF of $h(x)$ and $h(y)$ can be obtained from that of H_o and H_L, again, through the transformation of variables. The joint PDF reads as

$$p_2[h(x), h(y)] = \frac{L}{y-x} p_{H_o H_L}\left[\frac{yh(x) - xh(y)}{y-x}, \frac{(L-y)h(x) - (L-x)h(y)}{x-y}\right] \tag{1.31}$$

If H_o and H_L are independent, then $p_{H_oH_l}(H_o, H_L) = p_{H_o}(H_o)p_{H_l}(H_L)$. Under this condition, the one- and two-point PDFs of $h(x)$ read as

$$p_1[h(x)] = \frac{L^2}{x(L-x)} \int_{-\infty}^{\infty} p_{H_o}\left\{\frac{L}{L-x}[h(x) - Z_2]\right\} p_{H_L}\left[\frac{L}{x}Z_2\right] dZ_2 \quad (1.32)$$

$$p_2[h(x), h(y)] = \frac{L}{y-x} p_{H_o}\left[\frac{yh(x) - xh(y)}{y-x}\right]$$
$$\cdot p_{H_L}\left[\frac{(L-y)h(x) - (L-x)h(y)}{x-y}\right] \quad (1.33)$$

Alternatively, one can obtain the one-point PDF $p_1[h(x)]$ as the marginal PDF of the two-point (joint) PDF $p_2[h(x), h(y)]$ once the latter becomes available. The head covariance $C_h(x, y)$ can be evaluated with its definition given in Eq. (1.7) (for $m = l = 1$).

Next let us examine the special case that the boundary heads H_o and H_L obey normal (or Gaussian) distributions. (In this book, the terms "normal distribution" and "Gaussian distribution" are used interchangeably. See Chapter 2 for a review of the properties of normal distributions.) Then the PDF for H_o or H_L reads as

$$p_{H_i}(H_i) = \frac{1}{\sigma_{H_i}\sqrt{2\pi}} \exp\left[-\frac{(H_i - \langle H_i \rangle)^2}{2\sigma_{H_i}^2}\right] \quad (1.34)$$

where $H_i = H_o$ or H_L. The distribution of H_i is completely characterized by the first two moments: the mean $\langle H_i \rangle$ and the variance $\sigma_{H_i}^2$. We denote this normal distribution by $N(\langle H_i \rangle, \sigma_{H_i}^2)$. In theory, the boundary head H_i cannot exactly follow a normal distribution because it cannot take negative values in a saturated system. However, a normal distribution may be a good, practical approximation if the mean $\langle H_i \rangle$ is sufficiently larger than zero and the standard deviation σ_{H_i} is small compared to the mean. Figure 1.5 shows an example of normal boundary head distributions, where the heads are specified with the following statistics: $\langle H_o \rangle = 7$ [L] and $\sigma_{H_o}^2 = 0.5$ [L^2]; $\langle H_L \rangle = 3$ [L] and $\sigma_{H_L}^2 = 0.3$ [L^2].

It is well known that a linear function of a normally distributed variable also obeys a normal distribution. This can be verified for Z_1 and Z_2 by substituting Eq. (1.34) into Eq. (1.28), resulting in

$$p_{Z_1}[Z_1(x)] = \frac{1}{((L-x)/L)\sigma_{H_o}\sqrt{2\pi}} \exp\left[-\frac{(Z_1 - ((L-x)/L)\langle H_o \rangle)^2}{2(((L-x)/L)\sigma_{H_o})^2}\right]$$
$$p_{Z_2}[Z_2(x)] = \frac{1}{(x/L)\sigma_{H_L}\sqrt{2\pi}} \exp\left[-\frac{(Z_2 - (x/L)\langle H_L \rangle)^2}{2((x/L)\sigma_{H_L})^2}\right] \quad (1.35)$$

Thus, we see that Z_1 obeys the normal distribution

$$N\left(\frac{L-x}{L}\langle H_o \rangle, \frac{(L-x)^2}{L^2}\sigma_{H_o}^2\right)$$

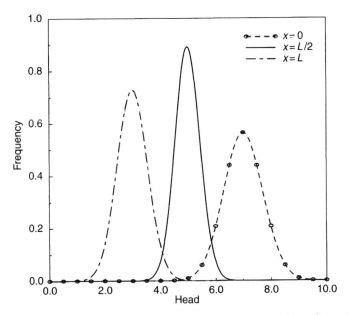

Figure 1.5. One-point PDFs of head at two boundaries and the center of one-dimensional domain.

and Z_2 obeys

$$N\left(\frac{x}{L}\langle H_L\rangle, \frac{x^2}{L^2}\sigma^2_{H_L}\right)$$

It is also well known that the sum of two independent normal variables with parameters (μ_1, σ_1^2) and (μ_2, σ_2^2) follows a normal distribution with parameters $(\mu_1 + \mu_2, \sigma_1^2 + \sigma_2^2)$ (see Section 2.2.5). Therefore, $h(x)$ satisfies

$$N\left(\frac{L-x}{L}\langle H_o\rangle + \frac{x}{L}\langle H_L\rangle, \frac{(L-x)^2}{L^2}\sigma^2_{H_o} + \frac{x^2}{L^2}\sigma^2_{H_L}\right)$$

The one-point PDF of $h(x)$ at the center of a one-dimensional domain with $L = 10$ [L] subject to random boundary heads is shown in Fig. 1.5. One may verify from Eq. (1.33) that

$$C_h(x, y) = \frac{(L-x)(L-y)}{L^2}\sigma^2_{H_o} + \frac{xy}{L^2}\sigma^2_{H_L}$$

1.3.3 Discussion

For the case of independent, normal H_o and H_L, the first two moments of head obtained with the PDF method are exactly the same as those in Eqs. (1.24) and (1.26) derived with the moment method. This is so because in this special case both methods are exact without any further approximation. For other cases where some type of approximations are required, the PDF method and the moment method may produce different results.

In this special case, the PDF method is relatively simple. But the moment method is still more straightforward. This example supports the common practice of directly deriving the statistical moments of dependent variables. With the moments, one may then construct (or approximate) the PDFs.

It is found that the head obeys the same type of distribution as the boundary heads when the latter are assumed to be normally distributed. However, this may not be true if the boundary heads are specified with a different type of distribution. For example, if each of H_o and H_L obeys a uniform distribution, $h(x)$ is generally not uniformly distributed. One may verify this with Eq. (1.32).

1.4 SATURATED FLOW WITH RANDOM COEFFICIENTS

In this example, we investigate the situation where hydraulic conductivity $K_S(x)$ is a random space function with known statistical moments. For the sake of clear presentation, we assume the forcing term $g(x)$ to be zero and the boundary conditions to be known with certainty. Hence, the one-dimensional flow equation can be rewritten from Eqs. (1.1) and (1.2) as

$$K_S(x)\frac{d^2h(x)}{dx^2} + \frac{dK_S(x)}{dx}\frac{dh(x)}{dx} = 0 \tag{1.36}$$

subject to

$$h(x) = H_o, \quad x = 0$$
$$h(x) = H_L, \quad x = L \tag{1.37}$$

One may derive moment equations for $h(x)$ in terms of the moments of $K_S(x)$ on the basis of Eqs. (1.36) and (1.37). As an alternative and maybe a better way, one may work with log-transformed hydraulic conductivity, $f(x) = \ln K_S(x)$. With $K_S = \exp(f)$, one may rewrite Eq. (1.36) as

$$\frac{d^2h(x)}{dx^2} + \frac{df(x)}{dx}\frac{dh(x)}{dx} = 0 \tag{1.38}$$

As $f(x)$ is a random space function, Eq. (1.38) becomes a stochastic differential equation. In this section, we discuss several methods for deriving moment equations for $h(x)$, from Eq. (1.38), in terms of the statistical moments of f.

1.4.1 Closure Approximation Method

As usual, one may decompose the random space function into its mean and fluctuation: $f(x) = \langle f(x) \rangle + f'(x)$. If $f(x)$ is second-order stationary, then the mean is $\langle f(x) \rangle \equiv \langle f \rangle$

and the covariance is $C_f(x, y) \equiv C_f(x - y)$. (This assumption will be relaxed in other chapters.) With $h(x) = \langle h(x) \rangle + h'(x)$, one has

$$\frac{d^2 \langle h(x) \rangle}{dx^2} + \frac{d^2 h'(x)}{dx^2} + \frac{df'(x)}{dx} \frac{d\langle h(x) \rangle}{dx} + \frac{df'(x)}{dx} \frac{dh'(x)}{dx} = 0 \qquad (1.39)$$

$$\begin{aligned} \langle h(x) \rangle + h'(x) = H_o, \quad x = 0 \\ \langle h(x) \rangle + h'(x) = H_L, \quad x = L \end{aligned} \qquad (1.40)$$

Taking expectation of Eqs. (1.39) and (1.40) yields

$$\frac{d^2 \langle h(x) \rangle}{dx^2} + \left\langle \frac{df'(x)}{dx} \frac{dh'(x)}{dx} \right\rangle = 0 \qquad (1.41)$$

$$\begin{aligned} \langle h(x) \rangle = H_o, \quad x = 0 \\ \langle h(x) \rangle = H_L, \quad x = L \end{aligned} \qquad (1.42)$$

Subtracting Eqs. (1.41) and (1.42) from Eqs. (1.39) and (1.40) gives the equation governing the head fluctuation,

$$\frac{d^2 h'(x)}{dx^2} + \frac{df'(x)}{dx} \frac{d\langle h(x) \rangle}{dx} + \frac{df'(x)}{dx} \frac{dh'(x)}{dx} - \left\langle \frac{df'(x)}{dx} \frac{dh'(x)}{dx} \right\rangle = 0 \qquad (1.43)$$

$$\begin{aligned} h'(x) = 0, \quad x = 0 \\ h'(x) = 0, \quad x = L \end{aligned} \qquad (1.44)$$

where the assumption that the boundary conditions are deterministic has been used.

Multiplying Eqs. (1.43) and (1.44) with $h'(y)$ at a different location and taking expectation yields

$$\frac{d^2 \langle h'(x) h'(y) \rangle}{dx^2} + \frac{d\langle h(x) \rangle}{dx} \frac{d\langle f'(x) h'(y) \rangle}{dx} + \left\langle \frac{df'(x)}{dx} \frac{dh'(x)}{dx} h'(y) \right\rangle = 0 \qquad (1.45)$$

$$\begin{aligned} \langle h'(x) h'(y) \rangle = 0, \quad x = 0 \\ \langle h'(x) h'(y) \rangle = 0, \quad x = L \end{aligned} \qquad (1.46)$$

Rewriting Eqs. (1.43) and (1.44) in terms of y, multiplying them with $f'(x)$, and taking expectation gives the equation governing $\langle f'(x) h'(y) \rangle$,

$$\frac{d^2 \langle f'(x) h'(y) \rangle}{dy^2} + \frac{d\langle h(y) \rangle}{dy} \frac{d\langle f'(x) f'(y) \rangle}{dy} + \left\langle f'(x) \frac{df'(y)}{dy} \frac{dh'(y)}{dy} \right\rangle = 0 \qquad (1.47)$$

$$\begin{aligned} \langle f'(x) h'(y) \rangle = 0, \quad y = 0 \\ \langle f'(x) h'(y) \rangle = 0, \quad y = L \end{aligned} \qquad (1.48)$$

It is seen from Eqs. (1.41) and (1.45) that the first moment $\langle h(x) \rangle$ depends on one unknown second moment involving $\langle f'(x)h'(y) \rangle$ and the second moment $\langle h'(x)h'(y) \rangle$ depends on two unknown moments: one being the second moment $\langle f'(x)h'(y) \rangle$ and the other being a third moment involving $\langle f'(x)h'(x)h'(y) \rangle$. The second moment $\langle f'(x)h'(y) \rangle$ is governed by Eq. (1.47), which is in terms of the input covariance $C_f(x, y) = \langle f'(x)f'(y) \rangle$ and an additional unknown (third) moment $\langle f'(x)f'(y)h'(y) \rangle$. We can certainly write a similar equation for the third moment. But we would find that the third moment is in terms of other third moments and an additional fourth moment. No matter how we try, we always have one more unknown than the number of equations. That is, the (exact) equation for the nth moment requires knowledge of the $(n + 1)$th moment. This is the so-called *closure problem*. The closure problem is usually treated by approximating the nth moment in terms of lower moments. This approximation is called a closure approximation.

One particular closure approximation is to assume the third cross central moment $\langle f'(x)f'(y)h'(y) \rangle$ to be zero, where the word "central" is used to denote that the moment is for mean-removed quantities (i.e., fluctuations). This treatment is exact if $h(x)$ obeys a Gaussian distribution, and it should be a good approximation if $h(x)$ is approximately Gaussian. This is so because f is assumed to be Gaussian, the product of Gaussian variables is still Gaussian, and the third central moment (and other higher odd central moments) of a Gaussian distribution is zero. Under this closure approximation, the equations for the first two moments become

$$\frac{d^2 \langle h(x) \rangle}{dx^2} + \left[\frac{\partial C_{fh}(x, y)}{\partial x \partial y} \right]_{x=y} = 0 \tag{1.49}$$

$$\begin{aligned} \langle h(x) \rangle &= H_o, \quad x = 0 \\ \langle h(x) \rangle &= H_L, \quad x = L \end{aligned} \tag{1.50}$$

$$\frac{\partial^2 C_{fh}(x, y)}{\partial y^2} = -\frac{d \langle h(y) \rangle}{dy} \frac{\partial C_f(x, y)}{\partial y} \tag{1.51}$$

$$\begin{aligned} C_{fh}(x, y) &= 0, \quad y = 0 \\ C_{fh}(x, y) &= 0, \quad y = L \end{aligned} \tag{1.52}$$

$$\frac{\partial^2 C_h(x, y)}{\partial x^2} = -\frac{d \langle h(x) \rangle}{dx} \frac{\partial C_{fh}(x, y)}{\partial x} \tag{1.53}$$

$$\begin{aligned} C_h(x, y) &= 0, \quad x = 0 \\ C_h(x, y) &= 0, \quad x = L \end{aligned} \tag{1.54}$$

where $C_{fh}(x, y) = \langle f'(x)h'(y) \rangle$ is the cross-covariance between log hydraulic conductivity and head. In Eq. (1.49), $[\partial^2 C_{fh}(x, y)/\partial x \partial y]_{x=y}$ denotes the second derivative of $C_{fh}(x, y)$ with respect to x and y but evaluated at $x = y$, and thus $[\partial^2 C_{fh}(x, y)/\partial x \partial y]_{x=y} \neq d^2 C_{fh}(x, x)/dx^2$.

It is seen that Eqs. (1.49)–(1.51) are coupled equations and need to be solved either simultaneously or iteratively. With the solution of $\langle h(x) \rangle$ and $C_{fh}(x, y)$, the head covariance $C_h(x, y)$ can be solved from Eqs. (1.53) and (1.54). This (weakly nonlinear) coupling is not desirable when solving the moment equations, especially considering that the original (stochastic) equation is linear. In the literature of flow in porous media, the coupling between $\langle h(x) \rangle$ and $C_{fh}(x, y)$ has usually been avoided by either assuming uniform mean flow (and hence providing the explicit, straightline solution for the mean head) or simply disregarding the second-order term in the mean equation. The treatment of discarding the second-order term in the mean equation would result in the zeroth-order approximation for the mean head. One may formally get rid of this coupling with an alternative approach.

1.4.2 Perturbative Expansion Method

Below we outline a perturbative moment equation approach, which has the above-mentioned property of avoiding the coupling between the first two moments. We first expand $h(x)$ into an infinite series:

$$h(x) = h^{(0)}(x) + h^{(1)}(x) + h^{(2)}(x) + \cdots \tag{1.55}$$

where $h^{(n)}$ is, in a statistical sense, a term of nth order in σ_f, i.e., $h^{(n)} = \mathbf{O}(\sigma_f^n)$, and σ_f is the standard deviation of f. Substituting this expansion and the usual decomposition of $f(x) = \langle f \rangle + f'(x)$ into Eqs. (1.37) and (1.38) yields

$$\frac{d^2}{dx^2} \left[h^{(0)}(x) + h^{(1)}(x) + h^{(2)}(x) + \cdots \right]$$
$$+ \frac{df'(x)}{dx} \frac{d}{dx} \left[h^{(0)}(x) + h^{(1)}(x) + h^{(2)}(x) + \cdots \right] = 0 \tag{1.56}$$

$$h^{(0)}(x) + h^{(1)}(x) + h^{(2)}(x) + \cdots = H_o, \quad x = 0$$
$$h^{(0)}(x) + h^{(1)}(x) + h^{(2)}(x) + \cdots = H_L, \quad x = L \tag{1.57}$$

Collecting terms at separate order, we have

$$\frac{d^2 h^{(0)}(x)}{dx^2} = 0 \tag{1.58}$$

$$h^{(0)}(x) = H_o, \quad x = 0$$
$$h^{(0)}(x) = H_L, \quad x = L \tag{1.59}$$

and

$$\frac{d^2 h^{(n)}(x)}{dx^2} = -\frac{dh^{(n-1)}(x)}{dx} \frac{df'(x)}{dx} \tag{1.60}$$

$$h^{(n)}(x) = 0, \quad x = 0$$
$$h^{(n)}(x) = 0, \quad x = L \tag{1.61}$$

where $n \geq 1$. In the above derivations, the zeroth-order term $h^{(0)}$ is required to absorb the entire deterministic boundary condition and thus the higher-order terms satisfy a homogeneous boundary condition. A different treatment of the boundary condition may be possible if real data suggest that the boundary terms should be decomposed into various order terms. The general approach can also be applied to situations with random boundary conditions (see Chapter 3).

Taking expectation of Eqs. (1.58) and (1.59) reveals that $h^{(0)}(x) = \langle h^{(0)}(x) \rangle$ and $[h^{(0)}(x)]' = 0$; the same procedure applied to Eqs. (1.60) and (1.61) for $n = 1$ gives that $\langle h^{(1)}(x) \rangle = 0$. Therefore, the mean head is $\langle h(x) \rangle = h^{(0)}(x)$ to *zeroth* or *first order* in σ_f and $\langle h(x) \rangle = h^{(0)}(x) + \langle h^{(2)}(x) \rangle$ to *second order*. For the head fluctuation, $h'(x) = h^{(1)}(x)$ to first order. Therefore, the head covariance is $C_h(x, y) = \langle h^{(1)}(x) h^{(1)}(y) \rangle$ to second order in σ_f (or first order in σ_f^2).

The zeroth-order mean head term is governed by the Laplace equation (1.58) subject to boundary condition (1.59). The first-order (in σ_f^2) head covariance $C_h(x, y)$ and the cross-covariance $C_{fh}(x, y) = \langle f'(x) h^{(1)}(y) \rangle$ can be constructed from Eqs. (1.60) and (1.61) for $n = 1$ with the same procedure for deriving Eqs. (1.45), (1.46) and (1.47), (1.48). The resulting covariance equations read as

$$\frac{\partial^2 C_{fh}(x, y)}{\partial y^2} = -\frac{dh^{(0)}(y)}{dy} \frac{\partial C_f(x, y)}{\partial y} \tag{1.62}$$

$$C_{fh}(x, y) = 0, \quad y = 0$$
$$C_{fh}(x, y) = 0, \quad y = L \tag{1.63}$$

$$\frac{\partial^2 C_h(x, y)}{\partial x^2} = -\frac{dh^{(0)}(x)}{dx} \frac{\partial C_{fh}(x, y)}{\partial x} \tag{1.64}$$

$$C_h(x, y) = 0, \quad x = 0$$
$$C_h(x, y) = 0, \quad x = L \tag{1.65}$$

These covariance equations look like Eqs. (1.51)–(1.54) except that in the new equations the mean head $\langle h(x) \rangle$ is replaced by the zeroth mean head $h^{(0)}(x)$. The mean head given in Eqs. (1.58) and (1.59) is no longer coupled with the cross-covariance equations (1.62) and (1.63). Hence, after solving for $h^{(0)}$, the deterministic equation governing C_{fh} is fully solvable. With C_{fh}, one can obtain C_h from Eqs. (1.64) and (1.65). If the second-order term is neglected in the mean equations (1.49) and (1.50), then the particular closure approximation discussed in Section 1.4.1 and the expansion approach yields identical results, up to first order in σ_f for the mean head and in σ_f^2 for the head covariance. Here we use different quantities σ_f (standard deviation) and σ_f^2 (variance) as reference when talking about the orders of $\langle h(x) \rangle$ and $C_h(x, y)$. We do so because $\langle h \rangle$ and C_h have different units, and we will follow this convention throughout the rest of the book.

Alternatively, one may write the head covariance as

$$\frac{\partial^4 C_h(x, y)}{\partial x^2 \partial y^2} = \frac{dh^{(0)}(x)}{dx} \frac{dh^{(0)}(y)}{dy} \frac{\partial^2 C_f(x, y)}{\partial x \, \partial y} \tag{1.66}$$

$$\begin{aligned}
C_h(x, y) &= 0, \quad x = 0, \quad y = 0 \\
C_h(x, y) &= 0, \quad x = 0, \quad y = L \\
C_h(x, y) &= 0, \quad x = L, \quad y = 0 \\
C_h(x, y) &= 0, \quad x = L, \quad y = L
\end{aligned} \tag{1.67}$$

which is obtained by multiplying Eqs. (1.60) and (1.61) for $n = 1$ with the same equations but in terms of y and then taking expectation.

The two formalisms are equivalent for the head covariances in that they give identical results. When analytical solutions can be obtained, it may be more beneficial to work with Eqs. (1.66) and (1.67). However, if numerical solutions are the only choice, it is generally more convenient to deal with Eqs. (1.62)–(1.65), especially when the domain is large (and in high space dimensions). If the domain is discretized into a mesh of N nodes, the full covariance of $C_h(x, y)$ consists of N^2 values. In the formalism based on Eqs. (1.66) and (1.67), one must solve the N^2 values of $C_h(x, y)$ simultaneously, while with the other formalism one may solve the problem sequentially: N unknowns for M times (where $1 \leq M \leq N$ depending on how much information needed). In addition, the sequential approach has some inherent parallel structures and may hence take advantage of some parallel algorithms. The details of numerical solutions of the covariance equations are discussed in Chapter 3.

In this moment equation approach, the zeroth- (or first-) order mean head can be corrected by higher-order terms. The second-order correction term $\langle h^{(2)}(x) \rangle$ for the mean head is obtained by directly taking expectation of Eqs. (1.60) and (1.61) for $n = 2$,

$$\frac{d^2 \langle h^{(2)}(x) \rangle}{dx^2} = -\frac{\partial}{\partial x} \frac{\partial}{\partial y} \left[C_{fh}(x, y) \right]_{x=y} \tag{1.68}$$

$$\begin{aligned}
\langle h^{(2)}(x) \rangle &= 0, \quad x = 0 \\
\langle h^{(2)}(x) \rangle &= 0, \quad x = L
\end{aligned} \tag{1.69}$$

It has recently been shown [e.g., Zhang, 1998] that for uniform mean flow the magnitude of this second-order correction term decreases as the domain size L increases and that it is zero for an unbounded domain.

It is seen from Eqs. (1.60) and (1.61) that the head terms are recursive. That is, the higher-order terms are in terms of lower-order ones. Hence, if desirable we may formally proceed to evaluate any arbitrarily higher-order terms (e.g., $\langle h^{(3)} \rangle$, $\langle h^{(4)} \rangle$, and so on) and higher joint moments (e.g., $\langle h^{(1)}(x)h^{(1)}(y)h^{(1)}(z) \rangle$, $\langle h^{(1)}(x)h^{(1)}(y)h^{(1)}(z)h^{(1)}(\chi) \rangle$, $\langle h^{(2)}(x)h^{(2)}(y) \rangle$, and so on). It is the formal advantage of this approach that one does not have to decide *a priori* at which order to stop and what terms to disregard. In theory, one may evaluate the contributions of each term and examine the convergence (or divergence) of the expansion. In practice, higher-order terms are, however, increasingly more difficult to evaluate. Therefore, except for some simple cases we usually stop at low orders by perturbing the infinite series (1.55). This perturbative procedure is meaningful only if the expansion series (1.55) converges. The convergence of this expansion cannot be judged by the magnitude of $h^{(n)}$ because it is a random quantity. Instead, one must work with the statistical moments of $h^{(n)}$. It is seen from Eqs. (1.64)–(1.67) that the variance of $h^{(1)}$ is of the order of σ_f^2, and it may be seen from Eqs. (1.60) and (1.61) with some algebraic manipulations that the variance of $h^{(2)}$ is of the order of σ_f^4. Hence, it is usually believed that the perturbation expansion converges in an asymptotic sense if σ_f^2 is smaller than unity. In practice, the validity of the expansion and the range of the σ_f^2 values for which the expansion remains valid are determined by other means such as Monte Carlo simulations.

Statistical moments for other flow variables may be derived with the head moments. For example, the flux moments may be obtained by rewriting Eq. (1.2) as

$$q(x) = -\exp[\langle f \rangle + f'(x)] \frac{d}{dx} \left[h^{(0)}(x) + h^{(1)}(x) + h^{(2)}(x) + \cdots \right]$$

$$= -K_G \left[1 + f'(x) + \frac{f'^2(x)}{2} + \cdots \right] \frac{d}{dx} \left[h^{(0)}(x) + h^{(1)}(x) + h^{(2)}(x) + \cdots \right] \tag{1.70}$$

where $K_G = \exp(\langle f \rangle)$ is the geometric mean hydraulic conductivity. Collecting terms at separate order, one has

$$q^{(0)}(x) = K_G J(x) \tag{1.71}$$

$$q^{(1)}(x) = -K_G \left[f'(x) \frac{dh^{(0)}(x)}{dx} + \frac{dh^{(1)}(x)}{dx} \right] \tag{1.72}$$

$$q^{(2)}(x) = -K_G \left[f'(x) \frac{dh^{(1)}(x)}{dx} + \frac{1}{2} f'^2(x) \frac{dh^{(0)}(x)}{dx} + \frac{dh^{(2)}(x)}{dx} \right] \tag{1.73}$$

where $J(x) = -dh^{(0)}(x)/dx$ is the negative of the (zeroth-order) mean hydraulic gradient. It may be shown that the mean flux is $\langle q \rangle = q^{(0)}$ to zeroth or first order in σ_f, $\langle q \rangle = q^{(0)} + \langle q^{(2)} \rangle$ to second order, and the flux fluctuation is $q' = q^{(1)}$ to first order. Therefore, to first order the mean flux $\langle q(x) \rangle$ and the flux covariance $C_q(x, y) = \langle q'(x) q'(y) \rangle$ are given as

$$\langle q(x) \rangle = K_G J(x) \tag{1.74}$$

$$C_q(x, y) = K_G^2 \left[J(x) J(y) C_f(x, y) - J(x) \frac{\partial}{\partial y} C_{fh}(x, y) \right.$$

$$\left. - J(y) \frac{\partial}{\partial x} C_{fh}(y, x) + \frac{\partial^2}{\partial x \partial y} C_h(x, y) \right] \tag{1.75}$$

The second-order correction term for the mean flux is given as

$$\langle q^{(2)}(x) \rangle = K_G \left[\frac{J(x)}{2} \sigma_f^2(x) - \frac{d}{dx} \langle h^{(2)}(x) \rangle - \frac{\partial}{\partial y} C_{fh}(x, y)|_{x-y} \right] \quad (1.76)$$

Another quantity of interest is the *effective hydraulic conductivity*, defined as

$$K^{ef}(x) = -\frac{\langle q(x) \rangle}{d\langle h(x) \rangle / dx} \quad (1.77)$$

Hence, to zeroth order (or first order) the effective conductivity is $K^{ef} = K_G$; to second order, it is

$$K^{ef}(x) = \frac{K_G J(x) + \langle q^{(2)}(x) \rangle}{J(x) - d\langle h^{(2)}(x) \rangle / dx} \quad (1.78)$$

The flux moments and the effective conductivity can be evaluated after the first few moments of head are available.

1.4.3 Green's Function Method

The stochastic differential equation (1.38) with the boundary condition (1.37) can be recast into an integral form with the aid of Green's function $G(x, \chi)$. In the case of one-dimensional flow, $G(x, \chi)$ is defined as the solution of the following equation:

$$\frac{\partial^2 G(x, \chi)}{\partial x^2} = -\delta(x - \chi) \quad (1.79)$$

subject to homogeneous boundary conditions (i.e., $G(x, \chi) = 0$ for $x = 0$ and L). In Eq. (1.79), $\delta(x - \chi)$ denotes the *Dirac delta function*, acting as a point source. It is a generalized function with the well-known properties $\delta(x - \chi) = 0$ for $x \neq \chi$, $\int \delta(x - \chi) \, dx = 1$, and more generally, $\int s(x)\delta(x - \chi) \, dx = s(\chi)$ for any function $s(x)$. A comprehensive review of the Dirac delta function is given in Section 2.4.1. Physically speaking, the Green's function $G(x, \chi)$ is the hydraulic head (pressure) response at point x in a field of unit hydraulic conductivity due to a source of unit strength at χ. Also, note that here $G(x, \chi)$ is a deterministic function. In the one-dimensional case, $G(x, \chi)$ is given by [e.g., Dagan, 1982a],

$$G(x, \chi) = H(\chi - x) \left(1 - \frac{\chi}{L} \right) x + H(x - \chi) \left(1 - \frac{x}{L} \right) \chi \quad (1.80)$$

where $H(x - \chi)$ is the so-called *Heaviside step function*,

$$\begin{aligned} H(x - \chi) &= 1, \quad x > \chi \\ H(x - \chi) &= 0, \quad x < \chi \end{aligned} \quad (1.81)$$

Multiplying Eq. (1.38) by $G(x, \chi)$, integrating over the domain Ω, and applying Green's (first) identity gives a stochastic integral equation for the head field [e.g., Dagan, 1982a;

Neuman and Orr, 1993],

$$h(x) = \int_{\Omega} \frac{df(\chi)}{d\chi} \frac{dh(\chi)}{d\chi} G(x, \chi) \, d\chi - \int_{\Gamma_D} H_i(\chi) \frac{\partial G(x, \chi)}{\partial \chi} n_\chi \, d\chi \tag{1.82}$$

where $H_i = H_o$ or H_L is the boundary head, Γ_D stands for the Dirichlet boundary (in this case, at $x = 0$ and L), and n_χ is an outward unit vector normal to the boundary (in this case, $n_\chi = -1$ at $\chi = 0$ and $n_\chi = 1$ at $\chi = L$). Under the given conditions, this equation can be written as

$$h(x) = \frac{L - x}{L} H_o + \frac{x}{L} H_L + \int_0^L \frac{df(\chi)}{d\chi} \frac{dh(\chi)}{d\chi} G(x, \chi) \, d\chi \tag{1.83}$$

Note that Eq. (1.83) is valid even for the case where H_o and H_L are given as random variables. However, we will still keep H_o and H_L deterministic for the sake of clear presentation.

Equation (1.83) is a stochastic equation because of the randomness in $f(x)$. If the statistical moments of $f(x)$ are given, one may derive the moments of $h(x)$ with the aid of Eq. (1.83). As in Section 1.4.1, let us decompose f and h into their respective mean and fluctuation: $f(x) = \langle f \rangle + f'(x)$ and $h(x) = \langle h(x) \rangle + h'(x)$. Here we also assume $f(x)$ to be second-order stationary so that $\langle f \rangle$ is a constant and $C_f(x, y) = C_f(x - y)$. Substituting these into Eq. (1.83) and taking expectation yields

$$\langle h(x) \rangle = \frac{L - x}{L} H_o + \frac{x}{L} H_L + \int_0^L \left\langle \frac{df'(\chi)}{d\chi} \frac{dh'(\chi)}{d\chi} \right\rangle G(x, \chi) \, d\chi \tag{1.84}$$

Subtracting Eq. (1.84) from Eq. (1.83) yields the equation for the head fluctuation,

$$h'(x) = \int_0^L \frac{df'(\chi)}{d\chi} \frac{d\langle h(\chi) \rangle}{d\chi} G(x, \chi) \, d\chi + \int_0^L \frac{df'(\chi)}{d\chi} \frac{dh'(\chi)}{d\chi} G(x, \chi) \, d\chi$$
$$- \int_0^L \left\langle \frac{df'(\chi)}{d\chi} \frac{dh'(\chi)}{d\chi} \right\rangle G(x, \chi) \, d\chi \tag{1.85}$$

Equation (1.85) can be used to express the head covariance $C_h(x, y)$. Multiplying Eq. (1.85) with $h'(y)$ and taking expectation gives

$$\langle h'(x)h'(y) \rangle = \int_0^L \left\langle \frac{df'(\chi)}{d\chi} h'(y) \right\rangle \frac{d\langle h(\chi) \rangle}{d\chi} G(x, \chi) \, d\chi$$
$$+ \int_0^L \left\langle \frac{df'(\chi)}{d\chi} \frac{dh'(\chi)}{d\chi} h'(y) \right\rangle G(x, \chi) \, d\chi \tag{1.86}$$

It is seen that we have encountered the closure problem in that the first moment depends on a second moment and the second moment depends on a third moment. This is the same problem as in Section 1.4.1. As a matter of fact, the stochastic integral equation (1.83) is merely a restatement of the stochastic differential equation (1.38); the integral equations (1.84), (1.85), and (1.86) exactly correspond to the differential equations (1.41) and (1.42),

(1.43) and (1.44), and (1.45) and (1.46), respectively. Therefore, we see that as they should be, the differential-equation-based method and the Green's-function-based moment method are equivalent.

As discussed in Section 1.4.1, a closure approximation may be used to close the system of moment equations. In Chapter 3, we will discuss a few closure approximations. Here, we elect to use the perturbative expansion introduced in Section 1.4.2 as the approach of attack. With the expansion (1.55), Eq. (1.83) can be rewritten as

$$
\begin{aligned}
h^{(0)}(x) &+ h^{(1)}(x) + h^{(2)}(x) + \cdots \\
&= \frac{L-x}{L}H_o + \frac{x}{L}H_L + \int_0^L \frac{df'(\chi)}{d\chi}\frac{d}{d\chi}\left[h^{(0)}(\chi) + h^{(1)}(\chi) + h^{(2)}(\chi) + \ldots\right] \\
&\cdot G(x,\chi)\,d\chi
\end{aligned}
\tag{1.87}
$$

Collecting terms at separate order yields

$$
h^{(0)}(x) = \frac{L-x}{L}H_o + \frac{x}{L}H_L
\tag{1.88}
$$

$$
h^{(1)}(x) = \int_0^L \frac{df'(\chi)}{d\chi}\frac{dh^{(0)}(\chi)}{d\chi}G(x,\chi)\,d\chi
\tag{1.89}
$$

$$
h^{(2)}(x) = \int_0^L \frac{df'(\chi)}{d\chi}\frac{dh^{(1)}(\chi)}{d\chi}G(x,\chi)\,d\chi
\tag{1.90}
$$

Higher-order terms can be given similarly. As discussed in Section 1.4.2, to first order in σ_f the mean head is $\langle h(x)\rangle = h^{(0)}(x)$ and the fluctuation is $h'(x) = h^{(1)}(x)$. The first order head covariance is $C_h(x,y) = \langle h^{(1)}(x)h^{(1)}(y)\rangle$ in terms of σ_f^2. The mean head may be corrected with higher-order terms such as $\langle h^{(2)}\rangle$.

The first-order head term can be rewritten as

$$
h^{(1)}(x) = -J\int_0^L \frac{df'(\chi)}{d\chi}G(x,\chi)\,d\chi
\tag{1.91}
$$

where $J(\chi) = -dh^{(0)}(\chi)/d\chi$. In the case under study, $J = (H_o - H_L)/L$. Integrating the right-hand side of Eq. (1.91) by parts leads to

$$
h^{(1)}(x) = -J\left[f'(\chi)G(x,\chi)\right]_{\chi=0}^L + J\int_0^L f'(\chi)\frac{\partial G(x,\chi)}{\partial\chi}\,d\chi
\tag{1.92}
$$

Substitution of Eq. (1.80) into Eq. (1.92) gives

$$
h^{(1)}(x) = J\int_0^x f'(\chi)\,d\chi - \frac{Jx}{L}\int_0^L f'(\chi)\,d\chi
\tag{1.93}
$$

which can be used to obtain the expressions for the head covariance $C_h(x, y)$ and the cross-covariance $C_{fh}(x, y)$ between $f(x)$ and $h(y)$,

$$C_h(x, y) = J^2 \int_0^x \int_0^y C_f(\chi - \eta) \, d\chi \, d\eta - \frac{J^2 y}{L} \int_0^x \int_0^L C_f(\chi - \eta) \, d\chi \, d\eta$$

$$- \frac{J^2 x}{L} \int_0^L \int_0^y C_f(\chi - \eta) \, d\chi \, d\eta + \frac{J^2 xy}{L^2} \int_0^L \int_0^L C_f(\chi - \eta) \, d\chi \, d\eta$$

$$(1.94)$$

$$C_{fh}(x, y) = J \int_0^y C_f(x - \eta) \, d\eta - \frac{Jy}{L} \int_0^L C_f(x - \eta) \, d\eta \qquad (1.95)$$

For a given $C_f(x - y)$, C_h and C_{fh} can be evaluated, respectively, from Eqs. (1.95) and (1.94) either analytically or numerically. One may verify that Eqs. (1.95) and (1.94) are indeed the solutions of Eqs. (1.62), (1.63) and (1.64), (1.65).

1.4.4 Solutions

As for the recharge covariance $C_w(r)$ in Section 1.2.3, we assume that the covariance of log hydraulic conductivity f takes the exponential form

$$C_f(r) = \sigma_f^2 \exp\left(-\frac{|r|}{\lambda_f}\right) \qquad (1.96)$$

where $r = x - y$, σ_f^2 is the variance and λ_f is the correlation scale of f. Although the process may be tedious, one can obtain the head covariance (and the head variance) $C_h(x, y)$ and the cross-covariance $C_{fh}(x, y)$ by integrating Eqs. (1.94) and (1.95). Rather than giving the general solutions here, we discuss two limiting cases.

The first case is that the log hydraulic conductivity is fully correlated in space so that $\lambda_f \to \infty$ and $C_f(r) = \sigma_f^2$. Under this condition, one obtains from Eqs. (1.95) and (1.94) the following results: $C_{fh}(x, y) \equiv 0$ and $C_h(x, y) \equiv 0$. The results are not surprising if one realizes that in this special case $f(x)$ is a constant, albeit random, throughout the domain. In such a situation the actual head is no longer random although the flux $q = -K_S \, dh/dx$ is.

The second case is that $f(x)$ is uncorrelated in space. This special case may be obtained from the exponential model by letting $\lambda_f \to 0$,

$$C_f(r) = S_o \delta(r) \qquad (1.97)$$

where S_o is a *spectral density*. This is the so-called *white noise* covariance function, which may be used to approximate the situation that the log hydraulic conductivity $f(x)$ varies rapidly in space. The magnitude of S_o is obtained from $S_o = \int_{-\infty}^{\infty} C(r) \, dr$, where $C(r)$ is the actual covariance to be approximated. In the case that $C(r)$ is given by $C_f(r)$ in Eq. (1.96), one has $S_o = 2\sigma_f^2 \lambda_f$. Therefore, the cross-covariance $C_{fh}(x, y)$ and the head covariance $C_h(x, y)$ are obtained as

$$C_{fh}(x, y) = 2J\sigma_f^2\lambda_f \left[H(y - x) - \frac{y}{L} \right] \tag{1.98}$$

$$C_h(x, y) = 2J^2\sigma_f^2\lambda_f \left[x\frac{L - y}{L} - (x - y)H(x - y) \right] \tag{1.99}$$

In obtaining the results (1.98) and (1.99) from Eqs. (1.94) and (1.95), we have to use some special properties of $\delta(x - y)$ and $H(x - y)$, which are given in Section 2.4.1.

Figure 1.6 shows $C_{fh}(x, y)$ and $C_h(x, y)$ for $x = L/2$ and $L = 10$ [L] as a function of y/L, where C_{fh} is normalized by $J^2\sigma_f^2\lambda_f$ and C_h is by $J^2\sigma_f^2\lambda_f^2$. The solutions corresponding to the white noise approximation are plotted as the solid curves, along with some curves obtained with the exponential covariance of different λ_f values. The results based on the exponential model are obtained by numerically solving the moment equations (1.62), (1.63) and (1.64), (1.65). The details of the numerical solutions are discussed in Section 3.4.1. It is seen that $C_{fh}(x, y)$ is negative for $y/L < 0.5$, (i.e., $x > y$) and positive for $y/L > 0.5$ (i.e., $x < y$), and $C_{fh}(x, y)$ is not continuous at $x = y$. The discontinuity at $x = y$ is due to the white noise approximation. For the exponential model, this sudden jump in C_{fh} at $x = y$ is replaced by a rapid yet smooth transition. The head covariance $C_h(x, y)$ is, however, continuous. It is of interest to note that for $x = L/2$, $C_h(x, y)$ as a function of y is symmetric about $y = L/2$, while $C_{fh}(x, y)$ is antisymmetric. It is seen from Fig. 1.6 that the simple white noise approximation captures well the general behaviors of the head covariances even for $\lambda_f/L = 0.1$ and that the approximation improves as the decrease of λ_f/L. At $\lambda_f/L = 0.01$, the results based on the white noise model match the exponential model results very closely. The behavior of $C_{fh}(x, y) < 0$ for $x > y$ and $C_{fh} > 0$ for $x < y$ is general regardless of C_f functions and can be explained with an argument based

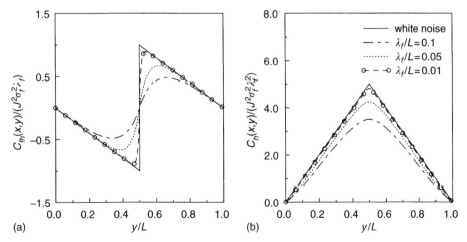

Figure I.6. The cross- and autocovariances of head in one-dimensional domain of two constant head boundaries based on the white noise covariance and the exponential covariance of various λ_f.

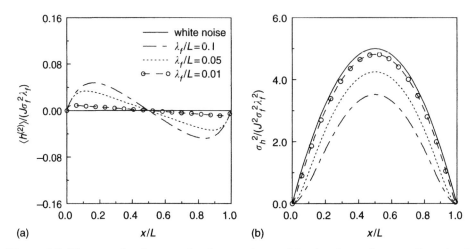

Figure I.7. The second-order mean head corrections and head variances in a one-dimensional domain of two constant head boundaries based on the white noise covariance and the exponential covariance of various λ_f.

on fluid continuity (see Dagan [1989, Fig. 3.7.2 and associated text] for the details of the argument).

The head variance is obtained from Eq. (1.99) as

$$\sigma_h^2(x) = 2J^2 \sigma_f^2 \lambda_f \frac{(L - x)x}{L} \tag{1.100}$$

It is easily seen from this expression that the maximum variance occurs at the domain center, which has the largest distance from the two known boundaries. The head variances are zero at these boundaries. One may also show that for a fixed J, the peak variance increases indefinitely with the size of domain L. In an unbounded, one-dimensional domain, the head variance may not be well defined except for some specific covariances such as a *hole function* covariance [e.g., Gutjahr and Gelhar, 1981], which has a zero *integral scale*. The head variances based on the exponential covariance are plotted in Fig. 1.7b for the case of $L = 10$ [L]. It is seen that as for the head covariance, the variance profiles approach the white noise approximation as λ_f/L decreases. In this case, the zeroth-order (or first-order) mean head is a straight line as given in Eq. (1.88). The second-order mean head correction term $\langle h^{(2)}(x) \rangle$ is shown in Fig. 1.7a for various λ_f. It is seen that the correction term is generally nonzero for a finite ratio of λ_f/L. However, the magnitude of this term approaches zero as $\lambda_f/L \to 0$. The mean head and the head variance can be used to construct confidence intervals for the head field as seen in Fig. 1.4.

I.4.5 Discussion

Contrary to the cases with a random forcing term or random boundary conditions only, the statistical moments of head cannot be solved without approximations of some type when the log hydraulic conductivity f (or hydraulic conductivity K_S) is a random space function. This is true whether a differential-equation-based or an intergal-equation-based method is used. In this section, we introduced two approximation methods. One is the method of closure approximations, which usually invokes some scheme to approximate certain high moments in terms of lower ones. The other is the perturbative expansion method, which avoids the closure problem but theoretically requires the variance of log hydraulic conductivity to be smaller than unity.

For the particular closure approximation invoked in Section 1.4.1, up to first order it produces the same results as the perturbation expansion approach. In general, the choice of closure approximations depends on the type of problem under study, one's physical insight into the problem and other factors. In Chapter 3, we will introduce a few closure approximations for the problem of saturated flow in multiple dimensions.

Some studies indicate that the perturbation approach is quite robust in that it produces satisfactory results for large σ_f^2 values. Zhang and Winter [1999] found a very good agreement for the head variance values obtained from a perturbation approach and from a Monte Carlo simulation study even for $\sigma_f^2 = 4.38$. It seems that the magnitude of σ_f^2 is not the only factor to determine the goodness of a perturbation approximation. The correlation structure of log hydraulic conductivity f, the size of the flow domain relative to the correlation scale of f, and the flow configuration and boundary conditions may all play a role in the accuracy of the perturbation approximation. We will discuss this issue in more detail in the rest of the book.

It is found that the differential-equation- (PDE) based approach and the Green's function approach are equivalent when the same closure approximation or the same perturbation expansion is used. However, the two techniques have their own advantages and disadvantages. The PDE-based approach is more straightforward and requires less mathematical manipulation. The resulting moment equations are PDEs, whose structures are similar or identical to the original (stochastic) PDE. The moments from the Green's function approach are usually in the form of integral equations (or integrodifferential equations, in some cases shown in Chapter 3). For the cases such as one-dimensional, steady-state flow, whose Green's function is simple and explicitly available, the moments can be expressed as integrals of the input covariance function $C_f(x, y)$. This sometimes leads to analytical solutions. Hence, the Green's-function-based approach is more suitable for obtaining analytical solutions for the moments of dependent variables. However, a Green's function is not always simple or available analytically. For example, in a rectangular two-dimensional bounded domain the Green's function needs to be expressed as an infinite Fourier sine series [Osnes, 1995]; in an irregular domain the Green's function may have to be computed numerically. If numerical integrations are needed to obtain the moments of dependent variables, one may find it more efficient to numerically solve the moment PDEs directly [Tartakovsky and Mirkov, 1999].

In summary, there are different techniques to derive moment equations (or expressions) from stochastic differential equations and to solve them. The choice of the techniques may depend on the problem on hand, previous research on the same topic or similar topics, and other factors. In other chapters, we will introduce more moment equation approaches, some of which may be general, while others may only be suitable for some special problems.

I.5 UNSATURATED FLOW IN RANDOM MEDIA

In the previous sections, we studied flow in saturated systems, where the pore space is fully occupied by water. In this section, we consider the situation where air and water coexist in the pore space. In a one-dimensional *unsaturated* system, the flow satisfies the continuity equation (1.1) and the slightly modified version of Darcy's law [e.g., Bear, 1972; Freeze and Cherry, 1979],

$$q(x) = -K[\psi(x), \dots] \frac{d\Phi(x)}{dx} \qquad (1.101)$$

where $q(x)$ is the specific discharge (flux), $\Phi(x) = \psi(x) + x$ is the total head, $\psi(x)$ is the pressure head, $K(x) = K_S(x)k_r[\psi(x), \dots]$ is the unsaturated hydraulic conductivity, $k_r[\psi(x), \dots]$ is the relative permeability which depends on the pressure head and soil properties at x, and x is directed vertically upward. Note that if x is along the horizontal direction, then $\Phi = \psi(x)$.

Substitution of Eq. (1.101) into Eq. (1.1) yields

$$K[\psi(x), \dots] \frac{d^2\psi(x)}{dx^2} + \frac{dK[\psi(x), \dots]}{dx} \left[\frac{d\psi(x)}{dx} + 1 \right] = 0 \qquad (1.102)$$

subject to appropriate boundary conditions. It is immediately seen that the coefficient K in Eq. (1.102) is a function of the dependent variable $\psi(x)$. Hence, this equation is *nonlinear* and needs to be solved by iterations. The problem of unsaturated flow is complicated by this nonlinearity. As for the saturated hydraulic conductivity $K_S(x)$, the unsaturated hydraulic conductivity $K(x)$ exhibits a high degree of spatial variability and can also be treated as a random space function. In turn, Eq. (1.102) becomes a nonlinear, stochastic differential equation.

I.5.I Moment Differential Equations

For the same reason as for saturated flow, let us work with the log-transformed unsaturated hydraulic conductivity $Y[\psi(x), \dots] = \ln K[\psi(x), \dots] = f(x) + \ln k_r[\psi(x), \dots]$ where $f(x) = \ln K_S(x)$. For convenience, we will write the log-transformed unsaturated hydraulic conductivity as $Y(x)$ with the understanding that it is a function of pressure head $\psi(x)$ and the soil properties yet to be defined. Since the variability of ψ depends on the variabilities of the soil properties and the variability of Y depends on those of ψ and the soil properties, we may express h and Y as infinite series in the following form: $Y(x) = Y^{(0)}(x) + Y^{(1)}(x) + Y^{(2)}(x) + \cdots$, and $\psi(x) = \psi^{(0)}(x) + \psi^{(1)}(x) + \psi^{(2)}(x) + \cdots$. It will be clear later that these terms are in terms of the variability of the soil properties. These expansions are similar to those for saturated flow in Section 1.4.2 except that unlike $\langle f \rangle$, $Y^{(0)}(x)$ is generally a function of space even if the log saturated hydraulic conductivity $f(x)$ and the soil properties are stationary. Substituting these expansions into Eq. (1.102) and collecting terms at separate order yields

$$\frac{d^2\psi^{(0)}(x)}{dx^2} + \frac{dY^{(0)}(x)}{dx}\left[\frac{d\psi^{(0)}(x)}{dx} + 1\right] = 0 \tag{1.103}$$

$$\frac{d^2\psi^{(1)}(x)}{dx^2} + \frac{dY^{(0)}(x)}{dx}\frac{d\psi^{(1)}(x)}{dx} = -J_1(x)\frac{dY^{(1)}(x)}{dx} \tag{1.104}$$

$$\frac{d^2\psi^{(2)}(x)}{dx^2} + \frac{dY^{(0)}(x)}{dx}\frac{d\psi^{(2)}(x)}{dx} = -J_1(x)\frac{dY^{(2)}(x)}{dx} - \frac{dY^{(1)}(x)}{dx}\frac{d\psi^{(1)}(x)}{dx} \tag{1.105}$$

subject to appropriate boundary conditions. Here,

$$J_1(x) = \frac{d\psi^{(0)}(x)}{dx} + 1$$

is the mean vertical gradient of total head. Equations for higher-order terms may be written similarly.

Since $Y^{(n)}$ is a function of $\psi^{(n)}$, Eqs.(1.103)–(1.105) are nonlinear. A direct application of the procedure developed in Section 1.4.2 will lead to nonlinear equations for head covariances. For example, multiplying Eq. (1.104) by $\psi^{(1)}$ and $Y^{(1)}$ at a different location and taking expectation leads to the following equations for $C_\psi(x, y) = \langle\psi^{(1)}(x)\psi^{(1)}(y)\rangle$ and $C_{Y\psi}(x, y) = \langle Y^{(1)}(x)\psi^{(1)}(y)\rangle$:

$$\frac{\partial^2 C_\psi(x, y)}{\partial x^2} + \frac{dY^{(0)}(x)}{dx}\frac{\partial C_\psi(x, y)}{\partial x} = -J_1(x)\frac{\partial C_{Y\psi}(x, y)}{\partial x} \tag{1.106}$$

$$\frac{\partial^2 C_{Y\psi}(x, y)}{\partial y^2} + \frac{dY^{(0)}(y)}{dy}\frac{\partial C_{Y\psi}(x, y)}{\partial y} = -J_1(y)\frac{\partial C_Y(x, y)}{\partial y} \tag{1.107}$$

which are nonlinearly coupled.

The unsaturated flow problem cannot be solved without knowing the *constitutive relationship* of $K[\psi(x), \dots]$ (or Y) versus $\psi(x)$. There are numerous models in the literature to describe the functional relationship. Owing to its simplicity, we choose the Gardner [1958] model for illustration,

$$K(x) = K_s(x)\exp[\alpha(x)\psi(x)] \tag{1.108}$$

where α is the soil parameter related to the pore size distribution. The complex yet more realistic models such as the van Genuchten [1980] model and the Brooks–Corey [1964] model will be introduced in Chapter 5. As mentioned earlier, both $f = \ln K_S$ and α may be treated as random space functions and are assumed second-order stationary. One may also decompose α into its mean and fluctuation: $\alpha = \langle\alpha\rangle + \alpha'$. With this, the log transformed hydraulic conductivity $Y(x)$ can be written as

$$Y(x) = f(x) + \alpha(x)\psi(x)$$
$$= \langle f \rangle + f'(x) + [\langle \alpha \rangle + \alpha'(x)][\psi^{(0)}(x) + \psi^{(1)}(x) + \psi^{(2)}(x) + \cdots] \qquad (1.109)$$

Hence, one has

$$Y^{(0)}(x) = \langle f \rangle + \langle \alpha \rangle \psi^{(0)}(x) \qquad (1.110)$$

$$Y^{(1)}(x) = f'(x) + \langle \alpha \rangle \psi^{(1)}(x) + \psi^{(0)}(x)\alpha'(x) \qquad (1.111)$$

$$Y^{(2)}(x) = \langle \alpha \rangle \psi^{(2)}(x) + \psi^{(1)}(x)\alpha'(x) \qquad (1.112)$$

With Eq. (1.111), one derives

$$
\begin{aligned}
C_Y(x, y) = & \, C_f(x, y) + \langle \alpha \rangle^2 C_\psi(x, y) + \psi^{(0)}(x)\psi^{(0)}(y)C_\alpha(x, y) \\
& + \langle \alpha \rangle [C_{f\psi}(x, y) + C_{f\psi}(y, x)] \\
& + \psi^{(0)}(y)C_{f\alpha}(x, y) + \psi^{(0)}(x)C_{f\alpha}(y, x) \\
& + \langle \alpha \rangle \psi^{(0)}(x)C_{\alpha\psi}(x, y) + \langle \alpha \rangle \psi^{(0)}(y)C_{\alpha\psi}(y, x) \qquad (1.113)
\end{aligned}
$$

$$C_{Y\psi}(x, y) = C_{f\psi}(x, y) + \langle \alpha \rangle C_\psi(x, y) + \psi^{(0)}(x)C_{\alpha\psi}(x, y) \qquad (1.114)$$

With Eqs. (1.113) and (1.114), one may confirm that the covariance equations (1.106) and (1.107) are nonlinear.

Alternatively, one may first substitute Eq. (1.111) into Eq. (1.103) and then rearrange some terms to obtain an explicit equation for $\psi^{(1)}(x)$,

$$
\begin{aligned}
\frac{d^2\psi^{(1)}(x)}{dx^2} & + b_1(x)\frac{d\psi^{(1)}(x)}{dx} \\
& = -J_1(x)\left\{ \frac{df'(x)}{dx} + \psi^{(0)}(x)\frac{d\alpha'(x)}{dx} + [J_1(x) - 1]\alpha'(x) \right\} \qquad (1.115)
\end{aligned}
$$

where $b_1(x) = [2J_1(x) - 1]\langle \alpha \rangle$. Multiplying it with $\psi^{(1)}(y)$ and taking expectation yields

$$
\begin{aligned}
\frac{\partial^2 C_\psi(x, y)}{\partial x^2} & + b_1(x)\frac{\partial C_\psi(x, y)}{\partial x} = -J_1(x)\left\{ \frac{\partial C_{f\psi}(x, y)}{\partial x} + \psi^{(0)}(x)\frac{\partial C_{\alpha\psi}(x, y)}{\partial x} \right. \\
& \left. + [J_1(x) - 1]C_{\alpha\psi}(x, y) \right\} \qquad (1.116)
\end{aligned}
$$

subject to appropriate boundary conditions. In a similar manner, one has

$$\frac{\partial^2 C_{f\psi}(x, y)}{\partial y^2} + b_1(y)\frac{\partial C_{f\psi}(x, y)}{\partial y} = -J_1(y)\left\{\frac{\partial C_f(x - y)}{\partial y} + \psi^{(0)}(y)\frac{\partial C_{f\alpha}(x - y)}{\partial y}\right.$$

$$\left. + [J_1(y) - 1]C_{f\alpha}(x - y)\right\} \tag{1.117}$$

$$\frac{\partial^2 C_{\alpha\psi}(x, y)}{\partial y^2} + b_1(y)\frac{\partial C_{\alpha\psi}(x, y)}{\partial y} = -J_1(y)\left\{\frac{\partial C_{\alpha f}(x - y)}{\partial y} + \psi^{(0)}(y)\frac{\partial C_{\alpha}(x - y)}{\partial y}\right.$$

$$\left. + [J_1(y) - 1]C_{\alpha}(x - y)\right\} \tag{1.118}$$

1.5.2 Solutions and Discussion

Once that $\psi^{(0)}$ is solved from the nonlinear equation (1.103) with Eq. (1.110), the covariance equations (1.116)–(1.118), which are linear, are fully solvable. It is thus seen that different procedures may lead to different equations. Although the set of equations (1.106) and (1.107) coupled with Eq. (1.113) should be equivalent to Eqs. (1.116)–(1.118) in terms of final results, the latter are more desirable because they are linear. Similarly, the second-order mean head correction term $\langle\psi^{(2)}(x)\rangle$ is governed by a linear equation and can be solved with these covariances.

It is often difficult to analytically solve these moment equations without further approximations such as mean gravity-dominated flow (by setting $J_1(x) = 1$) and unbounded domain. Under these simplified conditions, the mean equation (1.103) becomes trivial with the solution of $\psi^{(0)}(x)$ being a constant, and the head covariance depends on the separation vector $r = x - y$ rather than the actual locations of the two points. Hence the head field (and other flow variables) is stationary in space. Under these conditions, the head moment equations read as

$$\frac{d^2 C_{f\psi}(r)}{dr^2} - \langle\alpha\rangle\frac{dC_{f\psi}(r)}{dr} = \frac{dC_f(r)}{dr} + \psi^{(0)}\frac{dC_{f\alpha}(r)}{dr} \tag{1.119}$$

$$\frac{d^2 C_{\alpha\psi}(r)}{dr^2} - \langle\alpha\rangle\frac{dC_{\alpha\psi}(r)}{dr} = \frac{dC_{\alpha f}(r)}{dr} + \psi^{(0)}\frac{dC_{\alpha}(r)}{dr} \tag{1.120}$$

$$\frac{d^2 C_{\psi}(r)}{dr^2} + \langle\alpha\rangle\frac{dC_{\psi}(r)}{dr} = -\frac{dC_{f\psi}(r)}{dr} - \psi^{(0)}\frac{dC_{\alpha\psi}(r)}{dr} \tag{1.121}$$

Solutions for these equations may be derived analytically. When $C_f(r)$ and $C_{\alpha}(r)$ are specified as exponential covariances with the respective variance and correlation scale (σ_f^2, λ_f) and $(\sigma_\alpha^2, \lambda_\alpha)$, and when f and α are uncorrelated (i.e., $C_{f\alpha}(r) = 0$), the cross-covariances

$C_{f\psi}(r)$ and $C_{\alpha\psi}(r)$ read as

$$C_{f\psi}(r) = \frac{\sigma_f^2 \lambda_f}{1 - \langle\alpha\rangle^2 \lambda_f^2} \left\{ 2[H(r) - 1]\exp(-\langle\alpha\rangle|r|) \right.$$

$$\left. - [(2H(r) - 1) - \langle\alpha\rangle\lambda_f]\exp\left(-\frac{|r|}{\lambda_f}\right) \right\}$$

$$\sigma_{f\psi} = C_{f\psi}(0) = -\frac{\sigma_f^2 \lambda_f}{1 + \langle\alpha\rangle\lambda_f} \tag{1.122}$$

$$C_{\alpha\psi}(r) = \frac{\psi^{(0)}\sigma_\alpha^2 \lambda_\alpha}{1 - \langle\alpha\rangle^2 \lambda_\alpha^2} \left\{ 2[H(r) - 1]\exp(-\langle\alpha\rangle|r|) \right.$$

$$\left. - [(2H(r) - 1) - \langle\alpha\rangle\lambda_\alpha]\exp\left(-\frac{|r|}{\lambda_\alpha}\right) \right\}$$

$$\sigma_{\alpha\psi} = -\frac{\psi^{(0)}\sigma_\alpha^2 \lambda_\alpha}{1 + \langle\alpha\rangle\lambda_\alpha} \tag{1.123}$$

which are derived with the condition of $C_{f\psi}(-\infty) = C_{f\psi}(\infty) = 0$. One may verify these solutions by substituting them into Eqs. (1.119) and (1.120), respectively. These two covariance expressions are similar to Eqs. (39) and (41) of Zhang *et al.* [1998] but with some subtle yet important differences. We are using ψ to denote the pressure head, while in that work, h stands for the suction head (the negative of the pressure head). It is of interest to mention that the covariance $C_{f\psi}(r)$ or $C_{\alpha\psi}(r)$ in the present work is, however, not simply the negative of the counterpart covariance in Zhang *et al.* [1998]. The head covariance is the same as Eq. (45) of Zhang *et al.* [1998],

$$C_\psi(r) = \frac{\sigma_f^2 \lambda_f}{1 - \langle\alpha\rangle^2 \lambda_f^2} \left[\frac{\exp(-\langle\alpha\rangle|r|)}{\langle\alpha\rangle} - \lambda_f \exp\left(-\frac{|r|}{\lambda_f}\right) \right]$$

$$+ \frac{[\psi^{(0)}]^2\sigma_\alpha^2 \lambda_\alpha}{1 - \langle\alpha\rangle^2 \lambda_\alpha^2} \left[\frac{\exp(-\langle\alpha\rangle|r|)}{\langle\alpha\rangle} - \lambda_\alpha \exp\left(-\frac{|r|}{\lambda_\alpha}\right) \right]$$

$$\sigma_\psi^2 = \frac{\sigma_f^2 \lambda_f}{(1 + \langle\alpha\rangle\lambda_f)\langle\alpha\rangle} + \frac{[\psi^{(0)}]^2\sigma_\alpha^2 \lambda_\alpha}{(1 + \langle\alpha\rangle\lambda_\alpha)\langle\alpha\rangle} \tag{1.124}$$

Another covariance of interest is that of log unsaturated hydraulic conductivity Y, which can be easily obtained with the above expressions,

$$C_Y(r) = \frac{\sigma_f^2}{1 - \langle\alpha\rangle^2 \lambda_f^2} \left[\exp\left(-\frac{|r|}{\lambda_f}\right) - \langle\alpha\rangle\lambda_f \exp(-\langle\alpha\rangle|r|) \right]$$

$$+ \frac{[\psi^{(0)}]^2\sigma_\alpha^2}{1 - \langle\alpha\rangle^2 \lambda_\alpha^2} \left[\exp\left(-\frac{|r|}{\lambda_\alpha}\right) - \langle\alpha\rangle\lambda_\alpha \exp(-\langle\alpha\rangle|r|) \right]$$

$$\sigma_Y^2 = \frac{\sigma_f^2}{1 + \langle\alpha\rangle\lambda_f} + \frac{[\psi^{(0)}]^2\sigma_\alpha^2}{1 + \langle\alpha\rangle\lambda_\alpha} \tag{1.125}$$

It is seen that the head (auto-) covariance $C_\psi(r)$ and the covariance of log unsaturated conductivity $C_Y(r)$ are symmetric about $r = 0$, whereas the cross-covariance $C_{f\psi}(r)$ and $C_{\alpha\psi}(r)$ are neither symmetric nor antisymmetric. It can be seen from Eq. (1.124) that the head variance increases with $\sigma_f^2 \lambda_f$ and $\sigma_\alpha^2 \lambda_\alpha$. However, when the absolute value of $\psi^{(0)}$ is much greater than 1 (i.e., when the soil is dry), this variance is much more sensitive to the properties of the α field (σ_α^2 and λ_α) than to the properties of the saturated hydraulic conductivity (σ_f^2 and λ_f). This observation is also true for the variance of log unsaturated hydraulic conductivity σ_Y^2. It suggests that the variability of the pore size distribution parameter α should not be neglected even if it seems small compared to that of saturated hydraulic conductivity K_S. Furthermore, a better parameter for quantifying the variability of α is the *coefficient of variation*, defined as $C_{v_\alpha} = \sigma_\alpha / \langle \alpha \rangle$. A small σ_α does not necessarily mean that C_{v_α} is small when $\langle \alpha \rangle$ is also small.

As mentioned earlier, these solutions are derived by assuming mean gravity-dominated flow and unbounded domain. The condition of mean gravity-dominated flow requires the mean pressure head to be constant throughout the domain. This condition is certainly violated near the water table and near the land surface, where the pressure head moments are strongly location dependent [e.g., Zhang, 1999b]. Also, medium nonstationary features such as distinct geological layers and zones will render the flow field nonstationary. In Chapter 5, we will compare various methods for deriving moment partial differential equations under nonstationary flow conditions in multiple dimensions and discuss numerical techniques for solving these moment equations.

1.6 SCOPE OF THE BOOK

In this chapter, we have discussed saturated and unsaturated flows in a special situation of one space dimension. Flow in one dimension is often unrealistic and of little interest in applications. However, it serves the purpose of introducing stochastic differential equations and some methods for solving them. It is seen that random coefficients, random boundary conditions, and random forcing terms, or any one of these, may render the governing equations stochastic. Stochastic differential equations may be solved using either PDF methods or statistical moment methods. Although the complete solution would be the (joint) probability density functions, it is usually rather difficult to construct PDF equations and to solve them. Except for a very special case, we have concentrated on deriving the statistical moments of dependent variables, which may be expressed explicitly, governed by differential equations or given in terms of integral equations, depending on the nature of the problem. In general, the moment equations need to be derived under the assumption of small perturbations or with some kind of closure approximation. Although our development is for flow in one-space-dimensional, stationary media, some of these techniques are applicable to flow in multidimensional, nonstationary media, which is one of the main topics of the book.

In the rest of the book, we will apply some of these moment equation methods to various flows in porous media and introduce a few more techniques for deriving and solving moment equations. We start with the premise that the reader has a basic understanding of flow in porous media and has some familiarity with partial differential equations and probability concepts. Therefore, only a brief discussion will be given in the beginning of each chapter about the type of flow to be studied in that chapter. The fundamentals of the moment equation methods will, however, be given in a tutorial way so that no prior in-depth knowledge of stochastic processes is needed to comprehend most of the material in the book. For each method introduced, we present its potential and limitations and discuss solution techniques. Some of the methods are illustrated with examples. It is hoped that after studying the various moment equation methods the reader will be able to determine the appropriate method(s) and use it (them) for the problem of his/her interest, which may not necessarily be a problem of flow in porous media. For the reader who is interested in having hands-on experience with the moment equation methods and applying these techniques to real problems, a number of computer codes which numerically implement some of the derived moment equations are available from the author upon request.

The elements of probability theory and the fundamentals of some important concepts pertinent to stochastic processes (random fields) are reviewed in Chapter 2. Also presented are some key mathematical tools such as the Dirac delta function and Fourier and Laplace transforms, which are frequently used in the book. The reader familiar with the basic ideas in stochastic processes may wish to skip this chapter and consult it only when needed.

In Chapter 3, we systematically introduce a number of moment equation methods for multidimensional, steady-state flow in randomly heterogeneous porous media. The media can be spatially nonstationary in that the statistical moments of the medium properties may be location dependent. Medium nonstationarity may be caused by the presence of distinct geological zones, layers or facies or as a result of conditioning on available measurements. The methods discussed include perturbative expansion method, Green's function method, stationary and nonstationary spectral methods, adjoint state method, Adomian decomposition, and closure approximations. Also presented is the Monte Carlo simulation method, which is a complementary approach to the moment equation methods mentioned above and usually serves as an independent tool to validate or invalidate them. This chapter ends with some remarks on gas flow in porous media, conditional moment equations, and other techniques such as renormalization, renormalization group, and Feynman diagrams.

Chapter 4 is devoted to transient saturated flow in random porous media. Several moment equation methods are applied to this problem and some examples are used to illustrate how uncertainty in the input propagates with time and through the domain.

In Chapter 5, we study steady-state and transient unsaturated flow in random porous media of multiple space dimensions. We first give a brief introduction on unsaturated flow with some common constitutive models and on the spatial variabilities of unsaturated parameters. We then divide steady-state flow into two flow regimes: spatially nonstationary flow (Section 5.2) and gravity-dominated flow (Section 5.3). For the latter flow regime, some nice analytical (spectral analysis) or semi-analytical (fast Fourier transformation) methods may be used, whereas the flow nonstationarity calls for numerical solutions. Some examples are presented. It is found that flow boundaries, especially the water table boundary, play a crucial role for unsaturated flow and the effect of the water table should generally not be neglected. In Section 5.4, the approach of Kirchhoff transformation is discussed, which attempts to avoid or delay the linearization procedure required for formulating the moment equations. In Section 5.5, moment partial differential equations for transient, unsaturated flow in nonstationary (multiscale) media are presented with some illustrative examples.

Chapter 6 addresses two-phase flows in random porous media. A general introduction to two-phase flow is given first. Depending on the applications, two-phase flow may be divided into two categories: capillarity- (diffusion) dominated flow and advection-dominated flow. The former calls for an Eulerian approach, while the latter may be dealt with using a Lagrangian approach. Since almost all of the previously introduced methods are Eulerian, the derivation of the moment equations for two-phase flow is a straightforward application. However, these Eulerian moment equations are rather difficult to solve except for some special cases. It is found that the Lagrangian approach is particularly suitable for advection-dominated two-phase flow, which decomposes the problem of nonlinear displacement in random media into two subproblems: nonlinear dispacement in homogeneous media and linear advection in random media. Examples are included to illustrate both the Eulerian and Lagrangian approaches.

While the previous chapters deal with flows in nonfractured porous media, Chapter 7 is devoted to flow in fractured porous media. An introduction to flow in fractured porous media is first given, then a brief discussion of discrete fracture network models is presented. In this chapter, we concentrate on the continuum representation of fractured flow with a dual-porosity/double-permeability model. Moment partial differential equations are developed for saturated flow, solved numerically, and illustrated with some examples. This chapter ends with the derivation of moment equations for unsaturated flow in fractured porous media.

1.7 EXERCISES

The exercises are designed to either help understand some techniques introduced in the text or supplement the material conveyed in the chapter. Some of them are trivial, while others may be rather difficult (especially in other chapters). Even if the reader plans to skip some or all of the problems, he/she is encouraged to read all of them.

1. One usually decomposes a random variable, say, s, into its mean and fluctuation: $s = \langle s \rangle + s'$. Show that the mean of the fluctuation is zero.

2. The saturated hydraulic conductivity K_S has the unit of L/T, and the log transformed conductivity $f = \ln K_S$ is in terms of ln L/T. What unit does the variance of f have? Or does it have a unit at all?

3. The saturated hydraulic conductivity K_S [L/T] is related to the absolute (intrinsic) permeability k [L^2] by $K = (\rho g/\mu)k$, where ρ is the fluid density, g is the gravitational acceleration factor, and μ is the fluid viscosity. When ρ, g, and μ are known constants, show that the variance of the log-transformed absolute permeability $\ln k$ is equal to the variance of the log-transformed conductivity $\ln K_S$.

4. The linear integral scale for a random variable $s(x)$ is defined as

$$I_s = \frac{1}{\sigma_s^2} \int_0^\infty C_s(r)\, dr \qquad (1.126)$$

where $\sigma_s^{\prime 2}$ is the variance and C_s is the two-point covariance of the variable s. The integral scale may be interpreted as the average distance between two points beyond which $s(x)$ and $s(x + r)$ are only weakly correlated.

(a) Show that for the exponential covariance given in Eq. (1.96), the integral scale is equal to the parameter λ_f.

(b) Show that for the Gaussian covariance,

$$C_s(r) = \sigma_s^2 \exp\left(-\frac{r^2}{\lambda_s^2}\right) \tag{1.127}$$

one has $I_s = \sqrt{\pi}\lambda_s/2$.

(c) Show that for the hole function covariance,

$$C_s(r) = \sigma_s^2 \left[1 - \frac{|r|}{\lambda_s}\right]\exp\left(-\frac{|r|}{\lambda_s}\right) \tag{1.128}$$

one has $I_s = 0$.

5. The white noise covariance introduced in Eq. (1.97) may be used to approximate any covariance function C_s if the underlying random variable $s(x)$ varies rapidly in space. It is given in Section 1.4.4 that the white noise approximation reads as $C_s(r) = 2\sigma_s^2\lambda_s\delta(r)$ if it is used to approximate the exponential covariance. Show that for the Gaussian model (1.127), the white noise approximation is $C_s(r) = \sqrt{\pi}\sigma_s^2\lambda_s\delta(r)$.

6. If the white noise covariance is used to approximate the exponential covariance (1.21) for the normalized recharge $w(x)$, the four-fold integrations for obtaining the head covariance from Eq. (1.19) are relatively easy. The resulting head covariance reads as

$$C_h(x, y) = \frac{\sigma_w^2\lambda_w}{3L}\left[x(y - L)(x^2 + y^2 - 2Ly) + L(x - y)^3 H(x - y)\right] \tag{1.129}$$

where L is the size of the domain and $H(x - y)$ is the Heaviside step function.

(a) Try to recover this result by using the properties of the Dirac delta and the Heaviside step functions introduced in Section 2.4.1.

(b) Find out the head variance $\sigma_h^2(x)$ with Eq. (1.129).

(c) Try to plot the normalized head variance $\sigma_h^2/[\sigma_w^2 L^4]$ as a function of x/L for $\lambda_w = 0.1L$. Compare this head variance profile with the one in Fig. 1.3b for the case of fully correlated recharge and see which case has larger variances.

7. The normal probability density function (PDF) for the boundary head H_o or H_L is given by Eq. (1.34). Show that when the variance $\sigma_{H_i}^2 = 0$ (where $i = o$ or L), the PDF becomes $p_{H_i}(H_i) = \delta(H_i - \langle H_i \rangle)$, which mathematically states the simple fact that when the variance of H_i is zero the mean $\langle H_i \rangle$ is the only possible value that

H_i can take. Show that if both $\sigma^2_{H_o} = 0$ and $\sigma^2_{H_L} = 0$, the joint PDF of H_o and H_L is
$p_{H_o H_L}(H_o, H_L) = \delta(H_o - \langle H_o \rangle)\delta(H_L - \langle H_L \rangle)$.

8. In this chapter, one-dimensional steady-state flow is mainly considered. In the absence of internal source/sink the divergence free condition reads as $dq/dx = 0$ in the one-dimensional domain, which means that q must be a (random or deterministic) constant in space.

   (a) Show that under such conditions one has $C_q(x, y) = \sigma^2_q$, which states that q is fully correlated in space. If q is given at one boundary with certainty such as the case of vertical infiltration when the recharge rate at the top is assumed known, then one has $C_q \equiv 0$.
   (b) Derive the expressions of $C_q(x, y)$ for the following two cases: (1) saturated flow with random boundary heads and constant hydraulic conductivity K_S, as discussed in Section 1.3; (2) saturated flow with fully correlated random $K_S(x)$ and deterministic boundary heads, as discussed in Section 1.4.4.

2

STOCHASTIC VARIABLES AND PROCESSES

In this chapter, we review the elements of random variables and the fundamentals of some important concepts pertinent to stochastic processes. This is by no means a comprehensive review. Instead, the review is given for the purpose of better understanding the material presented in this book. For a comprehensive treatment of probability theory and stochastic processes, the reader is referred to standard texts such as Yaglom [1987], Papoulis [1991], and Ross [1997].

2.1 REAL RANDOM VARIABLES

A *random variable (stochastic variable)* U is defined by a set Ω of possible values (the *sample space*) and a *probability distribution P* over this set. In this book we will be concerned primarily with *real* (or *real-valued*) random variables for which Ω is a subset of the real numbers, and with collections of real random variables indexed by $\{1, 2, \ldots, n\}$ (*vector-valued* random variables) or by time and/or space (*stochastic processes*). Moreover, for the real random variables of interest to us, Ω will consist either of a finite or countably infinite subset of the real numbers, or of an open or closed, possibly infinite, interval on the real line. More generally, however, it is possible to define random variables whose values are complex numbers, members of finite sets such as {animal, vegetable, mineral}, probability distributions on the real line, or just about any other set.

The probability distribution P is defined on a collection of subsets of Ω, denoted by S. Formally, S must be a sigma-algebra of subsets, which means that it must be closed under countable unions and finite intersections, and P must satisfy certain conditions, of which the most important for our purposes are that $P(\phi) = 0$, where ϕ denotes the empty set, $P(\Omega) = 1$, and if $S_1 \subseteq S_2$, then $P(S_1) \leq P(S_2)$. It follows that $P(S)$ (which is shorthand for "the probability of the *event* $U \in S$") is between 0 and 1, inclusive, for any subset S belonging to Ω. The triple (Ω, S, P) is called a probability space. (For a more complete and mathematically accurate definition, consult one of the above-cited references.)

2.1.1 Continuous Random Variables

Let us first look at a real-valued random variable whose set of possible values is an interval on the real line. For example, if the random variable represents permeability at a particular location in a geological formation, the set of possible values is the half-infinite interval $(0, \infty)$. For porosity at the same position, the possible values are represented by $[0, 1]$. The probability distribution of such a random variable can be represented by its *cumulative (probability) distribution function* (CDF) defined by

$$F(u) = P(U \leq u) \tag{2.1}$$

for values of u in Ω and $P(U \leq u)$ stands for the probability of the event that the random variable U is less than or equal to u. It follows from the conditions defining a probability distribution that F must be a nondecreasing function of u taking values in the interval $[0, 1]$, and

$$\lim_{u \to -\infty} F(u) = 0 \quad \text{and} \quad \lim_{u \to \infty} F(u) = 1 \tag{2.2}$$

The probability that U lies between u and v is

$$P(u \leq X \leq v) = F(v) - F(u) \tag{2.3}$$

For the case that v only differs from u by an infinitesimal amount du, the probability that U lies between u and $u + du$ is $F(u + du) - F(u) = dF(u)$. If $F(u)$ is differentiable for all u, then U is a *continuous* (real) random variable, and its *probability density function* (PDF) $p(u)$, already mentioned in Chapter 1, can be defined as

$$p(u) = \frac{dF(u)}{du} \tag{2.4}$$

where $p(u)du$ is the probability that U lies between u and $u + du$. Because of Eq. (2.2), one has

$$p(-\infty) = p(\infty) = 0 \tag{2.5}$$

It is obvious that the CDF is related to the PDF via

$$F(u) = \int_{-\infty}^{u} p(u') \, du' \tag{2.6}$$

Note that the lower limit of the integral is appropriate for any PDF even though the probability space may not span from $-\infty$ to ∞. This is so because the PDF $p(u)$ is zero outside the probability space.

All probability questions about U can be answered either in terms of the PDF $p(u)$ or the CDF $F(u)$. For example, the probability that U lies between the interval $[a, b]$ is given by

$$P(a \leq U \leq b) = \int_{a}^{b} p(u) \, du = F(b) - F(a) \tag{2.7}$$

If $a = b$, then

$$P(U = a) = \int_a^a p(u)\,du = 0 \tag{2.8}$$

which states that the probability that a continuous random variable takes any particular value is zero.

Let us look at two common distributions for continuous random variables: *uniform* and *normal*. A random variable is said to be uniformly distributed over the interval (a, b) if its PDF is given by

$$p(u) = \begin{cases} \dfrac{1}{b-a}, & a < u < b \\ 0, & \text{otherwise} \end{cases} \tag{2.9}$$

The corresponding CDF is

$$F(u) = \begin{cases} 0, & a < u \\ \dfrac{u-a}{b-a}, & a < u < b \\ 1, & u \geq b \end{cases} \tag{2.10}$$

Such a distribution is shown in Fig. 2.1a.

A random variable is called normal, or *Gaussian*, if its PDF is given by

$$p(u) = \frac{1}{\sqrt{2\pi}\sigma} \exp\left[-\frac{(u-\mu)^2}{2\sigma^2}\right], \qquad -\infty < u < \infty \tag{2.11}$$

The normal (Gaussian) distribution is usually denoted by $N(\mu, \sigma^2)$, where μ is the mean and σ^2 is the variance of u. The mean and variance are two statistical moments of U, to be

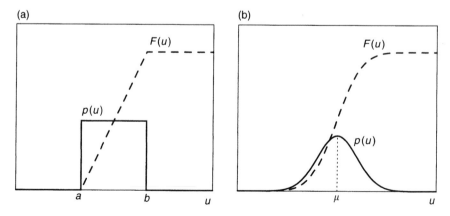

Figure 2.1. Probability density function $p(u)$ and cumulative distribution function $F(u)$: (a) uniform and (b) normal distributions.

defined later. As illustrated in Fig. 2.1b, the PDF is bell shaped and symmetric around μ. The CDF of a normal random variable is

$$F(u) = \frac{1}{\sqrt{2\pi}\sigma} \int_{-\infty}^{u} \exp\left[-\frac{(u'-\mu)^2}{2\sigma^2}\right] du' = \frac{1}{2} + \mathrm{erf}\left(\frac{u-\mu}{\sigma}\right) \qquad (2.12)$$

where erf is the *error function*

$$\mathrm{erf}(x) = \frac{1}{\sqrt{2\pi}} \int_{0}^{x} \exp\left(-\frac{y^2}{2}\right) dy \qquad (2.13)$$

2.1.2 Discrete Random Variables

Let us now look at real random variables whose sample space consists of a finite or countable infinite subset of the real numbers. For example, the number of fractures in a given volume may be any nonnegative integer. Such a random variable U is called a *discrete random variable*, and has a probability mass function p defined by

$$p(a) = P(U = a) \qquad (2.14)$$

for any a in the sample space of U. If the sample space is the set $\{u_1, u_2, \ldots\}$, then since P is a probability distribution,

$$\sum_{i=1}^{\infty} p(u_i) = 1 \qquad (2.15)$$

The cumulative distribution function $F(a)$ can be expressed with aid of $p(a)$,

$$F(a) = \sum_{\text{all } u_i \leq a} p(u_i) \qquad (2.16)$$

An example of such distributions is shown in Fig. 2.2. Although the probability mass function $p(u_i)$ is discontinuous, the cumulative distribution function $F(a)$ is a piecewise continuous function. Let us consider a laboratory column in which the same quantity of sand is packed repeatedly with the same procedure. In each packing experiment, a fixed point of the column (say, the center of the column, x_c) may be occupied by either a solid particle or a void space. We may denote the event of x_c being on a solid particle by $u = 0$ and that of x_c within a void by $u = 1$. After a large number of experiments, the probability that x_c lies within a void can be calculated by

$$P(U = 1) = \frac{\text{the number of experiments with the outcome of } U = 1}{\text{the total number of experiments}} \qquad (2.17)$$

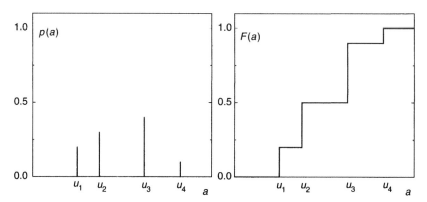

Figure 2.2. Probability mass function $p(a)$ and cumulative distribution function $F(a)$ for a discrete variable.

which, under some conditions, is equivalent to the familiar concept of porosity. Let us assume $p(1) = P(U = 1) = 0.3$, then $p(0) = 0.7$. In this case, the CDF of U is

$$F(u) = \begin{cases} 0, & u < 0 \\ 0.7, & 0 \leq u < 1 \\ 1, & u \geq 1 \end{cases} \tag{2.18}$$

It is seen that the probability mass function for a discrete random variable is conceptually different from the probability density function for a continuous random variable. However, one can avoid the dual notation by permitting the PDF to include *Dirac delta functions*, whose properties are introduced in Section 2.4.1. In general, the probability density function $p(u)$ can be written for a discrete random variable u as [Papoulis, 1991]

$$p(u) = \sum_{i=1}^{\infty} p(u_i)\delta(u - u_i) \tag{2.19}$$

For example, for the above example of sand columns, the PDF reads as

$$p(u) = 0.3\delta(u - 1) + 0.7\delta(u) \tag{2.20}$$

Then the probability mass function at u_i is interpreted as the area under the delta function at point u_i [e.g., Vanmarcke, 1983]. The density function (2.19) is usually called the *multimodal discrete PDF*. With this generalization, the CDF can be given consistently by Eq. (2.6) for both continuous and discrete random variables. Hence in the remainder of this chapter, we will not distinguish between continuous and discrete random variables.

2.1.3 Functions of a Random Variable

We are usually not only interested in a random variable U itself but also in a function $V = g(U)$ of the random variable. For example, a frequently used transformation, which has already been introduced in Chapter 1, is

$$f = \ln k \tag{2.21}$$

where k is the (intrinsic) permeability. If k is a random variable, then so is f. The probability distribution of f can be obtained from a knowledge of the distribution k, and vice versa. In general, the probability density function of $V = g(U)$ may be derived by equating the cumulative distributions, i.e., $F_V[g(u)] = F_U(u)$, and differentiating. When $g(U)$ is either monotonically increasing or decreasing, one has the following simple relationship:

$$p_V(v) = p_U(u) \left| \frac{du}{dv} \right| \equiv p_U(u) \left| \frac{dg^{-1}(v)}{dv} \right| \tag{2.22}$$

where $u = g^{-1}(v)$ is the *inverse transformation*. Here we have used the subscripts U and V to distinguish between the distributions for U and V. Equation (2.22) may be rewritten as $p_V(v)\,dv = \pm p_U(u)\,du$ (with the sign depending on whether $g(U)$ is a monotonically increasing or decreasing function), indicating that the probability that $V = g(U)$ lies between v and $v \pm dv$ is equal to the probability that U lies between u and $u + du$. When $g(U)$ is not a monotonic function, a more complex relation between $p_V(v)$ and $p_U(u)$ can be derived in a similar manner [Papoulis, 1991].

It is easy to show that if the PDF of f is known, the PDF of $k = \exp(f)$ can be expressed with Eq. (2.22) as

$$p_k(k) = \frac{1}{k} p_f(\ln k), \quad 0 < k < \infty \tag{2.23}$$

For the case that f is normal with the parameters $N(\mu_f, \sigma_f^2)$ given by Eq. (2.11), the PDF for k reads as

$$p_k(k) = \frac{1}{k\sqrt{2\pi}\,\sigma_f} \exp\left[-\frac{(\ln k - \mu_f)^2}{2\sigma_f^2} \right], \quad 0 < k < \infty \tag{2.24}$$

where μ_f and σ_f^2 are the mean and variance of the normal random variable f, respectively. A random variable with a probability density function of the form (2.24) is called *lognormal*.

It is seen from the above example that if U is normal, then $V = \exp(U)$ is lognormal, and, conversely, that if V is lognormal, $U = \ln V$ is normal. One can easily prove on the basis of Eq. (2.22) that if U is normal with the parameters $N(\mu, \sigma^2)$, then $V = \alpha U + \beta$ (where α and β are known constants) is also normal with the parameters $N(\alpha\mu + \beta, \alpha^2\sigma^2)$.

2.1.4 Statistical Moments

Although the probability density function $p_U(u)$ provides complete information about the random variable U, it is usually of great interest to obtain some *statistical moments* of U. For example, the *first moment* (i.e., the *mean* or *expectation*) of U can be derived with aid of $p_U(u)$,

$$\langle U \rangle \equiv E(U) = \int_{-\infty}^{\infty} u p_U(u)\,du \tag{2.25}$$

where both $\langle\ \rangle$ and $E(\ )$ indicate expectation. In this book, these two notations are used interchangeably.

Similarly, the expectation of any given function $V = g(U)$ of the random variable U is given by

$$\langle V \rangle \equiv E(V) \equiv \int_{-\infty}^{\infty} v p_V(v)\,dv \tag{2.26}$$

It appears that we must first find its PDF $p_V(v)$ in order to obtain the expectation $E(V)$. Finding $p_V(v)$ is not always straightforward for a nonmonotonical function $g(U)$. Fortunately, there is a basic theorem that avoids this procedure. That theorem reads as

$$E[g(U)] \equiv \int_{-\infty}^{\infty} g(u) p_U(u)\,du \tag{2.27}$$

which states mathematically that $E(V)$ for any given function $V = g(U)$ can be expressed directly in terms of the PDF $p_U(u)$ of U and the function $g(u)$. The proof of the theorem (2.27) shall not be given here, for which the reader is referred to Papoulis [1991].

With Eq. (2.27), one may define other moments of U by letting $g(U) = U^m$ where m is a nonnegative integer. For example, the *zeroth moment* for $m = 0$ is the integration of the PDF over the entire probability space, giving the value of one; the *second moment* for $m = 2$ is the *mean square* of U. One may also define the *central moments* of U by letting $g(U) = (U - \langle U \rangle)^m$, which are more commonly used for $m \geq 2$. The *first central moment* of U is the mean of the *fluctuation* $U' \equiv U - \langle U \rangle$, which is always zero. The *second central moment* is the variance of U,

$$\sigma_U^2 \equiv E(U'^2) = \langle U^2 \rangle - \langle U \rangle^2 \tag{2.28}$$

The square root of the variance is the *standard deviation* (σ_U) of U, which is a measure of the magnitude of the fluctuations of U about its mean $\langle U \rangle$. The *third central moment* is the *skewness* $s_U = E(U'^3)$, which measures the lack of symmetry of the distribution of U. Each *higher moment* gives some additional information about the *structure* of $p_U(u)$. As a matter of fact, it may need a complete knowledge of all moments to reconstruct $p_U(u)$.

One may want to prove that for the normal random variable with the PDF (2.11), the mean is μ, the variance is σ^2, and the skewness is zero. For a normal distribution, the first two central moments provide complete information because the higher moments can be expressed in terms of them,

$$E(U'^m) = \begin{cases} 0, & \text{for odd } m \\ 1 \cdot 3 \cdot 5 \cdots (m-1)\sigma^m, & \text{for even } m \end{cases} \tag{2.29}$$

For example, the *fourth central moment*, called the *kurtosis*, is $E(U'^4) = 3\sigma^4$.

2.1.5 Conditional Probability

We are often not only interested in the probability of a random variable or its functions but also in some statistical statements of the random variable for certain given conditions. For example, we may ask the following question: what is the probability that a randomly packed sand column has a permeability (k) less than a certain value k_o if it has been found that $k \in (k_1, k_2)$ (where $k_1 \le k_o \le k_2$)?

The *conditional distribution* $F(U|C)$ of a random variable, in the generic notation, U, for a given condition C is defined by

$$F(u|C) = P(U \le u|C) = \frac{P(U \le u, U \in C)}{P(U \in C)} \qquad (2.30)$$

where $P(U \in C)$ stands for the probability that U satisfies the condition C, $P(U \le u, U \in C)$ is the probability that $U \le u$ and $U \in C$. The conditional distribution has properties similar to the (unconditional) distribution defined in Eq. (2.1). The corresponding *conditional probability density function* can be defined as

$$p(u|C) = \frac{dF(u|C)}{du} \qquad (2.31)$$

The conditional distribution can be obtained if the unconditional CDF $F(u)$ or the unconditional PDF $p(u)$ is given. For instance, the conditional PDF of U under the condition $a < U \le b$ can be fully expressed in terms of the unconditional distributions,

$$p(u|a < U \le b) = \begin{cases} \dfrac{p(u)}{F(b) - F(a)}, & a < u \le b \\ 0, & \text{otherwise} \end{cases} \qquad (2.32)$$

Two such conditional PDFs are illustrated in Fig. 2.3.

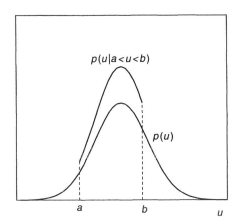
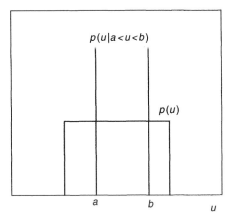

Figure 2.3. Sketch of conditional probability density functions.

2.1.6 Characteristic Functions and Cumulants

The *characteristic function* of a random variable U is defined as

$$\Phi(k) = \int_{-\infty}^{\infty} \exp(\iota k u) p(u)\, du \qquad (2.33)$$

where k is the wave number and $\iota = \sqrt{-1}$. The characteristic function $\Phi(k)$ of U is proportional to the *Fourier transform* of its probability density function. The properties of Fourier transforms are introduced in Section 2.4.4.

Since $p(u) \geq 0$, one has

$$|\Phi(k)| \leq \Phi(0) = 1 \qquad (2.34)$$

The PDF of U is given by the following *inverse formula* of $\Phi(k)$,

$$p(u) = \frac{1}{2\pi} \int_{-\infty}^{\infty} \exp(-\iota k u) \Phi(k)\, dk \qquad (2.35)$$

The fact that the PDF $p(u)$ can be determined with $\Phi(k)$ justifies the latter's name of characteristic function of U.

On the basis of Eq. (2.27), it is obvious that

$$\Phi(k) = E[\exp(\iota k U)] \qquad (2.36)$$

which may be expanded in the series

$$\Phi(k) = E\left[\sum_{m=0}^{\infty} \frac{(\iota k U)^m}{m!}\right] = \sum_{m=0}^{\infty} \frac{(\iota k)^m}{m!} E(U^m) \qquad (2.37)$$

Hence, if the series converges the characteristic function $\Phi(k)$ can be reconstructed with a knowledge of all the moments $E(U^m)$. In turn, the PDF can be obtained with Eq. (2.35). Differentiating both sides of Eq. (2.37) and evaluating at $k = 0$ leads to

$$E(U^m) = \frac{1}{\iota^m} \frac{d^m \Phi(k)}{dk^m}\bigg|_{k=0} \qquad (2.38)$$

Therefore, the moments of U can be obtained as the derivatives of its characteristic function. In particular, the mean and the mean square of U are

$$E(U) = \frac{1}{\iota} \frac{d\Phi(k)}{dk}\bigg|_{k=0} \qquad (2.39)$$

$$E(U^2) = -\frac{d^2 \Phi(k)}{dk^2}\bigg|_{k=0} \qquad (2.40)$$

The process of finding the moments can go on as long as the derivative is finite at $k = 0$.

Another quantity of particular interest is the logarithm of the characteristic function,

$$\Psi(k) = \ln \Phi(k) \tag{2.41}$$

which is called the *cumulant generating function*. Because of the fact that $\Psi(0) = 0$, the power series expansion about $k = 0$ has the form

$$\Psi(k) = \sum_{m=1}^{\infty} \frac{(\iota k)^m}{m!} \langle\langle U^m \rangle\rangle \tag{2.42}$$

where $\langle\langle U^m \rangle\rangle$ are called the *cumulants*, defined as

$$\langle\langle U^m \rangle\rangle = \frac{1}{\iota^m} \frac{d^m \Psi(k)}{dk^m}\bigg|_{k=0} \tag{2.43}$$

Throughout this book, the double brackets $\langle\langle \ \rangle\rangle$ are used for denoting cumulants. It can be found by comparing the two series for $\Phi(k)$ and $\Psi(k)$ that the cumulants of U are related to its statistical moments via [e.g., Beran, 1968]:

$$\langle\langle U^1 \rangle\rangle = E(U) = \langle U \rangle$$

$$\langle\langle U^2 \rangle\rangle = E[(U - \langle U \rangle)^2] = \sigma_U^2$$

$$\langle\langle U^3 \rangle\rangle = E[(U - \langle U \rangle)^3] = s_U \tag{2.44}$$

$$\langle\langle U^4 \rangle\rangle = E[(U - \langle U \rangle)^4] - 3\{E[(U - \langle U \rangle)^2]\}^2$$

$$\cdots$$

A striking feature of the cumulants is that higher cumulants contain information of decreasing significance, unlike higher moments. As a matter of fact, the cumulants $\langle\langle U^m \rangle\rangle$ are identically zero when $m > 2$ for a normal random variable.

2.2 JOINTLY DISTRIBUTED RANDOM VARIABLES

In Section 2.1, we have only discussed the probability distribution of a single random variable. We are also interested in probability statements concerning two or more random variables. Examples of joint random variables include the permeability k and porosity ϕ for one of the ensemble of sand columns packed repeatedly and randomly with the same quantity of sand, and the permeability values, $k(\mathbf{x}_1)$, $k(\mathbf{x}_2)$, and $k(\mathbf{x}_3)$, at three unmeasured locations of a given aquifer.

2.2.1 Joint Distributions

As for a single random variable, one may define the *joint cumulative distribution function* for any two random variables U_1 and U_2 by

$$F_{U_1 U_2}(u_1, u_2) = P(U_1 \leq u_1, U_2 \leq u_2) \tag{2.45}$$

where $P(U_1 \leq u_1, U_2 \leq u_2)$ stands for the probability that the random variable U_1 takes a value less than u_1 and U_2 less than u_2. The joint CDF is also a positive and monotonically increasing function and takes value between 0 and 1,

$$F_{U_1 U_2}(-\infty, u_2) = 0, \qquad F_{U_1 U_2}(u_1, -\infty) = 0, \qquad F_{U_1 U_2}(\infty, \infty) = 1 \qquad (2.46)$$

The *joint probability density function* $p_{U_1 U_2}(u_1, u_2)$ is defined as

$$p_{U_1 U_2}(u_1, u_2) = \frac{\partial^2 F_{U_1 U_2}(u_1, u_2)}{\partial u_1 \, \partial u_2} \qquad (2.47)$$

Conversely, one has

$$F_{U_1 U_2}(u_1, u_2) = \int_{-\infty}^{u_1} \int_{-\infty}^{u_2} p_{U_1 U_2}(u_1', u_2') \, du_1' \, du_2' \qquad (2.48)$$

Similarly, one may define the cumulative distribution function for a set of n random variables U_i as the probability that U_1 is less than u_1, U_2 is less than u_2, etc.,

$$F_U(u_1, u_2, \ldots, u_n) = P(U_1 \leq u_1, U_2 \leq u_2, \ldots, U_n \leq u_n) \qquad (2.49)$$

The *multivariate probability density function* is related to the CDF via

$$p_U(u_1, u_2, \ldots, u_n) = \frac{\partial^n F_U(u_1, u_2, \ldots, u_n)}{\partial u_1 \partial u_2 \cdots \partial u_n} \qquad (2.50)$$

$$F_U(u_1, u_2, \ldots, u_n) = \int_{-\infty}^{u_1} \int_{-\infty}^{u_2} \cdots \int_{-\infty}^{u_n} p_U(u_1', u_2', \ldots, u_n') \, du_1' \, du_2' \cdots du_n' \qquad (2.51)$$

2.2.2 Marginal Distributions

The cumulative distribution function $F_{U_1}(u_1)$ of U_1 alone can be obtained from the joint distribution $F_{U_1 U_2}(u_1, u_2)$ by letting $u_2 = \infty$,

$$F_{U_1}(u_1) = P(U_1 \leq u_1, U_2 \leq \infty) = F_{U_1 U_2}(u_1, \infty) \qquad (2.52)$$

$F_{U_1}(u_1)$ is called the *marginal CDF* of U_1. Similarly, the marginal CDF of U_2 is

$$F_{U_2}(u_2) = F_{U_1 U_2}(\infty, u_2) \qquad (2.53)$$

The *marginal probability density functions* are defined by

$$p_{U_1}(u_1) = \frac{\partial F_{U_1 U_2}(u_1, \infty)}{\partial u_1}, \qquad p_{U_2}(u_2) = \frac{\partial F_{U_1 U_2}(\infty, u_2)}{\partial u_2} \qquad (2.54)$$

It may be shown by differentiating Eq. (2.48) with respect to u_1 and u_2 and setting $u_2 = \infty$ and $u_1 = \infty$, respectively, that the marginal PDFs are given in terms of the joint PDF,

$$p_{U_1}(u_1) = \int_{-\infty}^{\infty} p_{U_1 U_2}(u_1, u_2)\, du_2, \qquad p_{U_2}(u_2) = \int_{-\infty}^{\infty} p_{U_1 U_2}(u_1, u_2)\, du_1 \qquad (2.55)$$

More generally, one has

$$p_U(u_1, u_2, \dots, u_{n-1}) = \int_{-\infty}^{\infty} p_U(u_1, u_2, \dots, u_n)\, du_n \qquad (2.56)$$

$$p_U(u_1, u_2, \dots, u_m)$$
$$= \int_{-\infty}^{\infty} \cdots \int_{-\infty}^{\infty} p_U(u_1, u_2, \dots, u_m, u_{m+1}, \dots, u_n) du_{m+1} \cdots du_n \qquad (2.57)$$

for $m \leq n$.

2.2.3 Functions of Random Variables

As for functions of a single random variable, we may be interested in the joint distribution of $V_1 = g_1(U_1, U_2)$ and $V_2 = g_2(U_1, U_2)$, which are functions of the random variables U_1 and U_2. The probability information about these new random variables V_1 and V_2 can be derived from a knowledge of the joint distribution of U_1 and U_2.

When g_1 and g_2 are single-valued functions of U_1 and U_2, they define a one-to-one transformation from the $U_1 U_2$-plane to the $V_1 V_2$-plane. That is, each point (u_1, u_2) in the $U_1 U_2$-plane corresponds to a unique point (v_1, v_2) in the $V_1 V_2$-plane. Under this condition, the joint PDF of V_1 and V_2 is related to that of U_1 and U_2 through the following simple relationship:

$$p_{V_1 V_2}(v_1, v_2) = p_{U_1 U_2}(u_1, u_2)|J(u_1, u_2)|^{-1} \qquad (2.58)$$

where $J(u_1, u_2)$ is the *Jacobian* of the transformation, defined with the *determinant*

$$J(u_1, u_2) = \begin{vmatrix} \dfrac{\partial g_1}{\partial u_1} & \dfrac{\partial g_1}{\partial u_2} \\[2mm] \dfrac{\partial g_2}{\partial u_1} & \dfrac{\partial g_2}{\partial u_2} \end{vmatrix} \qquad (2.59)$$

As for the case of a function of a single random variable, the relationship (2.58) can be derived by equating the cumulative distribution for (V_1, V_2) and (U_1, U_2) and then differentiating. In a similar manner, one may obtain a more general formula to relate the PDF of (V_1, V_2) to that of (U_1, U_2) when g_1 and g_2 are not single-valued functions of U_1 and U_2 [Papoulis, 1991]. The extension of Eqs. (2.58) and (2.59) to functions of n variables is straightforward.

2.2.4 Statistical Moments

As for a single random variable, the expectation of a function $V = g(U_1, U_2)$ can be expressed directly in terms of the function $g(U_1, U_2)$ and the joint PDF $p_{U_1 U_2}(u_1, u_2)$ of U_1 and U_2,

$$E[g(U_1, U_2)] = \int_{-\infty}^{\infty} \int_{-\infty}^{\infty} g(u_1, u_2) p_{U_1 U_2}(u_1, u_2) \, du_1 \, du_2 \qquad (2.60)$$

From Eq. (2.60), one may define the *joint statistical moments* of U_1 and U_2 by

$$E(U_1^m U_2^l) = \int_{-\infty}^{\infty} \int_{-\infty}^{\infty} u_1^m u_2^l p_{U_1 U_2}(u_1, u_2) \, du_1 \, du_2 \qquad (2.61)$$

where m and l are nonnegative integers. In particular, for $m = 1$ and $l = 0$ (or, $m = 0$ and $l = 1$), one has

$$\langle U_i \rangle \equiv E(U_i) = \int_{-\infty}^{\infty} u_i \, p_{U_i}(u_i) \, du_i \qquad (2.62)$$

which is the same as Eq. (2.25) given for a single random variable.

One may also define the *joint central moments* of U_1 and U_2 by

$$E(U_1'^m U_2'^l) = \int_{-\infty}^{\infty} \int_{-\infty}^{\infty} (u_1 - \langle U_1 \rangle)^m (u_2 - \langle U_2 \rangle)^l p_{U_1 U_2}(u_1, u_2) \, du_1 \, du_2 \qquad (2.63)$$

The case of particular interest is for $m = 2$ and $l = 0$ (or, $m = 0$ and $l = 2$), resulting in the variance, $\sigma_{U_i}^2$, of U_1 (or, U_2). Another quantity of great importance in this book is the *covariance* of U_1 and U_2, defined by

$$Cov(U_1, U_2) \equiv E(U_1' U_2')$$

$$= \int_{-\infty}^{\infty} \int_{-\infty}^{\infty} (u_1 - \langle U_1 \rangle)(u_2 - \langle U_2 \rangle) p_{U_1 U_2}(u_1, u_2) \, du_1 \, du_2$$

$$= E(U_1 U_2) - \langle U_1 \rangle \langle U_2 \rangle \qquad (2.64)$$

The covariance is an indication of how the two random variables U_1 and U_2 are correlated. In general, a positive covariance value indicates that U_2 tends to increase as U_1 increases, while a negative value indicates that U_2 tends to decrease as U_1 increases. Likewise, one may define the *correlation coefficient* of the two random variables U_1 and U_2 by the ratio

$$\rho = \frac{Cov(U_1, U_2)}{\sigma_{U_1} \sigma_{U_2}} \qquad (2.65)$$

which takes values from $[-1, 1]$. U_1 and U_2 are said to be *perfectly positively correlated* if $\rho = 1$, *perfectly negatively correlated* if $\rho = -1$, and *uncorrelated* if $\rho = 0$.

It should not be difficult to show the following properties for the first two moments of U_1 and U_2,

1. $E(U_1 + U_2) = E(U_1) + E(U_2)$,
2. $E(aU_1 + bU_2 + c) = aE(U_1) + bE(U_2) + c$,
3. $\text{Cov}(U, U) = \sigma_U^2$,
4. $\text{Cov}(U_1, U_2) = \text{Cov}(U_2, U_1)$,
5. $\text{Cov}(aU_1, bU_2 + c) = ab\text{Cov}(U_2, U_1)$,
6. $\text{Cov}(U_1, U_2 + U_3) = \text{Cov}(U_1, U_2) + \text{Cov}(U_1, U_3)$.

where a, b, and c are constants.

2.2.5 Independent Random Variables

The random variables U_1 and U_2 are said to be *independent* if and only if

$$p_{U_1 U_2}(u_1, u_2) = p_{U_1}(u_1) p_{U_2}(u_2) \tag{2.66}$$

which is equivalent to

$$F_{U_1 U_2}(u_1, u_2) = F_{U_1}(u_1) F_{U_2}(u_2) \tag{2.67}$$

It is easy to show that if U_1 and U_2 are independent, then one has $E(U_1 U_2) = E(U_1)E(U_2)$, and more generally, one has $E[g(U_1)h(U_2)] = E[g(U_1)]E[h(U_2)]$ for any functions g and h.

For two independent random variables U_1 and U_2, the PDF of their summation $V = U_1 + U_2$ is given as the *convolution* of their PDFs $p_{U_1}(u_1)$ and $p_{U_2}(u_2)$,

$$p_V(v) = \int_{-\infty}^{\infty} p_{U_1}(v - u_2) p_{U_2}(u_2) \, du_2 = \int_{-\infty}^{\infty} p_{U_1}(u_1) p_{U_2}(v - u_1) \, du_1 \tag{2.68}$$

It follows directly from Eq. (2.64) that $\text{Cov}(U_1, U_2) = 0$ if U_1 and U_2 are independent. That is, two independent random variables are uncorrelated. In general, two uncorrelated random variables may not necessarily be independent. However, for normal random variables, uncorrelatedness is equivalent to independence.

It is easy to show that for n independent random variables U_i, the mean and variance of $V = \sum_{i=1}^{n} U_i$ are

$$\langle V \rangle = \sum_{i=1}^{n} \langle U_i \rangle, \qquad \sigma_V^2 = \sum_{i=1}^{n} \sigma_{U_i}^2 \tag{2.69}$$

where $\langle U_i \rangle$ and $\sigma_{U_i}^2$ are the mean and variance of U_i. As a matter of fact, Eq. (2.69) only requires the weaker condition that the random variables U_i are mutually uncorrelated. If the random variables U_i are additionally normal, their summation V is also normal with the parameters $N(\sum \langle U_i \rangle, \sum \sigma_{U_i}^2)$.

2.2.6 Conditional Probability and Expectation

The conditional probability density function of the random variables U_1 and U_2, given $a < U_2 \leq b$, is defined by

$$p_{U_1 U_2}(u_1, u_2 | a < U_2 \leq b) = \begin{cases} \dfrac{p_{U_1 U_2}(u_1, u_2)}{F_{U_2}(b) - F_{U_2}(a)}, & a < u_2 \leq b \\ 0, & \text{otherwise} \end{cases} \tag{2.70}$$

Similarly, with the given condition of $a < U_1 \leq b$, the conditional PDF is given by

$$p_{U_1 U_2}(u_1, u_2 | a < U_1 \leq b) = \begin{cases} \dfrac{p_{U_1 U_2}(u_1, u_2)}{F_{U_1}(b) - F_{U_1}(a)}, & a < u_1 \leq b \\ 0, & \text{otherwise} \end{cases} \tag{2.71}$$

The conditional probability density function of the random variable U_1, given that U_2 was observed to take the value u_2, is defined by

$$p_{U_1}(u_1 | U_2 = u_2) = \frac{p_{U_1 U_2}(u_1, u_2)}{p_{U_2}(u_2)} = \frac{p_{U_1 U_2}(u_1, u_2)}{\int_{-\infty}^{\infty} p_{U_1 U_2}(u_1, u_2) \, du_1} \tag{2.72}$$

which can be simply denoted by $p_{U_1}(u_1 | u_2)$. The conditional cumulative probability distribution is obtained by integration,

$$F_{U_1}(u_1 | u_2) = \int_{-\infty}^{u_1} p_{U_1}(u_1' | u_2) \, du_1' \tag{2.73}$$

It is obvious that if U_1 and U_2 are independent, then $p_{U_1}(u_1 | u_2) = p_{U_1}(u_1)$.

Similar to Eq. (2.72), one has

$$p_{U_2}(u_2 | u_1) = \frac{p_{U_1 U_2}(u_1, u_2)}{p_{U_1}(u_1)} \tag{2.74}$$

Hence,

$$p_{U_1 U_2}(u_1, u_2) = p_{U_2}(u_2 | u_1) p_{U_1}(u_1) \tag{2.75}$$

$$p_{U_2}(u_2) = \int_{-\infty}^{\infty} p_{U_2}(u_2 | u_1) p_{U_1}(u_1) \, du_1 \tag{2.76}$$

Equation (2.76) is the so-called *total probability theorem*. Substituting Eqs. (2.75) and (2.76) into Eq. (2.72) leads to the well-known *Bayes' theorem*,

$$p_{U_1}(u_1 | u_2) = \frac{p_{U_2}(u_2 | u_1) p_{U_1}(u_1)}{\int_{-\infty}^{\infty} p_{U_2}(u_2 | u_1) p_{U_1}(u_1) \, du_1} \tag{2.77}$$

In the general case, the conditional PDF $p_U(u_1, u_2, \ldots, u_m | u_{m+1}, \ldots, u_n)$ is given as

$$p_U(u_1, u_2, \ldots, u_m | u_{m+1}, \ldots, u_n) = \frac{p_U(u_1, u_2, \ldots, u_m, u_{m+1}, \ldots, u_n)}{p_U(u_{m+1}, \ldots, u_n)} \quad (2.78)$$

for $m \leq n$.

The conditional PDFs can be used to define *conditional moments* of the random variables. For example, the *conditional mean* of U_1 for $U_2 = u_2$ is given by

$$\langle U_1 | u_2 \rangle = E(U_1 | u_2) = \int_{-\infty}^{\infty} u_1 p_{U_1}(u_1 | u_2) \, du_1 \quad (2.79)$$

and the *conditional variance* is

$$\sigma_{U_1|u_2}^2 = E[(U_1'|u_2)^2] = \int_{-\infty}^{\infty} (u_1 - \langle U_1 | u_2 \rangle)^2 p_{U_1}(u_1 | u_2) \, du_1 \quad (2.80)$$

where $U_1'|u_2 = U_1 - \langle U_1 | u_2 \rangle$. In general, the *conditional expectation* of a function $g(U_1, U_2)$ given $U_2 = u_2$ can be written as

$$E[g(U_1, U_2) | u_2] = \int_{-\infty}^{\infty} g(u_1, u_2) p_{U_1}(u_1 | u_2) \, du_1 \quad (2.81)$$

In the above, the conditional PDFs and the conditional moments are made to depend on a specific value of the random variable U_2, i.e., $U_2 = u_2$. The expressions for the conditional PDF and moments of U_1 are still valid if u_2 is replaced by U_2 (i.e., U_2 is left to be random). However, the conditional PDF of U_1 is a function of the random variable U_2, and the conditional moments of U_1 are functions of U_2. Hence, the conditional moments are random variables and are given by

$$\langle U_1 | U_2 \rangle = \int_{-\infty}^{\infty} u_1 p_{U_1}(u_1 | U_2) \, du_1 \quad (2.82)$$

$$\sigma_{U_1|U_2}^2 = \int_{-\infty}^{\infty} (u_1 - \langle U_1 | U_2 \rangle)^2 p_{U_1}(u_1 | U_2) \, du_1 \quad (2.83)$$

where $p_{U_1}(u_1 | U_2)$ is the conditional PDF of U_1 for a given (random) U_2. The conditional moments in terms of random variables are particularly useful when the conditioning variables are yet to be measured. Taking expectation of the conditional mean in Eq. (2.82) with respect to U_2 yields

$$E_{U_2}(\langle U_1 | U_2 \rangle) = \int_{-\infty}^{\infty} \left[\int_{-\infty}^{\infty} u_1 p_{U_1}(u_1 | U_2) \, du_1 \right] p_{U_2}(u_2) \, du_2$$

$$= \int_{-\infty}^{\infty} \int_{-\infty}^{\infty} u_1 p_{U_1 U_2}(u_1, u_2) \, du_2 \, du_1$$

$$= \langle U_1 \rangle \quad (2.84)$$

which is the unconditional mean of U_1. It can be shown with the same procedure that the expectation of the conditional variance $\sigma^2_{U_1|U_2}$ is the unconditional variance $\sigma^2_{U_1}$. In general, it can be shown that

$$E_{U_2}\left\{E_{U_1}\left[g(U_1, U_2)|U_2\right]\right\} = E\left[g(U_1, U_2)\right] \tag{2.85}$$

where the first (conditional) expectation is with respect to the random variable U_1 and the the second (unconditional) expectation is with the random variable U_2. Equation (2.85) provides an alternative and often more convenient way to compute the expectation of the function $g(U_1, U_2)$ of two random variables. This is done by first fixing U_2, obtaining the conditional expectation $h(u_2) = E_{U_1}[g(U_1, U_2)|U_2 = u_2]$, and then restoring the randomness to U_2 and taking unconditional expectation of $h(U_2)$, which is a function of U_2 only. Extending to multivariate variables is straightforward.

2.2.7 Joint Characteristic Functions

The *joint characteristic function* of two random variables U_1 and U_2 is defined as

$$\Phi(k_1, k_2) = \int_{-\infty}^{\infty}\int_{-\infty}^{\infty} \exp(\iota k_1 u_1 + \iota k_2 u_2) p_{U_1 U_2}(u_1, u_2)\, du_1\, du_2$$
$$= E[\exp(\iota k_1 U_1 + \iota k_2 U_2)] \tag{2.86}$$

Conversely, the joint PDF of U_1 and U_2 is given by the inverse formula of $\Phi(k_1, k_2)$,

$$p_{U_1 U_2}(u_1, u_2) = \frac{1}{(2\pi)^2}\int_{-\infty}^{\infty}\int_{-\infty}^{\infty} \exp(-\iota k_1 u_1 - \iota k_2 u_2)\Phi(k_1, k_2)\, dk_1\, dk_2 \tag{2.87}$$

Similarly, the *multivariate characteristic function* may be defined for n random variables as

$$\Phi(k_1, k_2, \ldots, k_n) = E\left[\exp\left(\sum_{i=1}^{n} \iota k_i U_i\right)\right] \tag{2.88}$$

The *marginal characteristic functions* can be given as

$$\Phi_{U_1}(k) = \Phi(k, 0) = E[\exp(\iota k U_1)]$$
$$\Phi_{U_2}(k) = \Phi(0, k) = E[\exp(\iota k U_2)] \tag{2.89}$$
$$\Phi(k_1, k_2, \ldots, k_m) = \Phi(k_1, k_2, \ldots, k_m, k_{m+1} = 0, \ldots, k_n = 0)$$

where $m \leq n$. It is easy to show that if U_1 and U_2 are independent, then one has

$$\Phi(k_1, k_2) = \Phi_{U_1}(k_1)\Phi_{U_2}(k_2) \tag{2.90}$$

If $U_1, U_2, \ldots, U_n$ are n mutually independent random variables with the respective characteristic functions $\Phi_{U_i}(k)$, one has

$$\Phi(k_1, k_2, \ldots, k_n) = \Phi_{U_1}(k_1)\Phi_{U_2}(k_2) \cdots \Phi_{U_n}(k_n) \tag{2.91}$$

The characteristic function of the sum $V = \sum_{i=1}^{n} U_i$ of these n independent variables is simply the product of the characteristic functions of the n variables,

$$\Phi_V(k) = \Phi_{U_1}(k)\Phi_{U_2}(k) \cdots \Phi_{U_n}(k) \tag{2.92}$$

As for the characteristic function of a single random variable discussed in Section 2.1.6, the multivariate characteristic function (2.88) can be expanded by *Taylor expansions* with respect to k_i (where $1 \leq i \leq n$),

$$\Phi(k_1, k_2, \ldots, k_n) = \sum_{M=0}^{\infty} \frac{(\iota k_1)^{m_1} (\iota k_2)^{m_2} \cdots (\iota k_n)^{m_n}}{m_1! m_2! \cdots m_n!} E(U_1^{m_1} U_2^{m_2} \cdots U_n^{m_n}) \tag{2.93}$$

where $M = m_1 + m_2 + \cdots + m_n$ and $E(U_1^{m_1} U_2^{m_2} \cdots U_n^{m_n})$ are the *multivariate moments* of $U_1, U_2, \ldots, U_n$, given as

$$E(U_1^{m_1} U_2^{m_2} \cdots U_n^{m_n}) = \frac{1}{\iota^M} \frac{\partial^M \Phi(k_1, k_2, \ldots, k_n)}{\partial k_1^{m_1} \partial k_2^{m_2} \cdots \partial k_n^{m_n}} \bigg|_{k_1=k_2=\cdots=k_n=0} \tag{2.94}$$

As a special case, the joint moments of U_1 and U_2 are given as

$$E(U_1^{m_1} U_2^{m_2}) = \frac{1}{\iota^{m_1+m_2}} \frac{\partial^{m_1+m_2} \Phi(k_1, k_2)}{\partial k_1^{m_1} \partial k_2^{m_2}} \bigg|_{k_1=k_2=0} \tag{2.95}$$

The *joint cumulant generating function* of U_1 and U_2 is defined as

$$\Psi(k_1, k_2) = \ln \Phi(k_1, k_2) \tag{2.96}$$

For n variables, the *multivariate cumulant generating function* is given as

$$\Psi(k_1, k_2, \ldots, k_n) = \ln \Phi(k_1, k_2, \ldots, k_n) \tag{2.97}$$

It is clear that as a special case, the cumulant generating function of the sum of n independent random variables is equal to the sum of the individual cumulant generating functions.

The multivariate cumulant generating function can be expanded in the following form:

$$\Psi(k_1, k_2, \ldots, k_n) = \sum_{M=1}^{\infty} \frac{(\iota k_1)^{m_1} (\iota k_2)^{m_2} \cdots (\iota k_n)^{m_n}}{m_1! m_2! \cdots m_n!} \langle\langle U_1^{m_1} U_2^{m_2} \cdots U_n^{m_n} \rangle\rangle \tag{2.98}$$

where $\langle\langle U_1^{m_1} U_2^{m_2} \cdots U_n^{m_n} \rangle\rangle$ are the Mth *multivariate cumulants*. As for the cumulants of a single random variable (see Section 2.1.6), the cumulants of n variables are somehow related to their moments. Although no simple formula can be given, the first few cumulants can be expressed as

$$\langle\langle U_i \rangle\rangle = \langle U_i \rangle \tag{2.99}$$

$$\langle\langle U_i U_j \rangle\rangle = E(U_i U_j) - \langle U_i \rangle\langle U_j \rangle = \text{Cov}(U_i, U_j) \tag{2.100}$$

$$\langle\langle U_i U_j U_l \rangle\rangle = E(U_i U_j U_l) - E(U_i U_j)\langle U_l \rangle - \langle U_i \rangle E(U_j U_l)$$
$$- \langle U_j \rangle E(U_i U_l) + 2\langle U_i \rangle\langle U_j \rangle\langle U_l \rangle \tag{2.101}$$

$$\langle\langle U_i U_j U_l U_m \rangle\rangle = E(U_i U_j U_l U_m) - 6\langle U_i \rangle\langle U_j \rangle\langle U_l \rangle\langle U_m \rangle - E(U_i U_j)E(U_l U_m)$$
$$- E(U_i U_l)E(U_j U_m) - E(U_i U_m)E(U_j U_l) - \langle U_i \rangle E(U_j U_l U_m)$$
$$- \langle U_j \rangle E(U_i U_l U_m) - \langle U_l \rangle E(U_i U_j U_m) - \langle U_m \rangle E(U_i U_j U_l)$$
$$+ 2[\langle U_i \rangle\langle U_j \rangle E(U_l U_m) + \langle U_i \rangle\langle U_l \rangle E(U_j U_m) + \langle U_i \rangle\langle U_m \rangle E(U_j U_l)$$
$$+ \langle U_j \rangle\langle U_l \rangle E(U_i U_m) + \langle U_j \rangle\langle U_m \rangle E(U_i U_l) + \langle U_l \rangle\langle U_m \rangle E(U_i U_j)] \tag{2.102}$$

These expressions are valid for any number of equal i, j, l, m. It can be verified that for $i = j = l = m$, Eqs. (2.99)–(2.102) reduce to Eq. (2.44). There are general procedures to express the higher cumulants in terms of moments, for which interested readers are referred to van Kampen [1981] and Gardiner [1985].

2.2.8 Multivariate Normal and Lognormal Random Variables

The joint distribution of two normal random variables is frequently used. The *bivariate probability density function* reads as

$$p_{U_1 U_2}(u_1, u_2) = \frac{1}{2\pi\sigma_1\sigma_2\sqrt{1-\rho^2}} \exp\left\{ \frac{-1}{1-\rho^2} \left[\frac{(u_1 - \mu_1)^2}{2\sigma_1^2} \right.\right.$$
$$\left.\left. - \rho\frac{(u_1 - \mu_1)(u_2 - \mu_2)}{\sigma_1\sigma_2} + \frac{(u_2 - \mu_2)^2}{2\sigma_2^2} \right] \right\} \tag{2.103}$$

where μ_i and σ_i^2 are the mean and variance of u_i, and ρ is the correlation coefficient. As for the univariate normal PDF of Eq. (2.11), the *bivariate normal PDF* may be denoted by $N(\mu_1, \sigma_1^2; \mu_2, \sigma_2^2; \rho)$. This density function has a single peak at $(u_1, u_2) = (\mu_1, \mu_2)$. A marginal distribution of this bivariate normal distribution is also a normal distribution, the same as Eq. (2.11) given for a single (univariate) normal random variable. For $\rho = 1$ (i.e., U_1 and U_2 are perfectly positively correlated), the bivariate PDF is reduced to

$$p_{U_1 U_2}(u_1, u_2) = p_{U_1}(u_1)\delta(u_1 - u_2) \tag{2.104}$$

where $p_{U_1}(u_1)$ is the univariate (marginal) PDF of U_1. In words, Eq. (2.104) states that the probability of $U_1 = U_2$ is one for $\rho = 1$. For $\rho = -1$, the bivariate PDF reads as

$$p_{U_1 U_2}(u_1, u_2) = p_{U_1}(u_1)\delta(u_1 + u_2) \tag{2.105}$$

which states mathematically that the probability that $U_1 = -U_2$ is one if U_1 and U_2 are perfectly negatively correlated.

The *bivariate lognormal PDF* of $V_1 = \exp(U_1)$ and $V_2 = \exp(U_2)$ can be easily obtained from Eq. (2.103) through transformation of variables (see Eq. (2.58)),

$$p_{V_1 V_2}(v_1, v_2) = \frac{1}{2\pi\sigma_1\sigma_2\sqrt{1-\rho^2}v_1 v_2} \exp\left\{\frac{-1}{1-\rho^2}\left[\frac{(\ln v_1 - \mu_1)^2}{2\sigma_1^2}\right.\right.$$
$$\left.\left. -\rho\frac{(\ln v_1 - \mu_1)(\ln v_2 - \mu_2)}{\sigma_1\sigma_2} + \frac{(\ln v_2 - \mu_2)^2}{2\sigma_2^2}\right]\right\} \tag{2.106}$$

where μ_i and σ_i^2 are the mean and variance of the normal random variable $U_i = \ln V_i$, and ρ is the correlation coefficient of U_1 and U_2.

The *first two moments* of the normal random variables (U_1, U_2) and the lognormal random variables $(V_1, V_2) = [\exp(U_1), \exp(U_2)]$ are related via

$$\langle V_i \rangle = \exp\left[\langle U_i \rangle + \tfrac{1}{2}\sigma_{U_i}^2\right] \tag{2.107}$$

$$\sigma_{V_i}^2 = \left[\exp(\sigma_{U_i}^2) - 1\right]\exp\left[2\langle U_i \rangle + \sigma_{U_i}^2\right] \tag{2.108}$$

$$\text{Cov}(V_1, V_2) = \{\exp[\text{Cov}(U_1, U_2)] - 1\}\exp\left[\langle U_1 \rangle + \langle U_2 \rangle + \tfrac{1}{2}\sigma_{U_1}^2 + \tfrac{1}{2}\sigma_{U_2}^2\right] \tag{2.109}$$

Conversely,

$$\langle U_i \rangle = 2\ln\langle V_i \rangle - \tfrac{1}{2}\ln[\langle V_i \rangle^2 + \sigma_{V_i}^2] \tag{2.110}$$

$$\sigma_{U_i}^2 = \ln\left[1 + \frac{\sigma_{V_i}^2}{\langle V_i \rangle^2}\right] \tag{2.111}$$

$$\text{Cov}(U_1, U_2) = \ln\left[1 + \frac{\text{Cov}(V_1, V_2)}{\langle V_1 \rangle \langle V_2 \rangle}\right] \tag{2.112}$$

Note that both the normal and lognormal distributions are completely characterized by their respective first two moments.

For a mixture of normal U_1 and lognormal $V_2 = \exp(U_2)$ variables, their bivariate PDF is obtained as

$$p_{U_1 V_2}(u_1, v_2) = \frac{1}{2\pi\sigma_1\sigma_2\sqrt{1-\rho^2}v_2} \exp\left\{\frac{-1}{1-\rho^2}\left[\frac{(u_1 - \mu_1)^2}{2\sigma_1^2}\right.\right.$$
$$\left.\left. -\rho\frac{(u_1 - \mu_1)(\ln v_2 - \mu_2)}{\sigma_1\sigma_2} + \frac{(\ln v_2 - \mu_2)^2}{2\sigma_2^2}\right]\right\} \tag{2.113}$$

where μ_i and σ_i^2 are, again, the mean and variance of the normal random variables U_1 and $U_2 = \ln V_2$, and ρ remains to be the correlation coefficient of the two normal random variables.

The multivariate PDF of the n normal random variables $U_1, U_2, \ldots, U_n$ is given by

$$p_U(u_1, u_2, \ldots, u_n) = \frac{|\det \mathbf{R}|^{1/2}}{(2\pi)^{n/2}} \exp\left[-\frac{1}{2}\sum_{i=1}^{n}\sum_{j=1}^{n} R_{ij}(u_i - \mu_i)(u_j - \mu_j)\right] \quad (2.114)$$

where R_{ij} are the elements and $\det \mathbf{R}$ is the determinant of the matrix $\mathbf{R}$, and the latter is the inverse of the $n \times n$ *covariance matrix* $\mathbf{C}$. The covariance matrix is defined as

$$C_{ij} = \text{Cov}(U_i, U_j) = E(U_i U_j) - \langle U_i \rangle\langle U_j \rangle \quad (2.115)$$

where $1 \le i, j \le n$. Higher (central) moments of the n normal random variables can be expressed in terms of the covariance. For example,

$$E(U_1' U_2' U_3' U_4') = \text{Cov}(U_1, U_2)\text{Cov}(U_3, U_4) + \text{Cov}(U_1, U_3)\text{Cov}(U_2, U_4)$$
$$+ \text{Cov}(U_1, U_4)\text{Cov}(U_2, U_3) \quad (2.116)$$

where $U_i' = U_i - \mu_i$.

The multivariate characteristic function of the n normal random variables is

$$\Phi(k_1, k_2, \ldots, k_n) = \exp\left(\iota\sum_{i=1}^{n} k_i \mu_i - \frac{1}{2}\sum_{i=1}^{n}\sum_{j=1}^{n} C_{ij}k_i k_j\right) \quad (2.117)$$

The *first cumulants* $\langle\langle U_i \rangle\rangle$ are equal to the means μ_i, the *second cumulants* $\langle\langle U_i U_j \rangle\rangle$ are the covariances C_{ij}, and the *higher cumulants* are identically zero. Therefore, the cumulants are particularly useful and convenient quantities for describing multivariate normal variables.

The conditional PDF $p_U(u_1, u_2, \ldots, u_m | u_{m+1}, \ldots, u_n)$ is also normal. The *conditional means and covariances* of U_i and U_j are given by

$$\langle U_i | u_{m+1}, \ldots, u_n \rangle = \mu_i + \sum_{k=m+1}^{n} a_{ik}(u_k - \mu_k) \quad (2.118)$$

$$\text{Cov}(U_i, U_j | u_{m+1}, \ldots, u_n) = \text{Cov}(U_i, U_j) - \sum_{k=m+1}^{n} a_{ik}\text{Cov}(U_k, U_j) \quad (2.119)$$

where $1 \le i, j \le m$, μ_i are the unconditional means, $\text{Cov}(U_i, U_j) = C_{ij}$ are the unconditional covariances, u_k are the measurements of U_k ($m + 1 \le k \le n$), and the coefficients

a_{ij} are solutions of the following linear equations:

$$\sum_{k=m+1}^{n} a_{ik}\text{Cov}(U_k, U_l) = \text{Cov}(U_i, U_l) \tag{2.120}$$

for $m + 1 \leq l \leq n$. The solutions of Eq. (2.120) are

$$a_{ik} = \sum_{l=m+1}^{n} \text{Cov}(U_i, U_l) R_{kl} \tag{2.121}$$

where R_{kl} are again the elements of the inverse of the covariance matrix. It is seen that the coefficients depend on the covariance but not on the actual measurements of the conditioning variables (i.e., $U_{m+1}, \ldots, U_n$). Hence, the conditional covariances (and variances) are not functions of the actual measurements of the conditioning variables, although the conditional means are.

Through transformation of variables, PDFs for n lognormal random variables or a mixture of normal and lognormal variables can be written similarly.

2.2.9 The Central Limit Theorem

Normal distributions are important in many fields of science and engineering because many random variables are empirically well approximated by normal distributions. The reason for the wide normality arises from the *central limit theorem*, which states that a random variable composed of the sum of many independent but arbitrarily distributed components is approximately normal.

For n independent random variables $U_1, U_2, \ldots, U_n$ with the respective mean $\langle U_i \rangle$ and variance $\sigma_{U_i}^2$, their sum $V = \sum_{i=1}^{n} U_i$ has the mean $\langle V \rangle = \sum_{i=1}^{n} \langle U_i \rangle$ and the variance $\sigma_V^2 = \sum_{i=1}^{n} \sigma_{U_i}^2$. The central limit theorem asserts that under certain general conditions, the probability density function of V approaches the normal PDF $N(\langle V \rangle, \sigma_V^2)$ as n increases. Some of the conditions are that the variances of U_i are larger than zero but finite and their third moments are bounded so that no single component would dominate the distribution of the summation as $n \to \infty$. The condition that U_i are independent is important. However, the central limit theorem is still valid when the random variables U_i are sufficiently weakly correlated. No smooth properties of the PDFs of U_i are required. Hence, the central limit theorem can be applied even when the random variables U_i are discrete and have the PDFs of the form (2.19). The extension of the central limit theorem to the bivariate distribution is that the joint PDF of two random variables V_1 and V_2, obtained by summing independent variables, will approach a bivariate normal PDF. Likewise, the central limit theorem can be generalized to multivariate distributions.

It follows from the central limit theorem that the product of n independent positive random variables, $W = \prod_{i=1}^{n} U_i$ with $U_i > 0$, is approximately lognormally distributed when n is large.

2.3 STOCHASTIC PROCESSES AND RANDOM FIELDS

A *random function* is an indexed collection of random variables. For example, when $S(t_i)$ stands for the stream stages at a particular location measured daily, $S(t)$ $(t \in (t_1, t_2, \ldots, t_n))$ may be regarded as a random function if the stream stages cannot be accurately predicted due to unknown effects; when $k(\mathbf{x})$ stands for the permeability at each point of one of the ensemble of randomly packed sand columns, $k(\mathbf{x})$ is a random function indexed with respect to spatial coordinates. When the index is time, the random function is usually called a *stochastic process*; when the index is spatial coordinates, it is called a *random space function*. However, in this book the word "stochastic" is used interchangeably with "random". Hence, a stochastic process may be a random function in *space*, in *time*, or in *space–time*. When the index $\mathbf{x}$ is multidimensional, a random function or a stochastic process is also called a *random field*. In this book, even one-dimensional random functions are sometimes referred to as random fields. As in the literature of flow in porous media, the three terms (random function, stochastic process, and random field) are used interchangeably in this book.

For the example of randomly packed sand columns, the sequence of permeability values $k(\mathbf{x})$ for one of these columns (a particular trial or experiment) constitutes a *sample function* or a *realization* of the stochastic process. The collection of all possible sequences of permeability values is referred to as the *ensemble* of the process. A *continuous random field*, such as the spatial distribution of permeability or pressure in an aquifer, contains an infinite and uncountable number of points so that $\mathbf{x} \in (\mathbf{x}_1, \mathbf{x}_2, \ldots, \mathbf{x}_n)$ with $n \to \infty$. All statistical questions about this random field U can be answered with a knowledge of the joint PDF $p_U[u(\mathbf{x}_1); u(\mathbf{x}_2); \ldots; u(\mathbf{x}_n)]$ as $n \to \infty$. A rigorous definition of the joint PDF of an infinite number of points (random variables) can be made with the aid of functional analysis [e.g., Beran, 1968]. In this book, we shall utilize a common, though less rigorous, treatment: a continuous random field is defined through the joint PDF $p_U[u(\mathbf{x}_1); u(\mathbf{x}_2); \ldots; u(\mathbf{x}_n)]$ for any set of an arbitrary, yet finite, number of points n.

2.3.1 Statistics of Stochastic Processes

Let us use the generic notation $U(\mathbf{x})$ to denote a stochastic process. When the stochastic process is with respect to space, $\mathbf{x}$ stands for $(x_1, \ldots, x_d)^T$ where d is the dimensionality; when it is given in time, $\mathbf{x}$ is to be replaced by t; for a space–time stochastic process, $\mathbf{x}$ can be replaced by $(\mathbf{x}, t)$. For a specific value of $\mathbf{x}$, $U(\mathbf{x})$ is a random variable with the cumulative distribution function (CDF) defined by

$$F_U(u, \mathbf{x}) = P[U(\mathbf{x}) \le u] \tag{2.122}$$

where $P[U(\mathbf{x}) \le u]$ is the probability that at the specific $\mathbf{x}$, $U(\mathbf{x})$ is less than u. It is seen that the CDF in Eq. (2.122) depends on $\mathbf{x}$. This CDF is called the *one-point distribution* of the stochastic process (random field) $U(\mathbf{x})$. The *one-point probability density function* of $U(\mathbf{x})$ is given by

$$p_U(u, \mathbf{x}) = \frac{\partial F_U(u, \mathbf{x})}{\partial u} \tag{2.123}$$

The *two-point CDF* of the process $U(\mathbf{x})$ is defined with the probability of the random variables $U(\mathbf{x}_1)$ and $U(\mathbf{x}_2)$,

$$F_U(u_1, \mathbf{x}_1; u_2, \mathbf{x}_2) = P[U(\mathbf{x}_1) \le u_1, U(\mathbf{x}_2) \le u_2] \tag{2.124}$$

The *two-point PDF* of the process reads as

$$p_U(u_1, \mathbf{x}_1; u_2, \mathbf{x}_2) = \frac{\partial^2 F_U(u_1, \mathbf{x}_1; u_2, \mathbf{x}_2)}{\partial u_1 \partial u_2} \tag{2.125}$$

Similarly, the *n-point* CDF and PDF of the process $U(\mathbf{x})$ can be given as

$$F_U(u_1, \mathbf{x}_1; u_2, \mathbf{x}_2; \ldots; u_n, \mathbf{x}_n) = P[U(\mathbf{x}_1) \le u_1, U(\mathbf{x}_2) \le u_2, \ldots, U(\mathbf{x}_n) \le u_n] \tag{2.126}$$

$$p_U(u_1, \mathbf{x}_1; u_2, \mathbf{x}_2; \ldots; u_n, \mathbf{x}_n) = \frac{\partial^n F_U(u_1, \mathbf{x}_1; u_2, \mathbf{x}_2; \ldots; u_n, \mathbf{x}_n)}{\partial u_1 \partial u_2 \cdots \partial u_n} \tag{2.127}$$

One may define conditional probability density functions from the joint PDFs of the stochastic process. For example, the conditional PDF of $U(\mathbf{x}_1)$, given that the process was measured at $\mathbf{x}_2$ with the measurement $U(\mathbf{x}_2) = u_2$, is

$$p_U(u_1, \mathbf{x}_1 | u_2, \mathbf{x}_2) = p_U[u_1, \mathbf{x}_1 | U(\mathbf{x}_2) = u_2] = \frac{p_U(u_1, \mathbf{x}_1; u_2, \mathbf{x}_2)}{p_U(u_2, \mathbf{x}_2)} \tag{2.128}$$

with

$$p_U(u_2, \mathbf{x}_2) = \int_{-\infty}^{\infty} p_U(u_1, \mathbf{x}_1; u_2, \mathbf{x}_2) \, du_1 \tag{2.129}$$

The conditional PDF of $U(\mathbf{x}_1)$ and $U(\mathbf{x}_2)$, given that the process took on the respective values u_3 and u_4 at $\mathbf{x}_3$ and $\mathbf{x}_4$, is

$$p_U(u_1, \mathbf{x}_1; u_2, \mathbf{x}_2 | u_3, \mathbf{x}_3; u_4, \mathbf{x}_4) = \frac{p_U(u_1, \mathbf{x}_1; u_2, \mathbf{x}_2; u_3, \mathbf{x}_3; u_4, \mathbf{x}_4)}{p_U(u_3, \mathbf{x}_3, u_4, \mathbf{x}_4)} \tag{2.130}$$

with

$$p_U(u_3, \mathbf{x}_3; u_4, \mathbf{x}_4) = \int_{-\infty}^{\infty} \int_{-\infty}^{\infty} p_U(u_1, \mathbf{x}_1; u_2, \mathbf{x}_2; u_3, \mathbf{x}_3; u_4, \mathbf{x}_4) \, du_1 \, du_2 \tag{2.131}$$

The statistical properties of a stochastic process (random field) are completely defined with all the multivariate probability density functions or cumulative distributions of the process. In reality, it is, however, extremely complicated and costly to infer the multivariate probability distribution functions. Hence, the properties of the stochastic process $U(\mathbf{x})$ are often studied through some characteristics of the multivariate distributions, the simplest and most widely used ones of which being their statistical moments. As for random variables, the various statistical moments of the process $U(\mathbf{x})$ are defined with the multivariate PDF $p_U(u_1, \mathbf{x}_1; u_2, \mathbf{x}_2; \ldots; u_n, \mathbf{x}_n)$,

$$E[U^{m_1}(\mathbf{x}_1)U^{m_2}(\mathbf{x}_2)\cdots U^{m_n}(\mathbf{x}_n)] = \int_{-\infty}^{\infty}\int_{-\infty}^{\infty}\cdots\int_{-\infty}^{\infty} u_1^{m_1}u_2^{m_2}\cdots u_n^{m_n}$$
$$\cdot\, p_U(u_1, \mathbf{x}_1; u_2, \mathbf{x}_2; \ldots; u_n, \mathbf{x}_n)\, du_1\, du_2\cdots du_n \tag{2.132}$$

where m_i are nonnegative integers. The first, and simplest, moment is the *mean* or *expected value*,

$$\langle U(\mathbf{x})\rangle = \int_{-\infty}^{\infty} u p_U(u, \mathbf{x})\, du \tag{2.133}$$

The expected value $\langle U(\mathbf{x})\rangle$, as a function of $\mathbf{x}$, describes the *trend* of the stochastic process $U(\mathbf{x})$. The *second moment* is given as

$$\langle U(\mathbf{x}_1)U(\mathbf{x}_2)\rangle = \int_{-\infty}^{\infty}\int_{-\infty}^{\infty} u_1 u_2 p_U(u_1, \mathbf{x}_1; u_2, \mathbf{x}_2)\, du_1\, du_2 \tag{2.134}$$

which is called the *autocorrelation function* of $U(\mathbf{x})$. One may define the *second central moment* as

$$C_U(\mathbf{x}_1, \mathbf{x}_2) = \langle U'(\mathbf{x}_1)U'(\mathbf{x}_2)\rangle = \langle U(\mathbf{x}_1)U(\mathbf{x}_2)\rangle - \langle U(\mathbf{x}_1)\rangle\langle U(\mathbf{x}_2)\rangle$$
$$= \int_{-\infty}^{\infty}\int_{-\infty}^{\infty} u_1 u_2 p_U(u_1, \mathbf{x}_2; u_2, \mathbf{x}_2)\, du_1\, du_2 - \langle U(\mathbf{x}_1)\rangle\langle U(\mathbf{x}_2)\rangle \tag{2.135}$$

where $U'(\mathbf{x}_i) = U(\mathbf{x}_i) - \langle U(\mathbf{x}_i)\rangle$ is the fluctuation of U at $\mathbf{x}_i$. A comparison of Eq. (2.135) with Eq. (2.64) reveals that $C_U(\mathbf{x}_1, \mathbf{x}_2)$ is the covariance of the random variable U at two (space/time) points. Hence, it is usually called the *autocovariance* of U. We also call it the *autocovariance function* (or, simply the *covariance function*) since $C_U(\mathbf{x}_1, \mathbf{x}_2)$ is generally a function of the locations.

Both the autocorrelation and autocovariance functions are symmetric, i.e., $\langle U(\mathbf{x}_1)U(\mathbf{x}_2)\rangle = \langle U(\mathbf{x}_2)U(\mathbf{x}_1)\rangle$ and $C_U(\mathbf{x}_1, \mathbf{x}_2) = C_U(\mathbf{x}_2, \mathbf{x}_1)$. Furthermore, both of them are *semi-positive definite*, namely

$$\sum_{i,j=1}^{n} a_i a_j \langle U(\mathbf{x}_i)U(\mathbf{x}_j)\rangle \geq 0, \qquad \sum_{i,j=1}^{n} a_i a_j C_U(\mathbf{x}_i, \mathbf{x}_j) \geq 0 \tag{2.136}$$

for any n points $\mathbf{x}_1, \mathbf{x}_2, \ldots, \mathbf{x}_n$ and any n real coefficients $a_1, a_2, \ldots, a_n$.

When $\mathbf{x}_1 = \mathbf{x}_2 = \mathbf{x}$, the autocovariance reduces to the *variance function*,

$$\sigma_U^2(\mathbf{x}) = C_U(\mathbf{x}, \mathbf{x}) = \int_{-\infty}^{\infty} u^2 p_U(u, \mathbf{x})\, du - \langle U(\mathbf{x})\rangle^2 \tag{2.137}$$

The *autocorrelation coefficient* is given by

$$\rho_U(\mathbf{x}_1, \mathbf{x}_2) = \frac{C_U(\mathbf{x}_1, \mathbf{x}_2)}{\sigma_U(\mathbf{x}_1)\sigma_U(\mathbf{x}_2)} \tag{2.138}$$

It is clear that the autocorrelation coefficient is also symmetric and semi-positive definite.

Similarly, the *cross-covariance* between the two stochastic processes (random fields) U and V can be defined as

$$C_{UV}(\mathbf{x}_1, \mathbf{x}_2) = \langle U'(\mathbf{x}_1)V'(\mathbf{x}_2)\rangle = \langle U(\mathbf{x}_1)V(\mathbf{x}_2)\rangle - \langle U(\mathbf{x}_1)\rangle\langle V(\mathbf{x}_2)\rangle$$

$$= \int_{-\infty}^{\infty}\int_{-\infty}^{\infty} uv p_{UV}(u, \mathbf{x}_1; v, \mathbf{x}_2)\, du\, dv - \langle U(\mathbf{x}_1)\rangle\langle V(\mathbf{x}_2)\rangle \tag{2.139}$$

where $p_{UV}(u, \mathbf{x}_1; v, \mathbf{x}_1)$ is the joint probability density function of $U(\mathbf{x}_1)$ and $V(\mathbf{x}_2)$. This cross-covariance function measures how the two processes are correlated in space and/or time. If $C_{UV}(\mathbf{x}_1, \mathbf{x}_2) = 0$, we say that U at $\mathbf{x}_1$ and V at $\mathbf{x}_2$ are uncorrelated. If the two processes U and V are uncorrelated, then $C_{UV}(\mathbf{x}_i, \mathbf{x}_j) \equiv 0$ for any i and j. One may as well define the *cross-correlation coefficient* as

$$\rho_{UV}(\mathbf{x}_1, \mathbf{x}_2) = \frac{C_{UV}(\mathbf{x}_1, \mathbf{x}_2)}{\sigma_U(\mathbf{x}_1)\sigma_V(\mathbf{x}_2)} \tag{2.140}$$

However, contrary to the autocovariance and the autocorrelation coefficient, $C_{UV}(\mathbf{x}_i, \mathbf{x}_j)$ and $\rho_{UV}(\mathbf{x}_1, \mathbf{x}_2)$ may be neither symmetric nor semi-positive definite.

It is seen that the autocovariance is usually a function of the two (space and/or time) locations. There are, of course, exceptions. The first example is the so called *white noise* process, whose values $U(\mathbf{x}_i)$ and $U(\mathbf{x}_j)$ are uncorrelated for every $\mathbf{x}_i \neq \mathbf{x}_j$. The autocovariance of a white noise process is given by

$$C_U(\mathbf{x}_i, \mathbf{x}_j) = S_o(\mathbf{x}_i)\delta(\mathbf{x}_i - \mathbf{x}_j) \tag{2.141}$$

where S_o is a spectral density. This covariance function has been introduced in Section 1.4.4 and can be used to approximate rapidly fluctuating stochastic processes. The second example is the process of *random constant*, in which U is independent of the location $\mathbf{x}$ but is a random variable. The autocovariance of this process is given by

$$C_U(\mathbf{x}_i, \mathbf{x}_j) = \sigma_U^2 = \text{const} \tag{2.142}$$

and its PDFs are given by

$$p_U(u, \mathbf{x}) = p_U(u) \tag{2.143}$$

$$p_U(u_1, \mathbf{x}_1; u_2, \mathbf{x}_2; \ldots; u_n, \mathbf{x}_n) = p_U(u_1)\delta(u_2 - u_1)\cdots\delta(u_n - u_1) \tag{2.144}$$

Another exception is the important class of *stationary (statistically homogeneous) processes*, which are frequently studied in many fields. We shall discuss the stationary processes separately in the next section.

A stochastic process (or a random field) $U(\mathbf{x})$ is called *normal* if the random variables $U(\mathbf{x}_1), U(\mathbf{x}_2), \ldots, U(\mathbf{x}_n)$ are jointly normal for any set of n points. A *normal (Gaussian) process* is completely characterized by its mean $\langle U(\mathbf{x}) \rangle = \mu_U(\mathbf{x})$ and autocovariance $C_U(\mathbf{x}_i, \mathbf{x}_j)$. The one-point PDF of the normal process is given as $N[\mu_U(\mathbf{x}), \sigma_U^2(\mathbf{x})]$, where $\sigma_U^2(\mathbf{x}) = C_U(\mathbf{x}, \mathbf{x})$. The two-point PDF of the process has the form of Eq. (2.103) with the parameters $N[\mu_U(\mathbf{x}_1), \sigma_U^2(\mathbf{x}_1); \mu_U(\mathbf{x}_2), \sigma_U^2(\mathbf{x}_2); \rho_U(\mathbf{x}_1, \mathbf{x}_2)]$, and the multivariate PDF has the form of Eq. (2.114) with appropriate parameters. The multivariate characteristic function of the normal stochastic process is

$$\Phi_U(k_1, \mathbf{x}_1; k_2, \mathbf{x}_2; \ldots; k_n, \mathbf{x}_n) = \exp\left[\iota \sum_{i=1}^n k_i \mu(\mathbf{x}_i) - \frac{1}{2} \sum_{i=1}^n \sum_{j=1}^n C_U(\mathbf{x}_i, \mathbf{x}_j) k_i k_j \right]$$

(2.145)

A (deterministic) function $V(\mathbf{x}) = g[U(\mathbf{x})]$ of a stochastic process $U(\mathbf{x})$ is also a stochastic process. The statistics of the new process $V(\mathbf{x})$ may be derived from those of $U(\mathbf{x})$ through transformation of variables (see Section 2.2.3), and vice versa. It is obvious that if $U(\mathbf{x})$ is normal, then $V(\mathbf{x}) = \exp(U)$ is a *lognormal stochastic process*; if $V(\mathbf{x})$ is lognormal, $U(\mathbf{x}) = \ln V$ is normal. The means, variances and autocovariances of the normal process $U(\mathbf{x})$ and the lognormal process $V(\mathbf{x})$ are related through the formulae (2.107)–(2.112).

2.3.2 Stationary Stochastic Processes

A stochastic process is called *strictly stationary* (or, *statistically homogeneous*) if the joint probability density function is invariant to a shift in $\mathbf{x}$, i.e.,

$$p_U(u_1, \mathbf{x}_1; u_2, \mathbf{x}_2; \ldots; u_n, \mathbf{x}_n) = p_U(u_1, \mathbf{x}_1 + \Delta; u_2, \mathbf{x}_2 + \Delta; \ldots; u_n, \mathbf{x}_n + \Delta) \quad (2.146)$$

for any Δ. Hence, the PDF depends on the points $\mathbf{x}_1, \mathbf{x}_2, \ldots, \mathbf{x}_n$ only through their differences $\mathbf{x}_2 - \mathbf{x}_1, \ldots, \mathbf{x}_n - \mathbf{x}_1$. In particular,

$$p_U(u, \mathbf{x}) = p_U(u) \tag{2.147}$$

$$p_U(u_1, \mathbf{x}_1; u_2, \mathbf{x}_2) = p_U(u_1, u_2; \mathbf{r}) \tag{2.148}$$

where $\mathbf{r} = \mathbf{x}_1 - \mathbf{x}_2$ is the *separation vector* of the two points $\mathbf{x}_1$ and $\mathbf{x}_2$. If $\mathbf{x}$ stands for space coordinates, the process is said to be *spatially stationary*; if $\mathbf{x}$ is replaced by time t, the process is *temporally stationary*; if $\mathbf{x}$ is a space–time index, the process is *spatially and temporally stationary*.

Second-order stationarity A stochastic process is called *second-order* (*weakly* or *wide-sense*) *stationary* if the first two moments of the process have the following properties:

$$\langle U(\mathbf{x}) \rangle = \langle U \rangle = \text{const} \tag{2.149}$$

$$\sigma_U^2(\mathbf{x}) = \text{const} \tag{2.150}$$

$$C_U(\mathbf{x}_1, \mathbf{x}_2) = C_U(\mathbf{r}) = C_U(-\mathbf{r}) \tag{2.151}$$

That is, the second-order stationarity requires the mean and the variance to be finite and constant and the two-point covariance to depend on the separation vector rather than the actual locations of the two points.

Two stochastic processes U and V are said to be *jointly second-order stationary* if the one-point moments (the means and variances) are constants for each of the two processes and their cross-covariance only depends on the separation vector of the two points,

$$C_{UV}(\mathbf{x}_1, \mathbf{x}_2) = C_{UV}(\mathbf{r}) \tag{2.152}$$

However, in general, $C_{UV}(\mathbf{r}) \neq C_{UV}(-\mathbf{r})$.

It is obvious that a strictly stationary process is second-order stationary if its first two moments exist. However, the converse is not necessarily true because the properties of the first two moments may not reveal any information about the higher moments. Normal (Gaussian) stochastic processes are an important exception because they are completely determined by the first two moments. It is because of this and the implication of the central limit theorem that Gaussian processes are frequently encountered and thoroughly studied in many fields of science and engineering. The first two moments play a crucial role in many applications even though the underlying processes may not be Gaussian. The importance of the first two moments is motivated by the following practical considerations: first, the first two moments usually have the most direct physical significance and often suffice to approximate confidence intervals for a stochastic process if it is not far from Gaussian; second, there are seldom enough measurements to infer higher statistical moments; finally, the higher moments are increasingly more difficult to come by for non-Gaussian stochastic processes from a theoretical point of view.

Examples of covariance functions The parameters needed to characterize the first two moments of a stationary process are significantly reduced because both the mean $\langle U \rangle$ and the variance σ_U^2 are constant and the (two-point) autocovariance depends on the separation vector rather than the actual locations of the two points. For some commonly used covariance functions, they are completely defined with a few parameters. The following are some examples of covariance functions commonly used in this book and in the literature:

1. *Exponential covariance,*

$$C_U(\mathbf{r}) = \sigma_U^2 \exp\left\{ -\left[\sum_{i=1}^{d} \frac{r_i^2}{\lambda_i^2} \right]^{1/2} \right\} \tag{2.153}$$

2. *Gaussian covariance,*

$$C_U(\mathbf{r}) = \sigma_U^2 \exp\left[-\frac{\pi}{4} \sum_{i=1}^{d} \frac{r_i^2}{\lambda_i^2} \right] \tag{2.154}$$

3. *Separate exponential covariance,*

$$C_U(\mathbf{r}) = \sigma_U^2 \exp\left[-\sum_{i=1}^{d} \frac{|r_i|}{\lambda_i}\right] \tag{2.155}$$

4. *Linear covariance,*

$$C_U(\mathbf{r}) = \begin{cases} \sigma_U^2 \prod_{i=1}^{d}\left[1 - \frac{|r_i|}{2\lambda_i}\right], & |r_i| \leq 2\lambda_i \\ 0, & \text{otherwise} \end{cases} \tag{2.156}$$

In the above, $\mathbf{r} = (r_1, \ldots, r_d)^T$, d is the number of (spatial and temporal) dimensions, and λ_i are some *length-scale parameters*, which may be related to the so-called *linear integral scales* of the process $U(\mathbf{x})$. The integral scales are defined, for each of the three directions in the case of $d = 3$, as

$$I_1 = \frac{1}{\sigma_U^2} \int_0^\infty C_U(r_1, 0, 0)\, dr_1$$

$$I_2 = \frac{1}{\sigma_U^2} \int_0^\infty C_U(0, r_2, 0)\, dr_2 \tag{2.157}$$

$$I_3 = \frac{1}{\sigma_U^2} \int_0^\infty C_U(0, 0, r_3)\, dr_3$$

I_i is the *directional integral scale*, which is a measure of the average distance between two points on the r_i axis beyond which $U(\mathbf{x})$ and $U(\mathbf{x} + \mathbf{r})$ (where $\mathbf{r} = r_i(\delta_{1i}, \ldots, \delta_{di})^T$, and $\delta_{ij} = 1$ for $i = j$ and $= 0$ for $i \neq j$) are only weakly correlated. One may also define a *global integral scale*, in three dimensions, as

$$I = \left[\frac{6}{\pi} \frac{1}{\sigma_U^2} \int_0^\infty \int_0^\infty \int_0^\infty C_U(r_1, r_2, r_3)\, dr_1\, dr_2\, dr_3\right]^{1/3} \tag{2.158}$$

which measures the radius of a sphere beyond which U are only weakly correlated. It can be shown that for the above four covariances, λ_i are the directional integral scales, i.e., $I_i = \lambda_i$.

A covariance function is said to be *isotropic* if it is invariant under rotation. Hence, the covariance only depends on the magnitude $r = |\mathbf{r}|$ of the separation vector. The exponential and Gaussian covariances are isotropic if and only if $\lambda_1 = \cdots = \lambda_d$. However, the separate exponential covariance and the linear covariance are always direction dependent and thus *anisotropic* even if $\lambda_1 = \cdots = \lambda_d$. It is of interest to note that the process with the linear covariance (2.156) may result from spatial averaging of an isotropic uncorrelated process (white noise), taking a rectangular box (rectangle or segment) of size $2\lambda_i$ moving parallel to coordinates as the averaging volume (area or length).

Those covariances that depend on $\sum_{i=1}^{d} r_i^2 / \lambda_i^2$ are referred to as the *ellipsoidal covariance functions*. The exponential and Gaussian covariances of Eqs. (2.153) and (2.154) are two examples of such functions. The ellipsoidal covariance functions can be made isotropic by the transformations $r_i' = r_i / \lambda_i$.

Parameter estimation and the ergodic hypothesis In theory, the statistical parameters of a stochastic process can be estimated from the ensemble of the process. For example, the mean can be estimated from n realizations $U(\mathbf{x}, \zeta_i)$ of the process $U(\mathbf{x})$ with

$$\langle U(\mathbf{x}) \rangle^s = \frac{1}{n} \sum_i^n U(\mathbf{x}, \zeta_i) \tag{2.159}$$

where ζ_i in $U(\mathbf{x}, \zeta_i)$ labels the ith realization of the process and $\langle \ \rangle^s$ stands for the *sample mean*. The sample mean provides a good estimate of $\langle U(\mathbf{x}) \rangle$ if a large number of realizations $U(\mathbf{x}, \zeta_i)$ are available. However, in many applications there is usually only one single realization available. An example of particular interest is the spatial distribution of permeability in a given geologic formation (see, e.g., Fig. 1.1). In such a situation, the ensemble is an abstract concept invoked to reflect the uncertainty in the description of the spatial structure of the given formation, rather than a set of equally likely formations. Hence, the parameters characterizing the stochastic process (permeability structure, in this case) must be obtained from the given realization (formation). In particular, the mean and the covariance of a stochastic process often have to be estimated from the space (or time) averages rather than ensemble averages.

The condition that allows us to equate the ensemble with the space averages is the *ergodic hypothesis*. The hypothesis requires that all states of the ensemble of a stochastic process are available in each realization of the process. If the space and ensemble means can be exchanged, the stochastic process $U(\mathbf{x})$ is called *mean-ergodic*. Stationarity is a necessary but not a sufficient condition for a process to be mean-ergodic. A counterexample is the process of random constant where the process takes a uniform value c_i through the domain in each realization and this value varies in different realizations. Hence, the space mean of a realization is c_i, while the ensemble mean is $(1/n) \sum_{i=1}^{n} c_i$ as $n \to \infty$. It is clear that the two means are not equal and thus the stationary process of random constant is not mean-ergodic. The necessary and sufficient condition for a stationary process to be mean-ergodic is that the variance of the space mean approaches zero as the averaging size increases. Let us briefly elaborate on this. The space mean of a stationary process $U(\mathbf{x})$ is defined as

$$\bar{U} = \frac{1}{V} \int_V U(\mathbf{x}) \, d\mathbf{x} \tag{2.160}$$

where V is the averaging volume (area or length). The space mean is a random variable with (ensemble) mean

$$\langle \bar{U} \rangle = \frac{1}{V} \int_V \langle U(\mathbf{x}) \rangle \, d\mathbf{x} = \langle U \rangle \tag{2.161}$$

Because of Eq. (2.161), $\bar{U}$ is said to be an *unbiased estimator* of $\langle U \rangle$. If the variance of $\bar{U}$, $\sigma_{\bar{U}}^2$, tends to zero as $V \to \infty$, then $\bar{U}$ approaches $\langle U \rangle$, in a probabilistic sense. The variance of $\bar{U}$ is given by

$$\sigma_{\bar{U}}^2 = E[(\bar{U} - \langle U \rangle)^2] = \frac{1}{V^2} \int_V \int_V C_U(\mathbf{x} - \boldsymbol{\chi}) \, d\mathbf{x} \, d\boldsymbol{\chi} \tag{2.162}$$

where $C_U(\mathbf{x} - \boldsymbol{\chi})$ is the (ensemble) autocovariance of the stationary process $U(\mathbf{x})$. Thus, the stationary process $U(\mathbf{x})$ is mean-ergodic if and only if

$$\lim_{V \to \infty} \frac{1}{V^2} \int_V \int_V C_U(\mathbf{x} - \boldsymbol{\chi}) \, d\mathbf{x} \, d\boldsymbol{\chi} = 0 \tag{2.163}$$

Let us first consider the case of one dimension. When the averaging domain is a line segment $[0, L_1]$, Eq. (2.162) can be written as

$$\sigma_{\bar{U}}^2 = \frac{1}{L_1^2} \int_0^{L_1} \int_0^{L_1} C_U(x_1 - \chi_1) \, dx_1 \, d\chi_1 \tag{2.164}$$

which can be simplified by a change of variables. Letting $r_1 = x_1 - \chi_1$, integrating with respect to x_1, and recognizing that $C_U(r_1)$ is an even function, leads to

$$\begin{aligned}
\sigma_{\bar{U}}^2 &= \frac{1}{L_1^2} \int_0^{L_1} C_U(r_1) \, dr_1 \int_0^{L_1 - r_1} dx_1 + \frac{1}{L_1^2} \int_{-L_1}^0 C_U(r_1) \, dr_1 \int_{-r_1}^{L_1} dx_1 \\
&= \frac{1}{L_1} \int_{-L_1}^{L_1} C_U(r_1) \left(1 - \frac{|r_1|}{L_1}\right) dr_1 \\
&= \frac{2}{L_1} \int_0^{L_1} C_U(r_1) \left(1 - \frac{r_1}{L_1}\right) dr_1
\end{aligned} \tag{2.165}$$

It may be seen from Eq. (2.163) that the mean-ergodicity of a process depends on the behavior of its autocovariance $C_U(\mathbf{r})$ for large $\mathbf{r}$. In many cases such as the four covariances (2.153)–(2.156), $C_U(\mathbf{r}) \to 0$ as $r \to \infty$. Under this condition, the variance of $\bar{U}$ in Eq. (2.165) may be well approximated by

$$\sigma_{\bar{U}}^2 \approx \frac{2}{L_1} \int_0^{L_1} C_U(r_1) \, dr_1 \tag{2.166}$$

By virtue of Eq. (2.157), one has that for $L_1 \to \infty$, $\sigma_{\bar{U}}^2 = (2I_1/L_1)\sigma_U^2 \to 0$. It is clear that when the averaging domain size is much larger than the integral scale I_1, the condition for mean-ergodicity is satisfied. However, if a process does not have a finite integral scale or a finite variance, $\sigma_{\bar{U}}^2$ may not tend to zero as $L_1 \to \infty$.

The condition for mean-ergodicity may be stated more generally as *Slutsky's theorem*, which asserts that a stationary process $U(\mathbf{x})$ is mean-ergodic if and only if

$$\lim_{L_1 \to \infty} \frac{1}{L_1} \int_0^{L_1} C_U(r_1) \, dr_1 = 0 \tag{2.167}$$

It is clear that Eq. (2.167) holds true if

$$\int_0^{\infty} C_U(r_1) \, dr_1 = I_1 \sigma_U^2 < \infty \tag{2.168}$$

or, if

$$C_U(r_1) = 0, \quad \text{as } r_1 \to \infty \tag{2.169}$$

Condition (2.168) is satisfied if the process $U(\mathbf{x})$ has a finite variance and a finite integral scale; condition (2.169) is satisfied if the process is uncorrelated for large separation r_1.

The conditions for mean-ergodicity can be generalized to multiple dimensions. When the averaging domain is a d-dimensional rectangular box $V = L_1 L_2 \cdots L_d$, similar to Eq. (2.165) the variance of $\bar{U}$ can be written in a general form as

$$\sigma_{\bar{U}}^2 = \frac{2^d}{L_1 L_2 \cdots L_d} \int_0^{L_1} \int_0^{L_2} \cdots \int_0^{L_d} C_U(r_1, r_2, \ldots, r_d)$$
$$\cdot \left(1 - \frac{r_1}{L_1}\right)\left(1 - \frac{r_2}{L_2}\right) \cdots \left(1 - \frac{r_d}{L_d}\right) dr_1 \, dr_2 \cdots dr_d \tag{2.170}$$

where the autocovariance is assumed to *quadrant symmetric*, a term coined by Vanmarcke [1983]. A covariance is quadrant symmetric if it is even with respect to each component r_i of the separation vector $\mathbf{r} = (r_1, r_2, \ldots, r_d)^T$: $C_U(r_1, \ldots, r_i, \ldots, r_d) = C_U(r_1, \ldots, -r_i, \ldots, r_d)$. It is obvious that the four covariances (2.153)–(2.156) possess this property. Hence, the *generalized form of Slutsky's theorem* is that a multidimensional stationary process (random field) $U(\mathbf{x})$ is mean-ergodic if and only if

$$\lim_{L_1, L_2, \ldots, L_d \to \infty} \frac{1}{L_1 L_2 \cdots L_d} \int_0^{L_1} \int_0^{L_2} \cdots \int_0^{L_d} C_U(r_1, r_2, \ldots, r_d) \, dr_1 \, dr_2 \cdots dr_d = 0 \tag{2.171}$$

It is also clear that Eq. (2.171) holds if

$$\int_0^{\infty} \int_0^{\infty} \cdots \int_0^{\infty} C_U(r_1, r_2, \ldots, r_d) \, dr_1 \, dr_2 \cdots dr_d < \infty \tag{2.172}$$

or, if

$$C_U(r_1, r_2, \ldots, r_d) = 0, \quad \text{as } r_1, r_2, \ldots, r_d \to \infty \tag{2.173}$$

One may define the *space covariance* of the stationary process $U(\mathbf{x})$ with a given mean $\langle U \rangle$ as

$$\overline{[U(\mathbf{x}) - \langle U \rangle][U(\mathbf{x} + \mathbf{r}) - \langle U \rangle]} = \frac{1}{V} \int_V [U(\mathbf{x}) - \langle U \rangle][U(\mathbf{x} + \mathbf{r}) - \langle U \rangle]\, d\mathbf{x} \quad (2.174)$$

which is an unbiased estimator of $C_U(\mathbf{r})$. If the space and ensemble covariances are exchangeable, the process $U(\mathbf{x})$ is said to be *covariance-ergodic*. In general, a condition similar to Eq. (2.167) or Eq. (2.171) is required for the fourth moment of $U(\mathbf{x})$ in order for a stationary process to be covariance-ergodic [e.g., Papoulis, 1991]. However, the condition (2.169) or (2.173) in terms of the autocovariance is sufficient for a normal, stationary process to be ergodic (in the mean, covariance, and any higher moments).

In applications for flow in porous formations, the validity of ergodicity cannot be examined rigorously because only a single realization of the formation under study is available and the (ensemble) covariance and fourth moment of the random field are not available beforehand. Therefore, a common approach is to presume ergodicity. After obtaining the moments of interest by space averaging, these space moments are substituted into the conditions (2.167)–(2.169) or (2.171)–(2.173) to check the validity of the ergodic hypothesis.

Estimating statistical parameters from real data has been a central problem in the applications of stochastic processes and has been subject to many studies. This topic is, however, beyond the scope of this book. The emphasis of the book is on how to derive the statistical moments of output flow variables such as hydraulic head, flux, and velocity on the basis of the statistical moments of input variables (medium properties such as permeability, porosity and unsaturated parameters, and boundary and initial conditions). The statistical moments of the input variables may be derived from measurements by the methods of statistical inference or more specifically, for the first two moments by correlation theories [e.g., Vanmarcke, 1983; Yaglom, 1987] and by geostatistical analyses [e.g., Journel and Huijbregts, 1978; Isaaks and Srivastava, 1989; Deutsch and Journel, 1998].

2.3.3 Nonstationary Stochastic Processes

A stochastic process is said to be spatially (or temporally) *nonstationary* if either its mean varies spatially (or temporally) or its autocovariance depends on the actual locations (or times). For example, distinct geological layers, zones, and facies may cause the permeability field (process) of a given geologic formation to be spatially nonstationary, and seasonal variations may render the infiltration rate of a vadose zone temporally nonstationary. A stationary stochastic process becomes nonstationary after conditioning on measurements. Furthermore, as shown in Chapter 1, the statistics of dependent flow variables such as hydraulic head, flux, and velocity are usually location dependent due to the effects of domain boundaries, spatial sinks/sources, or nonstationary medium features. Spatial nonstationarity is usually referred to as *statistical nonhomogeneity*. But in this book, the terms "nonstationarity" and "nonhomogeneity" are used interchangeably.

The first two moments of a nonstationary stochastic process (random field) can be expressed by Eqs. (2.133) and (2.134) as general space (or time) dependent functions.

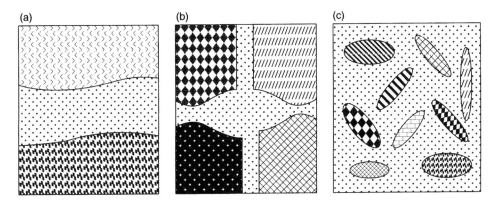

Figure 2.4. Sketch of geological layers (a), zones (b), and facies (c) with varying properties.

However, there exist some special forms of nonstationary processes whose statistical moments may exhibit some nice properties.

Zonal stationarity In many applications, a random field may be composed of a number of subfields. As illustrated in Fig. 2.4, a geological formation can include several distinct geological units (e.g., layers, zones, or facies), each of which may result from different geological processes. Although the permeability of each geological unit may be a stationary process (field), it has statistical moments different from its neighboring units. Therefore, the permeability of the entire formation is not a stationary field, although that of each unit is. Such a nonstationary random field is called, in one dimension, *piecewise stationary*, or more generally in multiple dimensions, *zonally stationary*.

Let us consider the permeability (or the log hydraulic conductivity) field in a formation of zonal stationarity. If the zones of stationary permeability can be identified on the basis of actual measurements of permeability (or other related rock properties), geophysical surveys, or some qualitative assessments, the statistical moments may be inferred from measurements separately at each zone. In this representation, well-defined large-scale features are treated deterministically while within-unit variabilities are modeled as random space functions.

When the geometries or boundaries of large-scale features cannot be very well defined due to incomplete knowledge about them, they may be themselves treated as random variables. Let us consider formations composed of M distinct features (e.g., layers, zones, or facies). The probability that a point $\mathbf{x}$ of the medium lies in feature F_i is $P_i(\mathbf{x}) = P[\mathbf{x} \in F_i]$. It is obvious that at any point $\sum_{i=1}^{M} P_i(\mathbf{x}) = 1$. An attribute at a point within the domain may be expressed by the following random variable:

$$Y(\mathbf{x}) = \sum_{i=1}^{M} I_i(\mathbf{x}) Y_i(\mathbf{x}) \qquad (2.175)$$

where $I_i(\mathbf{x})$ is the *geometry* or *indicator function* at $\mathbf{x}$ with $I_i = 1$ if $\mathbf{x} \in F_i$ and zero otherwise and with $\sum I_i(\mathbf{x}) = 1$, and Y_i is the attribute within feature F_i. For example,

$Y_1(\mathbf{x})$ may indicate the permeability of sand, $Y_2(\mathbf{x})$ may stand for the permeability of shale, and $Y_3(\mathbf{x})$ for that of clay, in a sand–shale–clay formation.

As both $I_i(\mathbf{x})$ and $Y_i(\mathbf{x})$ are random space functions, so is $Y(\mathbf{x})$. With the statistical properties of I_i and Y_i, it is, in principle, possible to obtain those of Y. For example, the expected value of $Y(\mathbf{x})$ is obtained as

$$\langle Y(\mathbf{x}) \rangle = \sum_{i=1}^{M} P_i(\mathbf{x}) \langle Y_i(\mathbf{x}) \rangle \tag{2.176}$$

upon assuming mutual independence of F_i. Other moments of $Y(\mathbf{x})$ are, however, not so easy to come by.

A special case of Eq. (2.175) has been utilized in stochastic analyses of subsurface flow and solute transport by several researchers such as Desbarats [1987], Rubin [1995], and Winter and Tartakovsky [2000]. These researchers considered *bimodal* heterogeneous media by letting $M = 2$. In this case, Eq. (2.175) reduces to

$$Y(\mathbf{x}) = I(\mathbf{x})Y_1(\mathbf{x}) + [1 - I(\mathbf{x})]Y_2(\mathbf{x}) \tag{2.177}$$

where $I(\mathbf{x})$ is equal to one for $Y = Y_1$ with probability $P(\mathbf{x})$ and equal to zero for $Y = Y_2$ with probability $1 - P(\mathbf{x})$. The statistical properties of $Y(\mathbf{x})$ can be derived in terms of those of $Y_1(\mathbf{x})$ and $Y_2(\mathbf{x})$. The moments of Y_1 and Y_2 can be regarded as conditional. For example, their expected values are given by

$$\langle Y_1(\mathbf{x}) \rangle = E[Y(\mathbf{x})|I(\mathbf{x}) = 1], \qquad \langle Y_2(\mathbf{x}) \rangle = E[Y(\mathbf{x})|I(\mathbf{x}) = 0] \tag{2.178}$$

and their autocovariances are given by

$$\begin{aligned} C_1(\mathbf{x}, \chi) &= E[Y(\mathbf{x})Y(\chi)|I(\mathbf{x}) = 1, I(\chi) = 1] - \langle Y_1(\mathbf{x}) \rangle \langle Y_1(\chi) \rangle \\ C_2(\mathbf{x}, \chi) &= E[Y(\mathbf{x})Y(\chi)|I(\mathbf{x}) = 0, I(\chi) = 0] - \langle Y_2(\mathbf{x}) \rangle \langle Y_2(\chi) \rangle \end{aligned} \tag{2.179}$$

These moments of Y_1 and Y_2 as well as those of $I(\mathbf{x})$ are assumed to be known as they may be inferred from measurements. The first two statistical moments of $Y(\mathbf{x})$ are expressed in terms of the above conditional moments [Rubin, 1995],

$$\langle Y(\mathbf{x}) \rangle = E[Y(\mathbf{x})|I(\mathbf{x}) = 1]P[I(\mathbf{x}) = 1] + E[Y(\mathbf{x})|I(\mathbf{x}) = 0]P[I(\mathbf{x}) = 0] \tag{2.180}$$

$$\begin{aligned} E[Y(\mathbf{x})Y(\chi)] = \ & E[Y(\mathbf{x})Y(\chi)|I(\mathbf{x}) = 1, I(\chi) = 1]P[I(\mathbf{x}) = 1, I(\chi) = 1] \\ & + E[Y(\mathbf{x})Y(\chi)|I(\mathbf{x}) = 1, I(\chi) = 0]P[I(\mathbf{x}) = 1, I(\chi) = 0] \\ & + E[Y(\mathbf{x})Y(\chi)|I(\mathbf{x}) = 0, I(\chi) = 1]P[I(\mathbf{x}) = 0, I(\chi) = 1] \\ & + E[Y(\mathbf{x})Y(\chi)|I(\mathbf{x}) = 0, I(\chi) = 0]P[I(\mathbf{x}) = 0, I(\chi) = 0] \end{aligned} \tag{2.181}$$

Assuming mutual independence of Y_1 and Y_2, the expected value and covariance of $Y(\mathbf{x})$ are obtained as

$$\langle Y(\mathbf{x}) \rangle = P(\mathbf{x}) \langle Y_1(\mathbf{x}) \rangle + [1 - P(\mathbf{x})] \langle Y_2(\mathbf{x}) \rangle \tag{2.182}$$

$$C_Y(\mathbf{x}, \boldsymbol{\chi}) = [P(\mathbf{x})P(\boldsymbol{\chi}) + C_I(\mathbf{x}, \boldsymbol{\chi})]C_1(\mathbf{x}, \boldsymbol{\chi})$$
$$+ \{[1 - P(\mathbf{x})][1 - P(\boldsymbol{\chi})] + C_I(\mathbf{x}, \boldsymbol{\chi})\}C_2(\mathbf{x}, \boldsymbol{\chi})$$
$$+ C_I(\mathbf{x}, \boldsymbol{\chi})[\langle Y_1(\mathbf{x})\rangle - \langle Y_2(\mathbf{x})\rangle][\langle Y_1(\boldsymbol{\chi})\rangle - \langle Y_2(\boldsymbol{\chi})\rangle] \tag{2.183}$$

where $C_I(\mathbf{x}, \boldsymbol{\chi}) = E[I(\mathbf{x})I(\boldsymbol{\chi})] - P(\mathbf{x})P(\boldsymbol{\chi})$ is the covariance of the indicator function I. It is clear that $\sigma_I^2(\mathbf{x}) = C_I(\mathbf{x}, \mathbf{x}) = P(\mathbf{x})[1 - P(\mathbf{x})]$. Hence, the variance of $Y(\mathbf{x})$ is

$$\sigma_Y^2(\mathbf{x}) = P(\mathbf{x})\sigma_1^2(\mathbf{x}) + [1 - P(\mathbf{x})]\sigma_2^2(\mathbf{x}) + P(\mathbf{x})[1 - P(\mathbf{x})][\langle Y_1(\mathbf{x})\rangle - \langle Y_2(\mathbf{x})\rangle]^2 \tag{2.184}$$

where $\sigma_1^2(\mathbf{x})$ and $\sigma_2^2(\mathbf{x})$ are the respective variance of $Y_1(\mathbf{x})$ and $Y_2(\mathbf{x})$. It is seen that $Y(\mathbf{x})$ is generally nonstationary as its moments are location dependent.

Some special cases of the bimodal media have been considered before. Desbarats [1987] studied sand–shale formations where $I(\mathbf{x})$ is assumed spatially stationary such that P becomes constant through the domain. In addition, the permeability is treated as a known constant, K_{ss} or K_{sh}, in the sand or shale, respectively. Hence, the permeability at any point can be written as $K(\mathbf{x}) = I(\mathbf{x})K_{ss} + [1 - I(\mathbf{x})]K_{sh}$, whose moments are

$$\langle K \rangle = PK_{ss} + (1 - P)K_{sh} \tag{2.185}$$

$$\sigma_K^2 = P(1 - P)(K_{ss} - K_{sh})^2 \tag{2.186}$$

$$C_K(\mathbf{r}) = C_I(\mathbf{r})(K_{ss} - K_{sh})^2 \tag{2.187}$$

where $\mathbf{r} = \mathbf{x} - \boldsymbol{\chi}$. In this special case, $K(\mathbf{x})$ is stationary. Note that the variance of $K(\mathbf{x})$ increases quadratically as the difference between K_{ss} and K_{sh}.

In the case considered by Rubin [1995], the attributes $Y_1(\mathbf{x})$ and $Y_2(\mathbf{x})$ are allowed to vary spatially as stationary random fields while $I(\mathbf{x})$ remains stationary. Then, Eqs. (2.182)–(2.184) become

$$\langle Y \rangle = P\langle Y_1 \rangle + (1 - P)\langle Y_2 \rangle \tag{2.188}$$

$$C_Y(\mathbf{r}) = [P^2 + C_I(\mathbf{r})]C_1(\mathbf{r}) + [(1 - P)^2 + C_I(\mathbf{r})]C_2(\mathbf{r}) + C_I(\mathbf{r})(\langle Y_1 \rangle - \langle Y_2 \rangle)^2 \tag{2.189}$$

$$\sigma_Y^2 = P\sigma_1^2 + (1 - P)\sigma_2^2 + P(1 - P)(\langle Y_1 \rangle - \langle Y_2 \rangle)^2 \tag{2.190}$$

Note that $Y(\mathbf{x})$ is stationary since $I(\mathbf{x})$ as well as $Y_1(\mathbf{x})$ and $Y_2(\mathbf{x})$ is stationary. The assumption of stationary $I(\mathbf{x})$ may be well justified for many situations such as interlaced sand–shale sequences [Desbarats, Fig. 3, 1987] and densely distributed fractures embedded in the matrix [Wels and Smith, Fig. 2, 1994]. However, the stationary assumption may be violated in the case of a domain of two disjoint units (or zones) with spatially varying or uncertain boundaries.

In the recent study by Winter and Tartakovsky [2000], the assumption of stationary $I(\mathbf{x})$ is relaxed, while the stationarity of $Y_i(\mathbf{x})$ is maintained. In their representation, P varies spatially and C_I depends on the actual locations of the two points $\mathbf{x}$ and $\boldsymbol{\chi}$ such that $Y(\mathbf{x})$ is spatially nonstationary,

$$\langle Y(\mathbf{x}) \rangle = P(\mathbf{x})\langle Y_1 \rangle + [1 - P(\mathbf{x})]\langle Y_2 \rangle \tag{2.191}$$

$$C_Y(\mathbf{x}, \boldsymbol{\chi}) = [P(\mathbf{x})P(\boldsymbol{\chi}) + C_I(\mathbf{x}, \boldsymbol{\chi})]C_1(\mathbf{r}) + \{[1 - P(\mathbf{x})][1 - P(\boldsymbol{\chi})]$$

$$+ C_I(\mathbf{x}, \boldsymbol{\chi})\}C_2(\mathbf{r}) + C_I(\mathbf{x}, \boldsymbol{\chi})(\langle Y_1 \rangle - \langle Y_2 \rangle)^2 \tag{2.192}$$

$$\sigma_Y^2(\mathbf{x}) = P(\mathbf{x})\sigma_1^2 + [1 - P(\mathbf{x})]\sigma_2^2 + P(\mathbf{x})[1 - P(\mathbf{x})](\langle Y_1 \rangle - \langle Y_2 \rangle)^2 \tag{2.193}$$

When $\mathbf{x}$ is deep within the unit F_1, $P(\mathbf{x}) \approx 1$, $\sigma_I^2(\mathbf{x}) \approx 0$ and $\sigma_Y^2(\mathbf{x}) \approx \sigma_1^2$. Near the boundary between the two units, the variability of $Y(\mathbf{x})$ may be large if $\langle Y_1 \rangle$ is significantly different from $\langle Y_2 \rangle$. It is thus seen that the attribute $Y(\mathbf{x})$ can be highly nonstationary even though the attribute $Y_i(\mathbf{x})$ in each unit is stationary.

Large-scale trending Another special form of nonstationarity is manifested in the mean of a stochastic process (random field). For example, the spatial distribution of permeability in an aquifer may exhibit a *trend* owing to downward or upward sand coarsening, or periodic sand–shale patterns; the infiltration rate in the vadose zone may vary periodically in time caused by seasonal effects. For such situations, Neuman and Jacobson [1984] and Rajaram and McLaughlin [1990] represented the hydrologic variability as a stationary random field superimposed on a large-scale, deterministic trend. That is, the nonstationarity of a stochastic process $U(\mathbf{x})$ is only manifested in the mean as a large-scale trend (or a global drift) whereas its fluctuations are stationary,

$$\langle U(\mathbf{x}) \rangle = \mu_U(\mathbf{x}) \tag{2.194}$$

$$C_U(\mathbf{x}, \boldsymbol{\chi}) = C_U(\mathbf{r}) \tag{2.195}$$

where $\mu_U(\mathbf{x})$ is a general function of space (and/or time), and $\mathbf{r} = \mathbf{x} - \boldsymbol{\chi}$.

Accounting for a spatially varying, large-scale trend in a random field reduces the variance of the fluctuations. Figure 2.5 shows a random log hydraulic conductivity Y field with periodically varying trend. If this trend is intentionally ignored to achieve (forced) stationarity, the resultant spatial variability in Y is artificially large with the variance equal to 5.12. After considering the periodic trend, the variability is significantly reduced with the variance equal to 0.87.

Although this form of medium nonstationarity is simple, many researchers have made use of this representation. Rehfeldt *et al.* [1992] found that the spatial variability in the log hydraulic conductivity at the Columbus Air Force Base in Mississippi could be better explained when a spatial trend in the mean log conductivity is included explicitly. Loaiciga *et al.* [1993] cited an example of trending in hydraulic conductivity observed in the Eye-Dashwa Lakes, Pluton, near Atitokan, Ontario.

Stationary increments There are some stochastic processes (random fields) that have unbounded variances and are thus not stationary but possess stationary increments. A simple

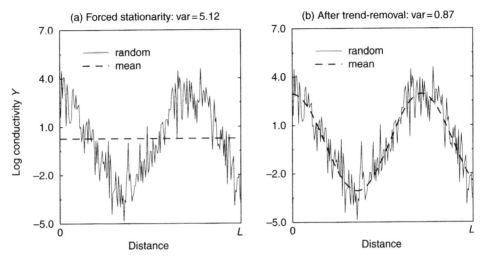

Figure 2.5. A random log hydraulic conductivity field with a large-scale trend: (a) ignoring the trend, and (b) accounting for it.

example of such processes can be constructed by the following integral of a one-dimensional stationary process $V(x)$ of mean $\langle V \rangle$, variance σ_V^2, and integral scale I_V [Dagan, 1989]:

$$U(x) = \int_0^x V(x')\,dx' \tag{2.196}$$

It is clear that the mean and fluctuation of $U(x)$ are given as

$$\langle U(x) \rangle = \langle V \rangle x \tag{2.197}$$

$$U'(x) = \int_0^x V'(x')\,dx' \tag{2.198}$$

The variance of $U(x)$ can be obtained as

$$\sigma_U^2(x) = 2\int_0^x (x - r')C_V(r')\,dr' = 2x\int_0^x C_V(r')\,dr' - 2\int_0^x r'C_V(r')\,dr' \tag{2.199}$$

Both the mean and the variance of $U(x)$ are functions of x. For large x, one has

$$\sigma_U^2(x) \approx 2x\sigma_V^2 I_V - \text{const} \tag{2.200}$$

It is thus seen that the variance of $U(x)$ increases indefinitely with x. However, the mean and variance of the increment $[U(x+r) - U(x)]$, given as

$$\langle [U(x+r) - U(x)] \rangle = \langle V \rangle r \tag{2.201}$$

$$\langle [U'(x+r) - U'(x)]^2 \rangle = 2 \int_0^r (r - r') C_V(r') \, dr' \tag{2.202}$$

do not depend on the actual locations of the two points x and $x + r$ but are only functions of the separation distance r. In addition, the variance of the increment $[U(x+r) - U(x)]$ is finite for a finite r. Such a process is called a *process of stationary increments*.

For any stochastic process $U(\mathbf{x})$, a *semi-variogram* may be defined as

$$\gamma_U(\mathbf{x}, \boldsymbol{\chi}) = \frac{1}{2} \langle [U'(\mathbf{x}) - U'(\boldsymbol{\chi})]^2 \rangle \tag{2.203}$$

The semi-variogram (or simply *variogram*) is the half of the variance of the increment $[U(\mathbf{x}) - U(\boldsymbol{\chi})]$ (i.e., the difference between the values of the general stochastic process U at two points $\mathbf{x}$ and $\boldsymbol{\chi}$). The process consists of stationary increments if the increment satisfies

$$\langle U(\mathbf{x}) - U(\boldsymbol{\chi}) \rangle = m_U(\mathbf{r})$$

$$\gamma_U(\mathbf{x}, \boldsymbol{\chi}) = \gamma_U(\mathbf{r}) \tag{2.204}$$

where $m_U(\mathbf{r})$ and $\gamma_U(\mathbf{r})$ are only functions of $\mathbf{r}$. It can be shown that under stationary increments,

$$m_U(\mathbf{r}_1 + \mathbf{r}_2) = m_U(\mathbf{r}_1) + m_U(\mathbf{r}_2) \tag{2.205}$$

The proof of Eq. (2.205) is left as an exercise at the end of this chapter. If $m_U(\mathbf{r})$ is a continuous function of $\mathbf{r}$, then Eq. (2.205) implies that $m_U(\mathbf{r})$ is a linear function of $\mathbf{r}$ [Yaglom, 1987],

$$m_U(\mathbf{r}) = c_1 r_1 + c_2 r_2 + \cdots + c_d r_d \tag{2.206}$$

where c_i are the coefficients, $\mathbf{r} = (r_1, \ldots, r_d)^T$ and d is the number of dimensions. It is clear that the mean value of the process (random field) $U(\mathbf{x})$ with stationary increments is given as

$$\langle U(\mathbf{x}) \rangle = c_0 + c_1 x_1 + c_2 x_2 + \cdots + c_d x_d \tag{2.207}$$

where $c_0 = \langle U(0) \rangle$ and $\mathbf{x} = (x_1, \ldots, x_d)^T$.

It follows from the definition (2.203) that if the variance and covariance of $U(\mathbf{x})$ exist,

$$\gamma_U(\mathbf{x}, \boldsymbol{\chi}) = \frac{1}{2} [\sigma_U^2(\mathbf{x}) + \sigma_U^2(\boldsymbol{\chi})] - C_U(\mathbf{x}, \boldsymbol{\chi}) \tag{2.208}$$

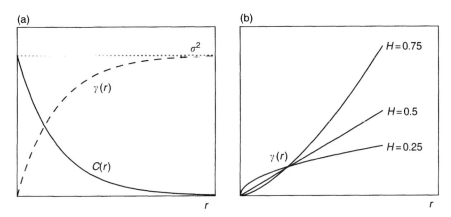

Figure 2.6. Variogram and covariance functions: (a) their relation for a stationary process, and (b) power law variogram models for a process without a finite variance.

Hence, when $U(\mathbf{x})$ is second-order stationary, the variogram is related to the covariance $C_U(\mathbf{r})$ in a simple way:

$$\gamma_U(\mathbf{r}) = \sigma_U^2 - C_U(\mathbf{r}) \tag{2.209}$$

It is seen that a stationary process consists also of stationary increments. As $r \to \infty$ (where $r = |\mathbf{r}|$), $C_U(\mathbf{r})$ generally tends to zero and thus $\gamma_U(\mathbf{r})$ approaches σ_U^2 (see Fig. 2.6a). In the terminology of geostatistics, the asymptotic constant is also called the *sill* of the variogram and the distance at which the variogram reaches this asymptotic value is called the *range*. However, if the process of interest does not have a finite variance, the variogram will never reach an asymptotic constant and the range is infinite (see Fig. 2.6b). In this case, the variogram is more general than the covariance because the former is well defined even if the latter is not.

Just as a covariance function must be semi-positive definite (see Eq. (2.136)), a variogram must be conditionally negative definite and be well behaved at $r \to \infty$ [Matheron, 1973; see also de Marsily, 1986, pp. 292–293]. In particular, γ_U for $r \to \infty$ must increase less rapidly than r^2, that is,

$$\lim_{r \to \infty} \frac{\gamma_U(\mathbf{r})}{r^2} = 0 \tag{2.210}$$

Self-similarity For a stochastic process with stationary increments, the variogram may take the following power law:

$$\gamma_U(\mathbf{r}) = ar^{2H} \tag{2.211}$$

where $0 < H < 1$ as per Eq. (2.210), $r = |\mathbf{r}|$, and a is a positive constant. H is the so-called *Hurst coefficient*. Three such power law variograms are shown in Fig. 2.6b. The case of

$H = 1$ is also possible, due to the special process of $U(\mathbf{x}) = V_o + V_1 x$, where $x = |\mathbf{x}|$ and V_o and V_1 are two random constants. It can be easily verified that for this special random field, $m_U(\mathbf{r}) - \langle V_1 \rangle_t$ and $\gamma_U(\mathbf{r}) = \frac{1}{2}\sigma_{V_1}^2 r^2$, where $\langle V_1 \rangle$ and $\sigma_{V_1}^2$ are the respective mean and variance of V_1.

There is a special property about the variogram (2.211): $\gamma_U(h\mathbf{r}) = c(h)\gamma_U(\mathbf{r})$, where $c(h) = ah^{2H}$ for any $h > 0$. That is, the form of this variogram is invariant under the similarity transformation $\mathbf{r} \to h\mathbf{r}$. The stochastic process (random field) that has the said property is called *self-similar*, or *self-affine*. When the increments of the process are self-similar the patterns of the incremental fluctuations appear similar regardless of the scale at which they are observed. In addition, this variogram is isotropic.

It is clear that there is no characteristic length, such as the range and the integral scale for a stationary field, associated with self-similar random fields. It can be shown that a stationary random field is not self-similar and that Eq. (2.211) is the only form for a self-similar variogram. The variogram of the form (2.211) has been used to represent the log hydraulic conductivity field whose variance increases with the scale of measurements (Hewett, 1986; Neuman, 1990, 1991, 1994; Ababou and Gelhar, 1990; Cushman, 1991, 1997; Kemblowski and Wen, 1993; Di Federico and Neuman, 1997, 1998; Hassan *et al.*, 1997; Molz *et al.*, 1997).

The power law behavior of $\gamma_U(\mathbf{r})$ in Eq. (2.211) implies that $U(\mathbf{x})$ is a random fractal field with dimension $D = d + 1 - H$, where d is the topological dimension of the random field. It has been shown [e.g., Molz *et al.*, 1997] that when $H < 0.5$, the increments of $U(\mathbf{x})$ are negatively correlated and $U(\mathbf{x})$ exhibits noisy behavior; when $H = 0.5$, the increments are uncorrelated; and when $H > 0.5$, the increments are positively correlated and $U(\mathbf{x})$ shows relatively long-range, smooth variations.

It can be verified [e.g., Hewett, 1986; Gelhar, 1993] that the spectral density of the fractal field $U(\mathbf{x})$ also has the power law form

$$S_U(\mathbf{k}) = Ak^{-\beta} \tag{2.212}$$

where $\beta = 2H + 1$, A is a normalization constant, $k = |\mathbf{k}|$ and $\mathbf{k}$ is the wave number vector. The wave number k is inversely related to a wavelength l by $k = 2\pi/l$. When $k \to 0$, $S_U \to \infty$ and thus the variance of U is infinite. Since $k = 2\pi/l$, $l \to \infty$ when $k \to 0$, which implies that the geological formation under consideration is unbounded. In reality, geological formations are always bounded, hence existing a maximum length, say L_{max}. Therefore, there exists a cut-off $k_{min} = 2\pi/L_{max}$ in the wave number. It is suggested by Zhan and Wheatcraft [1996] that the maximum length scale L_{max} may be determined by the no-flow boundaries in the formation. With such a cut-off, the spectral density reads as

$$S_U(\mathbf{k}) = \begin{cases} Ak^{-\beta}, & k \geq k_{min} \\ 0, & \text{otherwise} \end{cases} \tag{2.213}$$

After the low-frequency cut-off k_{min} is introduced, the spectral density and hence the variance are finite, and the truncated fractal field becomes stationary. Similarly, one may introduce an upper frequency cut-off k_{max} related to a minimum length scale L_{min}, say, the scale of the measurement support (sample volume).

Conditioning Nonstationarity may arise from a stationary process by conditioning on measurements. As for conditional moments of random variables, the conditional mean and covariance of the stationary process $U(\mathbf{x})$ for given n measurements at $\mathbf{x}_1, \mathbf{x}_2, \ldots, \mathbf{x}_n$ are given by

$$\langle U(\mathbf{x}) \rangle^c = \langle U \rangle + \sum_{i=1}^{n} a_i(\mathbf{x})[u(\mathbf{x}_i) - \langle U \rangle] \tag{2.214}$$

$$C_U^c(\mathbf{x}, \boldsymbol{\chi}) = C_U(\mathbf{x} - \boldsymbol{\chi}) - \sum_{i=1}^{n} a_i(\mathbf{x})C_U(\mathbf{x}_i - \boldsymbol{\chi}) \tag{2.215}$$

where $u(\mathbf{x}_i)$ is a specific value of U taken at $\mathbf{x}_i$, the moments with the superscript c are conditional ones, those without this superscript are the unconditional counterparts, and the coefficients a_i are solutions of the following linear equations:

$$\sum_{i=1}^{n} a_i(\mathbf{x})C_U(\mathbf{x}_i - \mathbf{x}_j) = C_U(\mathbf{x} - \mathbf{x}_j) \tag{2.216}$$

for $1 \leq j \leq n$.

It is seen that the conditional moments are generally nonstationary even though the unconditional moments are assumed to be stationary. This is so because the conditional moments depend on the distance relative to the measurement points. This dependency on the conditional moments stems from the weighting coefficients a_i. In addition to depending on the relative location to the measurement points, the conditional mean values are also dependent on the actual measured values at those points. This can be clearly seen from the special case that only one measurement point at $\mathbf{x}_o$ is used for conditioning. In this case, one has

$$a_1 = \frac{C_U(\mathbf{x} - \mathbf{x}_o)}{\sigma_U^2} = \rho_U(\mathbf{x} - \mathbf{x}_o) \tag{2.217}$$

$$\langle U(\mathbf{x}) \rangle^c = \langle U \rangle + \rho_U(\mathbf{x} - \mathbf{x}_o)[u(\mathbf{x}_o) - \langle U \rangle] \tag{2.218}$$

$$C_U^c(\mathbf{x}, \boldsymbol{\chi}) = C_U(\mathbf{x} - \boldsymbol{\chi}) - \rho_U(\mathbf{x} - \mathbf{x}_o)C_U(\mathbf{x}_o - \boldsymbol{\chi}) \tag{2.219}$$

It is clear that when $\mathbf{x} = \boldsymbol{\chi} = \mathbf{x}_o$, $a_1 = 1$. Hence, the conditional mean at the conditioning point $\mathbf{x}_o$ is exactly the measured value $u(\mathbf{x}_o)$ and the corresponding variance is zero. Since $\rho_U(\mathbf{x} - \mathbf{x}_o)$ has the same sign as $C_U(\mathbf{x}_o - \mathbf{x})$, the conditional variance at any point $\mathbf{x}$ is

$$C_U^c(\mathbf{x}, \mathbf{x}) = \sigma_U^2(\mathbf{x}) - |\rho_U(\mathbf{x} - \mathbf{x}_o)C_U(\mathbf{x}_o - \mathbf{x})| \leq \sigma_U^2(\mathbf{x}) \tag{2.220}$$

As a matter of fact, the conclusion that the conditional variance at any point **x** is less or equal to its unconditional variance holds true for any number of conditioning points.

It should be pointed out that the conditional moments are derived for a normal random field. Simple kriging yields the same system without necessarily assuming normality, but then the left-hand-sides of Eqs. (2.214) and (2.215) are minimum variance, unbiased estimates, not conditional moments. With this proviso, Eqs. (2.214)–(2.216) may not be limited to Gaussian random fields.

The formulae (2.214)–(2.216) can be generalized to the case that the unconditional statistics are nonstationary,

$$\langle U(\mathbf{x}) \rangle^c = \langle U(\mathbf{x}) \rangle + \sum_{i=1}^{n} a_i(\mathbf{x})[u(\mathbf{x}_i) - \langle U(\mathbf{x}) \rangle] \tag{2.221}$$

$$C_U^c(\mathbf{x}, \boldsymbol{\chi}) = C_U(\mathbf{x}, \boldsymbol{\chi}) - \sum_{i=1}^{n} a_i(\mathbf{x}) C_U(\mathbf{x}_i, \boldsymbol{\chi}) \tag{2.222}$$

with

$$\sum_{i=1}^{n} a_i(\mathbf{x}) C_U(\mathbf{x}_i, \mathbf{x}_j) = C_U(\mathbf{x}, \mathbf{x}_j) \tag{2.223}$$

for $1 \le j \le n$. Similar equations may be given for the case of two or more joint stochastic processes.

2.4 SOME MATHEMATICAL TOOLS

In this section, we shall introduce some mathematical tools such as the Dirac delta function, Fourier transform and Laplace transform, which are frequently used in the book for studying flow in porous media. For ease of reference, some important facts or properties of these tools are compiled.

2.4.1 Dirac Delta Functions

In Chapter 1 and the preceding sections of this chapter, we have encountered a nonordinary function, the *Dirac delta function*, and found the properties of this function useful in many derivations. In this subsection, we shall formally introduce this generalized function and review some of its properties.

The Dirac delta function $\delta(x)$, first introduced by the physicist P. Dirac, is usually defined by the relation

$$\int_{-\infty}^{\infty} \delta(x) u(x) \, dx = u(0) \tag{2.224}$$

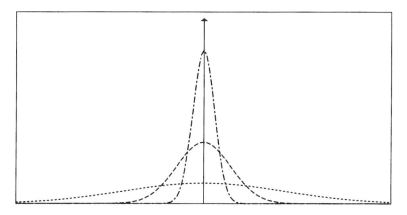

Figure 2.7. Construction of the Dirac delta function from a normal distribution.

for any function $u(x)$ that is continuous at $x = 0$. Mathematically speaking, the action of $\delta(x)$ on $u(x)$ is to pick out its value at $x = 0$. Similarly, $\delta(x - a)$ is defined with

$$\int_{-\infty}^{\infty} \delta(x - a)u(x)\,dx = u(a) \tag{2.225}$$

Physically speaking, the Dirac delta function $\delta(x - a)$ is a concentrated spike of unit area,

$$\delta(x - a) = 0, \quad \text{for } x \neq a$$

$$\int_{-\infty}^{\infty} \delta(x - a)\,dx = 1 \tag{2.226}$$

As illustrated in Fig. 2.7, this concentrated spike can be constructed by letting $\sigma^2 \to 0$ in the normal distribution of Eq. (2.11).

The derivative of $\delta(x - a)$ can be given with Eq. (2.226) and via integrating by parts,

$$\int_{-\infty}^{\infty} \frac{d\delta(x - a)}{dx}u(x)\,dx = \delta(x - a)u(x)\big|_{-\infty}^{\infty} - \int_{-\infty}^{\infty} \delta(x - a)\frac{du(x)}{dx}\,dx$$

$$= -\frac{du(x)}{dx}\bigg|_{x=a} \tag{2.227}$$

The ith derivative $\delta^{(i)}(x - a) = d^i\delta(x - a)/dx^i$ can be defined similarly as

$$\int_{-\infty}^{\infty} \delta^{(i)}(x - a)u(x) = (-1)^i u^{(i)}(a) \tag{2.228}$$

where $u^{(i)}(a) = d^i u(x)/dx^i$ is evaluated at $x = a$.

The following identities of the Dirac delta functions are commonly used:

$$x\delta(x) = 0 \tag{2.229}$$

$$u(x)\delta(x) = u(0)\delta(x) \tag{2.230}$$

$$\delta(x - a) = \delta(a - x) \tag{2.231}$$

$$\delta(bx) = \frac{1}{b}\delta(x) \tag{2.232}$$

$$\frac{d\delta[y - g(x)]}{dx} = -\frac{dg(x)}{dx}\frac{d\delta[y - g(x)]}{dy} \tag{2.233}$$

These identities can be verified by integrating with a test function. For example, since

$$\int_{-\infty}^{\infty} x\delta(x)v(x)\,dx = xv(x)|_{x=0} = 0 \tag{2.234}$$

for any ordinary function $v(x)$, Eq. (2.229) must be correct.

The Dirac delta function is related to the so-called *Heaviside step function* in a simple way,

$$H(x) = \int_{-\infty}^{x} \delta(x')\,dx' \tag{2.235}$$

The Heaviside step function is, however, an ordinary function taking values as follows:

$$H(x) = \begin{cases} 0, & x < 0 \\ 1, & x > 0 \end{cases} \tag{2.236}$$

and it is discontinuous at $x = 0$. Taking derivatives at both sides of Eq. (2.235) leads to

$$H'(x) = \delta(x) \tag{2.237}$$

The procedure that led to Eq. (2.237) is not valid in an ordinary sense because $H(x)$ is discontinuous at $x = 0$. Although $H'(x)$ does not exist at $x = 0$ in an ordinary sense, it is perfectly represented by the Dirac delta function $\delta(x)$. This may explain why $\delta(x)$ is called a generalized function.

The properties of the Heaviside step function are frequently used in many derivations. For example, the absolute value $|x|$ can be expressed by

$$|x| = x[2H(x) - 1] \tag{2.238}$$

Both the Dirac delta function and the Heaviside step function can be defined for multiple dimensions,

$$\delta(\mathbf{x}) = \delta(x_1)\delta(x_2)\cdots\delta(x_d) \tag{2.239}$$

$$H(\mathbf{x}) = H(x_1)H(x_2)\cdots H(x_d) \tag{2.240}$$

Their properties can be given in analogy to their one-dimensional counterparts.

2.4.2 Dirac Delta Representation of PDFs

With the Heaviside function $H(u)$, the cumulative distribution function (2.6) of a random variable U can be written as

$$F(u) = \int_{-\infty}^{u} p(u')\, du' = \int_{-\infty}^{\infty} H(u - u') p(u')\, du' \tag{2.241}$$

Hence, by the definition of expectation one has

$$F(u) = \langle H(u - U) \rangle \tag{2.242}$$

Differentiating both sides of Eq. (2.241) with respect to u and recognizing that $p(u')$ is not a function of u, leads to

$$p(u) = \langle \delta(u - U) \rangle \tag{2.243}$$

It can be shown that the Dirac delta representation is also applicable to a stochastic process $U(\mathbf{x})$ (as defined in Section 2.3),

$$F_U(u, \mathbf{x}) = \langle H[u - U(\mathbf{x})] \rangle \tag{2.244}$$

$$p_U(u, \mathbf{x}) = \langle \delta[u - U(\mathbf{x})] \rangle \tag{2.245}$$

Equation (2.245) has served as the basis of many stochastic approaches that aim to directly derive the governing equation for $p_U(u, \mathbf{x})$ [e.g., Pope, 1985, 1994; Klimenko and Bilger, 1999]. In such approaches, one first derives the equation governing $\delta[u - U(\mathbf{x})]$ and then takes the expectation of it.

The joint CDF and PDF of two stochastic processes $U(\mathbf{x})$ and $V(\mathbf{x})$ can be written similarly as

$$F_{UV}(u, \mathbf{x}; v, \mathbf{x}) = \langle H[u - U(\mathbf{x})] H[v - V(\mathbf{x})] \rangle \tag{2.246}$$

$$p_{UV}(u, \mathbf{x}; v, \mathbf{x}) = \langle \delta[u - U(\mathbf{x})] \delta[v - V(\mathbf{x})] \rangle \tag{2.247}$$

The expectation of $U(\mathbf{x})$ for given $V(\mathbf{x}) = v$, defined as

$$\langle U(\mathbf{x}) | V(\mathbf{x}) = v \rangle = \int_{-\infty}^{\infty} u\, p_{UV}[u, \mathbf{x} | V(\mathbf{x}) = v]\, du$$

$$= \frac{\int_{-\infty}^{\infty} u\, p_{UV}(u, \mathbf{x}; v, \mathbf{x})\, du}{p_V(v, \mathbf{x})} \tag{2.248}$$

can be expressed with the Dirac delta function by [Pope, Eq. (2.150), 1985],

$$\langle U(\mathbf{x}) | V(\mathbf{x}) = v \rangle\, p_V(v, \mathbf{x}) = \langle U(\mathbf{x}) \delta[v - V(\mathbf{x})] \rangle \tag{2.249}$$

The left-hand side of this equation involves a conditional expectation, whereas the right-hand side is an unconditional one. Equation (2.249) may be proven by multiplying each side by a test function $g(v)$ and integrating over v. Equation (2.249) can be generalized to multiple conditions of averaging, for example,

$$\langle U(\mathbf{x})|V(\mathbf{x}) = v, W(\mathbf{x}) = w\rangle \, p_{VW}(v, \mathbf{x}; w, \mathbf{x})$$
$$= \langle U(\mathbf{x})\delta[v - V(\mathbf{x})\delta[w - W(\mathbf{x})]\rangle \qquad (2.250)$$

Similarly, one may represent the multivariate CDF of Eq. (2.126) and PDF of Eq. (2.127) for a stochastic process with the Heaviside and Dirac delta functions,

$$F_U(u_1, \mathbf{x}_1; u_2, \mathbf{x}_2; \ldots; u_n, \mathbf{x}_n) = \langle H[u_1 - U(\mathbf{x}_1)]H[u_2 - U(\mathbf{x}_2)] \cdots H[u_n - U(\mathbf{x}_n)]\rangle \qquad (2.251)$$

$$p_U(u_1, \mathbf{x}_1; u_2, \mathbf{x}_2; \ldots; u_n, \mathbf{x}_n) = \langle \delta[u_1 - U(\mathbf{x}_1)]\delta[u_2 - U(\mathbf{x}_2)] \cdots \delta[u_n - U(\mathbf{x}_n)]\rangle \qquad (2.252)$$

2.4.3 Stochastic Continuity and Differentiation

A stochastic process (random field) $U(\mathbf{x})$ is called *mean square continuous* at a (space and/or time) point $\mathbf{x}$ if

$$\lim_{\Delta \mathbf{x} \to 0} \langle [U(\mathbf{x} + \Delta \mathbf{x}) - U(\mathbf{x})]^2\rangle = 0 \qquad (2.253)$$

The condition for the existence of Eq. (2.253) is that the autocorrelation function $\langle U(\mathbf{x} + \Delta \mathbf{x})U(\mathbf{x})\rangle$ is continuous in an ordinary sense at the point $\mathbf{x}$, that is,

$$\lim_{\Delta \mathbf{x} \to 0} \langle U(\mathbf{x} + \Delta \mathbf{x})U(\mathbf{x})\rangle = \langle U(\mathbf{x})U(\mathbf{x})\rangle \qquad (2.254)$$

If the autocorrelation function is continuous at any point of the field, the random field $U(\mathbf{x})$ is everywhere mean square continuous. When $U(\mathbf{x})$ is stationary, then it is mean square continuous at $\mathbf{x}$ if the autocovariance $C_U(\Delta \mathbf{x})$ is continuous for $\Delta \mathbf{x} \to 0$. Furthermore, the stationary random field is mean square continuous everywhere if it is mean square continuous at a point. Note that the mean square continuity of any random field $U(\mathbf{x})$ for all $\mathbf{x}$ does not mean that each realization $u(\mathbf{x})$ of this random field is a continuous function of $\mathbf{x}$.

If $U(\mathbf{x})$ is mean square continuous at $\mathbf{x}$, its mean is continuous at $\mathbf{x}$,

$$\lim_{\Delta \mathbf{x} \to 0} \langle U(\mathbf{x} + \Delta \mathbf{x})\rangle = \langle U(\mathbf{x})\rangle \qquad (2.255)$$

In turn, one has

$$\lim_{\Delta \mathbf{x} \to 0} \langle U(\mathbf{x} + \Delta \mathbf{x})\rangle = \langle \lim_{\Delta \mathbf{x} \to 0} U(\mathbf{x} + \Delta \mathbf{x})\rangle \qquad (2.256)$$

A stochastic process $U(\mathbf{x})$ is *mean square differentiable* at $\mathbf{x}$ if

$$\lim_{\Delta x_i \to 0} \left\langle \left[\frac{U(\mathbf{x} + \Delta \mathbf{x}) - U(\mathbf{x})}{\Delta x_i} - \frac{\partial U(\mathbf{x})}{\partial x_i}\right]^2\right\rangle = 0 \qquad (2.257)$$

where $i = 1, 2, \ldots, d$. The sufficient and necessary condition for Eq. (2.257) is that the partial derivatives of $\langle U(\mathbf{x} + \Delta\mathbf{x})U(\mathbf{x})\rangle$, i.e., $\partial^2 \langle U(\boldsymbol{\chi})U(\mathbf{x})\rangle / \partial\chi_i \partial x_i$, exist for $\boldsymbol{\chi} \to \mathbf{x}$. Therefore, one can show that

$$\left\langle \frac{\partial U(\mathbf{x})}{\partial x_i} \right\rangle = \left\langle \lim_{\Delta x_i \to 0} \frac{U(\mathbf{x} + \Delta\mathbf{x}) - U(\mathbf{x})}{\Delta x_i} \right\rangle$$

$$= \lim_{\Delta x_i \to 0} \frac{\langle U(\mathbf{x} + \Delta\mathbf{x})\rangle - \langle U(\mathbf{x})\rangle}{\Delta x_i}$$

$$= \frac{\partial \langle U(\mathbf{x})\rangle}{\partial x_i} \qquad (2.258)$$

That is, if the process $U(\mathbf{x})$ is mean square differentiable at the point $\mathbf{x}$, then the differentiation and expectation are *commutative*. As shown in Chapter 1, this property has been used extensively in deriving moment equations from original stochastic equations.

The basic equation (2.258) is valid for unconditional expectations, but it may not be valid for conditional expectations. The latter is so because the condition of averaging may also depend on $\mathbf{x}$. One may show this intuitively by writing

$$\left\langle \frac{\partial U(\mathbf{x})}{\partial x_i} \middle| V(\mathbf{x}) = v \right\rangle = \left\langle \lim_{\Delta x_i \to 0} \frac{U(\mathbf{x} + \Delta\mathbf{x}) - U(\mathbf{x})}{\Delta x_i} \middle| V(\mathbf{x}) = v \right\rangle$$

$$= \lim_{\Delta x_i \to 0} \frac{\langle U(\mathbf{x} + \Delta\mathbf{x})| V(\mathbf{x}) = v \rangle - \langle U(\mathbf{x})| V(\mathbf{x}) = v \rangle}{\Delta x_i} \qquad (2.259)$$

$$\frac{\partial \langle U(\mathbf{x})| V(\mathbf{x}) = v \rangle}{\partial x_i} = \lim_{\Delta x_i \to 0} \frac{\langle U(\mathbf{x} + \Delta\mathbf{x})| V(\mathbf{x} + \Delta\mathbf{x}) = v' \rangle - \langle U(\mathbf{x})| V(\mathbf{x}) = v \rangle}{\Delta x_i}$$

$$(2.260)$$

The right-hand sides of Eq. (2.259) and Eq. (2.260) may not be equal because the measurements v taken from a realization of the stochastic process $V(\mathbf{x})$ may not be continuous even if the stochastic process $U(\mathbf{x})$ and $V(\mathbf{x})$ are assumed to be mean square continuous.

A more rigorous proof of the noncommutativity of conditional expectation with differentiation may be given with the aid of the Dirac delta representation of conditional expectation of Eq. (2.249). This proof was outlined by Klimenko and Bilger [1999] for differentiation with respect to time. Let us first differentiate the quantity $U(\mathbf{x})\delta[v - V(\mathbf{x})]$ with respect to x_i,

$$\frac{\partial\{U(\mathbf{x})\delta[v - V(\mathbf{x})]\}}{\partial x_i} = \frac{\partial U(\mathbf{x})}{\partial x_i}\delta[v - V(\mathbf{x})] + U(\mathbf{x})\frac{\partial\delta[v - V(\mathbf{x})]}{\partial x_i}$$

$$= \frac{\partial U(\mathbf{x})}{\partial x_i}\delta[v - V(\mathbf{x})] - \frac{\partial}{\partial v}\left\{ U(\mathbf{x})\frac{\partial V(\mathbf{x})}{\partial x_i}\delta[v - V(\mathbf{x})] \right\} \quad (2.261)$$

In Eq. (2.261) we have utilized the multidimensional version of Eq. (2.233) and recognized that U is not a function of v. By taking unconditional expectation of Eq. (2.261) and utilizing

Eq. (2.249), we have

$$
\frac{\partial[\langle U(\mathbf{x})|V(\mathbf{x}) = v\rangle p_V(v, \mathbf{x})]}{\partial x_i}
$$

$$
= \left\langle \frac{\partial U(\mathbf{x})}{\partial x_i} \middle| V(\mathbf{x}) = v \right\rangle p_V(v, \mathbf{x}) - \frac{\partial}{\partial v} \left\{ \left\langle U(\mathbf{x}) \frac{\partial V(\mathbf{x})}{\partial x_i} \middle| V(\mathbf{x}) = v \right\rangle p_V(v, \mathbf{x}) \right\}
$$

(2.262)

which can be rewritten as

$$
\left\langle \frac{\partial U(\mathbf{x})}{\partial x_i} \middle| V(\mathbf{x}) = v \right\rangle p_V(v, \mathbf{x}) - \frac{\partial \langle U(\mathbf{x})|V(\mathbf{x}) = v\rangle}{\partial x_i} p_V(v, \mathbf{x})
$$

$$
= \langle U(\mathbf{x})|V(\mathbf{x}) = v\rangle \frac{\partial p_V(v, \mathbf{x})}{\partial x_i} + \frac{\partial}{\partial v} \left[\left\langle U(\mathbf{x}) \frac{\partial V(\mathbf{x})}{\partial x_i} \middle| V(\mathbf{x}) = v \right\rangle p_V(v, \mathbf{x}) \right]
$$

(2.263)

Since the right-hand side of Eq. (2.263) is generally nonzero, the conditional expectation and differentiation are usually not commutative. Therefore, additional conditions are needed to render the conditional expectation and differentiation commutative. However, this has not been recognized in the literature of flow in porous media. Many researchers have generalized the commutativity property of unconditional (conventional) expectation with differentiation to conditional expectations without realizing the approximations involved. This commutativity statement has been a key step in many conditional stochastic flow theories that derive conditional moment equations from the original stochastic equations.

It is obvious that when the condition of averaging (i.e., $V(\mathbf{x})$ in Eq. (2.263)) does not depend on x_i, we have

$$
\left\langle \frac{U(\mathbf{x})}{\partial x_i} \middle| V(\mathbf{x}) = v \right\rangle = \frac{\partial \langle U(\mathbf{x})|V(\mathbf{x}) = v\rangle}{\partial x_i}
$$

(2.264)

For example, if $U(\mathbf{x})$ stands for the hydraulic head $h(\mathbf{x}, t)$ and $V(\mathbf{x})$ denotes the log hydraulic conductivity $K_S(\mathbf{x})$ in a nondeformable, confined aquifer, the conditional expectation $\langle h(\mathbf{x}, t)|K_S(\mathbf{x})\rangle$ is commutative with differentiation with respect to time t. In the above example, if K_S is, in addition, a function of time (due to stress changes in a deformable formation or chemical reactions with the rock), then the conditional expectation $\langle h(\mathbf{x}, t)|K_S(\mathbf{x}, t)\rangle$ is not strictly commutative with time differentiation. Nevertheless, if $K_S(\mathbf{x}, t)$ varies slowly with time, then $\langle h(\mathbf{x}, t)|K_S(\mathbf{x}, t)\rangle$ might be approximately commutative with time differentiation and the associated error could be small. In general, we expect Eq. (2.264) to be a good approximation if the condition or conditions of averaging vary slowly with respect to x_i. However, a firm conclusion may not be stated without quantifying the two terms on the right-hand side of Eq. (2.263) and evaluating the effects of neglecting them in deriving conditional moment equations as done in previous studies.

As for the stochastic differentiation, one may define stochastic integration in a mean square sense. A stochastic process $U(\mathbf{x})$ is *mean square integrable* if the following limit exists in the mean square sense:

$$\int_\Omega U(\mathbf{x})\, d\mathbf{x} = \lim_{\Delta \mathbf{x}_i \to 0} \sum_i U(\mathbf{x}_i) \Delta \mathbf{x}_i \tag{2.265}$$

where Ω is the integration domain. The condition for Eq. (2.265) to exist is that

$$\int_\Omega \int_\Omega |\langle U(\mathbf{x}) U(\boldsymbol{\chi})\rangle|\, d\mathbf{x}\, d\boldsymbol{\chi} < \infty \tag{2.266}$$

2.4.4 Fourier Transform

The Fourier transform of a function $u(\mathbf{x})$ is given by

$$\hat{u}(\mathbf{k}) = \frac{1}{(2\pi)^d} \int \exp(-\iota \mathbf{k} \cdot \mathbf{x}) u(\mathbf{x})\, d\mathbf{x} \tag{2.267}$$

where $\mathbf{k} = (k_1, \ldots, k_d)^T$ is the wave number space vector (where d is the number of space dimensions), $\iota \equiv \sqrt{-1}$, and the integration is d-fold from $-\infty$ to ∞. The Fourier transform is usually denoted by $\mathbf{F}(u) \equiv \hat{u}$. The inverse Fourier transform is given by

$$u(\mathbf{x}) = \int \exp(\iota \mathbf{k} \cdot \mathbf{x}) \hat{u}(\mathbf{k})\, d\mathbf{k} \tag{2.268}$$

The notation for inverse is $\mathbf{F}^{-1}(\hat{u}) \equiv u$. It is clear that $\mathbf{F}^{-1}\mathbf{F}(u) = u(\mathbf{x})$. There are a number of extensive tables of Fourier transforms [e.g., Gradshteyn and Ryzhik, 1980].

The following are some important properties of the Fourier transform:

1. *Linearity.* For any constants a and b and any transformable functions u and v,

$$\mathbf{F}(au + bv) = a\hat{u} + b\hat{v} \tag{2.269}$$

$$\mathbf{F}^{-1}(a\hat{u} + b\hat{v}) = au + bv \tag{2.270}$$

2. *Transform of nth derivative.* If $\partial^{m_i} u(\mathbf{x})/\partial x_i^{m_i} \to 0$ as $\mathbf{x} \to \pm\infty$ for $m_i \leq (n-1)$, then

$$\mathbf{F}\left[\frac{\partial^n u(\mathbf{x})}{\partial x_1^{m_2} \partial x_2^{m_2} \cdots \partial x_d^{m_d}}\right] = (\iota)^n k_1^{m_1} k_2^{m_2} \cdots k_d^{m_d} \hat{u}(\mathbf{k}) \tag{2.271}$$

where $n = m_1 + m_2 + \cdots + m_d$.

3. *Fourier convolution.* One has

$$\mathbf{F}^{-1}(\hat{u}\hat{v}) = u * v(\mathbf{x}) \tag{2.272}$$

where $u * v(\mathbf{x})$ is the *Fourier convolution* of $u(\mathbf{x})$ and $v(\mathbf{x})$, defined as

$$u * v(\mathbf{x}) = \int u(\mathbf{x} - \boldsymbol{\chi})v(\boldsymbol{\chi}) \, d\boldsymbol{\chi} \qquad (2.273)$$

where the integration is d-fold from $-\infty$ to ∞. It is clear that $\mathbf{F}(u * v) = \hat{u}\hat{v}$ and $u * v(\mathbf{x}) = v * u(\mathbf{x})$.

Another quantity of common interest is the *complex conjugate* of the Fourier transform $\hat{u}(\mathbf{k})$, defined for a real function $u(\mathbf{x})$ by

$$\hat{u}^*(\mathbf{k}) \equiv \hat{u}(-\mathbf{k}) = \frac{1}{(2\pi)^d} \int \exp(\imath \mathbf{k} \cdot \mathbf{x}) u(\mathbf{x}) \, d\mathbf{x} \qquad (2.274)$$

It is seen that $\hat{u}^*(-\mathbf{k}) = \hat{u}(\mathbf{k})$. Some authors define the Fourier transform itself by Eq. (2.274) instead of Eq. (2.267). This is how the characteristic function is defined in Eq. (2.33) except for the factor $1/(2\pi)^d$. The corresponding inverse Fourier transform would be Eq. (2.35) without $1/(2\pi)^d$. To enhance the symmetry in the definition of Fourier transform, one may use $1/(2\pi)^{d/2}$ in both Eqs. (2.267) and (2.268) rather than putting the entire $1/(2\pi)^d$ into Eq. (2.267). One should pay special attention to the specific definitions of the Fourier transforms while looking up tables of Fourier transforms from different sources.

When $U(\mathbf{x})$ is a second-order stationary stochastic process (random field), the Fourier transform of its mean is equal to the mean of the Fourier transform of $U(\mathbf{x})$,

$$\langle \hat{U}(\mathbf{k}) \rangle = \frac{1}{(2\pi)^d} \int \exp(-\imath \mathbf{k} \cdot \mathbf{x}) \langle U(\mathbf{x}) \rangle \, d\mathbf{x} \qquad (2.275)$$

and the Fourier transform of its fluctuation is

$$\hat{U}'(\mathbf{k}) = \frac{1}{(2\pi)^d} \int \exp(-\imath \mathbf{k} \cdot \mathbf{x}) U'(\mathbf{x}) \, d\mathbf{x} \qquad (2.276)$$

The Fourier transform of the stationary covariance $C_U(\mathbf{r})$ is

$$S_U(\mathbf{k}) \equiv \hat{C}_U(\mathbf{k}) = \frac{1}{(2\pi)^d} \int \exp(-\imath \mathbf{k} \cdot \mathbf{r}) C_U(\mathbf{r}) \, d\mathbf{r} \qquad (2.277)$$

$S_U(\mathbf{k})$ is called the *spectral density* of the process $U(\mathbf{x})$. Conversely, one has

$$C_U(\mathbf{r}) = \int \exp(\imath \mathbf{k} \cdot \mathbf{r}) S_U(\mathbf{k}) \, d\mathbf{k} \qquad (2.278)$$

In particular, the variance $\sigma_U^2 = C_U(0)$ is given as

$$\sigma_U^2 = \int S_U(\mathbf{k}) \, d\mathbf{k} \qquad (2.279)$$

Equations (2.277) and (2.278) are the so-called *Wiener–Khinchine relations*.

Multiplying Eq. (2.276) by its complex conjugate at $\mathbf{k}'$ and taking expectation leads to

$$\langle \hat{U}'(\mathbf{k})\hat{U}'^*(\mathbf{k}')\rangle - \frac{1}{(2\pi)^{2d}} \int\int \exp(-\iota\mathbf{k}\cdot\mathbf{x} + \iota\mathbf{k}'\cdot\boldsymbol{\chi})\langle U'(\mathbf{x})U'(\boldsymbol{\chi})\rangle\, d\mathbf{x}\, d\boldsymbol{\chi}$$

$$= \frac{1}{(2\pi)^{2d}} \int \exp[-\iota(\mathbf{k}-\mathbf{k}')\cdot\boldsymbol{\chi}]\, d\boldsymbol{\chi} \int \exp(-\iota\mathbf{k}\cdot\mathbf{r})C_U(\mathbf{r})\, d\mathbf{r}$$

$$= \delta(\mathbf{k}-\mathbf{k}')S_U(\mathbf{k}) \tag{2.280}$$

because of the fact that $\int_{-\infty}^{\infty}\cdots\int_{-\infty}^{\infty}\exp[-\iota\mathbf{k}\cdot\boldsymbol{\chi}]\,d\boldsymbol{\chi} = (2\pi)^d\delta(\mathbf{k})$. It is seen from Eq. (2.280) that the Fourier transform of the fluctuation of a stationary stochastic process is a white noise process.

When $U(\mathbf{x})$ is a real stationary process, $C_U(\mathbf{r})$ is even and hence so is $S_U(\mathbf{k})$. Therefore, the Wiener–Khinchine relations (2.277) and (2.278) can be written as

$$S_U(\mathbf{k}) = \frac{1}{(2\pi)^d} \int \cos(\mathbf{k}\cdot\mathbf{r})C_U(\mathbf{r})\, d\mathbf{r} \tag{2.281}$$

$$C_U(\mathbf{r}) = \int \cos(\mathbf{k}\cdot\mathbf{r})S_U(\mathbf{k})\, d\mathbf{k} \tag{2.282}$$

One may express the spectral representation of the fluctuation of a stationary stochastic process in terms of the integral

$$Z_U(\mathbf{k}) = \int_{-\infty}^{\mathbf{k}} \hat{U}'(\mathbf{k}')\, d\mathbf{k}' \tag{2.283}$$

where $\hat{U}'(\mathbf{k}')$ is Fourier transform of $U'(\mathbf{x})$, given in Eq. (2.276). It follows from Eq. (2.283) that

$$dZ_U(\mathbf{k}) = \hat{U}'(\mathbf{k})\, d\mathbf{k} \tag{2.284}$$

Therefore, the fluctuation $U'(\mathbf{x})$ can be written as a *Fourier–Stieltjes integral*,

$$U'(\mathbf{x}) = \int \exp(\iota\mathbf{k}\cdot\mathbf{x})\, dZ_U(\mathbf{k}) \tag{2.285}$$

It follows from Eq. (2.280) that

$$\langle dZ_U(\mathbf{k})dZ_U^*(\mathbf{k}')\rangle = \delta(\mathbf{k}-\mathbf{k}')S_U(\mathbf{k})d\mathbf{k}\, d\mathbf{k}' \tag{2.286}$$

where dZ_U^* is the complex conjugate of dZ_U. With Eqs. (2.285) and (2.286), one can easily recover Eq. (2.278). It is obvious that $\langle dZ_U(\mathbf{k})\rangle = 0$.

2.4.5 Laplace Transform

Another transform of great importance is the *Laplace transform*,

$$\tilde{u}(s) = \int_0^\infty \exp(-st)u(t)\, dt \tag{2.287}$$

where s is a complex variable and t is a real one (e.g., time). The inverse of the Laplace transform is

$$u(t) = \frac{1}{2\pi\iota} \int_{\gamma-\iota\infty}^{\gamma+\iota\infty} \exp(st)\tilde{u}(s)\, ds \tag{2.288}$$

where γ is a real constant. As for the Fourier transform, it is sometimes convenient to use the operator notation L and L^{-1} for the Laplace transform and its inverse: $Lu = \tilde{u}$ and $L^{-1}\tilde{u} = u$.

There are some important properties of the Laplace transform:

1. *Linearity.* For any constants a and b and any transformable functions u and v,

$$L(au + bv) = a\tilde{u} + b\tilde{v} \tag{2.289}$$

$$L^{-1}(a\tilde{u} + b\tilde{v}) = au + bv \tag{2.290}$$

2. *Transform of derivatives.* One has

$$L\left[\frac{du(t)}{dt}\right] = s\tilde{u}(s) - u(0)$$

$$L\left[\frac{d^2u(t)}{dt^2}\right] = s^2\tilde{u}(s) - su(0) - \frac{du(t)}{dt}\bigg|_{s=0} \tag{2.291}$$

$$\cdots$$

$$L\left[u^{(n)}(t)\right] = sL\left[u^{(n-1)}(t)\right] - u^{(n-1)}(0)$$

3. *Transform of integrals.* One has

$$L\left[\int_0^t u(t')\, dt'\right] = \frac{1}{s}\tilde{u}(s)$$

$$L\left[\int_0^t \int_0^{t'} u(t'')\, dt''\, dt'\right] = \frac{1}{s^2}\tilde{u}(s) \tag{2.292}$$

4. *Laplace convolution.* One has

$$L^{-1}(\tilde{u}\tilde{v}) = u * v(t) \tag{2.293}$$

where $u * v(t)$ is the *Laplace convolution* of $u(t)$ and $v(t)$, defined as

$$u * v(t) = \int_0^t u(t - \tau)v(\tau) \, d\tau \tag{2.294}$$

It is clear that $L(u * v) = \tilde{u}\tilde{v}$ and $u * v(t) = v * u(t)$.

There exist extensive tables of Laplace transforms [e.g., Abramowitz and Stegun, 1970; Gradshteyn and Ryzhik, 1980]. In analogy to the one-dimensional Laplace transform pair, multidimensional Laplace transforms can be defined.

2.5 EXERCISES

1. Derive the mean, variance, and skewness of a random variable with the bimodal discrete probability density function $p(u) = (1 - \phi)\delta(u) + \phi\delta(u - 1)$.

2. If porosity ϕ is uniformly distributed over $(0.15, 0.45)$, calculate:

   (a) its cumulative distribution function;
   (b) the probability that (1) $\phi < 0.25$, (2) $\phi > 0.35$, (3) $0.2 < \phi < 0.4$;
   (c) the mean and variance of ϕ.

3. Show that a random variable U and a linear function $V = aU + b$ of U are perfectly correlated, where a and b are constants and $a \neq 0$.

4. Let V be the sum of two random variables U_1 and U_2.

   (a) Show that $\sigma_V^2 = \sigma_{U_1}^2 + 2\text{Cov}(U_1, U_2) + \sigma_{U_2}^2$.
   (b) Find out the condition under which σ_V^2 is, respectively, larger than, less than, or equal to the sum of the separate variances of U_1 and U_2, i.e., $\sigma_{U_1}^2 + \sigma_{U_2}^2$.

5. For a normal random variable U with the parameters $N(\mu_U, \sigma_U^2)$, its characteristic function is given as

$$\Phi_U(k) = \exp\left(\iota\mu_U k - \frac{1}{2}\sigma_U^2 k^2\right) \tag{2.295}$$

The central moments of the normal random variable can be obtained with Eq. (2.38) and by setting $\mu_U = 0$ in Eq. (2.295). Use this procedure to verify the results of Eq. (2.29).

6. If U is a normal random variable, $V = \exp(U)$ is lognormally distributed. The expectation of V^m (m being a positive integer) can be expressed via

$$E(V^m) = E[\exp(mU)] = \Phi_U\left(\frac{m}{\iota}\right) \tag{2.296}$$

where $\Phi_U(k)$ is the characteristic function of U, given by Eq. (2.295).

(a) Show that $E(V^m) = \exp(m\mu_U + (m^2/2)\sigma_U^2)$.

(b) Prove the relations (2.107) and (2.108) between the first two moments of U and V.

7. What units do the functions $\delta(x_1)$ and $H(x_1)$ have, respectively, if the unit of x_1 is $[L]$? How about $\delta(\mathbf{x})$ and $H(\mathbf{x})$ (where $\mathbf{x} = (x_1, x_2, \ldots, x_d)^T$ and d is the number of space dimensions)?

8. For a stochastic process (random field) $U(\mathbf{x})$ with stationary increments, one has $\langle U(\mathbf{x} + \mathbf{r}) - U(\mathbf{x}) \rangle = m_U(\mathbf{r})$, which only depends on the separation vector $\mathbf{r}$.

(a) Prove that $m_U(\mathbf{r}) = m_U(\mathbf{r}_1) + m_U(\mathbf{r}_2)$ if $\mathbf{r} = \mathbf{r}_1 + \mathbf{r}_2$.

(b) Show that $m_U(\mathbf{r})$ must be a linear function of $\mathbf{r}$.

3

STEADY-STATE SATURATED FLOW

3.1 INTRODUCTION

We consider steady-state, incompressible, single-phase fluid flow satisfying the following continuity equation and Darcy's law:

$$\nabla \cdot \mathbf{q}(\mathbf{x}) = g(\mathbf{x}) \tag{3.1}$$

$$q_i(\mathbf{x}) = -K_S(\mathbf{x})\frac{\partial h(\mathbf{x})}{\partial x_i} \tag{3.2}$$

subject to boundary conditions

$$h(\mathbf{x}) = H_B(\mathbf{x}), \quad \mathbf{x} \in \Gamma_D \tag{3.3}$$

$$\mathbf{q}(\mathbf{x}) \cdot \mathbf{n}(\mathbf{x}) = Q(\mathbf{x}), \quad \mathbf{x} \in \Gamma_N \tag{3.4}$$

where $\nabla = (\partial/\partial x_1, \ldots, \partial/\partial x_d)^T$ is the grad operator with respect to $\mathbf{x} = (x_1, \ldots, x_d)^T$ (where T denotes transpose), $\mathbf{q}$ is the specific discharge (flux) vector, $g(\mathbf{x})$ is the source/sink term (due to recharge, pumping, or injection; positive for source and negative for sink), $i = 1, \ldots, d$ (where d is the number of space dimensions), $h(\mathbf{x}) = P(\mathbf{x})/(\rho g) + x_3$ is the hydraulic head (where P is pressure, ρ density, g gravitational acceleration factor, and x_3 elevation), $K_S(\mathbf{x}) = k(\mathbf{x})\rho g/\mu$ is the saturated hydraulic conductivity (where k is intrinsic or absolute permeability and μ fluid viscosity), $H_B(\mathbf{x})$ is the prescribed head on Dirichlet boundary segments Γ_D, $Q(\mathbf{x})$ is the prescribed flux across Neumann boundary segments Γ_N, and $\mathbf{n}(\mathbf{x}) = (n_1, \ldots, n_d)^T$ is an outward unit vector normal to the boundary. In this work, K_S is assumed isotropic locally and is treated as a random space function. As already done in Section 1.4, we usually work with the log-transformed hydraulic conductivity $f = \ln K_S$. Note that the variance of the log hydraulic conductivity is the same as that of the log absolute permeability $\ln k$ because $f(\mathbf{x}) = \ln(\rho g/\mu) + \ln k(\mathbf{x})$ with ρ, g, and μ being known constants.

Substituting Eq. (3.2) into Eq. (3.1) and Eqs. (3.3) and (3.4), and utilizing $f(\mathbf{x}) = \ln K_S(\mathbf{x})$ yields

$$\frac{\partial^2 h(\mathbf{x})}{\partial x_i^2} + \frac{\partial f(\mathbf{x})}{\partial x_i}\frac{\partial h(\mathbf{x})}{\partial x_i} = -g(\mathbf{x})e^{-f(\mathbf{x})}$$

$$h(\mathbf{x}) = H_B(\mathbf{x}), \quad \mathbf{x} \in \Gamma_D \qquad n_i(\mathbf{x})\frac{\partial h(\mathbf{x})}{\partial x_i} = -Q(\mathbf{x})e^{-f(\mathbf{x})}, \quad \mathbf{x} \in \Gamma_N$$

(3.5)

Summation for repeated indices i is implied. As discussed in Chapter 1, the log-transformed hydraulic conductivity $f(\mathbf{x})$, the forcing term $g(\mathbf{x})$, the boundary head term $H_B(\mathbf{x})$, and the boundary flux term $Q(\mathbf{x})$ may be treated as stochastic processes. In turn, Eq. (3.5) becomes a set of stochastic partial differential equations. In this chapter, we introduce a number of methods for solving this set of stochastic equations.

3.2 PERTURBATIVE EXPANSION METHOD

We may first decompose the random variables into their respective mean and fluctuation,

$$f(\mathbf{x}) = \langle f(\mathbf{x})\rangle + f'(\mathbf{x}) \qquad g(\mathbf{x}) = \langle g(\mathbf{x})\rangle + g'(\mathbf{x})$$

$$H_B(\mathbf{x}) = \langle H_B(\mathbf{x})\rangle + H_B'(\mathbf{x}) \qquad Q(\mathbf{x}) = \langle Q(\mathbf{x})\rangle + Q'(\mathbf{x})$$

(3.6)

where angular brackets $\langle\ \rangle$ indicate a mathematical expectation (ensemble mean), and the primed quantities are the zero-mean fluctuations. Note that we have not made any assumption about these random variables (or functions) and their cross-correlations.

As done in Section 1.4.2, we may expand $h(\mathbf{x})$ into a formal series,

$$h(\mathbf{x}) = h^{(0)}(\mathbf{x}) + h^{(1)}(\mathbf{x}) + h^{(2)}(\mathbf{x}) + \cdots$$

(3.7)

However, unlike in Section 1.4.2, the $h^{(n)}$ term is not just in terms of σ_f^n but a function of some combination of σ_f, σ_g, σ_{H_B}, and σ_Q (where σ_s is the standard deviation of $s = f, g, H_B$, or Q). This is so beause the variability of $h(\mathbf{x})$ depends not only on the variability of $f(\mathbf{x})$ but also on the variabilities of $g(\mathbf{x})$, $H_B(\mathbf{x})$, and $Q(\mathbf{x})$.

With these definitions, Eq. (3.5) can be rewritten as

$$\left.\begin{aligned}
&\frac{\partial^2}{\partial x_i^2}\big[h^{(0)}(\mathbf{x}) + h^{(1)}(\mathbf{x}) + h^{(2)}(\mathbf{x}) + \cdots\big]\\
&+ \frac{\partial}{\partial x_i}[\langle f(\mathbf{x})\rangle + f'(\mathbf{x})]\frac{\partial}{\partial x_i}\big[h^{(0)}(\mathbf{x}) + h^{(1)}(\mathbf{x}) + h^{(2)}(\mathbf{x}) + \cdots\big]\\
&\quad = -\frac{\langle g(\mathbf{x})\rangle + g'(\mathbf{x})}{K_G(\mathbf{x})}\left[1 - f'(\mathbf{x}) + \frac{1}{2}f'^2(\mathbf{x}) + \cdots\right]\\
&h^{(0)}(\mathbf{x}) + h^{(1)}(\mathbf{x}) + h^{(2)}(\mathbf{x}) + \cdots = \langle H_B(\mathbf{x})\rangle + H_B'(\mathbf{x}), \quad \mathbf{x} \in \Gamma_D\\
&n_i(\mathbf{x})\frac{\partial}{\partial x_i}\big[h^{(0)}(\mathbf{x}) + h^{(1)}(\mathbf{x}) + h^{(2)}(\mathbf{x}) + \cdots\big]\\
&\quad = -\frac{\langle Q(\mathbf{x})\rangle + Q'(\mathbf{x})}{K_G(\mathbf{x})}\left[1 - f'(\mathbf{x}) + \frac{1}{2}f'^2(\mathbf{x}) + \cdots\right], \quad \mathbf{x} \in \Gamma_N
\end{aligned}\right\}$$

(3.8)

where $K_G(\mathbf{x}) = \exp[\langle f(\mathbf{x}) \rangle]$. Collecting terms at each separate order, we have

$$
\left.
\begin{aligned}
\frac{\partial^2 h^{(0)}(\mathbf{x})}{\partial x_i^2} + \frac{\partial \langle f(\mathbf{x}) \rangle}{\partial x_i} \frac{\partial h^{(0)}(\mathbf{x})}{\partial x_i} &= -\frac{\langle g(\mathbf{x}) \rangle}{K_G(\mathbf{x})} \\[2mm]
h^{(0)}(\mathbf{x}) &= \langle H_B(\mathbf{x}) \rangle, \quad \mathbf{x} \in \Gamma_D \\[2mm]
n_i(\mathbf{x}) \frac{\partial h^{(0)}(\mathbf{x})}{\partial x_i} &= -\frac{\langle Q(\mathbf{x}) \rangle}{K_G(\mathbf{x})}, \quad \mathbf{x} \in \Gamma_N
\end{aligned}
\right\}
\tag{3.9}
$$

$$
\left.
\begin{aligned}
&\frac{\partial^2 h^{(1)}(\mathbf{x})}{\partial x_i^2} + \frac{\partial \langle f(\mathbf{x}) \rangle}{\partial x_i} \frac{\partial h^{(1)}(\mathbf{x})}{\partial x_i} \\[2mm]
&= -\frac{\partial h^{(0)}(\mathbf{x})}{\partial x_i} \frac{\partial f'(\mathbf{x})}{\partial x_i} + \frac{\langle g(\mathbf{x}) \rangle}{K_G(\mathbf{x})} f'(\mathbf{x}) - \frac{g'(\mathbf{x})}{K_G(\mathbf{x})} \\[2mm]
&\quad h^{(1)}(\mathbf{x}) = H'_B(\mathbf{x}), \quad \mathbf{x} \in \Gamma_D \\[2mm]
&n_i(\mathbf{x}) \frac{\partial h^{(1)}(\mathbf{x})}{\partial x_i} = \frac{1}{K_G(\mathbf{x})} [-Q'(\mathbf{x}) + \langle Q(\mathbf{x}) \rangle f'(\mathbf{x})], \quad \mathbf{x} \in \Gamma_N
\end{aligned}
\right\}
\tag{3.10}
$$

and

$$
\left.
\begin{aligned}
&\frac{\partial^2 h^{(n)}(\mathbf{x})}{\partial x_i^2} + \frac{\partial \langle f(\mathbf{x}) \rangle}{\partial x_i} \frac{\partial h^{(n)}(\mathbf{x})}{\partial x_i} \\[2mm]
&= -\frac{\partial h^{(n-1)}(\mathbf{x})}{\partial x_i} \frac{\partial f'(\mathbf{x})}{\partial x_i} - \frac{(-1)^n}{n!} \frac{\langle g(\mathbf{x}) \rangle}{K_G(\mathbf{x})} [f'(\mathbf{x})]^n \\[2mm]
&\quad - \frac{(-1)^{n-1}}{(n-1)!} \frac{g'(\mathbf{x})}{K_G(\mathbf{x})} [f'(\mathbf{x})]^{n-1} \\[2mm]
&\quad h^{(n)}(\mathbf{x}) = 0, \quad \mathbf{x} \in \Gamma_D \\[2mm]
&n_i(\mathbf{x}) \frac{\partial h^{(n)}(\mathbf{x})}{\partial x_i} = -\frac{(-1)^n}{n!} \frac{\langle Q(\mathbf{x}) \rangle}{K_G(\mathbf{x})} [f'(\mathbf{x})]^n \\[2mm]
&\quad - \frac{(-1)^{n-1}}{(n-1)!} \frac{Q'(\mathbf{x})}{K_G(\mathbf{x})} [f'(\mathbf{x})]^{n-1}, \quad \mathbf{x} \in \Gamma_N
\end{aligned}
\right\}
\tag{3.11}
$$

where $n \geq 2$ and no summation for repeated indices n is implied.

It can be shown that $\langle h^{(0)} \rangle = h^{(0)}$, and $\langle h^{(1)} \rangle = 0$. Hence, the *mean head* is $\langle h \rangle = h^{(0)}$ to zeroth or first order in σ and $\langle h \rangle = h^{(0)} + \langle h^{(2)} \rangle$ to second order. The *head fluctuation* is $h' = h^{(1)}$ to first order. Therefore, the *head covariance* is $C_h(\mathbf{x}, \boldsymbol{\chi}) = \langle h^{(1)}(\mathbf{x}) h^{(1)}(\boldsymbol{\chi}) \rangle$ to first order in σ^2. Here, σ^2 is some combination of σ_f, σ_g, σ_{H_B}, and σ_Q.

In Chapter 1 and most previously published texts [e.g., Dagan, 1989; Gelhar, 1993], the log hydraulic conductivity is assumed stationary such that $\langle f \rangle$ is constant and the second term in Eqs. (3.9)–(3.11) would disappear. In this and subsequent chapters, $f(\mathbf{x})$ is mainly treated as a nonstationary random space function. Hence, the moment equations to be developed are applicable to flow in multiscale media, where random heterogeneities prevail at some small scale, while either deterministic or random geological structures and patterns exist at some larger scale. Such nonstationary medium features have been discussed in Section 2.3.3.

3.2.1 Moment Partial Differential Equations

The head covariance is obtained by multiplying Eq. (3.10) with $h^{(1)}$ at another location χ and then taking expectation,

$$
\left.
\begin{aligned}
\frac{\partial^2 C_h(\mathbf{x}, \chi)}{\partial x_i^2} &+ \frac{\partial \langle f(\mathbf{x}) \rangle}{\partial x_i} \frac{\partial C_h(\mathbf{x}, \chi)}{\partial x_i} \\
&= J_i(\mathbf{x}) \frac{\partial C_{fh}(\mathbf{x}, \chi)}{\partial x_i} + \frac{\langle g(\mathbf{x}) \rangle}{K_G(\mathbf{x})} C_{fh}(\mathbf{x}, \chi) - \frac{1}{K_G(\mathbf{x})} C_{gh}(\mathbf{x}, \chi) \\
C_h(\mathbf{x}, \chi) &= C_{H_B h}(\mathbf{x}, \chi), \quad \mathbf{x} \in \Gamma_D \\
n_i(\mathbf{x}) \frac{\partial C_h(\mathbf{x}, \chi)}{\partial x_i} &= \frac{1}{K_G(\mathbf{x})} [-C_{Qh}(\mathbf{x}, \chi) + \langle Q(\mathbf{x}) \rangle C_{fh}(\mathbf{x}, \chi)], \quad \mathbf{x} \in \Gamma_N
\end{aligned}
\right\}
$$
(3.12)

where $J_i(\mathbf{x}) = -\partial h^{(0)}(\mathbf{x})/\partial x_i$ is the negative of the (zeroth-order) mean hydraulic head gradient, $C_{fh}(\mathbf{x}, \chi) = \langle f'(\mathbf{x}) h^{(1)}(\chi) \rangle$ is the cross-covariance between log hydraulic conductivity and head, $C_{gh}(\mathbf{x}, \chi) = \langle g'(\mathbf{x}) h^{(1)}(\chi) \rangle$ is the cross-covariance between forcing term and head, $C_{H_B h}(\mathbf{x}, \chi) = \langle H_B'(\mathbf{x}) h^{(1)}(\chi) \rangle$ is the cross-covariance between boundary head term and head, and $C_{Qh}(\mathbf{x}, \chi) = \langle Q'(\mathbf{x}) h^{(1)}(\chi) \rangle$ is the cross-covariance between boundary flux term and head. These cross-covariances are yet to be given.

The cross-covariance C_{fh} is obtained by rewriting Eq. (3.10) in terms of χ, premultiplying it with $f'(\mathbf{x})$ and taking expectation,

$$
\left.
\begin{aligned}
\frac{\partial^2 C_{fh}(\mathbf{x}, \chi)}{\partial \chi_i^2} &+ \frac{\partial \langle f(\chi) \rangle}{\partial \chi_i} \frac{\partial C_{fh}(\mathbf{x}, \chi)}{\partial \chi_i} \\
&= J_i(\chi) \frac{\partial C_f(\mathbf{x}, \chi)}{\partial \chi_i} + \frac{\langle g(\chi) \rangle}{K_G(\chi)} C_f(\mathbf{x}, \chi) - \frac{1}{K_G(\chi)} C_{fg}(\mathbf{x}, \chi) \\
C_{fh}(\mathbf{x}, \chi) &= C_{fH_B}(\mathbf{x}, \chi), \quad \chi \in \Gamma_D \\
n_i(\chi) \frac{\partial C_{fh}(\mathbf{x}, \chi)}{\partial \chi_i} &= \frac{1}{K_G(\chi)} [-C_{fQ}(\mathbf{x}, \chi) + \langle Q(\chi) \rangle C_f(\mathbf{x}, \chi)], \quad \chi \in \Gamma_N
\end{aligned}
\right\}
$$
(3.13)

where $C_f(\mathbf{x}, \boldsymbol{\chi}) = \langle f'(\mathbf{x})f'(\boldsymbol{\chi})\rangle$ is the covariance of log hydraulic conductivity, $C_{fg}(\mathbf{x}, \boldsymbol{\chi}) = \langle f'(\mathbf{x})g'(\boldsymbol{\chi})\rangle$ is the cross-covariance between log hydraulic conductivity and recharge, $C_{fH_B}(\mathbf{x}, \boldsymbol{\chi}) = \langle f'(\mathbf{x})H'_B(\boldsymbol{\chi})\rangle$ is the cross-covariance between log hydraulic conductivity and boundary head term, and $C_{fQ}(\mathbf{x}, \boldsymbol{\chi}) = \langle f'(\mathbf{x})Q'(\boldsymbol{\chi})\rangle$ is the cross-covariance between log hydraulic conductivity and boundary flux term. These covariances may be given as input. Similarly, the other cross-covariances are given as functions of input statistics of f, g, H_B, and Q,

$$
\frac{\partial^2 C_{gh}(\mathbf{x}, \boldsymbol{\chi})}{\partial \chi_i^2} + \frac{\partial \langle f(\boldsymbol{\chi})\rangle}{\partial \chi_i} \frac{\partial C_{gh}(\mathbf{x}, \boldsymbol{\chi})}{\partial \chi_i}
$$

$$
= J_i(\boldsymbol{\chi})\frac{\partial C_{gf}(\mathbf{x}, \boldsymbol{\chi})}{\partial \chi_i} + \frac{\langle g(\boldsymbol{\chi})\rangle}{K_G(\boldsymbol{\chi})} C_{gf}(\mathbf{x}, \boldsymbol{\chi}) - \frac{1}{K_G(\boldsymbol{\chi})} C_g(\mathbf{x}, \boldsymbol{\chi})
$$

$$
C_{gh}(\mathbf{x}, \boldsymbol{\chi}) = C_{gH_B}(\mathbf{x}, \boldsymbol{\chi}), \quad \boldsymbol{\chi} \in \Gamma_D
$$

$$
n_i(\boldsymbol{\chi})\frac{\partial C_{gh}(\mathbf{x}, \boldsymbol{\chi})}{\partial \chi_i} = \frac{1}{K_G(\boldsymbol{\chi})}[-C_{gQ}(\mathbf{x}, \boldsymbol{\chi}) + \langle Q(\boldsymbol{\chi})\rangle C_{gf}(\mathbf{x}, \boldsymbol{\chi})], \quad \boldsymbol{\chi} \in \Gamma_N \tag{3.14}
$$

$$
\frac{\partial^2 C_{H_B h}(\mathbf{x}, \boldsymbol{\chi})}{\partial \chi_i^2} + \frac{\partial \langle f(\boldsymbol{\chi})\rangle}{\partial \chi_i} \frac{\partial C_{H_B h}(\mathbf{x}, \boldsymbol{\chi})}{\partial \chi_i}
$$

$$
= J_i(\boldsymbol{\chi})\frac{\partial C_{H_B f}(\mathbf{x}, \boldsymbol{\chi})}{\partial \chi_i} + \frac{\langle g(\boldsymbol{\chi})\rangle}{K_G(\boldsymbol{\chi})} C_{H_B f}(\mathbf{x}, \boldsymbol{\chi}) - \frac{1}{K_G(\boldsymbol{\chi})} C_{H_B g}(\mathbf{x}, \boldsymbol{\chi})
$$

$$
C_{H_B h}(\mathbf{x}, \boldsymbol{\chi}) = C_{H_B}(\mathbf{x}, \boldsymbol{\chi}), \quad \boldsymbol{\chi} \in \Gamma_D
$$

$$
n_i(\boldsymbol{\chi})\frac{\partial C_{H_B h}(\mathbf{x}, \boldsymbol{\chi})}{\partial \chi_i} = \frac{1}{K_G(\boldsymbol{\chi})}[-C_{H_B Q}(\mathbf{x}, \boldsymbol{\chi}) + \langle Q(\boldsymbol{\chi})\rangle C_{H_B f}(\mathbf{x}, \boldsymbol{\chi})], \quad \boldsymbol{\chi} \in \Gamma_N \tag{3.15}
$$

$$
\frac{\partial^2 C_{Qh}(\mathbf{x}, \boldsymbol{\chi})}{\partial \chi_i^2} + \frac{\partial \langle f(\boldsymbol{\chi})\rangle}{\partial \chi_i} \frac{\partial C_{Qh}(\mathbf{x}, \boldsymbol{\chi})}{\partial \chi_i}
$$

$$
= J_i(\boldsymbol{\chi})\frac{\partial C_{Qf}(\mathbf{x}, \boldsymbol{\chi})}{\partial \chi_i} + \frac{\langle g(\boldsymbol{\chi})\rangle}{K_G(\boldsymbol{\chi})} C_{Qf}(\mathbf{x}, \boldsymbol{\chi}) - \frac{1}{K_G(\boldsymbol{\chi})} C_{Qg}(\mathbf{x}, \boldsymbol{\chi})
$$

$$
C_{Qh}(\mathbf{x}, \boldsymbol{\chi}) = C_{QH_B}(\mathbf{x}, \boldsymbol{\chi}), \quad \boldsymbol{\chi} \in \Gamma_D
$$

$$
n_i(\boldsymbol{\chi})\frac{\partial C_{Qh}(\mathbf{x}, \boldsymbol{\chi})}{\partial \chi_i} = \frac{1}{K_G(\boldsymbol{\chi})}[-C_Q(\mathbf{x}, \boldsymbol{\chi}) + \langle Q(\boldsymbol{\chi})\rangle C_{Qf}(\mathbf{x}, \boldsymbol{\chi})], \quad \boldsymbol{\chi} \in \Gamma_N \tag{3.16}
$$

The zeroth- and first-order mean head is given by Eq. (3.9). The second-order *correction term* for the mean head is given by taking expectation of Eq. (3.11) with $n = 2$,

$$
\left.
\begin{aligned}
&\frac{\partial^2 \langle h^{(2)}(\mathbf{x}) \rangle}{\partial x_i^2} + \frac{\partial \langle f(\mathbf{x}) \rangle}{\partial x_i} \frac{\partial \langle h^{(2)}(\mathbf{x}) \rangle}{\partial x_i} \\
&= - \frac{\partial}{\partial x_i} \frac{\partial}{\partial \chi_i} C_{fh}(\mathbf{x}, \boldsymbol{\chi}) \bigg|_{\mathbf{x}=\boldsymbol{\chi}} - \frac{1}{2} \frac{\langle g(\mathbf{x}) \rangle}{K_G(\mathbf{x})} \sigma_f^2(\mathbf{x}) + \frac{1}{K_G(\mathbf{x})} \sigma_{fg}(\mathbf{x}) \\
&\qquad\qquad \langle h^{(2)}(\mathbf{x}) \rangle = 0, \quad \mathbf{x} \in \Gamma_D \\
&n_i(\mathbf{x}) \frac{\partial \langle h^{(2)}(\mathbf{x}) \rangle}{\partial x_i} = \frac{1}{K_G(\mathbf{x})} \left[-\frac{\langle Q(\mathbf{x}) \rangle}{2} \sigma_f^2(\mathbf{x}) + \sigma_{fQ}(\mathbf{x}) \right], \quad \mathbf{x} \in \Gamma_N
\end{aligned}
\right\}
\tag{3.17}
$$

where $\sigma_f^2(\mathbf{x}) = C_f(\mathbf{x}, \mathbf{x})$, $\sigma_{fg}(\mathbf{x}) = C_{fg}(\mathbf{x}, \mathbf{x})$, and $\sigma_{fQ}(\mathbf{x}) = C_{fQ}(\mathbf{x}, \mathbf{x})$. For *uniform mean flow* (e.g., $J_i(\mathbf{x}) \equiv J_i$) in unbounded domains of stationary media in the absence of sink/source, this second-order correction term is identically zero. However, for bounded domains it is generally nonzero.

The flux in Eq. (3.2) can be rewritten as

$$
\mathbf{q} = -K_G \left[1 + f' + \frac{f'^2}{2} + \cdots \right] \nabla \left[h^{(0)} + h^{(1)} + h^{(2)} + \cdots \right]
\tag{3.18}
$$

Collecting terms at each separate order, we have

$$
\mathbf{q}^{(0)}(\mathbf{x}) = -K_G(\mathbf{x}) \nabla h^{(0)}(\mathbf{x})
\tag{3.19}
$$

$$
\mathbf{q}^{(1)}(\mathbf{x}) = -K_G(\mathbf{x}) \left[f'(\mathbf{x}) \nabla h^{(0)}(\mathbf{x}) + \nabla h^{(1)}(\mathbf{x}) \right]
\tag{3.20}
$$

$$
\mathbf{q}^{(2)}(\mathbf{x}) = -K_G(\mathbf{x}) \left[f'(\mathbf{x}) \nabla h^{(1)}(\mathbf{x}) + \tfrac{1}{2} f'^2(\mathbf{x}) \nabla h^{(0)}(\mathbf{x}) + \nabla h^{(2)}(\mathbf{x}) \right]
\tag{3.21}
$$

It can be shown that the *mean flux* is $\langle \mathbf{q} \rangle = \mathbf{q}^{(0)}$ to zeroth or first order in σ, $\langle \mathbf{q} \rangle = \mathbf{q}^{(0)} + \langle \mathbf{q}^{(2)} \rangle$ to second order, and the *flux fluctuation* is $\mathbf{q}' = \mathbf{q}^{(1)}$ to first order. Therefore, to first order the mean flux and *flux covariance* are given as

$$
q_i^{(0)}(\mathbf{x}) = K_G(\mathbf{x}) J_i(\mathbf{x})
\tag{3.22}
$$

$$
\begin{aligned}
C_{q_{ij}}(\mathbf{x}, \boldsymbol{\chi}) = K_G(\mathbf{x}) K_G(\boldsymbol{\chi}) \Big[&J_i(\mathbf{x}) J_j(\boldsymbol{\chi}) C_f(\mathbf{x}, \boldsymbol{\chi}) - J_i(\mathbf{x}) \frac{\partial}{\partial \chi_j} C_{fh}(\mathbf{x}, \boldsymbol{\chi}) \\
&- J_j(\boldsymbol{\chi}) \frac{\partial}{\partial x_i} C_{fh}(\boldsymbol{\chi}, \mathbf{x}) + \frac{\partial^2}{\partial x_i \partial \chi_j} C_h(\mathbf{x}, \boldsymbol{\chi}) \Big]
\end{aligned}
\tag{3.23}
$$

The second-order mean flux correction term is given as

$$
\langle q_i^{(2)}(\mathbf{x}) \rangle = K_G(\mathbf{x}) \left[\frac{J_i(\mathbf{x})}{2} \sigma_f^2(\mathbf{x}) - \frac{\partial}{\partial x_i} \langle h^{(2)}(\mathbf{x}) \rangle - \frac{\partial}{\partial x_i} C_{fh}(\boldsymbol{\chi}, \mathbf{x}) \bigg|_{\boldsymbol{\chi}=\mathbf{x}} \right]
\tag{3.24}
$$

The mean head $\langle h \rangle$ and mean flux $\langle q_i \rangle$ can be used to estimate (or predict) the fields of head (pressure) and flux in a heterogeneous medium, and the corresponding (co)variance to evaluate the uncertainty (error) associated with the estimation (prediction). These first two moments can be used to construct confidence intervals for the pressure and flux fields. In addition, the head covariance, the flux covariance, the cross-covariance between log hydraulic conductivity and head, and other covariances obtained based on them may be used to derive (conditional) estimates of head, log hydraulic conductivity, and velocity from related field measurements by inverse methods or conditioning [Dagan, 1982a; Graham and McLaughlin, 1989a; Rubin, 1991; Zhang and Neuman, 1995a,b]. Since solute transport is controlled by the underlying velocity field, the statistical moments of the velocity field are essential for studying solute macrodispersion (field-scale dispersion) [see, e.g., Dagan, 1982b, 1984; Gelhar and Axness, 1983; Winter *et al.*, 1984; Neuman *et al.*, 1987; Kitanidis, 1988; Neuman, 1993; Zhang and Neuman, 1995a]. As will be seen in Chapter 6, the single-phase velocity moments also form the basis for stochastic analyses of two-phase displacement in random porous media.

3.3 GREEN'S FUNCTION METHOD

As discussed in Section 1.4.3, one may formulate the moment equations with the aid of *Green's function*. It is shown that there are two ways to do so. One is to first derive *exact* yet *unclosed* equations governing the flow moments with the help of *random* or *deterministic* Green's functions and then to close the equation set by some *closure approximations* or perturbation schemes. This approach has recently been undertaken intensively by Neuman and coworkers [Neuman and Orr, 1993; Neuman *et al.*, 1996; Tartakovsky and Neuman, 1998a,b; Guadagnini and Neuman, 1999a,b]. The other approach is to invoke the perturbation schemes from the onset and then to write moment equations on the basis of Green's functions. This approach has been extensively used in the literature of flow in porous media [e.g., Dagan, 1982a, 1989; Naff and Vecchia, 1986; Rubin and Dagan, 1988, 1989; Cheng and Lafe, 1991; Osnes, 1995; Hsu *et al.*, 1996]. The former approach yields a set of exact equations, which, however, cannot be solved without approximations, whereas the latter gives a set of approximated but solvable equations. It has been found that the two approaches yield identical results when consistent approximations are made, albeit at different stages. Since each approach has its own advantages and disadvantages, we introduce both of them in this section and make a detailed comparison.

3.3.1 Formal Integrodifferential Equations

In this subsection, we attempt to derive and analyze some formal equations governing the first two moments of head and flux. The development is mainly based on the work of Neuman's group [Neuman and Orr, 1993; Neuman *et al.*, 1996; Tartakovsky and Neuman, 1998a,b; Guadagnini and Neuman, 1999a,b].

Substitution of Eq. (3.2) into Eqs. (3.1) and (3.4) yields

$$\nabla \cdot [K_S(\mathbf{x}) \nabla h(\mathbf{x})] = -g(\mathbf{x}) \tag{3.25}$$

subject to boundary conditions

$$h(\mathbf{x}) - H_B(\mathbf{x}), \quad \mathbf{x} \in \Gamma_D \tag{3.26}$$

$$K_S(\mathbf{x})\nabla h(\mathbf{x}) \cdot \mathbf{n}(\mathbf{x}) = -Q(\mathbf{x}), \quad \mathbf{x} \in \Gamma_N \tag{3.27}$$

One may decompose K_S, g, H_B, and Q into their respective mean and fluctuation such that $K_S(\mathbf{x}) = \langle K_S(\mathbf{x}) \rangle + K_S'(\mathbf{x})$, $g(\mathbf{x}) = \langle g(\mathbf{x}) \rangle + g'(\mathbf{x})$, $H_B(\mathbf{x}) = \langle H_B(\mathbf{x}) \rangle + H_B'(\mathbf{x})$, and $Q(\mathbf{x}) = \langle Q(\mathbf{x}) \rangle + Q'(\mathbf{x})$. For simplicity, they are assumed statistically independent such that their cross-covariances $C_{pq} \equiv 0$ (where $p, q = K_S$, g, H_B, and Q but $p \neq q$). Likewise, one may decompose the random function $h(\mathbf{x})$ into its mean $\langle h(\mathbf{x}) \rangle$ and zero-mean fluctuation $h'(\mathbf{x})$. Taking expectation of Eqs. (3.25)–(3.27) yields

$$\nabla \cdot [\langle K_S(\mathbf{x}) \rangle \nabla \langle h(\mathbf{x}) \rangle - \mathbf{r}_f(\mathbf{x})] = -\langle g(\mathbf{x}) \rangle$$

$$\langle h(\mathbf{x}) \rangle = \langle H_B(\mathbf{x}) \rangle, \quad \mathbf{x} \in \Gamma_D \tag{3.28}$$

$$[\langle K_S(\mathbf{x}) \rangle \nabla \langle h(\mathbf{x}) \rangle - \mathbf{r}_f(\mathbf{x})] \cdot \mathbf{n}(\mathbf{x}) = -\langle Q(\mathbf{x}) \rangle, \quad \mathbf{x} \in \Gamma_N$$

where $\mathbf{r}_f(\mathbf{x}) = -\langle K_S'(\mathbf{x})\nabla h'(\mathbf{x}) \rangle$ is a cross-covariance between K_S and ∇h at $\mathbf{x}$. This covariance has been called the *residual flux* by Neuman and coworkers.

One may define a random Green's function $\mathcal{G}(\boldsymbol{\chi}, \mathbf{x})$ satisfying the following *Poisson equation*:

$$\nabla_{\boldsymbol{\chi}} \cdot [K_S(\boldsymbol{\chi})\nabla_{\boldsymbol{\chi}}\mathcal{G}(\boldsymbol{\chi}, \mathbf{x})] = -\delta(\boldsymbol{\chi} - \mathbf{x})$$

$$\mathcal{G}(\boldsymbol{\chi}, \mathbf{x}) = 0, \quad \boldsymbol{\chi} \in \Gamma_D \tag{3.29}$$

$$K_S(\mathbf{x})\nabla_{\boldsymbol{\chi}}\mathcal{G}(\boldsymbol{\chi}, \mathbf{x}) \cdot \mathbf{n}(\boldsymbol{\chi}) = 0, \quad \boldsymbol{\chi} \in \Gamma_N$$

where the operation of $\nabla_{\boldsymbol{\chi}}$ is with respect to $\boldsymbol{\chi}$, and $\delta(\boldsymbol{\chi} - \mathbf{x})$ is the d-dimensional Dirac delta function, whose properties have been discussed in Section 2.4.1.

Rewriting Eq. (3.25) in terms of $\boldsymbol{\chi}$, multiplying by $\mathcal{G}(\boldsymbol{\chi}, \mathbf{x})$ and integrating over the domain Ω leads to

$$\int_{\Omega} \left\{ \nabla_{\boldsymbol{\chi}} \cdot [K_S(\boldsymbol{\chi})\nabla_{\boldsymbol{\chi}}h(\boldsymbol{\chi})] + g(\boldsymbol{\chi}) \right\} \mathcal{G}(\boldsymbol{\chi}, \mathbf{x}) \, d\boldsymbol{\chi} = 0 \tag{3.30}$$

Applying *Green's identity* twice yields

$$\int_{\Omega} \left\{ \nabla_{\boldsymbol{\chi}} \cdot [K_S(\boldsymbol{\chi})\nabla_{\boldsymbol{\chi}}\mathcal{G}(\boldsymbol{\chi}, \mathbf{x})]h(\boldsymbol{\chi}) + g(\boldsymbol{\chi})\mathcal{G}(\boldsymbol{\chi}, \mathbf{x}) \right\} d\boldsymbol{\chi}$$

$$= \int_{\Gamma_D} K_S(\boldsymbol{\chi})\nabla_{\boldsymbol{\chi}}\mathcal{G}(\boldsymbol{\chi}, \mathbf{x}) \cdot \mathbf{n}(\boldsymbol{\chi})h(\boldsymbol{\chi}) \, d\boldsymbol{\chi}$$

$$- \int_{\Gamma_N} K_S(\boldsymbol{\chi})\nabla_{\boldsymbol{\chi}}h(\boldsymbol{\chi}) \cdot \mathbf{n}(\boldsymbol{\chi})\mathcal{G}(\boldsymbol{\chi}, \mathbf{x}) \, d\boldsymbol{\chi} \tag{3.31}$$

where the boundary condition in Eq. (3.29) has been utilized. With Eqs. (3.29), (3.26), and (3.27), one obtains from Eq. (3.31) the following solution for $h(\mathbf{x})$:

$$h(\mathbf{x}) = \int_{\Omega} g(\boldsymbol{\chi}) \mathcal{G}(\boldsymbol{\chi}, \mathbf{x}) \, d\boldsymbol{\chi} - \int_{\Gamma_N} Q(\boldsymbol{\chi}) \mathcal{G}(\boldsymbol{\chi}, \mathbf{x}) \, d\boldsymbol{\chi}$$
$$- \int_{\Gamma_D} K_S(\boldsymbol{\chi}) \nabla_{\boldsymbol{\chi}} \mathcal{G}(\boldsymbol{\chi}, \mathbf{x}) \cdot \mathbf{n}(\boldsymbol{\chi}) H_B(\boldsymbol{\chi}) \, d\boldsymbol{\chi} \tag{3.32}$$

It is a *random solution* because $h(\mathbf{x})$ is in terms of random quantities, although it is exact. As a matter of fact, Eq. (3.32) is merely a *restatement* of Eqs. (3.25)–(3.27).

Similarly, one may define a deterministic Green's function, $G(\boldsymbol{\chi}, \mathbf{x})$, as the solution of the following Poisson equation:

$$\nabla_{\boldsymbol{\chi}} \cdot [\langle K_S(\boldsymbol{\chi}) \rangle \nabla_{\boldsymbol{\chi}} G(\boldsymbol{\chi}, \mathbf{x})] = -\delta(\boldsymbol{\chi} - \mathbf{x})$$
$$G(\boldsymbol{\chi}, \mathbf{x}) = 0, \quad \boldsymbol{\chi} \in \Gamma_D \tag{3.33}$$
$$\langle K_S(\mathbf{x}) \rangle \nabla_{\boldsymbol{\chi}} G(\boldsymbol{\chi}, \mathbf{x}) \cdot \mathbf{n}(\boldsymbol{\chi}) = 0, \quad \boldsymbol{\chi} \in \Gamma_N$$

With the aid of this deterministic Green's function and in a way similar to that leading to Eq. (3.32), one obtains $h_o(\mathbf{x})$, the solution of Eq. (3.28) when $\mathbf{r}_f$ is set to be zero,

$$h_o(\mathbf{x}) = \int_{\Omega} \langle g(\boldsymbol{\chi}) \rangle G(\boldsymbol{\chi}, \mathbf{x}) \, d\boldsymbol{\chi} - \int_{\Gamma_N} \langle Q(\boldsymbol{\chi}) \rangle G(\boldsymbol{\chi}, \mathbf{x}) \, d\boldsymbol{\chi}$$
$$- \int_{\Gamma_D} \langle K_S(\boldsymbol{\chi}) \rangle \nabla_{\boldsymbol{\chi}} G(\boldsymbol{\chi}, \mathbf{x}) \cdot \mathbf{n}(\boldsymbol{\chi}) \langle H_B(\boldsymbol{\chi}) \rangle \, d\boldsymbol{\chi} \tag{3.34}$$

It is seen that $h_o(\mathbf{x})$ is a deterministic quantity.

One may recast Eq. (3.25) as [Neuman and Orr, 1993; Neuman *et al.*, 1996],

$$\mathcal{L}h(\mathbf{x}) + g(\mathbf{x}) = [L + \mathcal{R}]h(\mathbf{x}) + g(\mathbf{x}) = 0 \tag{3.35}$$

where $\mathcal{L} = \nabla \cdot [K_S(\mathbf{x})\nabla]$ and $\mathcal{R} = \nabla \cdot [K'_S(\mathbf{x})\nabla]$ are *stochastic operators*, and $L = \langle \mathcal{L} \rangle = \nabla \cdot [\langle K_S(\mathbf{x}) \rangle \nabla]$ is a *deterministic operator*. One may immediately recognize that $\mathcal{G}$ satisfies $\mathcal{L}\mathcal{G} + \delta = 0$ and G satisfies $LG + \delta = 0$, subject to *homogeneous boundary conditions*.

Rewrite Eq. (3.35) as

$$Lh(\mathbf{x}) = -g(\mathbf{x}) - \mathcal{R}h(\mathbf{x}) \tag{3.36}$$

Premultiplying Eq. (3.36) by G, and applying Green's identity twice to the left-hand side, yields [Neuman *et al.*, 1996, (A4)]

$$h(\mathbf{x}) = -L^{-1}g(\mathbf{x}) - L^{-1}\mathcal{R}h(\mathbf{x}) + T_{bc}(\mathbf{x}) \tag{3.37}$$

where T_{bc} is the *nonhomogeneous* boundary integral defined as

$$T_{bc}(\mathbf{x}) = - \int_{\Gamma_D} \langle K_S(\mathbf{x}') \rangle \nabla_{\mathbf{x}'} G(\mathbf{x}', \mathbf{x}) \cdot \mathbf{n}(\mathbf{x}') H_B(\mathbf{x}') \, d\mathbf{x}'$$
$$- \int_{\Gamma_N} G(\mathbf{x}', \mathbf{x}) \langle K_S(\mathbf{x}') \rangle [K_S(\mathbf{x}')]^{-1} Q(\mathbf{x}') \, d\mathbf{x}' \quad (3.38)$$

In the above, L^{-1} is the *inverse operator* defined as [Neuman *et al.*, 1996, (A5)]

$$L^{-1} g(\mathbf{x}) = - \int_{\Omega} G(\mathbf{x}', \mathbf{x}) g(\mathbf{x}') \, d\mathbf{x}' \quad (3.39)$$

so that $s = -L^{-1}g$ is the solution of $Ls + g = 0$, subject to homogeneous boundary conditions $H_B = Q = 0$. Similarly, one may define the inverse operator $\mathcal{L}^{-1}$ as

$$\mathcal{L}^{-1} g(\mathbf{x}) = - \int_{\Omega} \mathcal{G}(\mathbf{x}', \mathbf{x}) g(\mathbf{x}') \, d\mathbf{x}' \quad (3.40)$$

so that $s = -\mathcal{L}^{-1}g$ is the solution of $\mathcal{L}s + g = 0$, subject to homogeneous boundary conditions $H_B = Q = 0$. It is obvious that $LL^{-1} = I$ and $\mathcal{L}\mathcal{L}^{-1} = I$, where I is the *identity operator*. With this, one may verify that $(1 - \mathcal{L}^{-1}\mathcal{R})(1 + L^{-1}\mathcal{R}) = I$ [Neuman *et al.*, 1996, (A8)]. By rewriting Eq. (3.37) as

$$(1 + L^{-1}\mathcal{R})h(\mathbf{x}) = -L^{-1}g(\mathbf{x}) + T_{bc}(\mathbf{x}) \quad (3.41)$$

and operating with $(1 - \mathcal{L}^{-1}\mathcal{R})$, one obtains

$$h(\mathbf{x}) = (1 - \mathcal{L}^{-1}\mathcal{R})[-L^{-1}g(\mathbf{x}) + B(\mathbf{x}) + N(\mathbf{x})] \quad (3.42)$$

where

$$B(\mathbf{x}) = - \int_{\Gamma_D} \langle K_S(\mathbf{x}') \rangle \nabla_{\mathbf{x}'} G(\mathbf{x}', \mathbf{x}) \cdot \mathbf{n}(\mathbf{x}') H_B(\mathbf{x}') \, d\mathbf{x}'$$
$$- \int_{\Gamma_n} G(\mathbf{x}', \mathbf{x}) Q(\mathbf{x}') \, d\mathbf{x}' \quad (3.43)$$

$$N(\mathbf{x}) = - \int_{\Gamma_n} G(\mathbf{x}', \mathbf{x})[\langle K_S(\mathbf{x}') \rangle K_S(\mathbf{x}')^{-1} - 1] Q(\mathbf{x}') \, d\mathbf{x}'$$

One may obtain $\mathbf{r}_f(\mathbf{x}) = -\langle K_S'(\mathbf{x}) \nabla h'(\mathbf{x}) \rangle = -\langle K_S'(\mathbf{x}) \nabla h(\mathbf{x}) \rangle$ from Eq. (3.42) by premultiplying with $-K_S' \nabla$ and taking expectation. Recognizing that $h_o(\mathbf{x})$ in Eq. (3.34) is equal to $-L^{-1} \langle g(\mathbf{x}) \rangle + B(\mathbf{x})$, substituting the definitions of $\mathcal{L}^{-1}$, $\mathcal{R}$, B, and N, and applying

Green's identity, leads to [Neuman *et al.*, 1996, (11)–(13)]

$$\mathbf{r}_f(\mathbf{x}) = \int_\Omega \mathbf{a}(\mathbf{x}, \mathbf{x}') \nabla_{\mathbf{x}'} h_o(\mathbf{x}') \, d\mathbf{x}' + \int_{\Gamma_N} \mathbf{b}(\mathbf{x}, \mathbf{x}') \langle Q(\mathbf{x}') \rangle \, d\mathbf{x}'$$

$$+ \int_\Omega \int_{\Gamma_N} \mathbf{c}(\mathbf{x}, \mathbf{x}', \mathbf{x}'') \langle Q(\mathbf{x}'') \rangle \, d\mathbf{x}'' \, d\mathbf{x}' \tag{3.44}$$

where

$$\mathbf{a}(\mathbf{x}, \mathbf{x}') = \langle K_S'(\mathbf{x}) K_S'(\mathbf{x}') \nabla \nabla_{\mathbf{x}'}^T \mathcal{G}(\mathbf{x}', \mathbf{x}) \rangle \tag{3.45}$$

$$\mathbf{b}(\mathbf{x}, \mathbf{x}') = \langle K_S'(\mathbf{x}) K_S'(\mathbf{x}') \nabla \mathcal{G}(\mathbf{x}', \mathbf{x}) \rangle \langle K_S(\mathbf{x}') \rangle^{-1}$$

$$+ \langle K_S'(\mathbf{x}) K_S(\mathbf{x}')^{-1} \rangle \langle K_S(\mathbf{x}') \rangle \nabla G(\mathbf{x}', \mathbf{x}) \tag{3.46}$$

$$\mathbf{c}(\mathbf{x}, \mathbf{x}', \mathbf{x}'') = -\langle K_S'(\mathbf{x}) \nabla \mathcal{G}(\mathbf{x}', \mathbf{x}) \nabla_{\mathbf{x}'}^T K_S'(\mathbf{x}') \nabla_{\mathbf{x}'} G(\mathbf{x}'', \mathbf{x}')$$

$$\cdot [\langle K_S(\mathbf{x}'') \rangle K_S(\mathbf{x}'')^{-1} - 1] \rangle \tag{3.47}$$

Note that $G(\mathbf{x}', \mathbf{x})$ and $\mathcal{G}(\mathbf{x}', \mathbf{x})$ are the respective deterministic and random Green's functions defined earlier.

Now one may proceed to derive the head covariance $C_h(\mathbf{x}, \boldsymbol{\chi}) = \langle h'(\mathbf{x}) h'(\boldsymbol{\chi}) \rangle$. Subtracting Eq. (3.28) from Eqs. (3.25)–(3.27) yields

$$\nabla \cdot [K_S(\mathbf{x}) \nabla h'(\mathbf{x}) + K_S'(\mathbf{x}) \nabla \langle h(\mathbf{x}) \rangle + \mathbf{r}_f(\mathbf{x})] = -g'(\mathbf{x})$$

$$h'(\mathbf{x}) = H_B'(\mathbf{x}), \quad \mathbf{x} \in \Gamma_D \tag{3.48}$$

$$[K_S(\mathbf{x}) \nabla h'(\mathbf{x}) + K_S'(\mathbf{x}) \nabla \langle h(\mathbf{x}) \rangle + \mathbf{r}_f(\mathbf{x})] \cdot \mathbf{n}(\mathbf{x}) = -Q'(\mathbf{x}), \quad \mathbf{x} \in \Gamma_N$$

Following the same procedure leading to Eq. (3.32), one may derive the following expression for $h'(\mathbf{x})$ [Guadagnini and Neuman, 1999a, (A5)].

$$h'(\mathbf{x}) = \int_\Omega g'(\mathbf{x}') \mathcal{G}(\mathbf{x}', \mathbf{x}) \, d\mathbf{x}' - \int_\Omega \mathbf{r}_f(\mathbf{x}') \cdot \nabla_{\mathbf{x}'} \mathcal{G}(\mathbf{x}', \mathbf{x}) \, d\mathbf{x}'$$

$$- \int_\Omega K_S'(\mathbf{x}') \nabla_{\mathbf{x}'} \langle h(\mathbf{x}') \rangle \cdot \nabla_{\mathbf{x}'} \mathcal{G}(\mathbf{x}', \mathbf{x}) \, d\mathbf{x}'$$

$$- \int_{\Gamma_D} H_B'(\mathbf{x}') K_S(\mathbf{x}') \nabla_{\mathbf{x}'} \mathcal{G}(\mathbf{x}', \mathbf{x}) \cdot \mathbf{n}(\mathbf{x}') \, d\mathbf{x}'$$

$$- \int_{\Gamma_N} Q'(\mathbf{x}') \mathcal{G}(\mathbf{x}', \mathbf{x}) \, d\mathbf{x}' \tag{3.49}$$

The head covariance may be obtained by writing $h'(\chi)$ on the basis of Eq. (3.49), multiplying it with $h'(\mathbf{x})$ directly from Eq. (3.49), and taking expectation,

$$
\begin{aligned}
C_h(\mathbf{x}, \chi) = & \int_\Omega \int_\Omega \langle g'(\mathbf{x}')g'(\chi')\rangle \langle \mathcal{G}(\mathbf{x}', \mathbf{x})\mathcal{G}(\chi', \chi)\rangle \, d\mathbf{x}' \, d\chi' \\
& + \int_\Omega \int_\Omega \mathbf{r}_f^T(\mathbf{x}')\langle \nabla_{\mathbf{x}'}\mathcal{G}(\mathbf{x}', \mathbf{x})\nabla_{\chi'}^T\mathcal{G}(\chi', \chi)\rangle \mathbf{r}_f(\chi') \, d\mathbf{x}' \, d\chi' \\
& + \int_\Omega \int_\Omega \nabla_{\mathbf{x}'}^T\langle h(\mathbf{x}')\rangle \langle K_S'(\mathbf{x}')K_S'(\chi')\nabla_{\mathbf{x}'}\mathcal{G}(\mathbf{x}', \mathbf{x}) \\
& \quad \cdot \nabla_{\chi'}^T\mathcal{G}(\chi', \chi)\rangle \nabla_{\chi'}\langle h(\chi')\rangle \, d\mathbf{x}' \, d\chi' \\
& + \int_{\Gamma_D} \int_{\Gamma_D} \mathbf{n}^T(\mathbf{x}')\langle H_B'(\mathbf{x}')H_B'(\chi')\rangle \langle K_S(\mathbf{x}')K_S(\chi') \\
& \quad \cdot \nabla_{\mathbf{x}'}\mathcal{G}(\mathbf{x}', \mathbf{x})\nabla_{\chi'}^T\mathcal{G}(\chi', \chi)\rangle \mathbf{n}(\chi') \, d\mathbf{x}' \, d\chi' \\
& + 2 \int_\Omega \int_\Omega \mathbf{r}_f^T(\mathbf{x}')\langle K_S'(\chi')\nabla_{\mathbf{x}'}\mathcal{G}(\mathbf{x}', \mathbf{x})\nabla_{\chi'}^T\mathcal{G}(\chi', \chi)\rangle \nabla_{\chi'}\langle h(\chi')\rangle \, d\mathbf{x}' \, d\chi' \\
& + \int_{\Gamma_N} \int_{\Gamma_N} \langle Q'(\mathbf{x}')Q'(\chi')\rangle \langle \mathcal{G}(\mathbf{x}', \mathbf{x})\mathcal{G}(\chi', \chi)\rangle \, d\mathbf{x}' \, d\chi' \qquad (3.50)
\end{aligned}
$$

In deriving this expression, we have made use of the property that K_S, g, H_B, and Q are mutually uncorrelated and hence so are $\mathcal{G}$, g, H_B, and Q. Though exact, the head covariance (3.50) is unclosed in that it depends on unknown *higher moments* such as $\langle K_S'(\chi')\nabla_{\mathbf{x}'}\mathcal{G}(\mathbf{x}', \mathbf{x})\nabla_{\chi'}^T\mathcal{G}(\chi', \chi)\rangle$ and $\langle K_S(\mathbf{x}')K_S(\chi')\nabla_{\mathbf{x}'}\mathcal{G}(\mathbf{x}', \mathbf{x})\nabla_{\chi'}^T\mathcal{G}(\chi', \chi)\rangle$. It may be closed and become solvable by invoking some closure approximations or perturbation expansions. Rather than proceeding to approximate this *integral equation*, one may derive an alternative, *integrodifferential equation*, for the head covariance.

Multiplying Eq. (3.48) by $h'(\chi)$ and taking expectation yields

$$
\nabla \cdot [\langle K_S(\mathbf{x})\rangle \nabla C_h(\mathbf{x}, \chi) + \mathbf{d}(\mathbf{x}, \chi) + C_{Ksh}(\mathbf{x}, \chi)\nabla\langle h(\mathbf{x})\rangle] = -C_{gh}(\mathbf{x}, \chi)
$$
$$
C_h(\mathbf{x}, \chi) = C_{H_Bh}(\mathbf{x}, \chi), \quad \mathbf{x} \in \Gamma_D
$$
$$
[\langle K_S(\mathbf{x})\rangle \nabla C_h(\mathbf{x}, \chi) + \mathbf{d}(\mathbf{x}, \chi) + C_{Ksh}(\mathbf{x}, \chi)\nabla\langle h(\mathbf{x})\rangle] \cdot \mathbf{n}(\mathbf{x}) \qquad (3.51)
$$
$$
= -C_{Qh}(\mathbf{x}, \chi), \quad \mathbf{x} \in \Gamma_N
$$

where the third moment $\mathbf{d}(\mathbf{x}, \chi) = \langle K_S'(\mathbf{x})\nabla h'(\mathbf{x})h'(\chi)\rangle$ as well as the other second moments $C_{Ksh}(\mathbf{x}, \chi) = \langle K_S'(\mathbf{x})h'(\mathbf{x})\rangle$, $C_{gh}(\mathbf{x}, \chi) = \langle g'(\mathbf{x})h'(\mathbf{x})\rangle$, $C_{H_Bh}(\mathbf{x}, \chi) = \langle H_B'(\mathbf{x})h'(\mathbf{x})\rangle$, and $C_{Qh}(\mathbf{x}, \chi) = \langle Q'(\mathbf{x})h'(\mathbf{x})\rangle$ are given with the aid of Eq. (3.49),

$$
\begin{aligned}
\mathbf{d}^T(\mathbf{x}, \chi) = & -\int_\Omega \mathbf{r}_f^T(\chi')\nabla_{\chi'}\langle \mathcal{G}(\chi', \chi)K_S'(\mathbf{x})\nabla^T h'(\mathbf{x})\rangle \, d\chi' \\
& -\int_\Omega \nabla_{\chi'}^T\langle h(\chi')\rangle \langle K_S'(\chi')K_S'(\mathbf{x})\nabla_{\chi'}\mathcal{G}(\chi', \chi)\nabla^T h'(\mathbf{x})\rangle \, d\chi' \qquad (3.52)
\end{aligned}
$$

$$C_{K_S h}(\mathbf{x}, \chi) = -\int_\Omega \mathbf{r}_f^T(\chi') \nabla_{\chi'} \langle \mathcal{G}(\chi', \chi) K_S'(\mathbf{x}) \rangle \, d\chi'$$

$$-\int_\Omega \vee_{\chi'}^T \langle h(\chi') \rangle \langle K_S'(\mathbf{x}) K_S'(\chi') \nabla_{\chi'} \mathcal{G}(\chi', \chi) \rangle \, d\chi' \tag{3.53}$$

$$C_{gh}(\mathbf{x}, \chi) = \int_\Omega \langle g'(\mathbf{x}) g'(\chi') \rangle \langle \mathcal{G}(\chi', \chi) \rangle \, d\chi' \tag{3.54}$$

$$C_{H_B h}(\mathbf{x}, \chi) = -\int_{\Gamma_N} \langle H_B'(\mathbf{x}) H_B'(\chi') \rangle \langle K_S(\chi') \mathcal{G}(\chi', \chi) \rangle \cdot \mathbf{n}(\chi') \, d\chi' \tag{3.55}$$

$$C_{Qh}(\mathbf{x}, \chi) = -\int_{\Gamma_N} \langle Q'(\mathbf{x}) Q'(\chi') \rangle \langle \mathcal{G}(\chi', \chi) \rangle \, d\chi' \tag{3.56}$$

It is of interest to note that in analogy to Eq. (3.53), one obtains an alternative expression for the residual flux $\mathbf{r}_f$,

$$\mathbf{r}_f(\mathbf{x}) = \int_\Omega \langle K_S'(\mathbf{x}) \nabla \nabla_{\mathbf{x}'}^T \mathcal{G}(\mathbf{x}', \mathbf{x}) \rangle \mathbf{r}_f(\mathbf{x}') \, d\mathbf{x}' + \int_\Omega \mathbf{a}(\mathbf{x}', \mathbf{x}) \nabla_{\mathbf{x}'} \langle h(\mathbf{x}') \rangle \, d\mathbf{x}' \tag{3.57}$$

Compared to Eq. (3.44), Eq. (3.57) is implicit in $\mathbf{r}_f$ and has to be solved in an iterative manner.

In summary, the mean head is governed by Eq. (3.28) coupled with Eq. (3.44) or Eq. (3.57), and the head covariance is governed by Eq. (3.50) or Eq. (3.51) coupled with Eqs. (3.52)–(3.56). These moment equations are exact in that no approximations have been invoked. Although they may reveal some interesting qualitative features of the solutions, these moments are unclosed and cannot be solved without either approximations or high-resolution Monte Carlo simulations. These equations are essentially the restatement of the original set of partial differential equations. Hence, the significance of these equations is not their exactness (because the original equation set is also exact) but their provision of a device for further, workable approximations. In Section 3.10, we will introduce a few closure approximations to render these or similar moment equations workable. Below, we discuss a perturbative expansion scheme for these integrodifferential equations of head moments [Tartakovsky and Neuman, 1998a; Guadagnini and Neuman, 1999a].

3.3.2 Perturbed Integrodifferential Equations

As in Section 3.2, one may write $K_S(\mathbf{x}) = \exp[f(\mathbf{x})]$ and expand it as

$$K_S(\mathbf{x}) = K_G(\mathbf{x}) \left[1 + f'(\mathbf{x}) + \frac{[f'(\mathbf{x})]^2}{2} + \cdots \right]$$

Correspondingly, one may expand h as $h^{(0)} + h^{(1)} + h^{(2)} + \cdots$ and $\mathbf{r}_f$ as $\mathbf{r}_f^{(0)} + \mathbf{r}_f^{(1)} + \mathbf{r}_f^{(2)} + \cdots$. Other terms such as $\mathbf{a}$, $\mathbf{b}$, $\mathbf{c}$, $\mathbf{d}$, C_h, $C_{K_S h}$, C_{gh}, $C_{H_B h}$, and C_{Qh} can be expanded similarly. Substituting these expansions into Eqs. (3.28) and (3.44) and collecting separate order terms yields [Guadagnini and Neuman, 1999a]

$$
\left.
\begin{aligned}
\nabla \cdot [K_G(\mathbf{x})\nabla \langle h^{(0)}(\mathbf{x})\rangle] &= -\langle g(\mathbf{x})\rangle \\
\langle h^{(0)}(\mathbf{x})\rangle &= \langle H_B(\mathbf{x})\rangle, \quad \mathbf{x} \in \Gamma_D \\
[K_G(\mathbf{x})\rangle \nabla \langle h^{(0)}(\mathbf{x})\rangle] \cdot \mathbf{n}(\mathbf{x}) &= -\langle Q(\mathbf{x})\rangle, \quad \mathbf{x} \in \Gamma_N
\end{aligned}
\right\} \tag{3.58}
$$

$$
\left.
\begin{aligned}
\nabla \cdot \left\{ K_G(\mathbf{x}) \left[\nabla \langle h^{(2)}(\mathbf{x})\rangle + \frac{\sigma_f^2(\mathbf{x})}{2} \nabla \langle h^{(0)}(\mathbf{x})\rangle \right] - \mathbf{r}_f^{(2)}(\mathbf{x}) \right\} &= 0 \\
\langle h^{(0)}(\mathbf{x})\rangle = 0, \quad \mathbf{x} \in \Gamma_D & \\
\left\{ K_G(\mathbf{x}) \left[\nabla \langle h^{(2)}(\mathbf{x})\rangle + \frac{\sigma_f^2(\mathbf{x})}{2} \nabla \langle h^{(0)}(\mathbf{x})\rangle \right] - \mathbf{r}_f^{(2)}(\mathbf{x}) \right\} \cdot \mathbf{n}(\mathbf{x}) = 0, \quad \mathbf{x} \in \Gamma_N
\end{aligned}
\right\} \tag{3.59}
$$

where $\langle h^{(1)}\rangle \equiv 0$, and

$$
\mathbf{r}_f^{(2)}(\mathbf{x}) = \int_\Omega K_G(\mathbf{x})K_G(\mathbf{x}')C_f(\mathbf{x},\mathbf{x}')\nabla \nabla_{\mathbf{x}'}^T G^{(0)}(\mathbf{x}',\mathbf{x})\rangle \nabla_{\mathbf{x}'}\langle h^{(0)}(\mathbf{x}')\rangle \, d\mathbf{x}' \tag{3.60}
$$

In the above, $G^{(0)}(\mathbf{x}',\mathbf{x})$ is the mean Green's function satisfying the Poisson equation

$$
\begin{aligned}
\nabla_{\mathbf{x}'} \cdot [K_G(\mathbf{x}')\nabla_{\mathbf{x}'}G^{(0)}(\mathbf{x}',\mathbf{x})] &= -\delta(\mathbf{x}'-\mathbf{x}) \\
G^{(0)}(\mathbf{x}',\mathbf{x}) &= 0, \quad \mathbf{x}' \in \Gamma_D \\
K_G(\mathbf{x}')\nabla_{\mathbf{x}'}G^{(0)}(\mathbf{x}',\mathbf{x}) \cdot \mathbf{n}(\mathbf{x}') &= 0, \quad \mathbf{x}' \in \Gamma_N
\end{aligned} \tag{3.61}
$$

The same procedure applied to Eqs. (3.51)–(3.56) yields the head covariance [Guadagnini and Neuman, 1999a], up to first order in σ^2 (or, second order in σ),

$$
\begin{aligned}
\nabla \cdot [K_G(\mathbf{x})\nabla C_h(\mathbf{x},\boldsymbol{\chi}) + C_{Ksh}(\mathbf{x},\boldsymbol{\chi})\nabla \langle h^{(0)}(\mathbf{x})\rangle] &= -C_{gh}(\mathbf{x},\boldsymbol{\chi}) \\
C_h(\mathbf{x},\boldsymbol{\chi}) = C_{H_Bh}(\mathbf{x},\boldsymbol{\chi}), \quad \mathbf{x} \in \Gamma_D & \\
[K_G(\mathbf{x})\nabla C_h(\mathbf{x},\boldsymbol{\chi}) + C_{Ksh}(\mathbf{x},\boldsymbol{\chi})\nabla \langle h^{(0)}(\mathbf{x})\rangle] \cdot \mathbf{n}(\mathbf{x}) &= -C_{Qh}(\mathbf{x},\boldsymbol{\chi}), \quad \mathbf{x} \in \Gamma_N
\end{aligned} \tag{3.62}
$$

where

$$C_{K_S h}(\mathbf{x}, \boldsymbol{\chi}) = -\int_{\Omega} K_G(\mathbf{x}) K_G(\boldsymbol{\chi}') C_f(\mathbf{x}, \boldsymbol{\chi}') \nabla_{\chi'}^T \langle h^{(0)}(\boldsymbol{\chi}') \rangle \nabla_{\chi'} G^{(0)}(\boldsymbol{\chi}', \boldsymbol{\chi}) \, d\boldsymbol{\chi}' \quad (3.63)$$

$$C_{gh}(\mathbf{x}, \boldsymbol{\chi}) = \int_{\Omega} C_g(\mathbf{x}, \boldsymbol{\chi}') G^{(0)}(\boldsymbol{\chi}', \boldsymbol{\chi}) \, d\boldsymbol{\chi}' \quad (3.64)$$

$$C_{H_B h}(\mathbf{x}, \boldsymbol{\chi}) = -\int_{\Gamma_N} C_{H_B}(\mathbf{x}, \boldsymbol{\chi}') K_G(\boldsymbol{\chi}') G^{(0)}(\boldsymbol{\chi}', \boldsymbol{\chi}) \cdot \mathbf{n}(\boldsymbol{\chi}') \, d\boldsymbol{\chi}' \quad (3.65)$$

$$C_{Qh}(\mathbf{x}, \boldsymbol{\chi}) = -\int_{\Gamma_N} C_Q(\mathbf{x}, \boldsymbol{\chi}') G^{(0)}(\boldsymbol{\chi}', \boldsymbol{\chi}) \, d\boldsymbol{\chi}' \quad (3.66)$$

In principle, one may write integrodifferential equations for higher-order terms. However, like in the differential-equation-based approach discussed in Section 3.2, the higher-order equations are increasingly more difficult to solve. To the writer's knowledge, there has not yet been any attempt to solve the higher-order equations except for the special case of uniform mean flow in unbounded domains [e.g., Deng and Cushman, 1995, 1998; Hsu $et\,al.$, 1996; Hsu and Neuman, 1997; Hsu and Lamb, 2000]. These second-order studies will be discussed in Section 3.7.

It is worthwhile to mention that the above integrodifferential equations for the first two moments of head are equivalent to the partial differential equations (3.12)–(3.17) upon assuming the mutual independence of f, g, H_B, and Q. This is not surprising because the two approaches use the same perturbation expansions, though at different stages of derivation. However, the differential-equation-based approach is much more straightforward. As mentioned earlier, the advantage of the Green's function approach is not the derivation of the perturbed integrodifferential equations but the provision of a device for other workable approximations.

It is seen from Eq. (3.60) that $\mathbf{r}_f^{(2)}(\mathbf{x})$ is generally not proportional to the local hydraulic gradient (at $\mathbf{x}$) but is given as an integral involving, among other terms, the gradient of $h(\mathbf{x}')$ at all locations of the domain. Hence, the residual flux $\mathbf{r}_f^{(2)}$ is said to be $nonlocal$ and non-$Darcian$ [Neuman and Orr, 1993]. This in turn implies that the second-order mean head $\langle h^{(2)}(\mathbf{x}) \rangle$ and the mean flux $\langle \mathbf{q}^{(2)}(\mathbf{x}) \rangle$ are nonlocal. This nonlocality is not a unique feature of the Green's function approach (or the moment integrodifferential equations). As the moment partial differential equations (3.12)–(3.17) give exactly the same solutions as the integrodifferential equations upon assuming mutual independence of K_S, g, H_B, and Q in the former, these moment differential equations must also be nonlocal. The nonlocality of $C_{fh}(\mathbf{x}, \boldsymbol{\chi})$ in Eq. (3.13) (which is related to $\mathbf{r}_f^{(2)}(\mathbf{x})$) may not be as apparent as $\mathbf{r}_f^{(2)}(\mathbf{x})$ in Eq. (3.60). However, this can be made clear by recasting Eq. (3.13) into an integral form (see, e.g., Eq. (3.166)) with the aid of Green's function.

3.3.3 Alternative Procedure

If the perturbation scheme is used anyway, it may be more straightforward to do so from the onset. Below, we outline the procedure for deriving the integrodifferential moment equations with the alternative procedure.

Substituting the asymptotic expansions of $K_S = e^f$ and h as well as the decompositions of g, H_B, and Q into Eqs. (3.25)–(3.27) and separating terms at each order yields

$$
\left.
\begin{aligned}
\nabla \cdot [K_G(\mathbf{x})\nabla h^{(0)}(\mathbf{x})] &= -\langle g(\mathbf{x})\rangle \\[2mm]
h^{(0)}(\mathbf{x}) &= \langle H_B(\mathbf{x})\rangle, \quad \mathbf{x} \in \Gamma_D \\[2mm]
K_G(\mathbf{x})\nabla h^{(0)}(\mathbf{x}) \cdot \mathbf{n}(\mathbf{x}) &= -\langle Q(\mathbf{x})\rangle, \quad \mathbf{x} \in \Gamma_N
\end{aligned}
\right\}
\tag{3.67}
$$

$$
\left.
\begin{aligned}
\nabla \cdot \{K_G(\mathbf{x})[\nabla h^{(1)}(\mathbf{x}) + f'(\mathbf{x})\nabla h^{(0)}(\mathbf{x})]\} &= -g'(\mathbf{x}) \\[2mm]
h^{(1)}(\mathbf{x}) &= H_B'(\mathbf{x}), \quad \mathbf{x} \in \Gamma_D \\[2mm]
\{K_G(\mathbf{x})[\nabla h^{(1)}(\mathbf{x}) + f'(\mathbf{x})\nabla h^{(0)}(\mathbf{x})]\} \cdot \mathbf{n}(\mathbf{x}) &= -Q'(\mathbf{x}), \quad \mathbf{x} \in \Gamma_N
\end{aligned}
\right\}
\tag{3.68}
$$

$$
\left.
\begin{aligned}
\nabla \cdot \left\{K_G(\mathbf{x})\left[\nabla h^{(2)}(\mathbf{x}) + f'(\mathbf{x})\nabla h^{(1)}(\mathbf{x}) + \frac{f'^2(\mathbf{x})}{2}\nabla h^{(0)}(\mathbf{x})\right]\right\} &= 0 \\[2mm]
h^{(2)}(\mathbf{x}) = 0, \quad \mathbf{x} \in \Gamma_D& \\[2mm]
\left\{K_G(\mathbf{x})\left[\nabla h^{(2)}(\mathbf{x}) + f'(\mathbf{x})\nabla h^{(1)}(\mathbf{x}) + \frac{f'^2(\mathbf{x})}{2}\nabla h^{(0)}(\mathbf{x})\right]\right\}& \\[2mm]
\cdot \mathbf{n}(\mathbf{x}) = 0, \quad \mathbf{x} \in \Gamma_N&
\end{aligned}
\right\}
\tag{3.69}
$$

Equations for higher-order terms can be given similarly. Taking expectation of Eqs. (3.67) and (3.69) leads to Eqs. (3.58) and (3.59) directly. Multiplying Eq. (3.68) by $h'(\boldsymbol{\chi})$ and taking expectation yields

$$
\begin{aligned}
\nabla \cdot [K_G(\mathbf{x})\nabla C_h(\mathbf{x}, \boldsymbol{\chi}) + K_G(\mathbf{x})C_{fh}(\mathbf{x}, \boldsymbol{\chi})\nabla h^{(0)}(\mathbf{x})]\} &= -C_{gh}(\mathbf{x}) \\[2mm]
C_h(\mathbf{x}, \boldsymbol{\chi}) = C_{H_B h}(\mathbf{x}, \boldsymbol{\chi}), \quad \mathbf{x} \in \Gamma_D& \\[2mm]
[K_G(\mathbf{x})\nabla C_h(\mathbf{x}, \boldsymbol{\chi}) + C_{fh}(\mathbf{x}, \boldsymbol{\chi})\nabla h^{(0)}(\mathbf{x})] \cdot \mathbf{n}(\mathbf{x}) &= -C_{Qh}(\mathbf{x}), \quad \mathbf{x} \in \Gamma_N
\end{aligned}
\tag{3.70}
$$

where $C_{fh}(\mathbf{x}, \boldsymbol{\chi}) = \langle f'(\mathbf{x})h^{(1)}(\boldsymbol{\chi})\rangle$.

Rewriting Eq. (3.68) in terms of $\mathbf{x}'$, multiplying by the zeroth-order mean Green's function $G^{(0)}(\boldsymbol{\chi}, \mathbf{x})$ of Eq. (3.61), integrating over the domain Ω, and applying Green's identity, yields

$$
\begin{aligned}
h^{(1)}(\mathbf{x}) = &\int_\Omega g'(\mathbf{x}')G^{(0)}(\mathbf{x}', \mathbf{x})\, d\mathbf{x}' \\[2mm]
&- \int_\Omega K_G(\mathbf{x}')f'(\mathbf{x}')\nabla_{\mathbf{x}'}^T h^{(0)}(\mathbf{x}')\nabla_{\mathbf{x}'} G^{(0)}(\mathbf{x}', \mathbf{x})\, d\mathbf{x}' \\[2mm]
&- \int_{\Gamma_D} H_B'(\mathbf{x}')K_G(\mathbf{x}')\nabla_{\mathbf{x}'} G^{(0)}(\mathbf{x}', \mathbf{x}) \cdot \mathbf{n}(\mathbf{x}')\, d\mathbf{x}' \\[2mm]
&- \int_{\Gamma_N} Q'(\mathbf{x}')G^{(0)}(\mathbf{x}', \mathbf{x})\, d\mathbf{x}'
\end{aligned}
\tag{3.71}
$$

By expressing it in terms of χ, multiplying with $f'(\mathbf{x})$ and taking expectation, one obtains

$$K_G(\mathbf{x})C_{fh}(\mathbf{x}, \chi)$$

$$= -K_G(\mathbf{x}) \int_\Omega K_G(\chi')C_f(\mathbf{x}, \chi')\nabla^T_{\chi'}h^{(0)}(\chi')\nabla_{\chi'}G^{(0)}(\chi', \chi)\,d\chi' \qquad (3.72)$$

Comparing Eq. (3.72) with Eq. (3.63) reveals that $C_{Ksh}(\mathbf{x}, \chi) = K_G(\mathbf{x})C_{fh}(\mathbf{x}, \chi)$. With the same procedure, one obtains $C_{gh}(\mathbf{x}, \chi)$, $C_{H_Bh}(\mathbf{x}, \chi)$, and $C_{Qh}(\mathbf{x}, \chi)$, which are the same as Eqs. (3.64), (3.65), and (3.66), respectively. It is thus seen that the alternative yet simpler procedure leads to the same moment equations.

3.4 NUMERICAL SOLUTIONS TO THE MOMENT EQUATIONS

Analytical solutions of the statistical moments are available only for a few simple cases such as uniform mean flow in unbounded media and uniform mean flow in rectangular domains of stationary media, which are discussed in the next section. For more general situations with nonstationary material properties and/or in bounded domains of regular or irregular geometry, the problem usually needs to be solved numerically [Zhang, 1998; Guadagnini and Neuman, 1999a]. The solution is facilitated by recognizing that $h^{(0)}$, $\langle h^{(2)} \rangle$, C_{fh}, C_{gh}, C_{H_Bh}, C_{Qh}, and C_h are governed by the same type of equations but with different forcing terms on the right-hand side of each equation.

The moment partial differential or integrodifferential equations may be solved by conventional numerical techniques such as *finite differences*, *finite elements*, and *boundary elements*. In this section, we illustrate in detail how the finite difference and finite element methods can be used to solve the moment equations. The application of the boundary element method to solving moment integral equations can be found in the work of Cheng and Lafe [1991] for the special case of flow in a medium of known, uniform permeability but under random recharge and boundary conditions. In this section, we also briefly discuss the method of *mixed finite elements*.

3.4.1 Finite Differences

In this subsection, we show how to solve the moment partial differential equations in two dimensions by finite differences. For ease of discussion, we work on the special case that only the variability of log hydraulic conductivity $f(\mathbf{x})$ is significant so that the (cross) covariances involving g, H_B, and Q are zero. Under this condition, the number of moment equations is reduced significantly. For ease of reference, the resulting moment equations are reproduced from Section 3.2 as follows:

$$\left. \begin{array}{c} \dfrac{\partial^2 h^{(0)}(\mathbf{x})}{\partial x_i^2} + \dfrac{\partial \langle f(\mathbf{x}) \rangle}{\partial x_i}\dfrac{\partial h^{(0)}(\mathbf{x})}{\partial x_i} = -\dfrac{g(\mathbf{x})}{K_G(\mathbf{x})} \\[2mm] h^{(0)}(\mathbf{x}) = H_B(\mathbf{x}), \quad \mathbf{x} \in \Gamma_D \\[2mm] n_i(\mathbf{x})\dfrac{\partial h^{(0)}(\mathbf{x})}{\partial x_i} = 0, \quad \mathbf{x} \in \Gamma_N \end{array} \right\} \qquad (3.73)$$

$$\left.\begin{array}{c} \dfrac{\partial^2 C_{fh}(\mathbf{x}, \boldsymbol{\chi})}{\partial \chi_i^2} + \dfrac{\partial \langle f(\boldsymbol{\chi}) \rangle}{\partial \chi_i} \dfrac{\partial C_{fh}(\mathbf{x}, \boldsymbol{\chi})}{\partial \chi_i} \\[3mm] = J_i(\boldsymbol{\chi}) \dfrac{\partial C_f(\mathbf{x}, \boldsymbol{\chi})}{\partial \chi_i} + \dfrac{g(\boldsymbol{\chi})}{K_G(\boldsymbol{\chi})} C_f(\mathbf{x}, \boldsymbol{\chi}) \\[3mm] C_{fh}(\mathbf{x}, \boldsymbol{\chi}) = 0, \quad \boldsymbol{\chi} \in \Gamma_D \\[3mm] n_i(\boldsymbol{\chi}) \dfrac{\partial C_{fh}(\mathbf{x}, \boldsymbol{\chi})}{\partial \chi_i} = 0, \quad \boldsymbol{\chi} \in \Gamma_N \end{array}\right\} \quad (3.74)$$

$$\left.\begin{array}{c} \dfrac{\partial^2 C_h(\mathbf{x}, \boldsymbol{\chi})}{\partial x_i^2} + \dfrac{\partial \langle f(\mathbf{x}) \rangle}{\partial x_i} \dfrac{\partial C_h(\mathbf{x}, \boldsymbol{\chi})}{\partial x_i} \\[3mm] = J_i(\mathbf{x}) \dfrac{\partial C_{fh}(\mathbf{x}, \boldsymbol{\chi})}{\partial x_i} + \dfrac{g(\mathbf{x})}{K_G(\mathbf{x})} C_{fh}(\mathbf{x}, \boldsymbol{\chi}) \\[3mm] C_h(\mathbf{x}, \boldsymbol{\chi}) = 0, \quad \mathbf{x} \in \Gamma_D \\[3mm] n_i(\mathbf{x}) \dfrac{\partial C_h(\mathbf{x}, \boldsymbol{\chi})}{\partial x_i} = 0, \quad \mathbf{x} \in \Gamma_N \end{array}\right\} \quad (3.75)$$

$$\left.\begin{array}{c} \dfrac{\partial^2 \langle h^{(2)}(\mathbf{x}) \rangle}{\partial x_i^2} + \dfrac{\partial \langle f(\mathbf{x}) \rangle}{\partial x_i} \dfrac{\partial \langle h^{(2)}(\mathbf{x}) \rangle}{\partial x_i} \\[3mm] = -\dfrac{\partial}{\partial x_i} \dfrac{\partial}{\partial \chi_i} \left[C_{fh}(\mathbf{x}, \boldsymbol{\chi}) \right]_{\mathbf{x}=\boldsymbol{\chi}} - \dfrac{1}{2} \dfrac{g(\mathbf{x})}{K_G(\mathbf{x})} \sigma_f^2(\mathbf{x}) \\[3mm] \langle h^{(2)}(\mathbf{x}) \rangle = 0, \quad \mathbf{x} \in \Gamma_D \\[3mm] n_i(\mathbf{x}) \dfrac{\partial \langle h^{(2)}(\mathbf{x}) \rangle}{\partial x_i} = 0, \quad \mathbf{x} \in \Gamma_N \end{array}\right\} \quad (3.76)$$

In these equations, $g(\mathbf{x}) \equiv \langle g(\mathbf{x}) \rangle$ and $H_B(\mathbf{x}) \equiv \langle H_B(\mathbf{x}) \rangle$ while $Q(\mathbf{x})$ has further been set to zero. For nonzero Q, the numerical treatment of the boundary conditions can be found in Chapter 5 for the case of unsaturated flow. The numerical scheme outlined below can be easily extended to more general cases where some or all of the variabilities of recharge g, boundary head H_B, and boundary flux Q are considered.

The spatial derivatives may be discretized via the central-difference approximations,

$$\frac{\partial T(\mathbf{x}, \boldsymbol{\chi})}{\partial x_1} \approx \frac{T(x_1 + \Delta x_1^+, x_2; \chi_1, \chi_2) - T(x_1 - \Delta x_1^-, x_2; \chi_1, \chi_2)}{2\Delta x_1} \quad (3.77)$$

$$\frac{\partial^2 T(\mathbf{x}, \boldsymbol{\chi})}{\partial x_1^2} \approx \frac{T(x_1 + \Delta x_1^+, x_2; \chi_1, \chi_2) - T(x_1, x_2; \chi_1, \chi_2)}{\Delta x_1 \Delta x_1^+}$$

$$- \frac{T(x_1, x_2; \chi_1, \chi_2) - T(x_1 - \Delta x_1^-, x_2; \chi_1, \chi_2)}{\Delta x_1 \Delta x_1^-} \quad (3.78)$$

where T stands for $h^{(0)}$, $\langle h^{(2)} \rangle$, C_{fh} or C_h, and $\Delta x_1 = (\Delta x_1^+ + \Delta x_1^-)/2$. For homogeneous mesh grids, $\Delta x_1^+ = \Delta x_1^-$. With this, the left-hand side (LHS) of Eqs. (3.73)–(3.76) becomes

$$
\begin{aligned}
\text{LHS} = &\left[\frac{1}{\Delta x_1 \Delta x_1^+} + \frac{G_1(\mathbf{x})}{2\Delta x_1} \right] T(x_1 + \Delta x_1^+, x_2; \chi_1, \chi_2) \\
&\left[\frac{1}{\Delta x_1}\left(\frac{1}{\Delta x_1^+} + \frac{1}{\Delta x_1^-} \right) + \frac{1}{\Delta x_2}\left(\frac{1}{\Delta x_2^+} + \frac{1}{\Delta x_2^-} \right) \right] T(x_1, x_2; \chi_1, \chi_2) \\
&+ \left[\frac{1}{\Delta x_1 \Delta x_1^-} - \frac{G_1(\mathbf{x})}{2\Delta x_1} \right] T(x_1 - \Delta x_1^-, x_2; \chi_1, \chi_2) \\
&+ \left[\frac{1}{\Delta x_2 \Delta x_2^+} + \frac{G_2(\mathbf{x})}{2\Delta x_2} \right] T(x_1, x_2 + \Delta x_2^+; \chi_1, \chi_2) \\
&+ \left[\frac{1}{\Delta x_2 \Delta x_2^-} - \frac{G_2(\mathbf{x})}{2\Delta x_2} \right] T(x_1, x_2 - \Delta x_2^-; \chi_1, \chi_2)
\end{aligned}
\tag{3.79}
$$

where $G_1(\mathbf{x}) = \partial \langle f(\mathbf{x}) \rangle / \partial x_1$ and $G_2(\mathbf{x}) = \partial \langle f(\mathbf{x}) \rangle / \partial x_2$. The coefficients in LHS are functions of known quantities. For $h^{(0)}$ and $\langle h^{(2)} \rangle$, the dependency on χ vanishes. For a one-dimensional domain, the terms associated with x_2 disappear in the above. For the boundary conditions,

$$
T(\mathbf{x}, \chi) = T_R(\mathbf{x}), \quad \mathbf{x} \in \Gamma_D
$$

$$
\frac{n_1(\mathbf{x})}{\Delta x_1}[T(x_1 + \Delta x_1^+, x_2; \chi_1, \chi_2) - T(x_1 - \Delta x_1^-, x_2; \chi_1, \chi_2)]
$$

$$
+ \frac{n_2(\mathbf{x})}{\Delta x_2}[T(x_1, x_2 + \Delta x_2^+; \chi_1, \chi_2) - T(x_1, x_2 - \Delta x_2^-; \chi_1, \chi_2)] = 0,
$$

$$
\mathbf{x} \in \Gamma_N
\tag{3.80}
$$

For $T = h^{(0)}$, $T_R = H_B(\mathbf{x})$; otherwise, $T_R \equiv 0$. The right-hand side (RHS) of Eq. (3.73) for $h^{(0)}$ is the known function, $g(\mathbf{x})$, hence it can be solved readily. Equation (3.74) governing C_{fh} should be solved before doing so for $\langle h^{(2)} \rangle$ and C_h because the latter depend on the solution of C_{fh}. With these moments of head, the first two moments of flux can be calculated from Eqs. (3.22)–(3.24).

With these discretizations, the moment equations become

$$
\mathbf{AT} = \mathbf{R}
\tag{3.81}
$$

where $\mathbf{A}$ is the coefficient matrix, $\mathbf{T}$ is the solution vector for one of the four moments $h^{(0)}$, $\langle h^{(2)} \rangle$, C_{fh}, and C_h, and $\mathbf{R}$ is a vector containing information about the RHS of each equation and the boundary conditions. While the vector $\mathbf{R}$ is different for these four moments, the matrix $\mathbf{A}$ is exactly the same for all of them. These linear algebraic equations (LAEs) may be solved by, for instance, Gauss–Jordan elimination, lower–upper (LU) decomposition, successive over-relaxation, or the conjugate gradient method. Zhang [1998] and Zhang and Winter [1999] chose to solve the LAEs by LU decomposition with forward and back substitution [Press *et al.*, 1992]. It is obvious that the matrix **A** only needs to be decomposed once for each problem setup, though the substitution has to be performed as many times as the number of different RHS vectors.

For a specific grid, the equation governing $h^{(0)}$ or $\langle h^{(2)} \rangle$ only has to be solved once; that governing $C_h(\mathbf{x}, \chi)$ needs to be solved for each selected reference point χ. However, since the equation for $C_{fh}(\mathbf{x}, \chi)$ is written with respect to χ, it has to be solved as many times as the number of nodes on the grid for **x** in order to obtain the derivatives $\partial C_{fh}(\mathbf{x}, \chi)/\partial x_i$ (required for solving C_h and $\langle h^{(2)} \rangle$). As pointed out by Zhang [1998], it is because of this that the computational demand increases rapidly with the size of the domain.

This two-dimensional finite difference scheme has been implemented into a computer code called "STO-SAT". This code is capable of handling some complex situations such as the presence of sinks/sources in any arbitrary configuration, regularly nonuniform grids, and medium multiscale, nonstationary features. The code is available from the writer upon request.

3.4.2 Finite Elements

In this subsection, we outline how to solve the moment integrodifferential equations by the method of finite elements. For simplicity, the variabilities of recharge g and boundary terms H_B and Q are, again, not considered in the numerical implementation. An alternative treatment can be found in a recent work by Bonilla and Cushman [2000]. The development given below is mainly based on Guadagnini and Neuman [1999a].

For ease of reference, the moment integrodifferential equations developed in Section 3.3.2 are recalled for the case of deterministic sinks/sources and boundary conditions as follows:

$$\left.\begin{aligned} \nabla \cdot [K_G(\mathbf{x})\nabla \langle h^{(0)}(\mathbf{x})\rangle] &= -g(\mathbf{x}) \\ \langle h^{(0)}(\mathbf{x})\rangle &= H_B(\mathbf{x}), \quad \mathbf{x} \in \Gamma_D \\ [K_G(\mathbf{x})\rangle\nabla \langle h^{(0)}(\mathbf{x})\rangle] \cdot \mathbf{n}(\mathbf{x}) &= -Q(\mathbf{x}), \quad \mathbf{x} \in \Gamma_N \end{aligned}\right\} \tag{3.82}$$

$$\left.\begin{aligned} \nabla \cdot \left\{ K_G(\mathbf{x}) \left[\nabla \langle h^{(2)}(\mathbf{x})\rangle + \frac{\sigma_f^2(\mathbf{x})}{2} \nabla \langle h^{(0)}(\mathbf{x})\rangle \right] - \mathbf{r}_f^{(2)}(\mathbf{x}) \right\} &= 0 \\ \langle h^{(0)}(\mathbf{x})\rangle = 0, \quad \mathbf{x} \in \Gamma_D \\ \left\{ K_G(\mathbf{x}) \left[\nabla \langle h^{(2)}(\mathbf{x})\rangle + \frac{\sigma_f^2(\mathbf{x})}{2} \nabla \langle h^{(0)}(\mathbf{x})\rangle \right] - \mathbf{r}_f^{(2)}(\mathbf{x}) \right\} \\ \cdot \mathbf{n}(\mathbf{x}) = 0, \quad \mathbf{x} \in \Gamma_N \end{aligned}\right\} \tag{3.83}$$

$$\left.\begin{aligned} \nabla \cdot [K_G(\mathbf{x})\nabla C_h(\mathbf{x}, \chi) + C_{Ksh}(\mathbf{x}, \chi)\nabla \langle h^{(0)}(\mathbf{x})\rangle] &= 0 \\ C_h(\mathbf{x}, \chi) = 0, \quad \mathbf{x} \in \Gamma_D \\ [K_G(\mathbf{x})\nabla C_h(\mathbf{x}, \chi) + C_{Ksh}(\mathbf{x}, \chi)\nabla \langle h^{(0)}(\mathbf{x})\rangle] \cdot \mathbf{n}(\mathbf{x}) = 0, \quad \mathbf{x} \in \Gamma_N \end{aligned}\right\} \tag{3.84}$$

where

$$\mathbf{r}_f^{(2)}(\mathbf{x}) = \int_\Omega K_G(\mathbf{x}) K_G(\mathbf{x}') C_f(\mathbf{x}, \mathbf{x}') \nabla \nabla_{\mathbf{x}'}^T G^{(0)}(\mathbf{x}', \mathbf{x}) \nabla_{\mathbf{x}'} \langle h^{(0)}(\mathbf{x}') \rangle \, d\mathbf{x}' \tag{3.85}$$

$$C_{K_S h}(\mathbf{x}, \boldsymbol{\chi}) = - \int_\Omega K_G(\mathbf{x}) K_G(\boldsymbol{\chi}) C_f(\mathbf{x}, \boldsymbol{\chi}') \nabla_{\boldsymbol{\chi}'}^T \langle h^{(0)}(\boldsymbol{\chi}') \rangle \nabla_{\boldsymbol{\chi}'} G^{(0)}(\boldsymbol{\chi}', \boldsymbol{\chi}) \, d\boldsymbol{\chi}' \tag{3.86}$$

In the above, $G^{(0)}(\mathbf{x}', \mathbf{x})$ is the mean Green's function satisfying the Poisson equation

$$\nabla_{\mathbf{x}'} \cdot [K_G(\mathbf{x}') \nabla_{\mathbf{x}'} G^{(0)}(\mathbf{x}', \mathbf{x})] = -\delta(\mathbf{x}' - \mathbf{x})$$
$$G^{(0)}(\mathbf{x}', \mathbf{x}) = 0, \quad \mathbf{x}' \in \Gamma_D \tag{3.87}$$
$$K_G(\mathbf{x}') \nabla_{\mathbf{x}'} G^{(0)}(\mathbf{x}', \mathbf{x}) \cdot \mathbf{n}(\mathbf{x}') = 0, \quad \mathbf{x}' \in \Gamma_N$$

As discussed in Section 3.3.2, these integrodifferential equations are equivalent to the partial differential equations solved in the previous section by finite differences. Although these integrodifferential equations may be solved by finite differences as well, we introduce a *Galerkin* finite element scheme as an alternative.

One may first discretize the flow domain Ω into a grid of M elements and N nodes and denote the nodal solution of the mean head field as $h_j^{(n)}$ where $j = 1, 2, \ldots, N$ and $n = 0$ or 2. Then the head field can be interpolated with the aid of a *Lagrange interpolation* (basis, shape) *function* ψ_i as

$$\langle h^{(0)}(\mathbf{x}) \rangle = h_j^{(0)} \psi_j(\mathbf{x}), \qquad \langle h^{(2)}(\mathbf{x}) \rangle = h_j^{(2)} \psi_j(\mathbf{x}) \tag{3.88}$$

Summation for repeated indices is implied here and below if not stated otherwise. Applying *Galerkin orthogonalization* to the zeroth-order mean equation (3.82) yields for each node i

$$\int_\Omega \{\nabla \cdot [K_G(\mathbf{x}) \nabla \langle h^{(0)}(\mathbf{x}) \rangle] + g(\mathbf{x})\} \psi_i(\mathbf{x}) \, d\mathbf{x} = 0 \tag{3.89}$$

where $i = 1, 2, \ldots, N$. Using Green's identity, Eq. (3.89) can be rewritten as

$$\int_\Omega [K_G(\mathbf{x}) \nabla \langle h^{(0)}(\mathbf{x}) \rangle] \cdot \nabla \psi_i(\mathbf{x}) \, d\mathbf{x}$$
$$= \int_\Omega g(\mathbf{x}) \psi_i(\mathbf{x}) \, d\mathbf{x} + \int_\Gamma \psi_i(\mathbf{x}) [K_G(\mathbf{x}) \nabla \langle h^{(0)}(\mathbf{x}) \rangle] \cdot \mathbf{n}(\mathbf{x}) \, d\mathbf{x} \tag{3.90}$$

where Γ is the union of the Dirichlet Γ_D and the Neumann Γ_N boundaries. Substitution of the boundary conditions in Eq. (3.82) into Eq. (3.90) yields

$$\int_\Omega [K_G(\mathbf{x}) \nabla \langle h^{(0)}(\mathbf{x}) \rangle] \cdot \nabla \psi_i(\mathbf{x}) \, d\mathbf{x} = \int_\Omega g(\mathbf{x}) \psi_i(\mathbf{x}) \, d\mathbf{x} - \int_{\Gamma_N} Q(\mathbf{x}) \psi_i(\mathbf{x}) \, d\mathbf{x} \tag{3.91}$$

where $i = 1, 2, \ldots, N^*$, and N^* is the number of nodes which are not the Dirichlet (prescribed head) boundary Γ_D. For the $N^{**} = N - N^*$ nodes on Γ_D, Eq. (3.91) should be replaced by the appropriate Dirichlet boundary conditions through modifying the global matrix, to be introduced below. With the first equation of Eq. (3.88), Eq. (3.91) becomes

$$\left[\int_\Omega K_G(\mathbf{x})\nabla\psi_j(\mathbf{x}) \cdot \nabla\psi_i(\mathbf{x})\, d\mathbf{x}\right] h_j^{(0)} = \int_\Omega g(\mathbf{x})\psi_i(\mathbf{x})\, d\mathbf{x} - \int_{\Gamma_N} Q(\mathbf{x})\psi_i(\mathbf{x})\, d\mathbf{x}$$
$$- \left[\int_\Omega K_G(\mathbf{x})\nabla\psi_k(\mathbf{x}) \cdot \nabla\psi_i(\mathbf{x})\, d\mathbf{x}\right] h_k^{(0)} \quad (3.92)$$

where $j = 1, 2, \ldots, N^*$, $k = 1, 2, \ldots, N^{**}$, and $h_k^{(0)}$ are given as boundary conditions (i.e., $H_B(\mathbf{x})$). Equation (3.92) is the so-called *finite element equation* and can be written in a more compact form,

$$A_{ij}h_j^{(0)} = D_i \quad (3.93)$$

with

$$A_{ij} = \int_\Omega K_G(\mathbf{x})\nabla\psi_j(\mathbf{x}) \cdot \nabla\psi_i(\mathbf{x})\, d\mathbf{x} \quad (3.94)$$

$$D_i = \int_\Omega g(\mathbf{x})\psi_i(\mathbf{x})\, d\mathbf{x} - \int_{\Gamma_N} Q(\mathbf{x})\psi_i(\mathbf{x})\, d\mathbf{x} - A_{ik}h_k^{(0)} \quad (3.95)$$

where $i, j = 1, 2, \ldots, N^*$ and $k = 1, 2, \ldots, N^{**}$.

Applying the same procedure to Eq. (3.83) yields the following finite element equations for the second-order mean head term [Guadagnini and Neuman, 1999a],

$$A_{ij}h_j^{(2)} + B_{ij}h_j^{(0)} = P_i + S_i \quad (3.96)$$

with the new coefficients defined by

$$B_{ij} = \int_\Omega K_G(\mathbf{x})\frac{\sigma_f^2(\mathbf{x})}{2}\nabla\psi_j(\mathbf{x}) \cdot \nabla\psi_i(\mathbf{x})\, d\mathbf{x} \quad (3.97)$$

$$P_i = -A_{ik}h_k^{(2)} - B_{ik}h_k^{(0)} \quad (3.98)$$

$$S_i = \int_\Omega \mathbf{r}_f^{(2)}(\mathbf{x}) \cdot \nabla\psi_i(\mathbf{x})\, d\mathbf{x} \quad (3.99)$$

where $k = 1, 2, \ldots, N^{**}$ are the Dirichlet nodes (where $h_k^{(2)} = 0$ and $h_k^{(0)}$ are given by $H_B(\mathbf{x})$). Computation of the second-order residual flux $\mathbf{r}_f^{(2)}(\mathbf{x})$ in Eq. (3.99) requires evaluating the mean Green's function $G^{(0)}(\mathbf{x}', \mathbf{x})$ in Eq. (3.87).

In bounded domains of spatially varying $K_G(\mathbf{x})$, the mean Green's function may be evaluated numerically by placing an injection well of unit strength at node $\mathbf{x}$ of the finite element grid and computing $G^{(0)}(\mathbf{x}', \mathbf{x})$ at various points in the domain of $\mathbf{x}'$. Then the point source (reference point) is moved to a different location on the grid $\mathbf{x}$. For each reference point, the finite element equations for the mean Green's function read exactly as Eqs. (3.93)–(3.95) but with $g = \delta(\mathbf{x}' - \mathbf{x})$ and homogeneous boundary conditions. This system of equations needs to be solved N times (for N nodes on the grid $\mathbf{x}$) to obtain the $N \times N$ pairs of $G_{ij}^{(0)}$, which stands for the value of $G^{(0)}(\mathbf{x}', \mathbf{x})$ at node i on the grid $\mathbf{x}'$ due to a unit source at node j on the grid $\mathbf{x}$. It follows that $G^{(0)}(\mathbf{x}', \mathbf{x})$ is a $2d$-dimensional function with d being the space dimensionality. The computational effort for evaluating $G^{(0)}(\mathbf{x}', \mathbf{x})$ is similar to that for computing $C_{fh}(\mathbf{x}, \boldsymbol{\chi})$ in the partial-differential-equation-based finite difference scheme of Section 3.4.1.

The mean Green's function may be approximated as the weighted sum of its values at the $N \times N$ nodes on the $\mathbf{x}'$ and $\mathbf{x}$ grids via [Guadagnini and Neuman, 1999a],

$$G^{(0)}(\mathbf{x}', \mathbf{x}) = G_{ij}^{(0)} \psi_i(\mathbf{x}) \psi_j(\mathbf{x}') \tag{3.100}$$

With this, Guadagnini and Neuman [1999a, Appendix B] approximated the residual flux as

$$\mathbf{r}_f^{(2)}(\mathbf{x}^e) = K_G(e) \nabla \psi_i(\mathbf{x}^e) \sum_{e'=1}^{M'} K_G(e') C_f(e, e') G_{ij}^{ee'} h_l^{e'} \Theta_{jl}^{e'} \tag{3.101}$$

where e and e' stand for elements on the grids $\mathbf{x}$ and $\mathbf{x}'$, respectively, no summation is implied for repeated e, M' is the number of elements on the grid $\mathbf{x}'$, $K_G(e)$ is the geometric mean hydraulic conductivity in element e (assumed constant in each element), $C_f(e, e')$ is the covariance of log hydraulic conductivity between points located in elements e and e' (evaluated between the centers of these two elements), $h_l^{e'}$ is the value of $h_l^{(0)}$ at node l of element e' on the grid $\mathbf{x}'$, $G_{ij}^{ee'}$ is the mean Green's function at node i of element e due to a unit source at node j of element e', and

$$\Theta_{jl}^{e'} = \int_{\Omega(e')} \nabla \psi_j(\mathbf{x}'^{e'}) \cdot \nabla \psi_l(\mathbf{x}'^{e'}) \, d\mathbf{x}' \tag{3.102}$$

In Eq. (3.102), the integration is over element e' on the grid $\mathbf{x}'$. It follows that S_i of Eq. (3.99) can be evaluated with

$$S_i - \sum_{e=1}^{M} K_G(e) \Theta_{ik}^e \sum_{e'=1}^{M'} K_G(e') C_f(e, e') G_{kj}^{ee'} h_l^{e'} \Theta_{jl}^{e'} \tag{3.103}$$

where $j, k, l = 1, 2, \ldots, N^*$, and

$$\Theta_{ik}^e = \int_{\Omega(e)} \nabla \psi_i(\mathbf{x}^e) \cdot \nabla \psi_k(\mathbf{x}^e) \, d\mathbf{x} \tag{3.104}$$

By applying Galerkin orthogonalization to Eq. (3.84), one may derive the finite element equations for the head covariance as follows [Guadagnini and Neuman, 1999a]:

$$A_{ij} C_{jl'} + F_{ijl'} h_j^{(0)} = R_i \tag{3.105}$$

where $i, j = 1, 2, \ldots, N^*$, $C_{jl'}$ is the head covariance between node j on the grid $\mathbf{x}$ and node l' on the grid $\boldsymbol{\chi}$, the coefficients A_{ij} are given in Eq. (3.94), and

$$F_{ijl'} = \int_\Omega [C_{Ksh}(\mathbf{x})]_{l'} \nabla \psi_j(\mathbf{x}) \cdot \nabla \psi_i(\mathbf{x}) \, d\mathbf{x} \tag{3.106}$$

$$R_i = -A_{ik}C_{kl'} - F_{ikl'}h_k^{(0)} \tag{3.107}$$

where $k = 1, 2, \ldots, N^{**}$ are the Dirichlet nodes (where $C_{kl'} = 0$ in the case under consideration) and $[C_{Ksh}(\mathbf{x})]_{l'}$ is the cross-covariance C_{Ksh} between K_s at $\mathbf{x}$ and h at node l' on the grid $\boldsymbol{\chi}$. Similar to Eq. (3.101) the cross-covariance is evaluated from Eq. (3.86) by

$$[C_{Ksh}(\mathbf{x}^e)]_{l'} = -K_G(e) \sum_{e''=1}^{M''} K_G(e'') C_f(e, e'') G_{il'}^{e''e'} h_j^{e''} \Theta_{ij}^{e''} \tag{3.108}$$

where M'' is the number of elements e'' on the grid $\boldsymbol{\chi}'$, $G_{il'}^{e''e'}$ is the mean Green's function at node i on the grid $\boldsymbol{\chi}'$ due to a unit source at node l' on the grid $\boldsymbol{\chi}$, $h_j^{e''}$ is the value of $h_j^{(0)}$ of element e'' on the grid $\boldsymbol{\chi}'$, and

$$\Theta_{ij}^{e''} = \int_{\Omega(e'')} \nabla \psi_i(\mathbf{x}''^{e''}) \cdot \nabla \psi_j(\mathbf{x}''^{e''}) \, d\mathbf{x}'' \tag{3.109}$$

Flux moments may be approximated in a similar manner. The reader is referred to Guadagnini and Neuman [1999a] for such expressions. The flux moments can be computed after the head moments are available.

As for the moment partial differential equations, on a specific grid the equation governing $h^{(0)}$ or $\langle h^{(2)} \rangle$ only needs to be solved once; and that governing $C_h(\mathbf{x}, \boldsymbol{\chi})$ needs to be solved for each selected reference point $\boldsymbol{\chi}$. The integral equation for the cross-covariance $C_{Ksh}(\mathbf{x}, \boldsymbol{\chi})$ only needs to be solved for each selected $\boldsymbol{\chi}$, unlike the partial differential equation for $C_{fh}(\mathbf{x}, \boldsymbol{\chi})$ which has to be solved as many times as the number of nodes on the grid. However, as mentioned earlier, the computational effort for evaluating $G^{(0)}(\mathbf{x}', \mathbf{x})$ is similar to that for computing $C_{fh}(\mathbf{x}, \boldsymbol{\chi})$ in the partial-differential-equation-based-finite difference scheme. It follows that the computational efforts are roughly the same for the integrodifferential-equation-based finite element scheme and the partial-differential-equation-based finite difference scheme.

These finite element equations for the first two head moments are general, as no restriction on the shape of elements and on the choice of the basis functions $\psi_i(\mathbf{x})$ has been imposed yet. Guadagnini and Neuman [1999a,b] implemented these equations in a two-dimensional rectangular domain with square elements and using bilinear basis functions, and they illustrated their approach with a number of computational examples.

3.4.3 Mixed Finite Elements

In both the finite difference and finite element schemes discussed in the preceding subsections, the flux moments are obtained by numerically differentiating the moments of head. As pointed out by James and Graham [1999], this may lead to loss of accuracy in resulting flux moments if the head moments change rapidly over small distances within the domain and if the numerical grids are coarse. An alternative is to solve for head, flux, and their joint moments simultaneously by the method of mixed finite elements. The mixed finite element method has been used for deterministic simulations of groundwater flow problems [e.g., Durlofsky, 1994; Mose *et al.*, 1994]. It has been found that although it is computationally more demanding, the mixed finite element method results in greater accuracy of the flux approximations when the underlying permeability field varies rapidly in space. This feature may extend to the computation of the flux moments for flow in random media when the head moments vary rapidly. Below we briefly discuss the application of the mixed finite element method to the solution of statistical moment equations.

One may rewrite the flow partial differential equations (3.1)–(3.4) into the following equations:

$$\mathbf{q}(\mathbf{x}) + K_S(\mathbf{x})\nabla h(\mathbf{x}) = 0 \tag{3.110}$$

$$\nabla \cdot \mathbf{q}(\mathbf{x}) = g(\mathbf{x}) \tag{3.111}$$

$$h(\mathbf{x}) = H_B(\mathbf{x}), \quad \mathbf{x} \in \Gamma_D \tag{3.112}$$

$$\mathbf{q}(\mathbf{x}) \cdot \mathbf{n}(\mathbf{x}) = Q(\mathbf{x}), \quad \mathbf{x} \in \Gamma_N \tag{3.113}$$

Equations (3.110) and (3.111) are first-order partial differential equations. When the coefficient $K_S(\mathbf{x})$ and the boundary conditions are known, these equations are the starting point for the mixed finite element method, which simultaneously approximates the two variables h and $\mathbf{q}$ with different basis functions. The reader is referred to Durlofsky [1994] or Mose *et al.* [1994] for the deterministic set of mixed finite element equations.

When the hydraulic conductivity $K_S(\mathbf{x})$ is treated as a random space function, Eqs. (3.110)–(3.113) become a set of stochastic partial differential equations. For simplicity, we assume the source term $g(\mathbf{x})$ and the boundary conditions $H_B(\mathbf{x})$ and $Q(\mathbf{x})$ to be deterministic and known. As done in Section 3.2, we may decompose $f = \ln K_S$ as $f(\mathbf{x}) = \langle f(\mathbf{x}) \rangle + f'(\mathbf{x})$ and expand $h(\mathbf{x})$ and $\mathbf{q}(\mathbf{x})$ into formal series: $h(\mathbf{x}) = h^{(0)}(\mathbf{x}) + h^{(1)}(\mathbf{x}) + h^{(2)}(\mathbf{x}) + \cdots$ and $\mathbf{q}(\mathbf{x}) = \mathbf{q}^{(0)}(\mathbf{x}) + \mathbf{q}^{(1)}(\mathbf{x}) + \mathbf{q}^{(2)}(\mathbf{x}) + \cdots$. Substituting these into Eqs. (3.110)–(3.113) and collecting terms at each separate order leads to

$$\left. \begin{array}{c} \mathbf{q}^{(0)}(\mathbf{x}) + K_G(\mathbf{x})\nabla h^{(0)}(\mathbf{x}) = 0 \\ \nabla \cdot \mathbf{q}^{(0)}(\mathbf{x}) = g(\mathbf{x}) \\ h^{(0)}(\mathbf{x}) = H_B(\mathbf{x}), \quad \mathbf{x} \in \Gamma_D \\ \mathbf{q}^{(0)}(\mathbf{x}) \cdot \mathbf{n}(\mathbf{x}) = Q(\mathbf{x}), \quad \mathbf{x} \in \Gamma_N \end{array} \right\} \tag{3.114}$$

$$\left. \begin{array}{c} \mathbf{q}^{(1)}(\mathbf{x}) + K_G(\mathbf{x})[\nabla h^{(1)}(\mathbf{x}) + f'(\mathbf{x})\nabla h^{(0)}(\mathbf{x})] = 0 \\ \nabla \cdot \mathbf{q}^{(1)}(\mathbf{x}) = 0 \\ h^{(1)}(\mathbf{x}) = 0, \quad \mathbf{x} \in \Gamma_D \\ \mathbf{q}^{(1)}(\mathbf{x}) \cdot \mathbf{n}(\mathbf{x}) = 0, \quad \mathbf{x} \in \Gamma_N \end{array} \right\} \tag{3.115}$$

$$\left.\begin{array}{c} \mathbf{q}^{(2)}(\mathbf{x}) + K_G(\mathbf{x})\left[\nabla h^{(2)}(\mathbf{x}) + f'(\mathbf{x})\nabla h^{(1)}(\mathbf{x}) + \dfrac{f'^2(\mathbf{x})}{2}\nabla h^{(0)}(\mathbf{x})\right] = 0 \\[2mm] \nabla \cdot \mathbf{q}^{(2)}(\mathbf{x}) = 0 \\[2mm] h^{(2)}(\mathbf{x}) = 0, \quad \mathbf{x} \in \Gamma_D \\[2mm] \mathbf{q}^{(2)}(\mathbf{x}) \cdot \mathbf{n}(\mathbf{x}) = 0, \quad \mathbf{x} \in \Gamma_N \end{array}\right\} \quad (3.116)$$

Taking expectation of Eqs. (3.114)–(3.116) yields the mean equations for h and $\mathbf{q}$, up to second order in σ_f,

$$\left.\begin{array}{c} \langle \mathbf{q}^{(0)}(\mathbf{x}) \rangle + K_G(\mathbf{x})\nabla\langle h^{(0)}(\mathbf{x}) \rangle = 0 \\[2mm] \nabla \cdot \langle \mathbf{q}^{(0)}(\mathbf{x}) \rangle = g(\mathbf{x}) \\[2mm] \langle h^{(0)}(\mathbf{x}) \rangle = H_B(\mathbf{x}), \quad \mathbf{x} \in \Gamma_D \\[2mm] \langle \mathbf{q}^{(0)}(\mathbf{x}) \rangle \cdot \mathbf{n}(\mathbf{x}) = Q(\mathbf{x}), \quad \mathbf{x} \in \Gamma_N \end{array}\right\} \quad (3.117)$$

$$\left.\begin{array}{c} \langle \mathbf{q}^{(2)}(\mathbf{x}) \rangle + K_G(\mathbf{x})\nabla\langle h^{(2)}(\mathbf{x}) \rangle = -K_G\left[\langle f'(\mathbf{x})\nabla h^{(1)}(\mathbf{x}) \rangle + \dfrac{\sigma_f^2(\mathbf{x})}{2}\mathbf{J}(\mathbf{x})\right] \\[2mm] \nabla \cdot \langle \mathbf{q}^{(2)}(\mathbf{x}) \rangle = 0 \\[2mm] \langle h^{(2)}(\mathbf{x}) \rangle = 0, \quad \mathbf{x} \in \Gamma_D \\[2mm] \langle \mathbf{q}^{(2)}(\mathbf{x}) \rangle \cdot \mathbf{n}(\mathbf{x}) = 0, \quad \mathbf{x} \in \Gamma_N \end{array}\right\} \quad (3.118)$$

and $\langle h^{(1)} \rangle = \langle \mathbf{q}^{(1)} \rangle = 0$, where $\mathbf{J} = (J_1, \ldots, J_d)^T$ and $J_i(\mathbf{x}) = -\partial h^{(0)}(\mathbf{x})/\partial x_i$.

The first-order covariance equations of h and q_i are obtained by multiplying Eq. (3.115) with f', $h^{(1)}$, and $q_j^{(1)}$ at a different location and then taking expectation,

$$\left.\begin{array}{c} C_{q_i h}(\mathbf{x}, \boldsymbol{\chi}) + K_G(\mathbf{x})\dfrac{\partial C_h(\mathbf{x}, \boldsymbol{\chi})}{\partial x_i} = K_G(\mathbf{x})J_i(\mathbf{x})C_{fh}(\mathbf{x}, \boldsymbol{\chi}) \\[2mm] \dfrac{\partial C_{q_i h}(\mathbf{x}, \boldsymbol{\chi})}{\partial x_i} = 0 \\[2mm] C_h(\mathbf{x}, \boldsymbol{\chi}) = 0, \quad \mathbf{x} \in \Gamma_D \\[2mm] C_{q_i h}(\mathbf{x}, \boldsymbol{\chi})n_i(\mathbf{x}) = 0, \quad \mathbf{x} \in \Gamma_N \end{array}\right\} \quad (3.119)$$

$$\left.\begin{array}{c} C_{q_i q_j}(\mathbf{x}, \boldsymbol{\chi}) + K_G(\mathbf{x})\dfrac{\partial C_{q_j h}(\boldsymbol{\chi}, \mathbf{x})}{\partial x_i} = K_G(\mathbf{x})J_i(\mathbf{x})C_{fq_j}(\mathbf{x}, \boldsymbol{\chi}) \\[2mm] \dfrac{\partial C_{q_i q_j}(\mathbf{x}, \boldsymbol{\chi})}{\partial x_i} = 0 \\[2mm] C_{q_j h}(\boldsymbol{\chi}, \mathbf{x}) = 0, \quad \mathbf{x} \in \Gamma_D \\[2mm] C_{q_i q_j}(\mathbf{x}, \boldsymbol{\chi})n_i(\mathbf{x}) = 0, \quad \mathbf{x} \in \Gamma_N \end{array}\right\} \quad (3.120)$$

$$\left.\begin{array}{c} C_{fq_i}(\mathbf{x}, \boldsymbol{\chi}) + \dfrac{\partial C_{fh}(\mathbf{x}, \boldsymbol{\chi})}{\partial \chi_i} = K_G(\boldsymbol{\chi}) J_i(\boldsymbol{\chi}) C_f(\mathbf{x}, \boldsymbol{\chi}) \\[3mm] \dfrac{\partial C_{fq_i}(\mathbf{x}, \boldsymbol{\chi})}{\partial \chi_i} = 0 \\[3mm] C_{fh}(\mathbf{x}, \boldsymbol{\chi}) = 0, \quad \boldsymbol{\chi} \in \Gamma_D \\[2mm] C_{fq_i}(\mathbf{x}, \boldsymbol{\chi}) n_i(\boldsymbol{\chi}) = 0, \quad \boldsymbol{\chi} \in \Gamma_N \end{array}\right\} \qquad (3.121)$$

where $i, j = 1, \ldots, d$, summation for repeated indices is implied, $C_{q_i h}(\mathbf{x}, \boldsymbol{\chi}) = \langle q_i^{(1)}(\mathbf{x}) h^{(1)}(\boldsymbol{\chi}) \rangle$, $C_h(\mathbf{x}, \boldsymbol{\chi}) = \langle h^{(1)}(\mathbf{x}) h^{(1)}(\boldsymbol{\chi}) \rangle$, $C_{fh}(\mathbf{x}, \boldsymbol{\chi}) = \langle f'(\mathbf{x}) h^{(1)}(\boldsymbol{\chi}) \rangle$, and $C_{q_i q_j}(\mathbf{x}, \boldsymbol{\chi}) = \langle q_i^{(1)}(\mathbf{x}) q_j^{(1)}(\boldsymbol{\chi}) \rangle$. As an alternative, James and Graham [1999] derived equations for these mixed moments using the adjoint state equation method, which will be introduced in Section 3.8.

It should be recognized that the mixed moment equations (3.117)–(3.121) are the same type of equations as the original equations (3.110)–(3.113). This facilitates the solution of the moment equations. The zeroth-order mean head $\langle h^{(0)} \rangle$ and flux $\langle \mathbf{q}^{(0)} \rangle$ can be solved simultaneously from Eq. (3.117) by the method of mixed finite elements in the same manner as solving the deterministic flow equations. For the actual procedure of solving the latter equations, the reader is again referred to Durlofsky [1994] or Mose et al. [1994]. After obtaining $J_i(\mathbf{x})$ by differentiating $\langle h^{(0)} \rangle$, for a given reference point $\mathbf{x}$ a similar mixed finite element scheme (with respect to the grid $\boldsymbol{\chi}$) may be used to compute $C_{fq_i}(\mathbf{x}, \boldsymbol{\chi})$ and $C_{fh}(\mathbf{x}, \boldsymbol{\chi})$ simultaneously from Eq. (3.121). Then the moment pairs, $\langle h^{(2)}(\mathbf{x}) \rangle$ and $\langle q_i^{(2)}(\mathbf{x}) \rangle$, $C_h(\mathbf{x}, \boldsymbol{\chi})$ and $C_{q_i h}(\mathbf{x}, \boldsymbol{\chi})$, and $C_{q_i q_j}(\mathbf{x}, \boldsymbol{\chi})$ and $C_{q_j h}(\boldsymbol{\chi}, \mathbf{x})$, may be computed from Eqs. (3.118), (3.119), and (3.120), respectively, on the grid $\mathbf{x}$ by the method of mixed finite elements.

In this approach the head and flux moments are solved from the set of mixed moment equations (3.117)–(3.121), while in the finite difference scheme the head moments are solved from Eqs. (3.73)–(3.76) and the flux moments are postprocessed with Eqs. (3.22)–(3.24). Recognizing that the flux vector has d components in Eqs. (3.117)–(3.121) and that it may be easier to postprocess the head moments than to solve the flux moment equations, one should expect the set of mixed moment equations to be more demanding computationally. To the writer's knowledge, the covariance equations (3.119)–(3.121) have not been implemented by mixed finite elements. It is, therefore, unknown how a mixed finite element scheme actually compares with the finite difference and finite element schemes in terms of computational efficiency and accuracy in computing the flux moments. Future research is needed in this area.

3.4.4 Discussion

In comparison with the two- to three-decade-long research into numerically solving the original governing flow equations, the research into solving moment equations is still in its infancy. Research may reveal that the moment equations can be solved more efficiently. On the other hand, the problem of high computational effort for large domains may be alleviated with recent improvements in computer memory and speed, and with the availability of well-integrated massively parallel machines. The moment equations have inherent parallel structures. For example, Eq. (3.74) for $C_{fh}(\mathbf{x}; \boldsymbol{\chi})$ can be solved on different processes for different $\mathbf{x}$ because they are independent, and this is also true for $C_h(\mathbf{x}, \boldsymbol{\chi})$ with respect to each $(\boldsymbol{\chi})$. Recognition and utilization of this parallelism may speed up the computation significantly. This research topic is of paramount importance for the application of moment equation approaches to real-world problems.

3.4.5 Illustrative Examples

For ease of illustration, we consider rectangular domains of size L_1 by L_2 (Fig. 3.1). Unless stated otherwise, the left and right sides are specified as constant head boundaries: $h(x_1 = 0) = H_o$ and $h(x_1 = L_1) = H_L$. The lateral sides ($x_2 = 0$ and L_2) are no-flow boundaries. In the general equations given so far, the log hydraulic conductivity field is nonstationary, due to the presence of geological layers, zones, and facies or as a result of conditioning. Nonstationarity may manifest in two ways: the mean $\langle f \rangle$ may vary spatially; and the two-point covariance $C_f(\mathbf{x}, \boldsymbol{\chi})$ may depend on the actual locations of $\mathbf{x}$ and $\boldsymbol{\chi}$ rather than only on their separation distance. Some special cases of nonstationary random fields have been discussed in Section 2.3.3. In the following examples, we only consider stationary media and some special cases of nonstationary media. In the particular nonstationary cases to be considered, the log hydraulic conductivity $f(\mathbf{x})$ consists of a deterministic trend and a stationary fluctuation. That is to say, the mean $\langle f(\mathbf{x}) \rangle$ varies spatially and the two-point covariance $C_f(\mathbf{x} - \boldsymbol{\chi})$ only depends on the separation vector. It is of interest to note that although the numerical algorithm admits any covariance function, we consider only the exponential (2.152) and Gaussian (2.153) covariances in the following examples. These examples are computed with the finite-difference-based moment equation model STO-SAT, which can be obtained from the writer.

Stationary media The following examples concern flow in stationary media. The first is a comparison with the Monte Carlo simulation results published more than two decades

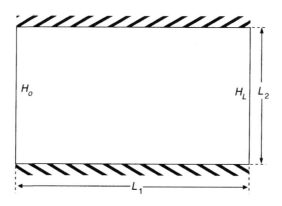

Figure 3.1. Sketch of a two-dimensional flow domain.

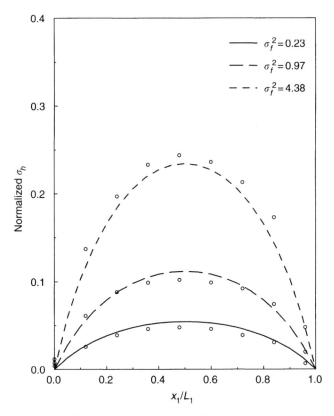

Figure 3.2. Comparison of head standard deviations from the moment equation approach and the Monte Carlo simulation study by Smith and Freeze [1979]. Curves, moment equation results; symbols, Monte Carlo results. (Adapted from Zhang and Winter [1999]. Copyright 1999 by the Society of Petroleum Engineers.)

ago by Smith and Freeze [1979]. Figure 3.2 compares the moment equation approach with the Monte Carlo results for flow in a rectangular domain of two constant head and two no-flow boundaries. The head standard deviations σ_h are normalized with respect to the head difference $(H_o - H_L)$ at the two opposite constant head boundaries and are a function of the normalized longitudinal distance x_1/L_1 passing the transverse center of the domain $(x_2 = L_2/2)$. In this case [Smith and Freeze, 1979, Figure 4], the domain size is $L_1 = 200$ [L] by $L_2 = 100$ [L] (where L in [ ] is an arbitrary length unit). Figure 3.2 shows three different cases: $\sigma_f^2 = 0.23$ and $\lambda_f = 18$ [L]; $\sigma_f^2 = 0.97$ and $\lambda_f = 17.5$ [L]; and $\sigma_f^2 = 4.38$ and $\lambda_f = 16.6$ [L]. Note that the variance values of $f = \ln K_S$ are different from those given in Smith and Freeze [1979] because it was defined as $f = \log K_S$ there. In the present study, the covariance function is assumed to be exponential for f. The head standard deviations are seen to be zero at the two constant head boundaries and to increase toward the center of the domain. Excellent agreements are found for the first two cases

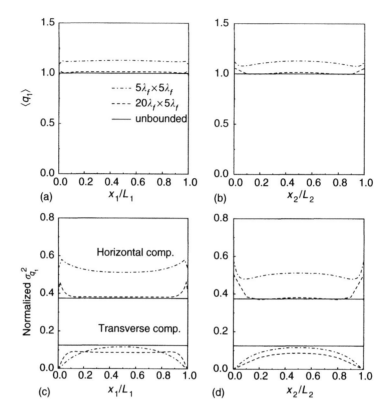

Figure 3.3. The first two moments of flux along longitudinal and transverse centerlines in rectangular domains: (a) and (b) mean flux $\langle q_1 \rangle$, (c) and (d) variance of longitudinal and transverse components of flux. (Adapted from Zhang and Winter [1999]. Copyright 1999 by the Society of Petroleum Engineers.)

even though the covariance function of f is slightly different in Smith and Freeze [1979]. The moment equations (3.74) and (3.75) are given up to first order, without accounting for higher-order correction terms. This is equivalent to theoretically requiring the variance σ_f^2 to be (much) less than one. Nevertheless, a good agreement is seen for the third case where $\sigma_f^2 = 4.38$ is much larger than one.

Figure 3.3 compares the first two moments of flux for rectangular domains with the analytical results for an unbounded domain. As in the previous example, the covariance takes the exponential form. The mean flux $\langle q_1 \rangle$ is normalized by the product of the (zeroth- or first-order) mean negative gradient J_1 and the geometric mean K_G of hydraulic conductivity, $J_1 K_G$, and the flux variance by $K_G^2 J_1^2 \sigma_f^2$. The results are expressed along the longitudinal ($x_2 = L_2/2$) and the transverse ($x_1 = L_1/2$) centerlines. Under the specific geometry and boundary conditions, the transverse component of mean flux is zero. It is seen from Figs. 3.3a and b that the mean flux $\langle q_1 \rangle$ tends to the unbounded domain limit as the domain size increases. It is well known that $\langle q_1 \rangle/(K_G J_1) = 1$ for two-dimensional (2-D), unbounded domains [e.g., Dagan, 1989]. For the flux variances, the 2-D unbounded domain limits are $\sigma_{q_1}^2/(K_G^2 J_1^2 \sigma_f^2) = \frac{3}{8}$ and $\sigma_{q_2}^2/(K_G^2 J_1^2 \sigma_f^2) = \frac{1}{8}$ [Rubin, 1990] (see also Section 3.5.1). It is seen from Figs. 3.3c and d that when the domain size increases, the variance of the longitudinal component q_1 of the flux quickly approaches the 2-D unbounded domain limit except near the boundaries. Along the transverse centerline, $\sigma_{q_1}^2$ is the largest at the no-flow lateral boundaries, decreases rapidly away from there, reaches its minimum shortly, then increases toward the domain center, and becomes stabilized there (Fig. 3.3d). Along the longitudinal (x_1) centerline, $\sigma_{q_1}^2$ decreases with distance from (except right at) the boundaries, and then becomes stabilized a few integral scales from there (Fig. 3.3c). There are numerical artifacts for both $\langle q_1 \rangle$ and $\sigma_{q_1}^2$ at the constant head (x_1) boundaries (Fig. 3.3c). This can be explained by recognizing that the derivatives with respect to x_1 are approximated by central differences at the interior of the domain and have to be approximated by either forward or backward differences at the x_1 boundaries. At the no-flow boundaries (along the transverse centerline), this numerical artifact disappears because the derivatives with respect to x_2 are zero and those with respect to x_1 can still be approximated by central differences. The variance $\sigma_{q_2}^2$ of the transverse component of the flux is zero at both the (longitudinal) constant head boundaries and the (transverse) no-flow boundaries and increases toward the domain center. At the domain center, $\sigma_{q_2}^2$ is closed to the unbounded limit when the domain is $5\lambda_f$ by $5\lambda_f$ (Figs. 3.3c and d). The variance becomes flatter near the domain center as the longitudinal dimension is increased to $20\lambda_f$. Overall, the mean and variance of flux are nonstationary near the boundaries (dependent on the distance from the boundaries) but become stationary after a few integral scales away from there. In general, at the domain center the variance of q_1 is larger than, and the variance of q_2 is smaller than, the respective counterpart for the case of the unbounded domain in which the variances are stationary.

It is worthwhile to note that the head standard deviation is location dependent and hence the head field is nonstationary in a bounded domain even though the medium is second-order stationary. Also, the head variance is always finite in a bounded domain, whereas it is infinite in a two-dimensional unbounded domain with a log hydraulic conductivity field characterized by a constant mean and an exponential covariance function [e.g., Dagan, 1989]. That is to say, the effects of the boundaries are always present in the head variance no matter how many correlation scales away from them for a domain of finite size. However, it is seen that the boundary impacts on the velocity field are only limited to within a few correlation scales.

Figures 3.4a and c show the flux covariance $C_{q_i}(\mathbf{x}, \boldsymbol{\chi})$ for the case of a $20\lambda_f$ by $5\lambda_f$ rectangular domain as in Fig. 3.3. The reference point $\boldsymbol{\chi}$ is selected at the middle of the domain, i.e., $\chi_i = L_i/2$. The covariances are normalized by $K_G^2 J_1^2 \sigma_f^2$, as for the variance of flux. The covariances for the longitudinal (q_1) and transverse (q_2) components are depicted along the longitudinal (x_1) and transverse (x_2) centerlines. The covariance of q_1 is the largest at and symmetric around $\mathbf{x} = \boldsymbol{\chi}$ where the covariance reduces to the variance, and it decays as the distance of $\mathbf{x}$ from $\boldsymbol{\chi}$ increases in both x_1 and x_2 directions (Fig. 3.4a). For the transverse component (q_2), the covariance decreases rapidly to below zero and then tends slowly toward zero along the x_1 direction (Fig. 3.4c). In this case, the integration of the covariance from 0 to L_1 is almost zero; in an unbounded domain, the integration is zero [Rubin, 1990] and the covariance is said to be a hole function. Along the x_2 direction, the

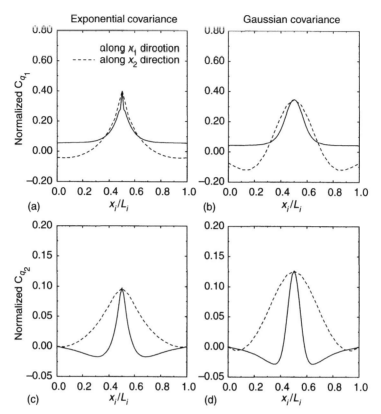

Figure 3.4. The covariance of flux along longitudinal and transverse centerlines in rectangular domains: (a) and (c) longitudinal and transverse components $C_{q_i}(\mathbf{x}, \chi)$ ($i = 1$ and 2) for an exponential covariance of log hydraulic conductivity f, (b) and (d) for a Gaussian covariance of f. (Adapted from Zhang and Winter [1999]. Copyright 1999 by the Society of Petroleum Engineers.)

covariance of q_2 decreases slowly toward zero in this case, while it is a hole function in the case of an unbounded domain. The difference is caused by the large contrast between the transverse dimension ($5\lambda_f$) and the longitudinal dimension ($20\lambda_f$). It is of interest to note that there is a *break point* on (or, *sudden jump* in the slope of) each half of the covariance C_{q_1} near $x_1 - \chi_1 = 0$ along the x_1 direction (Fig. 3.4a). This is a numerical artifact. The slope of the exponential covariance C_f is discontinuous at $x_1 - \chi_1 = 0$ and is approximated (regularized) to be zero there. This discontinuity and the approximation about it do not impact the behaviors of C_{fh} and C_h but affect that of $\partial C_{fh}/\partial x_i$ and hence that of C_{q_i} in Eq. (3.23). A similar numerical artifact exists for the case of a separate covariance C_f of Eq. (2.154). However, the break point disappears when a Gaussian function for C_f is used (Figs. 3.4b and d) because the Gaussian covariance is continuous in its slope at $\mathbf{x} - \chi = \mathbf{0}$ (and elsewhere). In addition, by comparing Figs. 3.4a and c with Figs. 3.4b and d one sees that even with the same variance and integral scale of log hydraulic conductivity f, the covariance form has some impacts on the velocity covariances, especially at and near $\mathbf{x} - \boldsymbol{\chi} = \mathbf{0}$ where the exponential and Gaussian covariances differ most.

Nonstationary media To illustrate the effect of a spatially variable mean hydraulic conductivity $\langle f \rangle$ on flow, we first consider the special case where $\langle f \rangle$ varies linearly according to

$$\langle f(\mathbf{x}) \rangle = a_0 + a_1 x_1 + a_2 x_2 \tag{3.122}$$

where a_i are known coefficients and, in particular, $a_0 = \langle f(0) \rangle$. Since $\langle f \rangle$ can vary in both the x_1 and x_2 directions, its direction may change spatially and may not be aligned with the mean hydraulic gradient.

Figure 3.5 shows the results for the case of linear trends parallel or normal to the mean flow in a square domain of size $10\lambda_f$ by $10\lambda_f$. Constant heads 100 and 90 are specified for

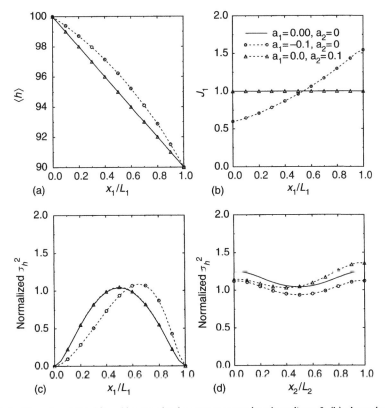

Figure 3.5. Mean head $\langle h \rangle$ (a) and longitudinal negative mean head gradient J_1 (b) along the longitudinal centerline ($x_2 = L_2/2$), normalized head variance σ_h^2 (c,d) along the longitudinal and transverse ($x_1 = L_1/2$) centerlines in a domain of size $10\lambda_f$ by $10\lambda_f$ with linear trends parallel or normal to the mean flow. (Adapted from Zhang [1998], published by the American Geophysical Union.)

the respective left and right boundaries, and the two lateral sides are no-flow boundaries. The covariance function of f takes the exponential form and the mean $\langle f(0) \rangle = 0$ at $\mathbf{x} = 0$. Figures 3.5a and b show the zeroth-order mean head $\langle h \rangle$ and the longitudinal negative mean head gradient J_1 along the longitudinal centerline ($x_2 = L_2/2$). In the key, a_1 and a_2 are defined in Eq. (3.122), and are the slopes of a linear trend along the x_1 and x_2 directions. Due to the specific geometry and boundary conditions, the (zeroth-order) mean head is a straight line along the longitudinal direction and the longitudinal negative mean head gradient is constant in the absence of any trend ($a_i = 0$); the mean head is constant along the transverse direction and the transverse mean gradient is zero in the whole domain. In the case of $a_1 = -0.1$ and $a_2 = 0$, $\langle f \rangle$ decreases linearly with the distance from $x_1 = 0$ in the longitudinal direction, $\langle h \rangle$ is generally larger than that without a trend (Fig. 3.5a), and J_1 increases exponentially with the distance from the left boundary (Fig. 3.5b). Physically speaking, the (pressure) head builds up in the left portion of the domain because the right portion with low hydraulic conductivity acts like a barrier to flow and the (mean) flow must occur from the left to the right with the specific geometry and boundaries. Mathematically speaking, Eq. (3.73) can be rewritten as $\partial J_1/\partial x_1 + a_1 J_1 = 0$ in the case of a linear trend parallel to the mean flow, which has the solution $J_1 = J_1(0) \exp(-a_1 x_1)$. In the case of $a_1 = 0$ and $a_2 = 0.1$, $\langle f \rangle$ increases with x_2 in the direction normal to the mean flow. The presence of a linear trend normal to the mean flow does not affect the mean head and the mean head gradients under the specific boundary conditions.

Figures 3.5c and d show the head variance along the longitudinal ($x_2 = L_2/2$) and transverse ($x_1 = L_1/2$) centerlines. The head variance $\sigma_h^2(\mathbf{x}) = C_h(x_1 = \chi_1, x_2 = \chi_2)$ is normalized with respect to $\lambda_f^2 \sigma_f^2$. In the absence of any trend in the mean log hydraulic conductivity $\langle f \rangle$, along the longitudinal centerline the head variance is zero at the left and right (constant head) boundaries due to the boundary constraint, increases toward the center of the domain, and reaches its peak there because the effect of constant head boundaries is minimal at the domain center. Along the transverse centerline the head variance is minimal at the center of the domain and increases toward the lateral (no-flow) boundaries (for this particular geometry). The effects of domain size and ratio between the longitudinal and transverse length of a rectangular domain on the head variance have recently been investigated by Osnes [1995] and Oliver and Christakos [1996] for the case without any trend. For the case of a linear trend with a negative slope in the x_1 direction ($a_1 = -0.1$ and $a_2 = 0$), the head variance profile along the longitudinal centerline is skewed toward the right portion of the domain. That is to say, the head variance is higher in the area where the negative mean head gradient is large. In this case, the head variances remain symmetric about the domain center along the transverse direction. In the presence of a linear trend normal to the mean flow ($a_1 = 0$ and $a_2 = 0.1$), the head variance along the longitudinal direction is the same as that in the case free of any trend; the variance along the transverse centerline is, however, skewed toward to the upper portion ($0.5L_2 < x_2 < L_2$) of the domain where $\langle f \rangle$ is larger. This is different from our earlier finding for the case of the linear trend parallel to the mean flow that the head variance is large in the area where $\langle f \rangle$ is smaller. This contradiction may be explained by realizing that the flow domain can be viewed as a series of layers normal or parallel to the mean flow in the case of a linear trend parallel or normal to it, respectively. In the case of flow normal to bedding, the flow is mainly controlled by the blocks with small (log) hydraulic conductivities; but in the case of flow parallel to bedding, the flow converges into the high-conductivity layers.

It is seen from Figs. 3.5c and d that the peak head variances are higher in the presence of trends in the mean log hydraulic conductivity. This is so because the head variance is normalized by $\lambda_f^2 \sigma_f^2$. Accounting for a spatially varying, large-scale trend in the log hydraulic conductivity field reduces the variance of log hydraulic conductivity. This justifies the small-variance assumption for many aquifers, although it makes the conductivity field nonstationary and significantly increases the mathematical complexity in the problem.

The special case of linear trends either parallel or normal to mean flow in unbounded domains has been well studied in the literature [e.g., Rubin and Seong, 1994; Li and McLaughlin, 1995; Indelman and Rubin, 1995]. Semi-analytical solutions have been found for this special case. The cases of quadratic and periodic trends in bounded domains have been investigated by Zhang [1998]. Physically speaking, trending in permeability may be caused by, for instance, downward or upward sand coarsening, or periodic sand–shale patterns.

Another nonstationary case of interest is the inclusion of distinct permeability zones or layers in an otherwise stationary permeability field. Although the numerical moment equation approach is able to handle any number of such zones in an arbitrary configuration, we illustrate the effects of such an inclusion with one higher or lower permeability zone in the center of the domain (see Fig. 3.6). As in the previous example, the fluctuations of f are assumed second-order stationary. Note that the treatment of different covariance functions for different zones is a straightforward extension in the numerical scheme.

Figure 3.7 shows the contours of the (first- and second-order) mean head $\langle h \rangle$, head standard deviation (STD $= \sigma_h$), and the mean flux vectors in a two-dimensional domain in the presence of a higher permeability zone where the mean log hydraulic conductivity is denoted by $\langle f_2 \rangle$. The contour levels are labeled by some integers and their corresponding values are listed in the key on the right-hand side of each plot. For example, in Fig. 3.7a

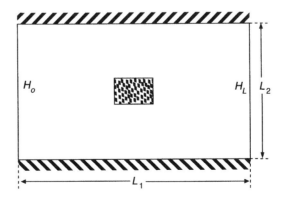

Figure 3.6. Sketch of a two-dimensional flow domain with a distinct permeability zone embedded in the center.

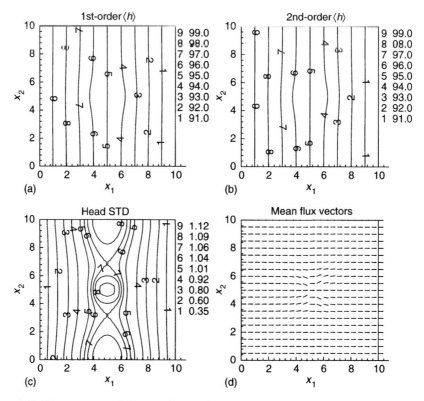

Figure 3.7. The contours of (first- and second-order) mean head $\langle h \rangle$, head standard deviation (STD), and vectors of mean flux in a domain of size $10\lambda_f$ by $10\lambda_f$ with a higher $\langle f_2 \rangle$ zone of $1\lambda_f$ by $1\lambda_f$ at the domain center.

contour level 9 corresponds to $\langle h \rangle = 99.0$. We let $\langle f_2 \rangle = 2$ in this case, while the background mean log conductivity is $\langle f_1 \rangle = 0$ (i.e., $K_G = 1$ [L/T]). The domain size is $10\lambda_f$ by $10\lambda_f$ and the inclusion zone is $1\lambda_f$ by $1\lambda_f$. The boundary conditions and other parameters are exactly the same as those for the case shown in Fig. 3.5.

As expected, the high permeability zone of $\langle f_2 \rangle = 2$ constitutes a fast fluid conduit, resulting in convergent flow toward this zone (see Fig. 3.7d). It is also seen from Figs. 3.7a and c that the mean head changes slowly in and around the high permeability inclusion zone, resulting in a reduced head gradient, and that the head standard deviation is enhanced there. This observation may be made clearer by looking at Fig. 3.8, which shows these moments along the longitudinal (x_1) and the transverse (x_2) centerlines. Also shown are the case of a lower permeability zone ($\langle f_2 \rangle = -2$) and the case of uniform mean permeability ($\langle f_2 \rangle = \langle f_1 \rangle = 1$). It is seen that the longitudinal negative mean head gradient J_1 is generally smaller outside of the low permeability zone of $\langle f_2 \rangle = -2$ (Fig. 3.8b). Although J_1 is larger in the low permeability zone, the longitudinal mean flux $\langle q_1 \rangle = K_G J_1$ is smaller

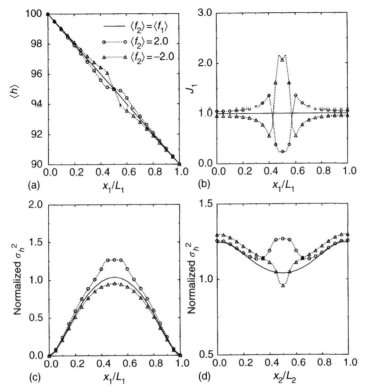

Figure 3.8. Mean head $\langle h \rangle$ (a) and longitudinal negative mean head gradient J_1 (b) along the longitudinal centerline ($x_2 = L_2/2$), and normalized head variance σ_h^2 (c,d) along the longitudinal and transverse ($x_1 = L_1/2$) centerlines in a domain of size $10\lambda_f$ by $10\lambda_f$ with a higher or lower $\langle f_2 \rangle$ zone of $1\lambda_f$ by $1\lambda_f$ at the domain center.

there than in the rest of the domain. It is also seen from Figs. 3.7b and 3.8c that for the case of $\langle f_2 \rangle = 2$, the normalized head variance is generally larger along the longitudinal and transverse centerlines than their counterparts without such a high permeability zone. For the case of $\langle f_2 \rangle = -2$, the head variance is lower along the longitudinal centerline and generally higher along the transverse centerline except at the lower permeability zone. That is, the nonstationary (multiscale) feature significantly impacts the behaviors of the statistical moments of flow. It may be of interest to note that a comparison between Fig. 3.7a and Fig. 3.7b reveals that the second order mean correction is not significant for the case studied.

Although only two types of medium nonstationary feature are shown here, the numerical moment equation approach is able to handle flow in media of more complex multiscale structures and patterns. A more complex configuration may be a combination of a number of such features just shown. The effect of flow nonstationarity caused by medium multiscale structures on solute transport was recently studied by Zhang *et al.* [2000b].

Nonuniform mean flow In the above, we looked at some examples of rectangular domain with two opposite constant head and two opposite no-flow boundaries, without any interior source/sink. However, injection and/or pumping wells are usually involved in oil/water production and aquifer remediation. The numerical moment equation approach can handle any number of wells in an arbitrary configuration. For ease of illustration, our examples only involve two wells.

We first show an example of one injection well and one pumping well in a square domain of closed boundary and of size $10\lambda_f$ by $10\lambda_f$. The injection well is located at $(2.5\lambda_f, 5\lambda_f)$ with the specified head of 100 [L]; the pumping well is located at $(7.5\lambda_f, 5\lambda_f)$ with the specified pumping rate of 0.833 [L^3/T]. In this case, the medium is assumed to be stationary. The covariance of f takes the Gaussian form with $\sigma_f^2 = 1$ and $\lambda_f = 1$ [L], and the geometric mean K_G is set to be 1 [L/T].

Figure 3.9 shows the contours of the (first- and second-order) mean head (pressure), the (second-order) head standard deviation (STD = σ_h), and the vectors of the mean flux $\langle \mathbf{q} \rangle$.

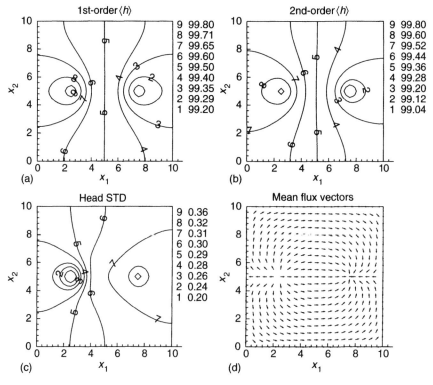

Figure 3.9. The contours of (first- and second-order) mean head $\langle h \rangle$, head standard deviation (STD), and vectors of mean flux for square domain of closed boundary and with injection well at $(2.5, 5)$ and production well at $(7.5, 5)$. (Adapted from Zhang and Winter [1999]. Copyright 1999 by the Society of Petroleum Engineers.)

The mean head can be used to estimate the spatial distribution of head (pressure) and the head standard deviation is a measure of uncertainty associated with the estimation. The mean head and the head standard deviation can be used to construct confidence intervals for head at each point in the field. The source (injector) and sink (producer) locations are apparent from the mean head contours and from the mean flux vectors (which are not drawn according to scale). The uncertainty associated with the head estimation is zero at the injection well with a specified head, increases with distance from it and is the largest at the producing well as one should expect (see Fig. 3.9a). A comparison of Figs. 3.9a and b reveals that the second-order head correction (the difference between the second- and first-order mean heads) is generally negative in this case, except at the (constant-head) injection well where it is zero, and has the largest absolute value at the producing well (of a constant rate).

Another case of great interest is a quarter of a five-spot system from a repeated, balanced pattern in stationary media. Hence the model domain has no-flow boundaries at the four sides, and consists of one injection well at the lower left corner and one pumping well at the upper right corner. This is exactly the setup in one of the cases for water flooding studied in Chapter 6. As in the previous case, the model domain size is $10\lambda_f$ by $10\lambda_f$; the injection well is maintained with the constant head of 100 [L] and the pumping well with the constant rate of 0.833 [L^3/T].

Figure 3.10 shows the contours of the first- and second-order mean head $\langle h \rangle$, the head standard deviation and the vectors of the mean flux $\langle \mathbf{q} \rangle$. Again, the mean flux vectors are not drawn to scale. The second moments of the flux can be found in Zhang and Winter [1999, Figure 7]. It is seen that the mean head varies rapidly near the wells and slowly elsewhere due to the nature of convergent flow (see Figs. 3.10a and b) and that the second-order head correction is the greatest at the production well (of constant rate), as in the previous case. The standard deviation of head is zero at the injection well (of constant pressure), increases with the distance from it, and reaches its maximum at the production well of constant rate (see Fig. 3.10c).

3.5 ANALYTICAL SOLUTIONS TO THE MOMENT EQUATIONS

As mentioned earlier, analytical solutions to the statistical moment equations of flow are only available for some special cases such as uniform mean flow in unbounded media [e.g., Dagan, 1989; Gelhar, 1993; Rubin, 1990; Zhang and Neuman, 1992] and uniform mean flow in rectangular domains of stationary media [Osnes, 1995]. In this section, we discuss a few analytical or semi-analytical solutions available in the literature.

3.5.1 Uniform Mean Flow In Stationary Media

For the case of uniform mean flow in spatially stationary media in the absence of sink/source and without considering the variability in boundary conditions, the (zeroth-order) mean head equation (3.73) or (3.67) reduces to $\partial^2 h^{(0)}(\mathbf{x})/\partial x_i = 0$ subject to boundary conditions.

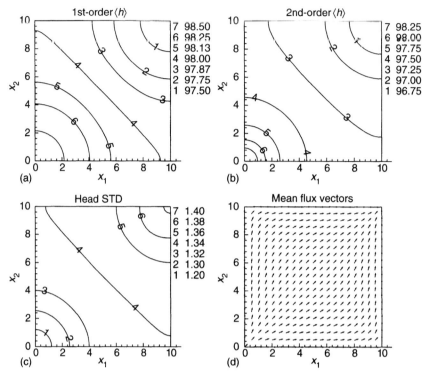

Figure 3.10. The contours of mean head $\langle h \rangle$, head standard deviation (STD), and vectors of mean flux for a quarter of a five-spot production system with injection well at the lower left corner and producing well at upper right. (Adapted from Zhang and Winter [1999]. Copyright 1999 by the Society of Petroleum Engineers.)

For uniform mean flow in stationary media, the equations for the head covariances can be simplified from Eqs. (3.74) and (3.75) as

$$
\left.\begin{aligned}
\frac{\partial^2 C_{fh}(\mathbf{x}, \boldsymbol{\chi})}{\partial \chi_i^2} &= J_i \frac{\partial C_f(\mathbf{x}, \boldsymbol{\chi})}{\partial \chi_i} \\
C_{fh}(\mathbf{x}, \boldsymbol{\chi}) &= 0, \quad \boldsymbol{\chi} \in \Gamma_D \\
n_i(\boldsymbol{\chi}) \frac{\partial C_{fh}(\mathbf{x}, \boldsymbol{\chi})}{\partial \chi_i} &= 0, \quad \boldsymbol{\chi} \in \Gamma_N
\end{aligned}\right\} \tag{3.123}
$$

$$
\left.\begin{aligned}
\frac{\partial^2 C_h(\mathbf{x}, \boldsymbol{\chi})}{\partial x_i^2} &= J_i \frac{\partial C_{fh}(\mathbf{x}, \boldsymbol{\chi})}{\partial x_i} \\
C_h(\mathbf{x}, \boldsymbol{\chi}) &= 0, \quad \mathbf{x} \in \Gamma_D \\
n_i(\mathbf{x}) \frac{\partial C_h(\mathbf{x}, \boldsymbol{\chi})}{\partial x_i} &= 0, \quad \mathbf{x} \in \Gamma_N
\end{aligned}\right\} \tag{3.124}
$$

These head covariances can be written with the aid of Green's function,

$$C_{fh}(\mathbf{x}, \boldsymbol{\chi}) = K_G J_i \int_\Omega C_f(\mathbf{x}, \boldsymbol{\chi}') \frac{\partial G^{(0)}(\boldsymbol{\chi}', \mathbf{x})}{\partial \chi_i'} \, d\boldsymbol{\chi}' \tag{3.125}$$

$$C_h(\mathbf{x}, \boldsymbol{\chi}) = K_G^2 J_i J_j \int_\Omega \int_\Omega C_f(\mathbf{x}', \boldsymbol{\chi}') \frac{\partial G^{(0)}(\boldsymbol{\chi}', \boldsymbol{\chi}) \, \partial G^{(0)}(\mathbf{x}', \mathbf{x})}{\partial \chi_i' \qquad \partial x_j'} \, d\boldsymbol{\chi}' \, d\mathbf{x}' \tag{3.126}$$

where $G^{(0)}(\mathbf{x}', \mathbf{x})$ is the mean Green's function defined as Eq. (3.61) with K_G as a constant. Note that Eqs. (3.125) and (3.126) can be directly obtained from Eq. (3.71) when the variabilities of g, H_B, and Q are not considered. Similarly, one may express the head variance as

$$\sigma_h^2(\mathbf{x}) = K_G^2 J_i J_j \int_\Omega \int_\Omega C_f(\mathbf{x}', \boldsymbol{\chi}') \frac{\partial G^{(0)}(\boldsymbol{\chi}', \mathbf{x}) \, \partial G^{(0)}(\mathbf{x}', \mathbf{x})}{\partial \chi_i' \qquad \partial x_j'} \, d\boldsymbol{\chi}' \, d\mathbf{x}' \tag{3.127}$$

Unbounded domain In an unbounded domain, the mean Green's function $G^{(0)}(\mathbf{x}', \mathbf{x})$ of Eq. (3.61) is given explicitly as [e.g., Arfken, 1985, Table 16.1]

$$G^{(0)}(\mathbf{x}', \mathbf{x}) = -\frac{1}{2\pi K_G} \ln(r) \tag{3.128}$$

in two space dimensions, and

$$G^{(0)}(\mathbf{x}', \mathbf{x}) = \frac{1}{4\pi K_G} \frac{1}{r} \tag{3.129}$$

in three space dimensions. Here, $\mathbf{r} = \mathbf{x}' - \mathbf{x}$ is the separation vector between the two points $\mathbf{x}'$ and $\mathbf{x}$, and $r = |\mathbf{r}|$ is the magnitude of this vector. With Green's function, the head covariances may be obtained by (either numerical or analytical) integrations.

For the case of the isotropic exponential covariance $C_f(\mathbf{r}) = \sigma_f^2 \exp(-r/\lambda_f)$ and the mean flow being aligned with x_1 such that $J_i = J_1 \delta_{i1}$, explicit analytical solutions are available. The mean head $h^{(0)}$ is given as a linear trend along x_1 but does not vary in the x_2 and x_3 directions. In this case, the two-point auto- and cross-covariances of head depend on the separation vector $\mathbf{r}$ rather than the actual locations of the two points. That is to say, the head fluctuation is stationary, although the mean head is not constant in space. Hence, the head field is said to be *incrementally stationary*. However, the head variance $\sigma_h^2(\mathbf{x}) = C_h(\mathbf{x}, \mathbf{x})$ may be unbounded in a one- or two-dimensional unbounded domain unless the integral scale of log hydraulic conductivity f is zero [Bakr et al., 1978]. The latter condition is satisfied if a hole function covariance similar to Eq. (1.128) is used. For the exponential covariance, the condition is not satisfied, since the integral scale is finite (being equal to λ_f). To circumvent the problem of an infinite head variance, one may consider the head variogram, already defined in Section 2.3.3 as

$$\Gamma_h(\mathbf{r}) = \sigma_h^2 - C_h(\mathbf{r}) \tag{3.130}$$

The advantage of this definition is that the head variogram is always finite for a finite r. When $r \to \infty$, $\Gamma_h(\mathbf{r}) = \sigma_h^2$.

In two dimensions, the cross-covariance $C_{fh}(\mathbf{r})$ and the head variogram $\Gamma_h(\mathbf{r})$ are given as [Dagan, 1985]

$$C_{fh}(\mathbf{r}) = \lambda_f \sigma_f^2 \frac{J_1 r_1'}{r'^2} [(1 + r') \exp(-r') - 1] \tag{3.131}$$

$$\Gamma_h(\mathbf{r}) = J_1^2 \lambda_f^2 \sigma_f^2 \left\{ \left(\frac{r_1'^2}{r'^2} - \frac{1}{2} \right) \left[\frac{1}{2} + \frac{(3 + 3r' + r'^2) - 3}{r'^2} \exp(-r') \right] \right.$$
$$\left. - \text{Ei}(-r') + \ln r' + \exp(-r') - 0.4228 \right\} \tag{3.132}$$

where $\mathbf{r}' = \mathbf{r}/\lambda_f$, $r_i' = r_i/\lambda_f$, $r_i = x_i - \chi_i$, and Ei is the exponential integral function. Figure 3.11 shows these head moments along the r_1 and r_2 directions. It is seen that the second moments of head are anisotropic. The cross-covariance C_{fh} is antisymmetric along the mean flow (r_1) direction about $r_1 = 0$, while it is identically zero along r_2 (i.e., $r_1 = 0$). The head variogram Γ_h is generally larger in the mean flow direction than normal to it. In both directions, the head variogram increases with the separation distance without bound. It can be seen from Eq. (3.132) that when $r' \to \infty$, $\Gamma_h \to \frac{1}{2} J_1^2 \lambda_f^2 \sigma_f^2 \ln(r')$. Hence, the

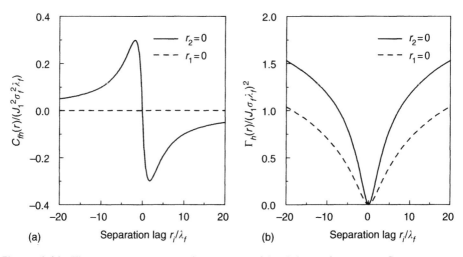

Figure 3.11. The cross-covariance and variogram of head for uniform mean flow in a two-dimensional, unbounded domain as functions of directional separation lag.

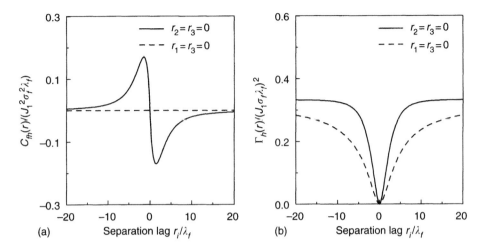

Figure 3.12. The cross-covariance and variogram of head for uniform mean flow in a three-dimensional, unbounded domain as functions of directional separation lag.

head variance is indeed unbounded for a two-dimensional unbounded domain with the exponential covariance $C_f(\mathbf{r})$.

In three dimensions, the head moments read as [Dagan, 1985]

$$C_{fh}(\mathbf{r}) = \lambda_f \sigma_f^2 \frac{J_1 r_1'}{r'^3} [(2 + 2r' + r'^2) \exp(-r') - 2] \tag{3.133}$$

$$\Gamma_h(\mathbf{r}) = J_1^2 \lambda_f^2 \sigma_f^2 \left[\frac{1}{3} - \frac{1}{r'} + \frac{4 + r_1'^2}{r'^3} - \frac{12 r_1'^2}{r'^5} \right.$$
$$\left. + \exp(-r') \left(-\frac{1}{r'} - \frac{4 - r_1'^2}{r'^2} - \frac{4 - 5r_1'^2}{r'^3} + \frac{12 r_1'^2}{r'^4} + \frac{12 r_1'^2}{r'^5} \right) \right] \tag{3.134}$$

Figure 3.12 depicts the behaviors of these moments as functions of the directional separation lag. It is seen that the cross-covariance is antisymmetric along the mean flow (r_1) direction and that the head variogram approaches a constant as the separation distance increases. It is of interest to note that the head variogram reaches the constant much faster in the mean flow direction than normal to it. This can be explained by the fact that for the uniform (unidirectional) mean flow, the head field is more strongly correlated normal to the mean flow than along it. It can be verified from Eq. (3.134) that the head variance is $\sigma_h^2 = \frac{1}{3} J_1^2 \lambda_f^2 \sigma_f^2$.

Substitution of these head covariances into Eq. (3.23) leads to explicit expressions for the flux covariances. The flux components are second-order stationary in that the means and the variances are constant and the covariances only depend on the separation vector for uniform mean flow in an unbounded domain of stationary media. The two-dimensional expressions are given as [Rubin, 1990]

$$
\frac{C_{q_{11}}(\mathbf{r})}{K_G^2 J_1^2 \sigma_f^2} = \frac{1}{2} \left\{ 2e^{-r'} + 4 \left[-\frac{r_1'^2}{r_2'^2} e^{-r'} + \left(\frac{1+r'}{r'^2} e^{-r'} - \frac{1}{r'^2} \right) \left(1 - 2\frac{r_1'^2}{r'^2} \right) \right] \right.
$$

$$
+ \frac{r_2'^2}{r'^2} \left[\beta \left(\frac{6}{r'^4} - \frac{6 + 6r' + 3r'^2 + r'^3}{r'^4} e^{-r'} \right) - \frac{1+r'}{r'^2} e^{-r'} + \frac{1}{r'^2} \right]
$$

$$
+ \frac{r_1'^2}{r'^2} \left[\beta \left(\frac{18 + 18r' + 9r'^2 + 3r'^3 + r'^4}{r'^4} e^{-r'} - \frac{18}{r'^4} \right) \right.
$$

$$
\left. + \frac{1 + r' + r'^2}{r'^2} e^{-r'} - \frac{1}{r'^2} \right]
$$

$$
+ \frac{4}{r'^2} \left(\frac{1}{2} + \frac{3 + 3r' + r'^2}{r'^2} e^{-r'} \right) \left(1 - 5\frac{r_1'^2}{r'^2} + 4\frac{r_1'^4}{r'^4} \right)
$$

$$
\left. + \frac{8r_1'^2}{r'^3} \left(1 - \frac{r_1'^2}{r'^2} \right) \left(\frac{6}{r'^3} - \frac{6 + 6r' + 3r'^2 + r'^3}{r'^3} e^{-r'} \right) \right\} \tag{3.135}
$$

$$
\frac{C_{q_{22}}(\mathbf{r})}{K_G^2 J_1^2 \sigma_f^2} = \frac{1}{2} \left\{ \frac{r_1'^2}{r'^2} \left[\beta \left(\frac{6}{r'^4} - \frac{6 + 6r' + 3r'^2 + r'^3}{r'^4} e^{-r'} \right) - \frac{1+r'}{r'^2} e^{-r'} + \frac{1}{r'^2} \right] \right.
$$

$$
+ \frac{r_2'^2}{r'^2} \left[\beta \left(\frac{18 + 18r' + 9r'^2 + 3r'^3 + r'^4}{r'^4} e^{-r'} - \frac{18}{r'^4} \right) \right.
$$

$$
\left. + \frac{1 + r' + r'^2}{r'^2} e^{-r'} - \frac{1}{r'^2} \right]
$$

$$
- \frac{4}{r'^2} \left(\frac{1}{2} + \frac{3 + 3r' + r'^2}{r'^2} e^{-r'} \right) \left(1 - 5\frac{r_2'^2}{r'^2} + 4\frac{r_2'^4}{r'^4} \right)
$$

$$
\left. + \frac{8r_2'^2}{r'^3} \left(\frac{r_2'^2}{r'^2} - 1 \right) \left(\frac{6}{r'^3} - \frac{6 + 6r' + 3r'^2 + r'^3}{r'^3} e^{-r'} \right) \right\} \tag{3.136}
$$

$$
\frac{C_{q_{12}}(\mathbf{r})}{K_G^2 J_1^2 \sigma_f^2} = \frac{r_1' r_2'}{r'} \left(\frac{2}{r'^3} - \frac{2 + 2r' + r'^2}{r'^3} e^{-r'} \right)
$$

$$
+ \frac{1}{2} \beta r_1' r_2' \left(\frac{4}{r'^4} - \frac{72}{r'^6} + \frac{72 + 72r' + 32r'^2 + 8r'^3 + r'^4}{r'^6} e^{-r'} \right)
$$

$$
+ \frac{r_1' r_2'}{2r'^2} \left(\frac{2 + 2r' + r'^2}{r'^2} e^{-r'} - \frac{2}{r'^2} \right) \tag{3.137}
$$

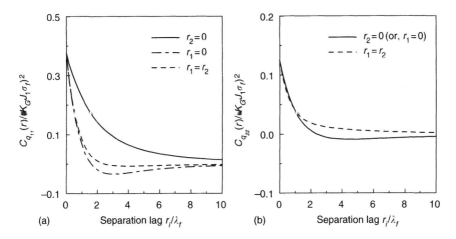

Figure 3.13. The flux covariances for uniform mean flow in a two-dimensional, unbounded domain as functions of directional separation lag.

where $\beta = (r_1'^2 - r_2'^2)/r'^2$. The flux variances are obtained upon taking the limits of $C_{q_{11}}(\mathbf{r})$ and $C_{q_{22}}(\mathbf{r})$ as $r' \to 0$,

$$\sigma_{q_1}^2 = \tfrac{3}{8} K_G^2 J_1^2 \sigma_f^2 \qquad \sigma_{q_2}^2 = \tfrac{1}{8} K_G^2 J_1^2 \sigma_f^2 \tag{3.138}$$

The flux covariances as functions of the directional separation lag are depicted in Fig. 3.13. It is seen that both $C_{q_{11}}$ and $C_{q_{22}}$ are anisotropic. The covariance $C_{q_{11}}$ of the longitudinal component q_1 is nonnegative along the r_1 (i.e., $r_2 = 0$) direction, that is, the fluxes are positively correlated in the mean flow direction. It is found that along r_2 (i.e., $r_1 = 0$) $C_{q_{11}}$ is a hole-type covariance whose integral scale is zero. In other directions such as $r_1 = r_2$ (Fig. 3.13a), $C_{q_{11}}$ may have a finite integral scale. The covariance $C_{q_{22}}$ of the transverse component is found to be a hole-type function with zero integral scale along the mean flow direction or normal to it; however, it has a finite integral scale in other directions such as $r_1 = r_2$ (Fig. 3.13b). Though not shown, the cross flux covariance $C_{q_{12}}$ is zero parallel or normal to the mean flow, and it is nonzero and symmetric about $r = 0$ in other directions.

The three-dimensional expressions read as [Zhang and Neuman, 1992],

$$\frac{C_{q_{11}}(\mathbf{r})}{K_G^2 J_1^2 \sigma_f^2} = -\frac{1}{r'^3} - \frac{36 + 6r_1'^2}{r'^5} + \frac{360 r_1'^2 + 15 r_1'^4}{r'^7} - \frac{420 r_1'^4}{r'^9}$$

$$+ e^{-r'} \left(1 + \frac{2}{r'} + \frac{7 - 2r_1'^2}{r'^2} + \frac{19 - 12 r_1'^2}{r'^3} \right.$$

$$+ \frac{36 - 54 r_1'^2 + r_1'^4}{r'^4} + \frac{36 - 174 r_1'^2 + 10 r_1'^4}{r'^5}$$

$$\left. - \frac{360 r_1'^2 - 55 r_1'^4}{r'^6} - \frac{360 r_1'^2 - 195 r_1'^4}{r'^7} + \frac{420 r_1'^4}{r'^8} + \frac{420 r_1'^4}{r'^9} \right) \tag{3.139}$$

$$\frac{C_{q_{kk}}(\mathbf{r})}{K_G^2 I_1^2 \sigma_f^2} = \frac{1}{r'^3} - \frac{12 + 3r_1'^2 + 3r_k'^2}{r'^5} + \frac{60r_1'^2 + 15r_1'^2 r_k'^2 + 60r_k'^2}{r'^7}$$

$$- \frac{420r_1'^2 r_k'^2}{r'^9} + e^{-r'}\left(\frac{1}{r'^2} + \frac{5 - r_1'^2 - r_k'^2}{r'^3}\right.$$

$$+ \frac{12 - 7r_1'^2 - 7r_k'^2 + r_1'^2 r_k'^2}{r'^4} + \frac{12 - 27r_1'^2 - 27r_k'^2 + 10r_1'^2 r_k'^2}{r'^5}$$

$$- \frac{60r_1'^2 + 60r_k'^2 - 55r_1'^2 r_k'^2}{r'^6} - \frac{60r_1'^2 + 60r_k'^2 - 195r_1'^2 r_k'^2}{r'^7}$$

$$\left. + \frac{420r_1'^2 r_k'^2}{r'^8} + \frac{420r_1'^2 r_k'^2}{r'^9}\right) \quad k = 2, 3 \tag{3.140}$$

$$\frac{C_{q_{1k}}(\mathbf{r})}{K_G^2 J_1^2 \sigma_f^2} = r_1' r_k' \left[-\frac{3}{r'^5} + \frac{180 + 15r_1'^2}{r'^7} - \frac{420r_1'^2}{r'^9} \right.$$

$$- e^{-r'}\left(\frac{1}{r'^2} + \frac{6}{r'^3} + \frac{27 - r_1'^2}{r'^4} + \frac{87 - 10r_1'^2}{r'^5} + \frac{180 - 55r_1'^2}{r'^6}\right.$$

$$\left.\left. + \frac{180 - 195r_1'^2}{r'^7} - \frac{420r_1'^2}{r'^8} - \frac{420r_1'^2}{r'^9}\right)\right] \quad k = 2, 3 \tag{3.141}$$

$$\frac{C_{q_{23}}(\mathbf{r})}{K_G^2 J_1^2 \sigma_f^2} = r_2' r_3' \left[-\frac{3}{r'^5} + \frac{60 + 15r_1'^2}{r'^7} - \frac{420r_1'^2}{r'^9} - e^{-r'}\left(\frac{1}{r'^3} + \frac{7 - r_1'^2}{r'^4}\right.\right.$$

$$\left.\left. + \frac{27 - 10r_1'^2}{r'^5} + \frac{60 - 55r_1'^2}{r'^6} + \frac{60 - 195r_1'^2}{r'^7} - \frac{420r_1'^2}{r'^8} - \frac{420r_1'^2}{r'^9}\right)\right]$$

$$\tag{3.142}$$

Though not shown here, the three-dimensional flux covariances behave similarly to their two-dimensional counterparts [Zhang and Neuman, 1992; Rubin and Dagan, 1992]. The flux variances are

$$\sigma_{q_1}^2 = \frac{8}{15} K_G^2 J_1^2 \sigma_f^2 \qquad \sigma_{q_2}^2 = \frac{1}{15} K_G^2 J_1^2 \sigma_f^2 \tag{3.143}$$

Analytical or semi-analytical solutions are also available for other covariance functions [Hsu, 1999] and for the anisotropic form of the exponential covariance $C_f(\mathbf{r})$ [Dagan, 1989; Rubin and Dagan, 1992; Russo, 1995a].

Rectangular domain Now we consider the case of uniform mean flow in a two-dimensional, rectangular domain of stationary media, schematically shown in Fig. 3.1.

For such a case, the mean Green's function is much more complicated. In 2-D, it may be given as a Fourier sine series [Osnes, 1995],

$$G^{(0)}(\mathbf{x}', \mathbf{x}) = \frac{1}{K_G} \sum_{n=1}^{\infty} a_n(x_2; x_1', x_2') \sin\left(\frac{n\pi x_1}{L_1}\right) \tag{3.144}$$

where

$$a_n(x_2; x_1', x_2') = c_n \cosh\left[\frac{n\pi(x_2' - L_2)}{L_1}\right] \cosh\left(\frac{n\pi x_2}{L_1}\right), \quad x_2 \leq x_2'$$

$$a_n(x_2; x_1', x_2') = c_n \cosh\left(\frac{n\pi x_2'}{L_1}\right) \cosh\left[\frac{n\pi(x_2 - L_2)}{L_1}\right], \quad x_2 > x_2' \tag{3.145}$$

$$c_n = \frac{\sin(n\pi x_1'/L_1)}{n\pi \sinh(n\pi L_2/L_1)}$$

With Green's function, the head covariances can be evaluated from Eqs. (3.125) and (3.126) for a given covariance function of $C_f(\mathbf{r})$. In a bounded domain, Green's function depends on the actual locations of $\mathbf{x}$ and $\mathbf{x}'$. Hence, even though the covariance $C_f(\mathbf{r})$ is stationary the head fluctuation is generally nonstationary in the presence of boundaries. Recently, Osnes [1995] derived closed-form expressions of $C_{fh}(\mathbf{x}, \boldsymbol{\chi})$ and $C_h(\mathbf{x}, \boldsymbol{\chi})$ for the separated exponential covariance

$$C_f(\mathbf{r}) = \sigma_f^2 \exp\left(-\frac{|r_1|}{\lambda_f} - \frac{|r_2|}{\lambda_f}\right) \tag{3.146}$$

Note that this covariance function is anisotropic. The resulting expressions for the head covariances read as

$$C_{fh}(\mathbf{x}, \boldsymbol{\chi}) = J_1 \sigma_f^2 \frac{\lambda_f^2}{L_1} \sum_{n=1}^{\infty} g_n\left(\frac{x_1}{\lambda_f}, \frac{x_2}{\lambda_f}, \frac{\chi_1}{\lambda_f}, \frac{\chi_2}{\lambda_f}, \frac{L_1}{\lambda_f}, \frac{L_2}{\lambda_f}\right) \tag{3.147}$$

$$C_h(\mathbf{x}, \boldsymbol{\chi}) = J_1^2 \sigma_f^2 \frac{\lambda_f^4}{L_1^2} \sum_{n=1}^{\infty} \sum_{m=1}^{\infty} f_{nm}\left(\frac{x_1}{\lambda_f}, \frac{x_2}{\lambda_f}, \frac{\chi_1}{\lambda_f}, \frac{\chi_2}{\lambda_f}, \frac{L_1}{\lambda_f}, \frac{L_2}{\lambda_f}\right) \tag{3.148}$$

where g_n and f_{nm} are two infinite series, given in Osnes [1995]. It is obvious that these expressions need to be evaluated numerically. It may take 100 terms to achieve convergence. Similarly, semi-analytical solutions can be obtained for the flux covariances [Osnes, 1996]. The behaviors of these covariances are similar to those shown in Figs. 3.2 and 3.3 computed with a numerical approach for the exponential function of $C_f(\mathbf{r})$.

It is seen that Green's function becomes very complex even for a domain of simple geometry. Analytical expressions of Green's function may not be available for more general geometrical shapes.

Effective conductivity Similar to the one-dimensional effective hydraulic conductivity discussed in Section 1.4.2, in two and three dimensions the effective hydraulic conductivity is defined as the tensor that relates the expected values of flux and hydraulic head gradient via the following relationship:

$$\langle \mathbf{q}(\mathbf{x}) \rangle = -\mathbf{K}^{ef}(\mathbf{x}) \nabla \langle h(\mathbf{x}) \rangle \tag{3.149}$$

Exact expressions for $\mathbf{K}^{ef}$ are difficult to derive except for some special cases. There are two exact results available in the literature. The first one is that the effective conductivity is equal to the geometric mean K_G for uniform mean flow in two-dimensional, unbounded domains of lognormal, statistically isotropic conductivity fields [Matheron, 1967]. The second exact result is that the effective conductivity is bounded by the harmonic mean $K_H \equiv \langle K_S^{-1} \rangle^{-1}$ and the arithmetic mean $K_A \equiv \langle K_S \rangle$ [e.g., Matheron, 1967], namely,

$$K_H \leq K^{ef} \leq K_A \tag{3.150}$$

For a lognormally distributed conductivity field, $K_H = K_G \exp(-\sigma_f^2/2)$, and $K_A = K_G \exp(\sigma_f^2/2)$. It is thus seen that these bounds are wide apart when the variance of log hydraulic conductivity is large. Next we discuss some approximate results for some specific cases.

It is seen from Eq. (3.22) that the zeroth-order (or first-order) effective hydraulic conductivity is

$$\mathbf{K}^{ef}(\mathbf{x}) = K_G(\mathbf{x})\mathbf{I} \tag{3.151}$$

where $\mathbf{I}$ is the identity tensor. That is, the effective conductivity is equal to the geometric mean hydraulic conductivity, to first order in σ_f. To second order in σ_f, the effective conductivity tensor is given by

$$K_G(\mathbf{x})J_i(\mathbf{x}) + \langle q_i^{(2)}(\mathbf{x}) \rangle = K_{ij}^{ef}(\mathbf{x}) \left[J_j(\mathbf{x}) - \frac{\partial \langle h^{(2)}(\mathbf{x}) \rangle}{\partial x_j} \right] \tag{3.152}$$

where $J_i = -\partial h^{(0)}/\partial x_i$ is the negative of the mean head gradient along the x_i direction, and $\langle h^{(2)}(\mathbf{x}) \rangle$ and $\langle q_i^{(2)}(\mathbf{x}) \rangle$ are the second-order mean head and directional flux, given by Eqs. (3.17) and (3.24), respectively. Many studies were devoted to deriving explicit expressions for the effective conductivity under various conditions. In this subsection, we only discuss some results pertinent to uniform mean flows.

As discussed earlier in this chapter, under uniform mean flow in an unbounded domain the second-order head correction term is identically zero. By substituting Eq. (3.24) into Eq. (3.152), one obtains

$$K_{ij}^{ef}(\mathbf{x})J_j(\mathbf{x}) = K_G(\mathbf{x}) \left[\delta_{ij} + \frac{1}{2}\sigma_f^2(\mathbf{x})\delta_{ij} - \alpha_{ij}(\mathbf{x}) \right] J_j(\mathbf{x}) \tag{3.153}$$

where

$$\alpha_{ij}(\mathbf{x})J_j(\mathbf{x}) = \left.\frac{\partial}{\partial x_i}C_{fh}(\boldsymbol{\chi},\mathbf{x})\right|_{\boldsymbol{\chi}=\mathbf{x}} \tag{3.154}$$

It is thus seen that the evaluation of the effective conductivity under uniform mean flow in an unbounded domain reduces to the computation of the α_{ij} terms. The latter has been the focus of many studies. The α_{ij} terms can be evaluated either by differentiating $C_{fh}(\boldsymbol{\chi},\mathbf{x})$ or directly with $\langle f'(\mathbf{x})\partial h^{(1)}(\mathbf{x})/\partial x_i\rangle$. When the log hydraulic conductivity f field is spatially stationary, the mean gradient J_1 is constant, the head fluctuation field is stationary and the effective conductivity becomes independent of location under the above mentioned condition of uniform mean flow. Gelhar and Axness [1983], Neuman and Depner [1988], and Dagan [1989] studied the case of ellipsoidal, anisotropic covariances C_f. The general result from these three and other studies is that $\mathbf{K}^{ef}$ has a tensorial structure whose components in the principle axes of anisotropy of log hydraulic conductivity f are given in the following form:

$$K_{ii}^{ef} = K_G\left[1 + \sigma_f^2\left(\tfrac{1}{2} - F_i\right)\right] \tag{3.155}$$

where

$$F_i = \frac{1}{\sigma_f^2}\int \frac{k_i^2}{k^2}S_f(\mathbf{k})\,d\mathbf{k} \tag{3.156}$$

In Eq. (3.156), $\mathbf{k} = (k_1,\ldots,k_d)^T$ is the wave number space vector, d is the number of space dimensions, and S_f is the spectral density of f, defined as the Fourier transform of C_f in Eq. (2.277). It can be shown with Eq. (2.279) that

$$\sum_{i=1}^{d} F_i = 1 \tag{3.157}$$

It has been found [e.g., Indelman and Abramovich, 1994a] that up to second order in σ_f, the results for the effective conductivity do not depend on the shape of the covariance C_f but only on its variance σ_f^2 and the ratios of the correlation lengths λ_i. Some particular results are given as follows [Gelhar and Axness, 1983]:

1. When the covariance of f is isotropic, all F_i are identical on the basis of Eq. (3.157) and are thus equal to $1/d$. The effective conductivity is obtained from Eq. (3.155) as [e.g., Gutjahr et al., 1978]

$$K_{ii}^{ef} = K_G\left[1 + \sigma_f^2\left(\frac{1}{2} - \frac{1}{d}\right)\right] \tag{3.158}$$

2. In a two-dimensional domain, the values of F_i are determined by the ratio $e = \lambda_2/\lambda_1$ as

$$F_1 = 1 - F_2 = \frac{e}{1 + e} \tag{3.159}$$

3. For the case of a three-dimensional axisymmetric covariance C_f with $\lambda_1 = \lambda_2$, the values of F_i are given as functions of $e = \lambda_3/\lambda_1$,

$$F_3 = \frac{1}{1-e^2}\left[1 - \frac{e}{\sqrt{1-e^2}}\tan^{-1}\left(\frac{\sqrt{1-e^2}}{e}\right)\right], \quad \text{for } e < 1$$

$$F_3 = \frac{1}{1-e^2}\left[1 - \frac{e}{2(e^2-1)}\ln\frac{e+\sqrt{e^2-1}}{e-\sqrt{e^2-1}}\right], \quad \text{for } e \geq 1$$

(3.160)

and $F_1 = F_2 = (1 - F_3)/2$, where $\tan^{-1}(\ )$ stands for the arctangent function.

The result of Eq. (3.155) has been generalized by regarding the expression within the brackets there as the first two terms in a series expansion of $\exp[\sigma_f^2(\frac{1}{2} - F_i)]$, namely,

$$K_{ii}^{ef} = K_G \exp\left[\left(\tfrac{1}{2} - F_i\right)\sigma_f^2\right]$$

(3.161)

This generalization, usually referred to as the *Landau–Lifshitz conjecture*, is well known in the literature of flow in porous media [e.g., Matheron, 1967; Gelhar and Axness, 1983]. The conjecture is rigorously valid in a one-dimensional domain where it yields the harmonic mean K_H of $K_S(\mathbf{x})$; it gives the exact result of the geometric mean K_G in a two-dimensional domain of lognormally distributed, statistically stationary, isotropic conductivities. For uniform mean flow in three-dimensional domains of lognormal, statistically isotropic conductivity fields, the conjecture has been shown to be valid to fourth order in σ_f by Dagan [1993],

$$K_{ii}^{ef} = K_G \exp\left[1 + \left(\tfrac{1}{2} - \tfrac{1}{3}\right)\sigma_f^2 + \tfrac{1}{2}\left(\tfrac{1}{2} - \tfrac{1}{3}\right)^2\sigma_f^4\right]$$

(3.162)

However, De Wit [1995] made one step further in evaluating the effective conductivity up to sixth order in σ_f and found that at the sixth order, K_{ii}^{ef} includes terms in addition to the σ_f^6 term contained in the expansion of Eq. (3.161). It is found that the additional terms depend on the shape of the covariance C_f. King [1987] and Noetinger [1994] evaluated Eq. (3.161) with the combined *field theoretic* and Green's function method. Using the technique of partial resummation of perturbation series, Noetinger [1994] confirmed Eq. (3.161) up to all orders for an uncorrelated isotropic lognormal conductivity field. The formula, however, remains an approximation, since not all terms were retained during resummation. Therefore, the three-dimensional version of the conjecture (3.161) may not be rigorously correct. Nevertheless, the error in neglecting the correction terms at sixth order in σ_f is extremely small (less than 1%) for $\sigma_f^2 < 3$. Numerical simulations have shown that the conjecture holds at least up to $\sigma_f^2 = 7$ for uniform mean flow in three-dimensional domains of lognormal, statistically isotropic conductivities [Dykaar and Kitanidis, 1992; Neuman and Orr, 1993]. Although it has been found that the statistical anisotropy in f renders the effective conductivity dependent on the shape of the covariance C_f at fourth order in σ_f, this effect is also small when σ_f^2 is small [Indelman and Abramovich, 1994a; De Wit, 1995].

The results discussed above are pertinent to unbounded domains. In bounded domains, the second-order mean head correct term $\langle h^{(2)}(\mathbf{x})\rangle$ is nonzero and location dependent, hence its gradients are generally not zero [Zhang, 1998]. Therefore, even under uniform (zeroth-order) mean flow, Eq. (3.152) cannot reduce to Eq. (3.153) without invoking additional approximations on $\langle h^{(2)}(\mathbf{x})\rangle$. However, the second-order correction term is usually small under uniform mean flow and decreases as the domain size increases [Zhang, 1998]. This may justify the utilization of Eq. (3.153) instead of Eq. (3.152) in evaluating the effective conductivity in bounded domains. This is exactly the approach undertaken by Paleologos *et al.* [1996]. These authors found that the effective conductivity in bounded domains of stationary media can also be expressed by Eq. (3.155) and be conjectured as Eq. (3.161) but with the F_i terms evaluated by numerical quadratures.

In bounded domains, the effective conductivity K_{ij}^{ef} is generally not a constant but depends on the location from the boundaries. However, for uniform mean flows this space dependency usually dies out within a few integral scales from the boundaries. When it varies spatially, the effective conductivity defined in Eq. (3.149) is sometimes referred to as the *apparent conductivity*, which emphasizes its local nature. In this book, however, we use these two terms interchangeably.

Another concept of great interest is the *equivalent conductivity* defined as the conductivity of a fictitious homogeneous medium that conveys the same discharge as the actual, heterogeneous one for a given pressure head drop. This definition is similar to the one commonly used for experimentally measuring hydraulic conductivity values. It is also called the *upscaled* or *block* conductivity when it is used to replace a block of heterogeneous media by an equivalent, homogeneous one in numerical simulations.

3.5.2 Flow Subject to Recharge

In this subsection, we consider the case of flow in a heterogeneous formation of stationary media subject to recharge. As seen in Section 1.2 for flow in homogeneous media with random recharge, the mean head is not linear in space but exhibits a quadratic trend. Hence, the mean head gradient is not constant and the head fluctuations become nonstationary.

The flow is assumed to happen at a *regional scale*, at which the horizontal extent of the formation is much larger than the vertical extent [Dagan, 1989]. Hence, the vertical component of the flow may be negligible and the flow can be taken as horizontal. The flow may then be well approximated by the two-dimensional version of Eqs. (3.1)–(3.4) with the hydraulic conductivity replaced by *transmissivity* $T(\mathbf{x})$. This two-dimensional representation can be either for flow in a *confined aquifer* or in an *unconfined aquifer* subject to *Dupuit's hypothesis* [e.g., de Marsily, 1986].

For the rectangular domain shown in Fig. 3.1 with two parallel constant heads (H_o and H_L) and two parallel no-flow boundaries and with uniform (deterministic) recharge $g(\mathbf{x}) = g$ normal to the flow domain, the (zeroth-order) mean head is solved from Eq. (3.73) as

$$h^{(0)}(\mathbf{x}) = H_o + \frac{x_1}{L_1}(H_L - H_o) + \frac{g}{2K_G}(L_1 - x_1)x_1 \qquad (3.163)$$

where K_G is the geometric mean of transmissivity. The mean negative head gradients are $J_2(\mathbf{x}) = 0$ and

$$J_1(\mathbf{x}) = \frac{H_o - H_L}{L_1} - \frac{gL_1}{2K_G} + \frac{g}{K_G}x_1 \qquad (3.164)$$

The latter can be written as

$$J_1(\mathbf{x}) = J_1(\mathbf{x}_o)\left[1 + \frac{\beta(x_1 - x_{o,1})}{\lambda_f}\right] \qquad (3.165)$$

where $\mathbf{x}_o = (x_{o,1}, x_{o,2})^T$ is a reference point, $\beta = g\lambda_f/[K_G J_1(\mathbf{x}_o)]$ is the normalized recharge, and λ_f is the integral scale of log-transformed transmissivity $f = \ln T$. It is seen that under the specific boundary conditions, the mean flow is unidirectional though nonuniform.

The cross-covariance $C_{fh}(\mathbf{x}, \boldsymbol{\chi})$ and the head covariance $C_h(\mathbf{x}, \boldsymbol{\chi})$ are given as

$$C_{fh}(\mathbf{x}, \boldsymbol{\chi}) = K_G \int_\Omega J_i(\boldsymbol{\chi}')C_f(\mathbf{x}, \boldsymbol{\chi}')\frac{\partial G^{(0)}(\boldsymbol{\chi}', \boldsymbol{\chi})}{\partial \chi_i'}d\boldsymbol{\chi}' \qquad (3.166)$$

$$C_h(\mathbf{x}, \boldsymbol{\chi}) - K_G^2 \int_\Omega \int_\Omega J_i(\mathbf{x}')J_j(\boldsymbol{\chi}')C_f(\mathbf{x}', \boldsymbol{\chi}')$$
$$\cdot \frac{\partial G^{(0)}(\mathbf{x}', \mathbf{x})}{\partial x_i'}\frac{\partial G^{(0)}(\boldsymbol{\chi}', \boldsymbol{\chi})}{\partial \chi_j'}d\boldsymbol{\chi}'\,d\mathbf{x}' \qquad (3.167)$$

which are obtained in a manner similar to Eqs. (3.125) and (3.126). Note that the only difference between Eqs. (3.166) and (3.167) and Eqs. (3.125) and (3.126) lies in the spatial dependency of the mean gradient $J_i(\mathbf{x})$. In Eqs. (3.166) and (3.167), $G^{(0)}(\boldsymbol{\chi}, \boldsymbol{\chi})$ is again the mean Green's function given, for example, by Eq. (3.128) for unbounded, two-dimensional domains and by Eq. (3.144) for rectangular domains.

The two covariances may be obtained by substituting the mean Green's function and the explicit expression for $J_i(\mathbf{x})$ in Eq. (3.165) into Eqs. (3.166) and (3.167). With this procedure, Rubin and Dagan [1987] derived analytical solutions of $C_{fh}(\mathbf{x}, \boldsymbol{\chi})$ and the head variogram $\Gamma_h(\mathbf{x}, \boldsymbol{\chi})$ for an exponential covariance of C_f. The cross-covariance reads as

$$C_{fh}(\mathbf{x}, \boldsymbol{\chi}) = C_{fh}^{(I)} + C_{fh}^{(II)} \tag{3.168}$$

where the first term has the same form as Eq. (3.131) but with the space-dependent mean head gradient $J_1(\boldsymbol{\chi})$,

$$C_{fh}^{(I)}(\mathbf{x}, \boldsymbol{\chi}) = \lambda_f \sigma_f^2 \frac{J_1(\mathbf{x}) r_1'}{r'^2}[(1 + r') \exp(-r') - 1] \tag{3.169}$$

and the second term is a new one, reflecting the presence of the quadratic trend owing to recharge,

$$C_{fh}^{(II)}(\mathbf{x}, \boldsymbol{\chi}) = \lambda_f \sigma_f^2 J_1(\mathbf{x}_o)\beta \left[L + \frac{r_1'^2}{r'}\frac{dL}{dr'} \right]$$

$$L = \exp(-r')\left(1 + \frac{3}{r'} + \frac{3}{r'^2} \right) - \frac{3}{r'^2} \tag{3.170}$$

In the above, $r_1' = (x_i - \chi_i)/\lambda_f$ and $r' = \sqrt{r_1'^2 + r_2'^2}$. The first term $C_{fh}^{(I)}$ is antisymmetric as shown in Fig. 3.11, whereas the second term $C_{fh}^{(II)}$ is symmetric with the maximum covariance at $r' = 0$. The head variogram is more complex. Rubin and Dagan [1987] evaluated the *isotropic average*, $\bar{\Gamma}_h = (1/2\pi) \int_0^{2\pi} \Gamma_h \, d\theta$ (where θ is the angle between $\mathbf{r}'$ and r_1'), of the true, anisotropic variogram Γ_h. The expression for $\bar{\Gamma}_h$ is

$$\bar{\Gamma}_h(\mathbf{x}, \boldsymbol{\chi}) = \bar{\Gamma}_h^{(I)} + \bar{\Gamma}_h^{(II)} \tag{3.171}$$

where

$$\bar{\Gamma}_h^{(I)} = \tfrac{1}{2}J_1^2(\bar{\mathbf{x}})\lambda_f^2 \sigma_f^2 \left[-\mathrm{Ei}(-r') + \ln r' + \exp(-r') - 0.4228 \right] \tag{3.172}$$

$$\bar{\Gamma}_h^{(II)} = J_1^2(\mathbf{x}_o)\beta^2\lambda_f^2\sigma_f^2 \{ 0.5649 + \exp(-r') \left[\tfrac{3}{32}r'^2 + \tfrac{13}{32}r' + \tfrac{25}{32} \right]$$

$$+ \tfrac{7}{64}r'^2 + \tfrac{3}{8}[\mathrm{Ei}(-r') - \ln r'] \} \tag{3.173}$$

In the above, Ei is the exponential integral function and $\bar{\mathbf{x}} = (\mathbf{x} + \boldsymbol{\chi})/2$.

It is seen that both C_{fh} and $\bar{\Gamma}_h$ are nonstationary in that they do not only depend on the separation vector $\mathbf{r} = \mathbf{x} - \boldsymbol{\chi}$ but also on the actual locations of $\mathbf{x}$ and $\boldsymbol{\chi}$. The relative contribution of the recharge component (i.e., the second term in Eqs. (3.168) and (3.171), respectively) is proportional to the normalized recharge β.

The flux covariances can be derived by substituting these head moments into Eq. (3.23). The flux covariances are given explicitly along the mean flow direction x_1

by Rubin and Bellin [1994],

$$\frac{C_{q_{11}}(\mathbf{x}, \chi)}{K_G^2 J_1^2(\mathbf{x}_o)\sigma_f^2} = \left(1 + \beta\frac{x_1}{\lambda_f}\right)\left(1 + \beta\frac{\chi_1}{\lambda_f}\right)\exp(-r')$$

$$+ \frac{1}{4}\left[\left(1 + \beta\frac{x_1}{\lambda_f}\right)^2 + \left(1 + \beta\frac{\chi_1}{\lambda_f}\right)^2\right]$$

$$\cdot \left[\exp(-r')\left(-2 + \frac{6}{r'^2} + \frac{18}{r'^3} + \frac{18}{r'^4}\right) - \frac{3}{r'^2} - \frac{18}{r'^4}\right]$$

$$+ \frac{\beta^2}{32r'^2}\left[-\exp(-r')(96 + 96r' + 62r'^2 + 30r'^3 + 3r'^4)\right.$$

$$\left. + 96 + 30r'^2\right] \tag{3.174}$$

$$\frac{C_{q_{22}}(\mathbf{x}, \chi)}{K_G^2 J_1^2(\mathbf{x}_o)\sigma_f^2} = \frac{1}{4}\left[\left(1 + \beta\frac{x_1}{\lambda_f}\right)^2 + \left(1 + \beta\frac{\chi_1}{\lambda_f}\right)^2\right]$$

$$\cdot \left[\frac{18 - r'^2 - \exp(-r')\left(18 + 18r' + 8r'^2 + 2r'^3\right)}{r'^4}\right.$$

$$\left. + \beta^2\frac{-12 + 7r'^2 + \exp(-r')\left(12 + 12r' + 7r'^2 + 3r'^3\right)}{32r'^2}\right] \tag{3.175}$$

The flux variances are obtained by taking the limit $\mathbf{r} = \mathbf{x} - \chi \to 0$ [Rubin and Bellin, 1994],

$$\frac{\sigma_{q_{11}}^2(\mathbf{x})}{K_G^2 J_1^2(\mathbf{x}_o)\sigma_f^2} = \frac{3}{8}\left(1 + \beta\frac{x_1}{\lambda_f}\right)^2 + \frac{\beta^2}{2}$$

$$\frac{\sigma_{q_{22}}^2(\mathbf{x})}{K_G^2 J_1^2(\mathbf{x}_o)\sigma_f^2} = \frac{1}{8}\left(1 + \beta\frac{x_1}{\lambda_f}\right)^2 + \frac{\beta^2}{4} \tag{3.176}$$

It is clear that Eq. (3.176) reduces to Eq. (3.138) in the absence of recharge, i.e., when $\beta = 0$.

Li and Graham [1998] derived similar, closed-form expressions for these covariances with Fourier transform techniques for flow in an unbounded, two-dimensional domain subject to random recharge. In their work, the log hydraulic conductivity (transmissivity) and recharge are assumed to be uncorrelated random fields characterized by the hole-type covariance functions of Mizell *et al.* [1982].

3.5.3 Flow in Linearly Trending Media

We now consider the case where the mean log hydraulic conductivity field exhibits a linear trend $\langle f(\mathbf{x})\rangle = a_0 + a_1 x_1 + a_2 x_2$ but the fluctuation $f'(\mathbf{x})$ is stationary. This case has been treated in Section 3.4.1 with a numerical approach for flow in bounded domains. Here we show how the same problem can be attacked analytically for unbounded domains. For simplicity, only the case that the trend is aligned with the mean flow (along the x_1 direction) such that $a_2 = 0$ is considered, although the procedure outlined below is applicable to a linear trend in an arbitrary direction. For unidirectional mean flow (with the mean head gradient $J_i = J_1\delta_{1i}$) in an unbounded domain of such media in the absence of sink/source, the mean head equation (3.73) reduces to

$$\frac{\partial^2 h^{(0)}(\mathbf{x})}{\partial x_i^2} + a_1 \frac{\partial h^{(0)}(\mathbf{x})}{\partial x_1} = 0 \tag{3.177}$$

It follows from Eq. (3.177) that the mean head gradient $J_1(\mathbf{x}) = -\partial h^{(0)}/\partial x_1$ is given as

$$J_1(\mathbf{x}) = J_0 \exp(-a_1 x_1) \tag{3.178}$$

where $J_0 = J_1(0)$ is known. It is seen that the mean gradient is nonuniform in space, though unidirectional. The covariances $C_{fh}(\mathbf{x}, \boldsymbol{\chi})$ and $C_h(\mathbf{x}, \boldsymbol{\chi})$ may be expressed with the general equation (3.71) for head fluctuation as

$$C_{fh}(\mathbf{x}, \boldsymbol{\chi}) = \int_\Omega K_G(\boldsymbol{\chi}') J_i(\boldsymbol{\chi}') C_f(\mathbf{x}, \boldsymbol{\chi}') \frac{\partial G^{(0)}(\boldsymbol{\chi}', \boldsymbol{\chi})}{\partial \chi_i'} d\boldsymbol{\chi}' \tag{3.179}$$

$$C_h(\mathbf{x}, \boldsymbol{\chi}) = \int_\Omega \int_\Omega K_G(\mathbf{x}') K_G(\boldsymbol{\chi}') J_i(\mathbf{x}') J_j(\boldsymbol{\chi}') C_f(\mathbf{x}', \boldsymbol{\chi}')$$
$$\cdot \frac{\partial G^{(0)}(\mathbf{x}', \mathbf{x})}{\partial x_i'} \frac{\partial G^{(0)}(\boldsymbol{\chi}', \boldsymbol{\chi})}{\partial \chi_j'} d\boldsymbol{\chi}' d\mathbf{x}' \tag{3.180}$$

where $K_G(\mathbf{x}) = \exp(a_0 + a_1 x_1)$ and $G^{(0)}(\mathbf{x}', \mathbf{x})$ is the mean Green's function satisfying

$$\nabla_{\mathbf{x}'} \cdot [K_G(\mathbf{x}') \nabla_{\mathbf{x}'} G^{(0)}(\mathbf{x}', \mathbf{x})] = -\delta(\mathbf{x}' - \mathbf{x}) \tag{3.181}$$

in an unbounded domain. Equation (3.181) can be rewritten as

$$\frac{\partial^2 G^{(0)}(\mathbf{x}', \mathbf{x})}{\partial x_i'^2} + a_1 \frac{\partial G^{(0)}(\mathbf{x}', \mathbf{x})}{\partial x_i'} = -\frac{\delta(\mathbf{x}' - \mathbf{x})}{K_G(\mathbf{x}')} \tag{3.182}$$

The solution of Eq. (3.182) is given via $G^{(0)}(\mathbf{x}', \mathbf{x}) = F(\mathbf{x}', \mathbf{x}) \exp(-\frac{1}{2} a_1 x_1')$, where F satisfies the so-called *modified Helmholtz equation*,

$$\frac{\partial^2 F(\mathbf{x}', \mathbf{x})}{\partial x_i'^2} - \frac{a_1^2}{4} F(\mathbf{x}', \mathbf{x}) = -\frac{\delta(\mathbf{x}' - \mathbf{x})}{K_G(\mathbf{x}')} \exp\left(\frac{1}{2} a_1 x_1'\right) \tag{3.183}$$

The solution of $F(\mathbf{x}', \mathbf{x})$ and hence that of $G^{(0)}(\mathbf{x}', \mathbf{x})$ can be obtained on the basis of Green's function $G_F(\mathbf{x}', \mathbf{x})$ for the modified Helmholtz equation,

$$\frac{\partial^2 G_F(\mathbf{x}', \mathbf{x})}{\partial x_i'^2} - \frac{a_1^2}{4} G_F(\mathbf{x}', \mathbf{x}) = -\delta(\mathbf{x}' - \mathbf{x}) \tag{3.184}$$

given as [Arfken, 1985, Table 16.1]

$$G_F(\mathbf{x}', \mathbf{x}) = \frac{1}{2\pi} K_0\left(\frac{1}{2}a_1 r\right)$$

$$G_F(\mathbf{x}', \mathbf{x}) = \frac{1}{4\pi r} \exp\left(-\frac{1}{2}a_1 r\right)$$

(3.185)

for 2-D and 3-D unbounded domains, respectively. In the above, K_0 is the *modified Bessel function* of order zero, $r = |\mathbf{r}|$, and $\mathbf{r} = \mathbf{x}' - \mathbf{x}$. With Eq. (3.185), the mean Green's function $G^{(0)}(\mathbf{x}', \mathbf{x})$ is given as

$$G^{(0)}(\mathbf{x}', \mathbf{x}) = \frac{1}{2\pi} K_0\left(\frac{1}{2}a_1 r\right) \frac{\exp\left(-\frac{1}{2}a_1 r_1\right)}{K_G(\mathbf{x})}$$

$$G^{(0)}(\mathbf{x}', \mathbf{x}) = \frac{1}{4\pi r} \exp\left(-\frac{1}{2}a_1 r\right) \frac{\exp\left(-\frac{1}{2}a_1 r_1\right)}{K_G(\mathbf{x})}$$

(3.186)

for 2-D and 3-D, respectively, where $r_1 = x_1' - x_1$.

On the basis of Eq. (3.178), it can be shown that $K_G(\mathbf{x}) J_1(\mathbf{x}) = K_G(0) J_1(0)$, which means that the mean flux is both uniform and unidirectional, although the mean head gradient is not uniform. With this and via integrating by parts, Eqs. (3.179) and (3.180) can be simplified to

$$C_{fh}(\mathbf{x}, \chi) = K_G(0) J_1(0) \int_\Omega \frac{C_f(\mathbf{x}, \chi')}{\partial \chi_1'} G^{(0)}(\chi', \chi) \, d\chi'$$

(3.187)

$$C_h(\mathbf{x}, \chi) = [K_G(0) J_1(0)]^2 \int_\Omega \int_\Omega \frac{\partial^2 C_f(\mathbf{x}', \chi')}{\partial x_1' \partial \chi_1'} G^{(0)}(\mathbf{x}', \mathbf{x}) G^{(0)}(\chi', \chi) \, d\chi' \, dx'$$

(3.188)

Equations equivalent to Eqs. (3.187) and (3.188) have been derived by Rubin and Seong [1994] and Indelman and Rubin [1995] with slightly different procedures. In particular, the equations in Indelman and Rubin [1995] are more general in that the linear trend is along an arbitrary direction relative to the mean flow.

Equations (3.187) and (3.188) may be evaluated analytically or semi-analytically with Eq. (3.186) for certain forms of the input covariance C_f [Rubin and Seong, 1994; Indelman and Rubin, 1995]. With these results, other moments such as the second-order mean head correction term in Eq. (3.17) and the flux moments in Eqs. (3.23) and (3.24) can be evaluated. In particular, it has been found that the second-order mean head correction is identically zero in an unbounded domain of linear trend parallel to the mean flow [e.g., Rubin and Seong, 1994]. However, this correction term is generally nonzero in a bounded domain [Zhang, 1998].

In Section 3.6.2, we will illustrate how the problem of flow in trending media can be solved by a nonstationary spectral method and will provide some closed-form results. It should be noted that the mean Green's function $G^{(0)}(\mathbf{x}', \mathbf{x})$ is difficult to derive for flow in bounded domains. The complexity of such a Green's function, if available, usually prevents us from obtaining simple analytical or semi-analytical solutions of C_{fh} and C_h. For flow in bounded domains of trending media, the numerical moment equation methods already discussed in Section 3.4 provide a viable approach.

3.5.4 Convergent Flow

In Section 3.4.1 the case of nonuniform mean flow caused by the presence of wells is studied by the finite difference moment equation approach. In this subsection, we approach the problem of convergent flow in random media analytically. Figure 3.14 shows a confined aquifer pumped by a fully penetrating well of radius r_w. Being consistent with the shallow water approximation and the fully penetrating well condition, the flow domain considered is horizontal and two-dimensional. The boundary conditions may be given as $h = H_w$ at $r = r_w$ and $h = H_L$ at $r = r_L$, where r is the radial distance from the center of the well. On the basis of fluid continuity the pumping rate per unit thickness is given as

$$
\begin{aligned}
Q_w &= r \int_0^{2\pi} K_S(r, \theta) \frac{\partial h(r, \theta)}{\partial r} \, d\theta \\
&= r_w \int_0^{2\pi} K_S(r_w, \theta) \left. \frac{\partial h(r, \theta)}{\partial r} \right|_{r=r_w} d\theta
\end{aligned}
\tag{3.189}
$$

where r and θ are the coordinates of a polar coordinate system, whose origin is at the center of the well. Owing to the randomness in the K_S field, it is not possible to impose deterministic conditions for H_w, Q_w, and H_L simultaneously. Out of these three conditions, H_L is relatively less important because it drops out when the flow domain is sufficiently large and thus can be treated as unbounded, as done in many studies. However, the treatment

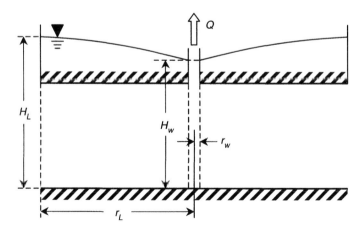

Figure 3.14. Two-dimensional cross-section of flow toward a fully penetrating well in a confined aquifer.

of the conditions at the pumping well location is crucial in the solutions of the problem. One way is to specify the pumping rate per unit thickness, Q_w, as a deterministic constant [e.g., Dagan, 1982a; Naff, 1991, Desbarats, 1992, Indelman and Abramovich, 1994b; Sanchez-Vila, 1997]. In this case, H_w becomes a random variable. Another, and probably more realistic, way is to treat the head H_w at the well deterministically [e.g., Indelman et al., 1996; Sanchez-Vila et al., 1999]. Then the pumping rate Q_w is random.

Specified pumping rate at the well We first consider the case of a specified pumping rate Q_w at the well and with a constant H_L at the distant boundary $r = r_L$. When the well radius r_w is much smaller than the domain size r_L and the integral scale λ_f of the log conductivity (or log transmissivity), the well may be approximated by a point source at $\mathbf{x}_w$ of strength Q_w. Therefore, the governing equation for $h_r = h - H_L$ can be written from Eq. (3.5) as

$$\frac{\partial^2 h_r(\mathbf{x})}{\partial x_i^2} + \frac{\partial f(\mathbf{x})}{\partial x_i}\frac{\partial h_r(\mathbf{x})}{\partial x_i} = -Q_w \delta(\mathbf{x} - \mathbf{x}_w)e^{-f(\mathbf{x})}$$

$$h_r = 0, \quad r = r_L \tag{3.190}$$

where $r = |\mathbf{x} - \mathbf{x}_w|$. As usual, $h_r(\mathbf{x})$ may be expanded into a formal series: $h_r(\mathbf{x}) = h_r^{(0)}(\mathbf{x}) + h_r^{(1)}(\mathbf{x}) + h_r^{(2)}(\mathbf{x}) + \cdots$. The equations for the head terms can be written from Eqs. (3.9)–(3.11) as

$$\frac{\partial^2 h_r^{(0)}(\mathbf{x})}{\partial x_i^2} = -\frac{Q_w}{K_G}\delta(\mathbf{x} - \mathbf{x}_w)$$

$$h_r^{(0)} = 0, \quad r = r_L \tag{3.191}$$

$$\frac{\partial^2 h_r^{(1)}(\mathbf{x})}{\partial x_i^2} = J_i(\mathbf{x})\frac{\partial f'(\mathbf{x})}{\partial x_i} + \frac{Q_w}{K_G}\delta(\mathbf{x} - \mathbf{x}_w)f'(\mathbf{x})$$

$$h_r^{(1)} = 0, \quad r = r_L \tag{3.192}$$

and so on, where $J_i = -\partial h_r^{(0)}/\partial x_i$ and $K_G = \exp(\langle f \rangle)$. In this case, the equations for the relative head $h_r^{(n)}(\mathbf{x})$ are given under the condition of a stationary $f(\mathbf{x}) = \ln K_S(\mathbf{x})$ field. It can be verified with Eq. (3.191) that Eq. (3.192) is equivalent to

$$\frac{\partial^2 h_r^{(1)}(\mathbf{x})}{\partial x_i^2} = \frac{\partial}{\partial x_i}\left[J_i(\mathbf{x})f'(\mathbf{x})\right] \tag{3.193}$$

The zeroth-order mean head $h_r^{(0)}(\mathbf{x})$ can be expressed with the aid of the mean Green's function $G^{(0)}$,

$$h_r^{(0)}(\mathbf{x}) = \int_\Omega \frac{Q_w}{K_G}\delta(\mathbf{x}' - \mathbf{x}_w)[K_G G^{(0)}(\mathbf{x}', \mathbf{x})]\, d\mathbf{x}' = Q_w G^{(0)}(\mathbf{x}_w, \mathbf{x}) \tag{3.194}$$

where $G^{(0)}$ is defined by Eq. (3.61). Similarly, the first-order head fluctuation can be written as

$$h_r^{(1)}(\mathbf{x}) = -\int_\Omega \frac{\partial}{\partial x_i'} \left[J_i(\mathbf{x}') f'(\mathbf{x}') \right] K_G G^{(0)}(\mathbf{x}', \mathbf{x}) \, d\mathbf{x}'$$

$$= K_G \int_\Omega J_i(\mathbf{x}') f'(\mathbf{x}') \frac{\partial G^{(0)}(\mathbf{x}', \mathbf{x})}{\partial x_i'} \, d\mathbf{x}' \qquad (3.195)$$

Alternatively, Eq. (3.195) can be written directly from Eq. (3.71) by using the specific conditions in the present case. The cross- and autocovariances of head can be expressed with Eq. (3.195). For example, the autocovariance of head reads as

$$C_{h_r}(\mathbf{x}, \boldsymbol{\chi}) = K_G^2 \int_\Omega \int_\Omega J_i(\mathbf{x}') J_j(\boldsymbol{\chi}') C_f(\mathbf{x}', \boldsymbol{\chi}')$$

$$\cdot \frac{\partial G^{(0)}(\mathbf{x}', \mathbf{x})}{\partial x_i'} \frac{\partial G^{(0)}(\boldsymbol{\chi}', \boldsymbol{\chi})}{\partial \chi_j'} \, d\mathbf{x}' \, d\boldsymbol{\chi}' \qquad (3.196)$$

The solution to Eq. (3.191) is the well-known *Thiem's formula* [e.g., de Marsily, 1986],

$$h_r^{(0)}(\mathbf{x}) = \frac{Q_w}{2\pi K_G} \ln\left(\frac{r}{r_L}\right) \qquad (3.197)$$

With the mean Green's function $G^{(0)}$, the first-order head covariance can, in principle, be evaluated from Eq. (3.196). However, closed-form solutions for the head covariance or variance are difficult to obtain except for some special cases. In the case of an unbounded domain as considered by Dagan [1982a], the mean Green's function is given by Eq. (3.128). Approximate solutions for the head variance were derived by Dagan [1982a] at locations either very close to the well or far away from it,

$$\sigma_{h_r}^2(r) = \frac{Q_w^2 \sigma_f^2}{4\pi^2 K_G^2} \ln\left(\frac{r}{\lambda_f}\right), \quad r \ll \lambda_f$$

$$\sigma_{h_r}^2(r) = \frac{c Q_w^2 \sigma_f^2}{16\pi^3 K_G^2 r^2} \left[\ln\left(\frac{r}{r_w}\right) + \ln\left(\frac{r}{\lambda_f}\right) \right], \quad r \gg \lambda_f \qquad (3.198)$$

where $c = 2\pi \lambda_f^2$ for an isotropic exponential covariance C_f with the integral scale λ_f.

Specified head at the well When the head H_w at the well is known with certainty, the pumping rate per unit thickness Q_w becomes a random quantity. Therefore, the governing equation for $h_r = h - H_L$ can be written from Eq. (3.5) as

$$\frac{\partial^2 h_r(\mathbf{x})}{\partial x_i^2} + \frac{\partial f(\mathbf{x})}{\partial x_i} \frac{\partial h_r(\mathbf{x})}{\partial x_i} = 0$$

$$h_r = H_w - H_L, \quad r = r_w \tag{3.199}$$

$$h_r = 0, \quad r = r_L$$

Similar to Eqs. (3.191) and (3.192), the (relative) head may be decomposed into terms of various orders, and these terms are governed by the following equations:

$$\left.\begin{aligned} \frac{\partial^2 h_r^{(0)}(\mathbf{x})}{\partial x_i^2} &= 0 \\ h_r^{(0)} = H_w - H_L, \quad r &= r_w \\ h_r^{(0)} = 0, \quad r &= r_L \end{aligned}\right\} \tag{3.200}$$

$$\left.\begin{aligned} \frac{\partial^2 h_r^{(1)}(\mathbf{x})}{\partial x_i^2} &= \frac{\partial}{\partial x_i}\left[J_i(\mathbf{x}) f'(\mathbf{x})\right] \\ h_r^{(1)} = 0, \quad r &= r_w \\ h_r^{(1)} = 0, \quad r &= r_L \end{aligned}\right\} \tag{3.201}$$

It is well known the solution of the zeroth-order mean head in Eq. (3.200) is

$$h_r^{(0)}(\mathbf{x}) = (H_w - H_L) - \frac{H_w - H_L}{\ln(r_L/r_w)} \ln\left(\frac{r}{r_w}\right) \tag{3.202}$$

where $r = |\mathbf{x} - \mathbf{x}_w|$ and $\mathbf{x}_w$ is again the center of the well. The first-order head fluctuation term can be written as Eq. (3.195) in terms of the mean Green's function, and the head covariance can be expressed as Eq. (3.196). The mean Green's function in a circular domain bounded at $r = r_w$ and $r = r_L$ is given as [e.g., Sanchez-Vila, 1997]

$$G^{(0)}(\mathbf{x}', \mathbf{x}) = \frac{\ln(r/r_w)\ln(r_L/r')}{2\pi \ln(r_L/r_w)}$$

$$+ \sum_{n=1}^{\infty} \frac{(r^n/r_w^n - r^{-n}/r_w^{-n})\cos n(\theta' - \theta)}{2\pi n(r_L^n/r_w^n - r_L^{-n}/r_w^{-n})}\left[\left(\frac{r_L}{r'}\right)^n - \left(\frac{r_L}{r'}\right)^{-n}\right] \tag{3.203}$$

valid for $r < r'$ with a similar expression for $r > r'$ but with r and r' exchanged. In Eq. (3.203), $r = |\mathbf{x} - \mathbf{x}_w|$ and $r' = |\mathbf{x}' - \mathbf{x}_w|$.

The head covariance or variance can, in principle, be evaluated from Eq. (3.196) with Eqs. (3.202) and (3.203), although there are no closed-form results available in the literature. Instead, a number of recent studies were devoted to investigate effective or equivalent conductivities under radial flow. Next we discuss some of the results.

Effective and equivalent conductivities The problem of seeking some type of effective or equivalent conductivities under converging or diverging flows is of great practical importance because most field methods to obtain hydraulic parameters involve flow towards sinks and/or from sources. As in Section 3.5.1 for uniform mean flow, the effective hydraulic conductivity is defined in Eq. (3.149) as the tensor that relates the expected values of flux and hydraulic head gradient. The equivalent conductivity K^{eq} is defined as the conductivity of a fictitious homogeneous medium that conveys the same discharge Q as the actual, heterogeneous formation.

Effective conductivities under converging flow in random media were first studied by Shvidler [1964]. It was found that unlike in uniform mean flow, a single effective conductivity that depends only on the statistical properties of the hydraulic conductivity field cannot be defined. Instead, the effective conductivity generally depends on the distance from the pumping well. Dagan [1989] obtained the effective conductivities both close to the well and far from it with simple arguments. His analysis is based on the observation that close to the well, $K_S(r, \theta)$ in Eq. (3.189) can be regarded as homogeneous and is thus independent of θ. With this argument, Eq. (3.189) can be written as

$$\frac{Q_w}{2\pi r} = K_S \frac{dh}{dr} \tag{3.204}$$

When the pumping rate per unit thickness Q_w is given as a deterministic quantity, ensemble averaging of Eq. (3.204) yields

$$\frac{d\langle h \rangle}{dr} = \left\langle \frac{1}{K_S} \right\rangle \frac{Q_w}{2\pi r} = \frac{1}{K_H} \frac{Q_w}{2\pi r} \tag{3.205}$$

After defining a radial specific discharge $q_r = Q_w/(2\pi r) = -K_S(dh/dr)$, one has

$$q_r = -K_H \frac{d\langle h \rangle}{dr} \tag{3.206}$$

It is thus seen that the effective conductivity at and near the well is equal to the harmonic mean K_H of the hydraulic conductivity. If the head at the well $h(r_w)$ is known with certainty, then the pumping rate Q_w and the radial specific discharge q_r are random. As such, the ensemble averages of q_r and K_S are related by

$$\langle q_r \rangle = -\langle K_S \rangle \frac{dh}{dr} \tag{3.207}$$

Hence the effective conductivity becomes the arithmetic mean $K_A = \langle K_S \rangle$. These limiting results are only rigorously valid close to the well, i.e., $r \to r_w$. Dagan [1989] also argued that the effective conductivity should be equal to the geometric mean K_G far away from the well on the ground so that the two-dimensional, horizontal flow should be almost uniform in the mean there.

The above limiting results have been verified by various studies [e.g., Naff, 1991; Neuman and Orr, 1993; Indelman et al., 1996]. Furthermore, some recent studies have derived full-range, closed-form solutions for effective conductivity as a function of the radial distance from the well for unbounded [Indelman and Abramovich, 1994b] and bounded

[Sanchez-Vila, 1997] domains. It has been found that for the case of a given pumping rate at the well, the effective conductivity (transmissivity) is a monotonic increasing function of the radial distance r, going from the harmonic mean K_H for small r to the geometric mean K_G for large r. The radial distance at which the asymptotic value of K_G is attained depends on the covariance function of $f = \ln K_S$: 1.5–2 integral scales for the Gaussian model and 3–5 for the exponential model. It is also found that although the effective transmissivity K^{ef} depends upon the choice of boundary conditions and the covariance models of f, it is approximately equal to the geometric mean K_G at a few integral scales away from the boundaries.

The equivalent conductivity (transmissivity) may be defined, more specifically with Thiem's formula (3.197), as

$$K^{eq} = \frac{Q_w \ln(r_w/r_L)}{2\pi(h_L - h_w)} \tag{3.208}$$

where h_w and h_L are the heads at r_w and r_L, respectively. Thus, for the problem under consideration the equivalent conductivity is the value that best suits Thiem's formula. This definition corresponds to the transmissivity determined in a pumping test where h_w, h_L, and Q_w are observed. For the case with known heads h_w and h_L and random Q_w, the equivalent conductivity may be estimated with the statistical moments of Q_w. The latter depend on the statistical properties of the medium on the basis of Eq. (3.189). It is seen from Eq. (3.208) that K^{eq} also depends on the radii r_w and r_L. In this aspect, K^{eq} is regarded as a *weighted average* of the conductivity values in the area bounded by two circles with radii r_w and r_L. Hence, K^{eq} is an *upscaled quantity* rather than a *point value*. Recently, Indelman et al. [1996] and Sanchez-Vila et al. [1999] have evaluated K^{eq} for the case of radial flow up to second order in the standard deviation of the log hydraulic conductivity, σ_f.

3.6 SPECTRAL METHODS

Spectral methods offer an alternative, convenient way to derive and solve moment equations, especially for the case of stationary independent and dependent variable fluctuations. These methods have been extensively used in many fields of science and engineering [e.g., Lumley and Panofsky, 1964; Granger and Hatanaka, 1964; Beran, 1968; Priestley, 1981]. The stationary spectral method has been successfully applied to the problem of stochastic flow in porous media [e.g., Bakr et al., 1978; Gutjahr et al., 1978; Gutjahr and Gelhar, 1981; Mizell et al., 1982; Gelhar and Axness, 1983; Naff and Vecchia, 1986; Loaiciga et al., 1993, 1994]. Recently, a nonstationary generalization of the spectral method has been introduced for solving porous flow problems by Li and McLaughlin [1991, 1995]. In this section, we discuss these two spectral methods and some corresponding key results.

3.6.1 Stationary Spectral Method

The stationary spectral method is strictly applicable only to the situation where both the independent ($f(\mathbf{x}) = \ln K_S(\mathbf{x})$) and dependent (the head fluctuation $h'(\mathbf{x})$) variables are spatially stationary. As discussed in previous sections, this condition is equivalent to uniform mean flow in unbounded domains of stationary media. Under this condition, the fluctuations f' and h' can be expressed by the following stochastic *Fourier–Stieltjes integral representations* [Lumley and Panofsky, 1964; Bakr *et al.*, 1978]:

$$f'(\mathbf{x}) = \int \exp(\iota \mathbf{k} \cdot \mathbf{x}) \, dZ_f(\mathbf{k}) \tag{3.209}$$

$$h'(\mathbf{x}) = \int \exp(\iota \mathbf{k} \cdot \mathbf{x}) \, dZ_h(\mathbf{k}) \tag{3.210}$$

where $\mathbf{k} = (k_1, \ldots, k_d)^T$ is the wave number space vector (where d is the number of space dimensions), $\iota \equiv \sqrt{-1}$, and $dZ_f(\mathbf{k})$ and $dZ_h(\mathbf{k})$ are the complex Fourier increments of the fluctuations at $\mathbf{k}$. The integration is d-fold from $-\infty$ to ∞. As discussed in Section 2.4.4, the stochastic Fourier–Stieltjes integral has the following properties, using dZ_f as an example:

$$\langle dZ_f(\mathbf{k}) \rangle = 0$$
$$\langle dZ_f(\mathbf{k}) \, dZ_f^*(\mathbf{k}') \rangle = S(\mathbf{k})\delta(\mathbf{k} - \mathbf{k}') \, d\mathbf{k} \, d\mathbf{k}' \tag{3.211}$$

where dZ_f^* is the complex conjugate of dZ_f, and $S_f(\mathbf{k})$ is the *spectrum* (i.e., spectral density function) of f if it is integrable. The first equation of (3.211) indicates zero mean for the random increment dZ_f and the second one states the so-called *orthogonality property* of dZ_f. The covariance $C_f(\mathbf{r})$ is related to the spectrum by

$$C_f(\mathbf{r}) = \int \exp(\iota \mathbf{k} \cdot \mathbf{r}) S_f(\mathbf{k}) \, d\mathbf{k}$$
$$S_f(\mathbf{k}) = \frac{1}{(2\pi)^d} \int \exp(-\iota \mathbf{k} \cdot \mathbf{r}) C_f(\mathbf{r}) \, d\mathbf{r} \tag{3.212}$$

For example, for the three-dimensional anisotropic, exponential covariance $C_f(\mathbf{r})$,

$$C_f(\mathbf{r}) = \sigma_f^2 \exp\left[-\left(\frac{r_1^2}{\lambda_1^2} + \frac{r_2^2}{\lambda_2^2} + \frac{r_3^2}{\lambda_3^2}\right)^{1/2}\right] \tag{3.213}$$

its spectrum is

$$S_f(\mathbf{k}) = \frac{\sigma_f^2 \lambda_1 \lambda_2 \lambda_3}{\pi^2 (1 + \lambda_1^2 k_1^2 + \lambda_2^2 k_2^2 + \lambda_3^2 k_3^2)^2} \tag{3.214}$$

where λ_i is the correlation scale in the r_i direction.

For the situation of uniform mean flow in unbounded domains of stationary media, the head fluctuation may be written from Eq. (3.10), to first order, as

$$\frac{\partial^2 h'(\mathbf{x})}{\partial x_i^2} = J_1 \frac{\partial f'(\mathbf{x})}{\partial x_1} \tag{3.215}$$

where J_1 is the negative of the mean hydraulic head gradient and the mean flow has been assumed to be aligned with x_1.

Substituting Eqs. (3.209) and (3.210) into Eq. (3.215) yields

$$dZ_h(\mathbf{k}) = -\frac{\iota J_1 k_1}{k^2} dZ_f(\mathbf{k}) \tag{3.216}$$

where $k^2 = \sum_{i=1}^{d} k_i^2$. The spectrum $S_h(\mathbf{k})$ of head can be obtained by multiplying Eq. (3.216) by its complex conjugate, taking expectation and utilizing the orthogonality property of Eq. (3.211),

$$S_h(\mathbf{k}) = \frac{J_1^2 k_1^2}{k^4} S_f(\mathbf{k}) \tag{3.217}$$

If the right-hand side of Eq. (3.217) is integrable, the head covariance $C_h(\mathbf{r})$ can be obtained via the inverse Fourier transformation,

$$C_h(\mathbf{r}) = \int \exp(\iota \mathbf{k} \cdot \mathbf{r}) \frac{J_1^2 k_1^2}{k^4} S_f(\mathbf{k}) \, d\mathbf{k} \tag{3.218}$$

The head variance is

$$\sigma_h^2 = \int \frac{J_1^2 k_1^2}{k^4} S_f(\mathbf{k}) \, d\mathbf{k} \tag{3.219}$$

Similarly, the spectrum of $S_{fh}(\mathbf{k})$ is obtained as

$$S_{fh}(\mathbf{k}) = \frac{\iota J_1 k_1}{k^2} S_f(\mathbf{k}) \tag{3.220}$$

which is related to the cross-covariance $C_{fh}(\mathbf{r})$ via

$$C_{fh}(\mathbf{r}) = \int \exp(\iota \mathbf{k} \cdot \mathbf{r}) \frac{\iota J_1 k_1}{k^2} S_f(\mathbf{k}) \, d\mathbf{k} \tag{3.221}$$

Note that the spectra S_{fh} and S_h can be derived directly from the stationary version of Eqs. (3.123) and (3.124) by Fourier transforms.

The cross-covariance of head may be evaluated with Eq. (3.221) for a given $S_f(\mathbf{k})$. For the isotropic version of the exponential covariance C_f in Eq. (3.213) (i.e., $\lambda_1 = \lambda_2 = \lambda_3 = \lambda_f$), the three-dimensional (auto- and cross-) covariances can be integrated analytically [Bakr *et al.*, 1978], resulting in expressions equivalent to Eqs. (3.133) and (3.134). However, it is difficult to derive fully analytical covariance expressions for the case of anisotropic covariance $C_f(\mathbf{r})$. Instead, only the head variance is usually evaluated. For example, for the special case of axis-symmetric covariance C_f where $\lambda_1 = \lambda_2 = \lambda$, the head variance is obtained by substituting Eq. (3.213) into Eq. (3.220) with $\lambda_1 = \lambda_2 = \lambda$ and then integrating [Naff and Vecchia, 1986, (18); Dagan, 1989, (3.7.16)],

$$\frac{\sigma_h^2}{J_1^2 \lambda^2 \sigma_f^2} = \frac{e^2}{4\gamma^2} \left[1 + \frac{1 - 2e^2}{e^2} \left(\frac{e}{\gamma} \right) \tan^{-1} \left(\frac{\gamma}{e} \right) \right] \tag{3.222}$$

where $e = \lambda_3 / \lambda$ measures the statistical anisotropy of the medium, $\gamma^2 = 1 - e^2$, and $\tan^{-1}( )$ stands for the arctangent function. For $\gamma^2 < 0$, γ becomes a complex variable. However, σ_h^2 is always real because so is $(e/\gamma) \tan^{-1}(\gamma/e)$. This is obvious for $\gamma^2 \geq 0$. For $\gamma^2 < 0$, it can be shown that $(e/\gamma) \tan^{-1}(\gamma/e) = e \ln(2e^2 - 1 + 2e\sqrt{e^2 - 1})/(2\sqrt{e^2 - 1})$. Note that the result of Eq. (3.222) is specific for the situations where the mean flow is aligned with the horizontal (x_1) direction (i.e., parallel to formation bedding). Figure 3.15a shows the head variance as a function of the anisotropy ratio e. It is seen that the head variance increases with e monotonically without bound. There are some interesting limiting cases: for the case of $e \to 0$, which implies heterogeneity in the vertical (x_3) direction only (i.e., perfectly stratified formation), $\sigma_h^2 = 0$; for the case of $e = 1$ (i.e., isotropic three-dimensional heterogeneity), $\sigma_h^2 = J_1^2 \lambda^2 \sigma_f^2 / 3$; and for the case of $e \to \infty$, which implies heterogeneity in the horizontal directions only and is thus equivalent to the case of isotropic, two-dimensional heterogeneity, the head variance is unbounded.

Naff and Vecchia [1986, (44) and (45)] also derived a semi-analytical expression for the head variance, with a hybrid spectral and Green's function method, for the case of a semi-infinite, confined aquifer (i.e., $L_1 = L_2 \to \infty$ while L_3 is finite).

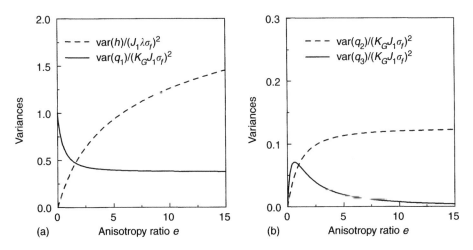

Figure 3.15. The head and flux variances of uniform mean flow in a three-dimensional unbounded domain as functions of anisotropic ratio e.

Flux moments can be derived as well by the spectral method. Applying Fourier–Stieltjes representations to the first-order flux fluctuation equation (3.20) leads to

$$dZ_{q_i}(\mathbf{k}) = K_G[\delta_{i1}J_i dZ_f(\mathbf{k}) - \iota k_i dZ_h(\mathbf{k})]$$

$$= K_G J_1 \left(\delta_{i1} - \frac{k_1 k_i}{k^2} \right) dZ_f(\mathbf{k}) \tag{3.223}$$

Hence, the spectrum of flux is

$$S_{q_{ij}}(\mathbf{k}) = K_G^2 J_1^2 \left(\delta_{i1} - \frac{k_1 k_i}{k^2} \right) \left(\delta_{j1} - \frac{k_1 k_j}{k^2} \right) S_f(\mathbf{k}) \tag{3.224}$$

For the isotropic exponential covariance $C_f(\mathbf{r})$, the flux covariances obtained by inverting this spectrum are equivalent to those in Eqs. (3.135)–(3.143) derived earlier. For the anisotropic exponential covariance, only semi-analytical flux covariances are available [Rubin and Dagan, 1992; Russo, 1995a]. However, fully explicit solutions for the flux variances can be derived for the case of horizontal mean flow in axis-symmetric formations [Russo, 1995a, (13)],

$$\frac{\sigma_{q_{11}}^2}{K_G^2 J_1^2 \sigma_f^2} = 1 + \frac{19e^2 - 10e^4}{16(\gamma^2)^2} - \frac{e(13 - 4e^2)}{16(\gamma^2)^2} \frac{\sin^{-1}(\gamma)}{\gamma} \tag{3.225}$$

$$\frac{\sigma_{q_{22}}^2}{K_G^2 J_1^2 \sigma_f^2} = \frac{e^2 + 2e^4}{16(\gamma^2)^2} + \frac{e(1 - 4e^2)}{16(\gamma^2)^2} \frac{\sin^{-1}(\gamma)}{\gamma} \tag{3.226}$$

$$\frac{\sigma_{q_{33}}^2}{K_G^2 J_1^2 \sigma_f^2} = \frac{-3e^2}{4(\gamma^2)^2} + \frac{e(1 + 2e^2)}{4(\gamma^2)^2} \frac{\sin^{-1}(\gamma)}{\gamma} \tag{3.227}$$

where $\sin^{-1}(\gamma)$ stands for the arcsine function. Like the head variance, the flux variances are always real because the term $\sin^{-1}(\gamma)/\gamma$ that may involve complex variables turns out to be real. It can be shown that for $\gamma^2 = 1 - e^2 < 0$, $\sin^{-1}(\gamma)/\gamma = -\ln(e - \sqrt{e^2 - 1})/\sqrt{e^2 - 1}$. These variances as functions of e are depicted in Fig. 3.15. It is seen that $\sigma_{q_1}^2/(K_G^2 J_1^2 \sigma_f^2)$ decreases monotonically with the increase of e and approaches the constant of 0.375 for $e \gg 1$ (Fig. 3.15a), $\sigma_{q_2}^2/(K_G^2 J_1^2 \sigma_f^2)$ increases monotonically with e and approaches constant of 0.125 for $e \gg 1$ (Fig. 3.15b), but $\sigma_{q_3}^2/(K_G^2 J_1^2 \sigma_f^2)$ is not a monotonic function of e and approaches zero for large e. The two transverse components have different variances when λ_2 and λ_3 are different. The following limiting cases of these variances can be either read from the figure or derived directly from Eqs. (3.225)–(3.227): for the infinite limit of $e \to \infty$, the well-known results for two-dimensional, isotropic formations are recovered, i.e., $\sigma_{q_1}^2 = 3K_G^2 J_1^2 \sigma_f^2/8$, $\sigma_{q_2}^2 = K_G^2 J_1^2 \sigma_f^2/8$, and $\sigma_{q_3}^2 = 0$; for the case of $e \to 0$, $\sigma_{q_1}^2 = K_G^2 J_1^2 \sigma_f^2$, and $\sigma_{q_2}^2 = \sigma_{q_3}^2 = 0$; and for the case of $e = 1$, the results for three-dimensional, isotropic formations are obtained, i.e., $\sigma_{q_1}^2 = 8K_G^2 J_1^2 \sigma_f^2/15$, and $\sigma_{q_2}^2 = \sigma_{q_3}^2 = K_G^2 J_1^2 \sigma_f^2/15$.

3.6.2 Nonstationary Spectral Method

As found earlier, the head fluctuation is generally nonstationary in the case of finite domains, medium nonstationarity, or nonuniform mean flow. For such a situation, the stationary spectral method may no longer be valid.

For the case that the mean log hydraulic conductivity $\langle f(\mathbf{x}) \rangle$ varies spatially but the fluctuation of f is stationary, the spectral representation of h' in Eq. (3.210) has been generalized by Li and McLaughlin [1991, 1995] through a transfer function

$$h'(\mathbf{x}) = \int \phi_{hf}(\mathbf{x}, \mathbf{k}) \, dZ_f(\mathbf{k}) \tag{3.228}$$

The transfer function $\phi_{hf}(\mathbf{x}, \mathbf{k})$ can be derived by substituting Eqs. (3.209) and (3.228) into the first-order head fluctuation equation (3.10). In the absence of sink/source and under deterministic boundary conditions, the resulting equation is

$$\left.\begin{array}{c} \displaystyle\int \left[\frac{\partial^2 \phi_{hf}(\mathbf{x}, \mathbf{k})}{\partial x_i^2} + \frac{\partial \langle f(\mathbf{x}) \rangle}{\partial x_i} \frac{\partial \phi_{hf}(\mathbf{x}, \mathbf{k})}{\partial x_i} - J_i(\mathbf{x}) \frac{\partial \exp(\iota \mathbf{k} \cdot \mathbf{x})}{\partial x_i} \right] dZ_f(\mathbf{k}) = 0 \\[3mm] \displaystyle\int \phi_{hf}(\mathbf{x}, \mathbf{k}) \, dZ_f(\mathbf{k}) = 0, \quad \mathbf{x} \in \Gamma_D \\[3mm] \displaystyle\int n_i(\mathbf{x}) \frac{\partial \phi_{hf}(\mathbf{x}, \mathbf{k})}{\partial x_i} \, dZ_f(\mathbf{k}) = 0, \quad \mathbf{x} \in \Gamma_N \end{array}\right\} \tag{3.229}$$

Multiplying Eq. (3.229) with its complex conjugate and taking expectation leads to

$$\left.\begin{array}{c} \displaystyle\int \left| \frac{\partial^2 \phi_{hf}(\mathbf{x}, \mathbf{k})}{\partial x_i^2} + \frac{\partial \langle f(\mathbf{x}) \rangle}{\partial x_i} \frac{\partial \phi_{hf}(\mathbf{x}, \mathbf{k})}{\partial x_i} - J_i(\mathbf{x}) \frac{\partial \exp(\iota \mathbf{k} \cdot \mathbf{x})}{\partial x_i} \right|^2 S_f(\mathbf{k}) \, d\mathbf{k} = 0 \\[3mm] \displaystyle\int |\phi_{hf}(\mathbf{x}, \mathbf{k})|^2 S_f(\mathbf{k}) \, d\mathbf{k} = 0, \quad \mathbf{x} \in \Gamma_D \\[3mm] \displaystyle\int \left| n_i(\mathbf{x}) \frac{\partial \phi_{hf}(\mathbf{x}, \mathbf{k})}{\partial x_i} \right|^2 S_f(\mathbf{k}) \, d\mathbf{k} = 0, \quad \mathbf{x} \in \Gamma_N \end{array}\right\} \tag{3.230}$$

where the vertical bars indicate absolute value. The necessary and sufficient condition for Eq. (3.230) to hold for any arbitrary $S_f(\mathbf{k})$ is

$$\frac{\partial^2 \phi_{hf}(\mathbf{x}, \mathbf{k})}{\partial x_i^2} + \frac{\partial \langle f(\mathbf{x}) \rangle}{\partial x_i} \frac{\partial \phi_{hf}(\mathbf{x}, \mathbf{k})}{\partial x_i} - J_i(\mathbf{x}) \frac{\partial \exp(\iota \mathbf{k} \cdot \mathbf{x})}{\partial x_i} = 0$$

$$\phi_{hf}(\mathbf{x}, \mathbf{k}) = 0, \quad \mathbf{x} \in \Gamma_D \tag{3.231}$$

$$n_i(\mathbf{x}) \frac{\partial \phi_{hf}(\mathbf{x}, \mathbf{k})}{\partial x_i} = 0, \quad \mathbf{x} \in \Gamma_N$$

This equation describes a well-posed boundary value problem with a unique solution. Hence, the generalized spectral representation (3.228) is unique [Li and McLaughlin, 1995].

The head covariance can then be obtained with Eq. (3.228) through multiplying it with its complex conjugate, and taking expectation and utilizing the orthogonality property of Eq. (3.211),

$$C_h(\mathbf{x}, \chi) = \iint \phi_{hf}(\mathbf{x}, \mathbf{k}) \phi_{hf}^*(\chi, \mathbf{k}') \langle dZ_f(\mathbf{k}) dZ_f^*(\mathbf{k}') \rangle \, d\mathbf{k} \, d\mathbf{k}'$$

$$= \int \phi_{hf}(\mathbf{x}, \mathbf{k}) \phi_{hf}^*(\chi, \mathbf{k}) S_f(\mathbf{k}) \, d\mathbf{k} \tag{3.232}$$

The cross-covariance C_{fh} is obtained similarly as

$$C_{fh}(\mathbf{x}, \chi) = \int \exp(\iota \mathbf{k} \cdot \mathbf{x}) \phi_{hf}^*(\chi, \mathbf{k}) S_f(\mathbf{k}) \, d\mathbf{k} \tag{3.233}$$

These covariances reduce to Eqs. (3.220) and (3.221) for stationary head fluctuations. It should be emphasized that in the generalized spectral representation the fluctuation of independent variable (f', in this case) must be stationary although the dependent variable (h, in this case) can be nonstationary. This restrictive requirement is, however, not necessary in the partial differential equation and the Green's-function-based approaches discussed in the previous sections.

In general, the transfer function $\phi_{hf}(\mathbf{x}, \mathbf{k})$ needs to be evaluated numerically from Eq. (3.231). For some special cases such as uniform mean flow in semi-infinite, confined aquifers of stationary media [Li and McLaughlin, 1991] and in unbounded domains of linear trending permeability, analytical solutions may, however, be derived. Below we illustrate the nonstationary spectral approach with an example involving a linear trend in the mean log hydraulic conductivity parallel to the mean flow and compare this nonstationary approach with the stationary spectral approach.

Linear trend aligned with mean flow Consider, again, the case of a log hydraulic conductivity field with the mean $\langle f(\mathbf{x}) \rangle = a_0 + a_1 x_1$ but with stationary fluctuation $f'(\mathbf{x})$ in a two-dimensional unbounded domain. This case has been studied analytically with the Green's function approach in Section 3.5.3. The mean head gradient $J_1(\mathbf{x})$ is given as $J_1(\mathbf{x}) = J_0 \exp(-a_1 x_1)$ in Eq. (3.178), where $J_0 = J_1(0)$. Then the equation governing the transfer function can be written as

$$\frac{\partial^2 \phi_{hf}(\mathbf{x}, \mathbf{k})}{\partial x_i^2} + a_1 \frac{\partial \phi_{hf}(\mathbf{x}, \mathbf{k})}{\partial x_1} = \iota k_1 J_1(\mathbf{x}) \exp(\iota \mathbf{k} \cdot \mathbf{x}) \tag{3.234}$$

It can be verified [Li and McLaughlin, 1995] that the solution of Eq. (3.234) is

$$\phi_{hf}(\mathbf{x}, \mathbf{k}) = -\frac{\iota J_1(\mathbf{x}) k_1}{k^2 + \iota a_1 k_1} \exp(\iota k_i x_i) \tag{3.235}$$

With the transfer function in Eq. (3.235), the head fluctuation can be obtained from Eq. (3.228) as

$$h'(\mathbf{x}) = -\int \frac{\iota J_1(\mathbf{x})k_1}{k^2 + \iota a_1 k_1} \exp(\iota k_i x_i) \, dZ_f(\mathbf{k}) \tag{3.236}$$

The head covariances are obtained from Eqs. (3.232) and (3.233) as

$$C_h(\mathbf{x}, \boldsymbol{\chi}) = J_1(\mathbf{x}) J_1(\boldsymbol{\chi}) \int \exp(\iota k_i r_i) \frac{k_1^2}{k^4 + a_1^2 k_1^2} S_f(\mathbf{k}) \, d\mathbf{k} \tag{3.237}$$

$$C_{fh}(\mathbf{x}, \boldsymbol{\chi}) = J_1(\boldsymbol{\chi}) \int \exp(\iota k_i r_i) \frac{\iota k_1}{k^2 - \iota a_1 k_1} S_f(\mathbf{k}) \, d\mathbf{k} \tag{3.238}$$

where $r_i = \chi_i - x_i$. It is obvious that these head covariances reduce to the stationary expressions (3.218) and (3.221) in the absence of a linear trend (i.e., $a_1 = 0$).

For the log hydraulic conductivity covariance of the following form [Mizell *et al.*, 1982]:

$$C_f(\mathbf{r}) = \sigma_f^2 \left[\frac{\pi r}{4\lambda_f} K_1 \left(\frac{\pi r}{4\lambda_f} \right) - \left(\frac{\pi r}{4\lambda_f} \right)^2 K_0 \left(\frac{\pi r}{4\lambda_f} \right) \right] \tag{3.239}$$

$$S_f(\mathbf{k}) = \frac{2\sigma_f^2 \alpha^2 k^2}{\pi (k^2 + \alpha^2)} \tag{3.240}$$

where $\alpha = \pi/(4\lambda_f)$ and K_j is the modified Bessel function of order j, the head covariances are obtained by integrations as [Li and McLaughlin, 1995],

$$C_h(\mathbf{x}, \boldsymbol{\chi}) = \frac{2}{\pi} \sigma_f^2 \alpha^2 J_1(\mathbf{x}) J_1(\boldsymbol{\chi}) M_1(\mathbf{r}, a_1) \tag{3.241}$$

$$C_{fh}(\mathbf{x}, \boldsymbol{\chi}) = \frac{2}{\pi} \sigma_f^2 \alpha^2 J_1(\boldsymbol{\chi})[-M_1(\mathbf{r}, a_1) + M_2(\mathbf{r}, a_1)] \tag{3.242}$$

where M_1 and M_2 are defined as

$$M_1(\mathbf{r}, a_1) = \int_{-\infty}^{\infty} \int_{-\infty}^{\infty} \frac{k_1^2 k^2 \cos(k_1 r_1 + k_2 r_2)}{(k^4 + a_1^2 k_1^2)(k^2 + \alpha^2)^3} \, dk_1 \, dk_2 \tag{3.243}$$

$$M_2(\mathbf{r}, a_1) = \int_{-\infty}^{\infty} \int_{-\infty}^{\infty} \frac{k_1^2 k^4 \sin(k_1 r_1 + k_2 r_2)}{(k^4 + a_1^2 k_1^2)(k^2 + \alpha^2)^3} \, dk_1 \, dk_2 \tag{3.244}$$

In general, the covariance expressions, however, must be evaluated numerically. In addition, since the integrals contain oscillating functions, Li and McLaughlin [1995] found it necessary to use a special analytically based integration procedure in order to obtain satisfactory results. For the special situation $\mathbf{x} = \boldsymbol{\chi}$, they were able to derive explicit solutions via *Mathematica* [Wolfram, 1991],

$$\frac{\sigma_h^2(\mathbf{x})}{J_1(\mathbf{x})^2 \lambda_f^2 \sigma_f^2} = \frac{1}{a_1^2 \lambda_f^2} - \frac{3\pi^6}{4a_1^2 \lambda_f^2 s^5} + \frac{24\pi^2}{s^4} - \frac{3\pi^4}{4a_1^2 \lambda_f^2 s^3} - \frac{3\pi^2}{4a_1^2 \lambda_f^2 s}$$

$$+ \frac{\pi^2}{a_1^2 \lambda_f^2 s^2} + \frac{384\pi a_1^2 \lambda_f^2}{s^5} \sin^{-1}\left(\frac{\pi}{4|a_1 \lambda_f|}\right) \qquad (3.245)$$

$$\frac{\sigma_{fh}(\mathbf{x})}{J_1(\mathbf{x}) \lambda_f \sigma_f^2} = -\left[\frac{1}{a_1 \lambda_f} - \frac{3\pi^6}{4a_1 \lambda_f s^5} + \frac{24\pi^2 a_1 \lambda_f}{s^4} - \frac{3\pi^4}{4a_1 \lambda_f s^3} \right.$$

$$\left. - \frac{3\pi^2}{4a_1 \lambda_f s} + \frac{\pi^2}{a_1 \lambda_f s^2} + \frac{384\pi a_1^3 \lambda_f^3}{s^5} \sin^{-1}\left(\frac{\pi}{4|a_1 \lambda_f|}\right) \right] \qquad (3.246)$$

where $s = (16a_1^2 \lambda_f^2 - \pi^2)^{1/2}$.

Evaluation of local stationarity Although the stationary spectral representation of Eq. (3.210) is strictly valid only for stationary head fluctuations, it has been used to approximate the situations with nonstationary head fluctuations by invoking the argument of *local stationarity* [e.g., Loaiciga *et al.*, 1993, 1994; Gelhar, 1993]. Below, we introduce and evaluate this treatment of nonstationarity.

For the case of a linear trend aligned with the mean flow in an unbounded domain, the first-order head fluctuation equation can be written from Eq. (3.10) as

$$\frac{\partial^2 h'(\mathbf{x})}{\partial x_i^2} + a_1 \frac{\partial h'(\mathbf{x})}{\partial x_1} = J_1(\mathbf{x}) \frac{\partial f'(\mathbf{x})}{\partial x_1} \qquad (3.247)$$

The head fluctuation is nonstationary because the mean head gradient $J_1(\mathbf{x})$ is spatially varying. The stationary spectral representation may be applied to this equation if the assumption of local stationarity is invoked. This assumption implies that $J_1(\mathbf{x})$ does not change significantly over the correlation scale of h so that J_1 may be treated as a local constant and h' may be regarded stationary in a local sense. Since the spatial variability of J_1 is caused by the presence of the linear trend in this case, it is expected that the notion of local stationarity is only meaningful when the product of a_1 and the correlation scale of head is small relative to one [Gelhar, 1993]. Under this condition, applying the spectral representations (3.209) and (3.210) to Eq. (3.247) yields

$$dZ_h(\mathbf{k}) = -\frac{\iota k_1 J_1(\mathbf{x})}{k^2 - \iota a_1 k_1} dZ_f(\mathbf{k}) \qquad (3.248)$$

This leads to the head covariances

$$C_h(\mathbf{x}, \boldsymbol{\chi}) = J_1(\mathbf{x}) J_1(\boldsymbol{\chi}) \int \exp(\iota k_i r_i) \frac{k_1^2}{k^4 + a_1^2 k_1^2} S_f(\mathbf{k}) \, d\mathbf{k} \qquad (3.249)$$

$$C_{fh}(\mathbf{x}, \boldsymbol{\chi}) = J_1(\boldsymbol{\chi}) \int \exp(\iota k_i r_i) \frac{\iota k_1}{k^2 + \iota a_1 k_1} S_f(\mathbf{k}) \, d\mathbf{k} \qquad (3.250)$$

Note that the dependency of the auto- and cross-covariances of head on the actual locations enters through the mean head gradients.

It is seen that the head covariance of Eq. (3.249) is identical to Eq. (3.237) obtained with the nonstationary spectral representation but the cross-covariance of Eq. (3.250) differs with Eq. (3.238) in the sign of the denominator term containing a_1. It is obvious that the nonstationary and the locally stationary approaches yield identical results in the absence of the linear trend (i.e., $a_1 = 0$). When $a_1 \neq 0$, although the stationary spectral approach correctly predicts the head covariance it fails to capture the qualitative behavior of $C_{fh}(\mathbf{x}, \boldsymbol{\chi})$ due to the incorrect sign in front of the a_1 term [Li and McLaughlin, 1995]. The error in the cross-covariance has an important implication for conditioning log hydraulic conductivities on the head measurements. This error may also migrate into other moments such as $\langle h^{(2)} \rangle$ in Eq. (3.17), $\langle q^{(2)} \rangle$ in Eq. (3.24), and K^{ef} defined in Eq. (3.149), all of which contain the C_{fh} term.

As one should expect, the magnitude of the error caused by the local stationarity assumption diminishes as the decrease of a_1. More generally, for a strongly nonstationary flow field (but with stationary log hydraulic conductivity fluctuations), the nonstationary spectral approach is necessary; for a weakly nonstationary flow field the much simpler, locally stationary spectral approach might be able to provide adequate approximations. When the fluctuations of the log hydraulic conductivity are nonstationary, the partial differential equation and the Green's-function-based methods discussed in the previous sections are appropriate.

3.6.3 Application to Flow in Semiconfined Aquifers

In this subsection we illustrate how the spectral methods are applied to saturated flow in *semiconfined aquifers*. An aquifer is called semiconfined if its confining beds are not truly impermeable and flow (leakage) occurs through these beds, as illustrated in Fig. 3.16. Flow is essentially horizontal in a shallow semiconfined aquifer if the semiconfined aquifer is horizontal, the lower layer is impermeable, and the head is constant above the upper semipermeable layer. Flow in such a situation satisfies the following governing equation [e.g., Bear, 1972]:

$$\frac{\partial}{\partial x_i} \left[K_S(\mathbf{x}) \frac{\partial h(\mathbf{x})}{\partial x_i} \right] = K_S(\mathbf{x}) \alpha^2 [h(\mathbf{x}) - h^*] \tag{3.251}$$

where α is the so-called leakage factor and h^* is the head in the aquifer above the semipermeable layer. Summation for repeated indices i is implied with $i = 1, 2$. In this study, both α and h^* are assumed to be deterministic and constant, while $K_S(\mathbf{x})$ is assumed to be a random space function of second-order stationarity. Equation (3.251) can be rewritten in terms of the relative head $h_r(\mathbf{x}) = h(\mathbf{x}) - h^*$,

$$\frac{\partial}{\partial x_i} \left[K_S(\mathbf{x}) \frac{\partial h_r(\mathbf{x})}{\partial x_i} \right] - K_S(\mathbf{x}) \alpha^2 h_r(\mathbf{x}) = 0 \tag{3.252}$$

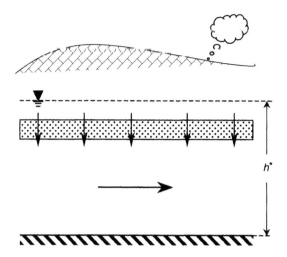

Figure 3.16. Two-dimensional cross-section of flow in a semiconfined aquifer.

subject to appropriate boundary conditions. It is seen that the relative head is governed by the modified Helmholtz equation, already encountered in Section 3.5.3. As usual, the relative head may be expanded into a formal series: $h_r(\mathbf{x}) = h_r^{(0)}(\mathbf{x}) + h_r^{(1)}(\mathbf{x}) + h_r^{(2)}(\mathbf{x}) + \cdots$. It can be shown, as done in Section 3.2, that the head terms satisfy the following equations subject to appropriate boundary conditions:

$$\frac{\partial^2 h_r^{(0)}(\mathbf{x})}{\partial x_i^2} - \alpha^2 h_r^{(0)}(\mathbf{x}) = 0 \qquad (3.253)$$

$$\frac{\partial^2 h_r^{(1)}(\mathbf{x})}{\partial x_i^2} - \alpha^2 h_r^{(1)}(\mathbf{x}) = J_i(\mathbf{x}) \frac{\partial f'(\mathbf{x})}{\partial x_i} \qquad (3.254)$$

and

$$\frac{\partial^2 h_r^{(n)}(\mathbf{x})}{\partial x_i^2} - \alpha^2 h_r^{(n)}(\mathbf{x}) = -\frac{\partial h_r^{(n-1)}(\mathbf{x})}{\partial x_i} \frac{\partial f'(\mathbf{x})}{\partial x_i} \qquad (3.255)$$

where $J_i(\mathbf{x}) = -\partial h_r^{(0)}(\mathbf{x})/\partial x_i$, $n \geq 2$ and no summation for repeated indices n is implied.

Let us first consider the solution of the (zeroth-order) mean head in Eq. (3.253) in a two-dimensional, horizontal bounded domain of two parallel constant head and two parallel no-flow boundaries (see Fig. 3.1). It can be verified that

$$\langle h_r^{(0)}(\mathbf{x})\rangle = \frac{H_o \exp(-\alpha L_1) - H_L}{\exp(-\alpha L_1) - \exp(\alpha L_1)} \exp(\alpha x_1)$$

$$+ \frac{H_L - H_o \exp(\alpha L_1)}{\exp(-\alpha L_1) - \exp(\alpha L_1)} \exp(-\alpha x_1) \qquad (3.256)$$

The mean head gradient is given as

$$J_1(\mathbf{x}) = A\exp(\alpha x_1) + B\exp(-\alpha x_1) \qquad (3.257)$$

and $J_2 = 0$, where $A = -\alpha[H_o\exp(-\alpha L_1) - H_L]/[\exp(-\alpha L_1) - \exp(\alpha L_1)]$ and $B = \alpha[H_l - H_o\exp(\alpha L_1)]/[\exp(-\alpha L_1) - \exp(\alpha L_1)]$. Hence, the first-order head fluctuation equation (3.254) can be written as

$$\frac{\partial^2 h_r^{(1)}(\mathbf{x})}{\partial x_i^2} - \alpha^2 h_r^{(1)}(\mathbf{x}) = [A\exp(\alpha x_1) + B\exp(-\alpha x_1)]\frac{\partial f'(\mathbf{x})}{\partial x_1} \qquad (3.258)$$

The head fluctuation can be expressed as Eq. (3.71) with the the mean Green's function replaced by the one for the modified Helmholtz equation, and then the head covariances C_{fh_r} and C_h can be written similar to Eqs. (3.179) and (3.180). In passing, for an unbounded domain, Green's function is given by Eq. (3.185). Alternatively, the head fluctuation can be expressed in terms of a spectral representation. Since the mean head J_1, in general, varies spatially, the head fluctuation is nonstationary even in an unbounded domain. Hence, the nonstationary spectral method is appropriate.

It can be shown that the transfer function $\phi_{hf}(\mathbf{x}, \mathbf{k})$ of Eq. (3.228) satisfies the following modified Helmholtz equation:

$$\frac{\partial^2 \phi_{hf}(\mathbf{x}, \mathbf{k})}{\partial x_i^2} - \alpha^2 \phi_{hf}(\mathbf{x}, \mathbf{k}) = \iota k_1[A\exp(\alpha x_1) + B\exp(-\alpha x_1)]\exp(\iota k_i x_i) \qquad (3.259)$$

subject to appropriate boundary conditions. With the solution of $\phi_{hf}(\mathbf{x}, \mathbf{k})$, the head covariances can be evaluated from Eqs. (3.232) and (3.233). For an unbounded domain, $\phi_{hf}(\mathbf{x}, \mathbf{k})$ is given by

$$\phi_{hf}(\mathbf{x}, \mathbf{k}) = -\left[\frac{\iota A k_1}{k^2 - 2\iota\alpha k_1}\exp(\alpha x_1) + \frac{\iota B k_1}{k^2 + 2\iota\alpha k_1}\exp(-\alpha x_1)\right]\exp(\iota k_i x_i) \qquad (3.260)$$

Hence, the head fluctuation can be expressed as

$$h_r^{(1)}(\mathbf{x}) = -\int\left[\frac{\iota A k_1}{k^2 - 2\iota\alpha k_1}\exp(\alpha x_1) + \frac{\iota B k_1}{k^2 + 2\iota\alpha k_1}\exp(-\alpha x_1)\right]\exp(\iota k_i x_i)\, dZ_f(\mathbf{k})$$
$$(3.261)$$

Recently, Zhu [1998] and Zhu and Sykes [2000] studied this shallow semiconfined flow problem with the stationary spectral method. It is already seen that the assumption of stationary flow is generally violated in the presence of leakage. However, when $\alpha \ll 1$, the mean gradient in Eq. (3.257) is approximately constant, i.e., $J_1(\mathbf{x}) \approx J_o$. Hence, the equation governing the transfer function ϕ_{hf} becomes

$$\frac{\partial^2 \psi_{hf}(\mathbf{x}, \mathbf{k})}{\partial x_i^2} - \alpha^2 \phi_{hf}(\mathbf{x}, \mathbf{k}) \approx \iota J_o k_1 \exp(\iota k_i x_i)] \tag{3.262}$$

and the corresponding solution for an unbounded domain is

$$\phi_{hf}(\mathbf{x}, \mathbf{k}) = -\frac{\iota J_o k_1}{k^2 + \alpha^2} \exp(\iota k_i x_i) \tag{3.263}$$

In turn, the head fluctuation is represented via Eq. (3.228) as

$$h_r^{(1)}(\mathbf{x}) = -\int \frac{\iota J_o k_1}{k^2 + \alpha^2} \exp(\iota k_i x_i) \, dZ_f(\mathbf{k}) \tag{3.264}$$

Equation (3.264) can be obtained by directly applying the Fourier–Stieltjes integral representations (3.209) and (3.210) to Eq. (3.254) if $J_1(\mathbf{x}) \approx J_o$ in the latter. If the local stationarity assumptions, introduced in the previous subsection, is invoked, one has

$$h_r^{(1)}(\mathbf{x}) = -\int \frac{\iota J_1(\mathbf{x}) k_1}{k^2 + \alpha^2} \exp(\iota k_i x_i) \, dZ_f(\mathbf{k}) \tag{3.265}$$

A comparison between Eqs. (3.265) and (3.261) reveals that although the spatial variability is accommodated in the local stationarity approach, the resulting spectral representation is substantially different from that obtained with the nonstationary approach. It is expected that the differences among the stationary representation Eq. (3.264), the locally stationary representation of Eq. (3.265), and the nonstationary representation of Eq. (3.261) increase with the increase of α.

With Eq. (3.264), Zhu [1998] and Zhu and Sykes [2000] derived some closed-form expressions for the head and flux covariances. However, it should be emphasized that such results are strictly valid only if $\alpha \to 0$ or $\alpha x_1 \to 0$. For a nonnegligible leakage factor α, the nonstationary spectral representation of Eq. (3.261) needs to be used. For flow in bounded semiconfined aquifers, the numerical moment equation methods introduced in Section 3.4 may be appropriate.

3.7 HIGHER-ORDER CORRECTIONS

In the preceding sections, only first-order results are obtained. By first order, we mean that for the mean quantities (first moments) such as $\langle h \rangle$ and $\langle q_i \rangle$, terms higher than the order of σ_f are not evaluated and are hence neglected; for the (co)variances (second moments) such as C_h and $C_{q_{ij}}$, only terms up to σ_f^2 are considered. As discussed in Section 1.4.2, we use different quantities σ_f (standard deviation) and σ_f^2 (variance) as references when talking about the orders of $\langle h \rangle$ and C_h. This convention is chosen because $\langle h \rangle$ and C_h have different units. In this section, we discuss the effects of higher-order terms on the first two moments of flow quantities.

Since almost all higher-order studies are performed for uniform mean flows in unbounded domains, the focus of this section shall be such flows. For uniform mean flow in an unbounded domain of stationary media in the absence of sink/source, the equations for the head terms $h^{(n)}$ can be written from Eqs. (3.9)–(3.11) as

$$\frac{\partial^2 h^{(1)}(\mathbf{x})}{\partial x_i^2} = J \frac{\partial f'(\mathbf{x})}{\partial x_1} \tag{3.266}$$

$$\frac{\partial^2 h^{(n)}(\mathbf{x})}{\partial x_i^2} = -\frac{\partial h^{(n-1)}(\mathbf{x})}{\partial x_i} \frac{\partial f'(\mathbf{x})}{\partial x_i} \tag{3.267}$$

where $J = -\partial h^{(0)}(\mathbf{x})/\partial x_1$ is the magnitude of the negative of the (zeroth-order) mean head gradient and is given as a constant, and the mean flow is assumed to be aligned with x_1. Note that summation for repeated indices is implied. The second-order mean head correction term is governed by the following equation:

$$\frac{\partial^2 \langle h^{(2)}(\mathbf{x}) \rangle}{\partial x_i^2} = -\left\langle \frac{\partial h^{(1)}(\mathbf{x})}{\partial x_i} \frac{\partial f'(\mathbf{x})}{\partial x_i} \right\rangle \tag{3.268}$$

It has been shown in previous sections that under uniform mean flow in unbounded domains the solution to Eq. (3.268) is $\langle h^{(2)} \rangle = 0$. It can be shown that the expected value of the third-order head term $h^{(3)}$ is zero if the log hydraulic conductivity f is Gaussian (see Eq. (3.273) below). Hence, the solution of $\langle h \rangle = h^{(0)}$ as being a linear trend is correct up to third order in σ_f for uniform mean flows in unbounded domains.

Up to third order in σ_f, the head fluctuation reads as

$$h'(\mathbf{x}) = h^{(1)}(\mathbf{x}) + h^{(2)}(\mathbf{x}) + h^{(3)}(\mathbf{x}) \tag{3.269}$$

The first-order head term can be expressed with the aid of Green's function,

$$h^{(1)}(\mathbf{x}) = -J \int G(\mathbf{x}', \mathbf{x}) \frac{\partial f'(\mathbf{x}')}{\partial x_1'} d\mathbf{x}' \tag{3.270}$$

where Green's function G is the solution of

$$\nabla_{\mathbf{x}'}^2 G(\mathbf{x}', \mathbf{x}) = -\delta(\mathbf{x}' - \mathbf{x}) \tag{3.271}$$

Notice that Green's function $G(\mathbf{x}', \mathbf{x})$ is different from the mean Green's function $G^{(0)}(\mathbf{x}', \mathbf{x})$ defined in Eq. (3.61). However, for a stationary medium, these two Green's functions are related in a simple way: $G(\mathbf{x}', \mathbf{x}) = K_G G^{(0)}(\mathbf{x}', \mathbf{x})$. Equation (3.270) may be directly written from Eq. (3.71) by using the specific conditions under consideration. By setting $n = 2$ in Eq. (3.267), substituting Eq. (3.270) and using the Green's function representation, one obtains [e.g., Hsu *et al.*, 1996],

$$h^{(2)}(\mathbf{x}) = J \iint G(\mathbf{x}', \mathbf{x}) \frac{\partial G(\mathbf{x}'', \mathbf{x}')}{\partial x_i'} \frac{\partial f'(\mathbf{x}')}{\partial x_i'} \frac{\partial f'(\mathbf{x}'')}{\partial x_1''} dx' \, dx'' \qquad (3.272)$$

Similarly, one has

$$h^{(3)}(\mathbf{x}) = -J \iiint G(\mathbf{x}', \mathbf{x}) \frac{\partial G(\mathbf{x}'', \mathbf{x}')}{\partial x_i'} \frac{\partial G(\mathbf{x}''', \mathbf{x}'')}{\partial x_j''} \frac{\partial f'(\mathbf{x}')}{\partial x_i'}$$

$$\cdot \frac{\partial f'(\mathbf{x}'')}{\partial x_j''} \frac{\partial f'(\mathbf{x}''')}{\partial x_1'''} dx' \, dx'' \, dx''' \qquad (3.273)$$

When f is Gaussian, it is seen that the mean of the third- (and higher odd-) order head term is zero. Note that the mean of an odd-order head term is not necessarily zero if f is not Gaussian. One may write the head covariance C_h expression up to second order in σ_f^2 (fourth order in σ_f) as

$$C_h^{[2]}(\mathbf{x}, \boldsymbol{\chi}) = \langle h^{(1)}(\mathbf{x}) h^{(1)}(\boldsymbol{\chi}) \rangle + \langle h^{(3)}(\mathbf{x}) h^{(1)}(\boldsymbol{\chi}) \rangle$$

$$+ \langle h^{(2)}(\mathbf{x}) h^{(2)}(\boldsymbol{\chi}) \rangle + \langle h^{(1)}(\mathbf{x}) h^{(3)}(\boldsymbol{\chi}) \rangle \qquad (3.274)$$

where the properties of Gaussian f' have been invoked such that terms including the odd moments of f' are zero. Note that the superscript $[n]$ denotes the overall order to which the head covariance is evaluated, whereas the superscript (n) stands for the order of each term. In Eq. (3.274), the first term on the right-hand side is the first-order head covariance evaluated so far and the remaining are the second-order correction terms.

As seen in Eqs. (3.19)–(3.21), the flux can be also expanded into terms $\mathbf{q}^{(n)}$ of various orders. Here we write explicitly such terms up to fourth order in σ_f,

$$\mathbf{q}^{(0)}(\mathbf{x}) = K_G \mathbf{J} \qquad (3.275)$$

$$\mathbf{q}^{(1)}(\mathbf{x}) = -K_G \left[\nabla h^{(1)}(\mathbf{x}) - \mathbf{J} f'(\mathbf{x}) \right] \qquad (3.276)$$

$$\mathbf{q}^{(2)}(\mathbf{x}) = -K_G \left[\nabla h^{(2)}(\mathbf{x}) + f'(\mathbf{x}) \nabla h^{(1)}(\mathbf{x}) - \frac{\mathbf{J}}{2} f'^2(\mathbf{x}) \right] \qquad (3.277)$$

$$\mathbf{q}^{(3)}(\mathbf{x}) = -K_G \left[\nabla h^{(3)}(\mathbf{x}) + f'(\mathbf{x}) \nabla h^{(2)}(\mathbf{x}) + \frac{f'^2(\mathbf{x})}{2} \nabla h^{(1)}(\mathbf{x}) - \frac{\mathbf{J}}{6} f'^3(\mathbf{x}) \right] \qquad (3.278)$$

$$\mathbf{q}^{(4)}(\mathbf{x}) = -K_G \left[\nabla h^{(4)}(\mathbf{x}) + f'(\mathbf{x}) \nabla h^{(3)}(\mathbf{x}) + \frac{f'^2(\mathbf{x})}{2} \nabla h^{(2)}(\mathbf{x}) \right.$$

$$\left. + \frac{f'^3(\mathbf{x})}{6} \nabla h^{(1)}(\mathbf{x}) - \frac{\mathbf{J}}{24} f'^4(\mathbf{x}) \right] \qquad (3.279)$$

where $\mathbf{J} = (J, 0, 0)^T$. It is seen that these flux terms can be expressed with the explicit representations of the head terms in Eqs. (3.270)–(3.273).

It can be shown that for a Gaussian f, $\langle \mathbf{q}^{(1)} \rangle = \langle \mathbf{q}^{(3)} \rangle = 0$ and

$$\langle \mathbf{q}^{(2)} \rangle = -K_G \left[\langle f'(\mathbf{x}) \nabla h^{(1)}(\mathbf{x}) \rangle - \frac{\mathbf{J}}{2} \sigma_f^2 \right] \tag{3.280}$$

$$\langle \mathbf{q}^{(4)} \rangle = -K_G \left[\langle f'(\mathbf{x}) \nabla h^{(3)}(\mathbf{x}) \rangle + \frac{1}{2} \langle f'^2(\mathbf{x}) \nabla h^{(2)}(\mathbf{x}) \rangle \right.$$
$$\left. + \frac{1}{6} \langle f'^3(\mathbf{x}) \nabla h^{(1)}(\mathbf{x}) \rangle - \frac{\mathbf{J}}{8} \sigma_f^4 \right] \tag{3.281}$$

and higher-order terms can be written similarly. In the above, the properties of Gaussian f have been used. Dagan [1993] evaluated Eqs. (3.280) and (3.281) analytically and obtained, for isotropic covariances C_f, the following general results:

$$\langle \mathbf{q}^{(2)} \rangle = K_G \sigma_f^2 \mathbf{J} \left(\frac{1}{2} - \frac{1}{d} \right) \tag{3.282}$$

$$\langle \mathbf{q}^{(4)} \rangle = K_G \sigma_f^4 \frac{\mathbf{J}}{2} \left(\frac{1}{2} - \frac{1}{d} \right)^2 \tag{3.283}$$

where d is the space dimensionality. As mentioned in Section 3.5.1, the results in Eqs. (3.282) and (3.283) agree with, up to fourth order in σ_f, the Landau–Lifshitz conjecture of Eq. (3.161). De Wit [1995] evaluated the sixth-order flux term $\langle \mathbf{q}^{(6)} \rangle$ for uniform mean flow in lognormal, statistically isotropic conductivity fields and obtained the following result:

$$\langle \mathbf{q}^{(6)} \rangle = K_G \sigma_f^6 \frac{\mathbf{J}}{6} \left(\frac{1}{2} - \frac{1}{d} \right)^3 + T(C_f, d) \tag{3.284}$$

where $T(C_f, d)$ is in general a function of the covariance C_f and the dimensionality d. It is found that for $d = 3$ the T term is nonzero, whereas $T \equiv 0$ for $d = 1$ and 2. This is consistent with the well-known fact that the Landau–Lifshitz conjecture is correct in one and two dimensions.

With the expected values of the flux terms, the flux fluctuations of various orders are obtained as $\mathbf{q}^{(n)}(\mathbf{x}) - \langle \mathbf{q}^{(n)} \rangle$. Then, the covariances of flux, up to second order in σ_f^2, are given as [Hsu et al., 1996; Hsu and Neuman, 1997]

$$C_{q_{ij}}^{[2]}(\mathbf{r}) = C_{q_{ij}}^{[1]}(\mathbf{r}) + C_{q_{ij}}^{(2)}(\mathbf{r}) \tag{3.285}$$

$$C_{q_{ij}}^{(2)}(\mathbf{r}) = C_{q_{ij}}^{*}(\mathbf{r}) + C_{q_{ij}}^{**}(\mathbf{r}) \tag{3.286}$$

where $\mathbf{r} = \mathbf{x} - \chi$, $C_{q_{ij}}^{[1]}(\mathbf{r})$ is the first-order flux covariance given in Eq. (3.23) and evaluated in the preceding sections, $C_{q_{ij}}^{(2)}(\mathbf{r})$ is a second-order correction, $C_{q_{ij}}^{*}(\mathbf{r})$ involves only first-order (in σ_f) head fluctuations, $h^{(1)}$, and $C_{q_{ij}}^{**}(\mathbf{r})$ involves higher-order head fluctuations, $h^{(2)}$ and $h^{(3)}$. The explicit expressions of $C_{q_{ij}}^{*}(\mathbf{r})$ and $C_{q_{ij}}^{**}(\mathbf{r})$ are tedious and are hence not given here. The reader is referred to Hsu et al. [1996, (A16) and (A20)–(A27)].

A number of recent studies evaluated the second-order correction $C_{q_{ij}}^{(2)}(\mathbf{r})$ for various cases of uniform mean flows in unbounded domains using different approaches. Deng and Cushman [1995] evaluated the second-order correction term numerically with a fast Fourier transform for three-dimensional, uniform mean flow and concluded that the contribution of $C_{q_{ij}}^{**}(\mathbf{r})$ should be negligible. Hsu $et\ al.$ [1996] evaluated the $C_{q_{ij}}^{*}(\mathbf{r})$ term analytically and the $C_{q_{ij}}^{**}(\mathbf{r})$ term numerically in two dimensions. The latter authors found that $C_{q_{ij}}^{*}(\mathbf{r})$ and $C_{q_{ij}}^{**}(\mathbf{r})$ are of similar order but opposite sign and thus argued that the conclusion by Deng and Cushman [1995] on the effects of head fluctuation terms higher than $h^{(1)}$ could be premature. This assertion was later confirmed by Hsu and Neuman [1997], who evaluated the second-order correction terms for uniform mean flows in two- and three-dimensional anisotropic media, and by Deng and Cushman [1998], who revisited their previous study. More recently, Hsu and Lamb [2000] presented an analytical, closed-form expression for the second-order correction term $C_{q_{ij}}^{**}(\mathbf{r})$ involving higher-order head fluctuations. Illustrated in Fig. 3.17 are the effects of second-order corrections to the flux covariances in both two- and

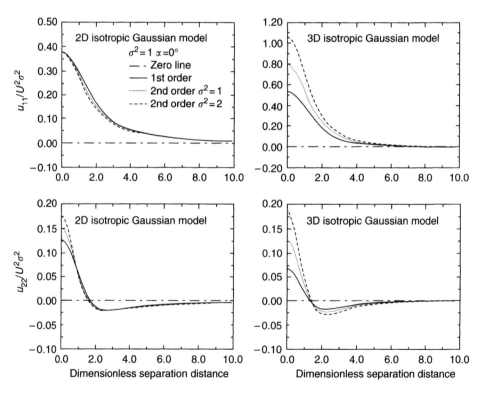

Figure 3.17. First- and second-order two- and three-dimensional velocity covariance ($u_{ii} = C_{q_{ii}}/\phi^2$) in isotropic media along the longitudinal direction for $\sigma_f^2 = 1$ and 2. (Adapted from Hsu and Neuman [1997]. Copyright 1997 by the American Geophysical Union.)

three-dimensional isotropic media with a Gaussian covariance C_f. It is seen that velocity variances (at lag zero) are generally larger when approximated consistently to second order in σ_f^2 than to first order. However, the second-order effects are more pronounced in three than in two dimensions. The ratio between second- and first-order variance approximations is larger in three than in two dimensions, and larger for the transverse than for longitudinal flux. In particular, the ratio for the longitudinal flux variances is 1.5 and 2 in three dimensions for $\sigma_f^2 = 1$ and 2, respectively, while the corresponding two-dimensional ratios are about 1. The transverse flux variance ratios are 1.9 and 2.8 for $\sigma_f^2 = 1$ and 2, respectively, in three dimensions, whereas the two-dimensional counterparts are only 1.2 and 1.4.

It is seen from Fig. 3.17 that the second-order effects increase with σ_f^2. However, it is not guaranteed that the second-order results are always better than the first-order counterparts, depending on the validity of the perturbative expansions invoked. For small variances of log hydraulic conductivity (or transmissivity), i.e., $\sigma_f^2 \ll 1$, it is well known that the perturbative approaches are valid and the expansions converge; it is thus expected that the second-order approximations are more accurate than the first-order ones. On the other hand, the effects of second- and perhaps higher-order corrections are small when σ_f^2 is small. For extremely large variances, the convergence of the expansions and hence the validity of the perturbative approaches are questionable. Therefore, it is expected that the second- and higher-order studies are most useful for moderate variances. However, a firm conclusion on the validity range of perturbative approaches cannot be reached without innovative theoretical studies or carefully designed, high-resolution Monte Carlo simulations.

The fact that second-order studies are mathematically involved and tedious has so far restricted the solutions of second-order equations to some simple cases of uniform mean flow in unbounded domains. In situations involving bounded domains and/or nonuniform mean flow, a fully numerical moment equation approach may be appropriate.

3.8 SPACE-STATE METHOD

In the preceding sections, it is seen that numerical moment equation approaches are indispensable for solving stochastic flow problems with various complications such as finite and complex domain boundaries, the presence of sinks and/or sources, and the appearance of medium nonstationary features or multiscale structures. In these numerical approaches, the equations governing the first two statistical moments of flow quantities are derived with perturbative techniques, and then the moment equations are solved by numerical schemes such as finite differences and finite elements. In this section, we introduce an alternative numerical approach, in which the original governing equation is first discretized on specified grids using finite differences or finite elements, and then the resulting (discretized) equation, the so-called *space-state equation*, is used to derive the statistical moments of the flow quantities. This approach is often termed the *vector space-state approach* and is usually combined with *sensitivity analyses* [Dettinger and Wilson, 1981; Hoeksema and Kitanidis, 1984].

3.8.1 Space-State Equations

Numerical discretization of the original (stochastic) partial differential equation (3.5) (or any other linear equations) using finite differences or finite elements usually leads to the following system of linear algebraic equations:

$$\mathbf{ch} = \mathbf{b}, \qquad \mathbf{h} = \mathbf{c}^{-1}\mathbf{b} \tag{3.287}$$

where $\mathbf{h} = (h_1, h_2, \ldots, h_N)^T$ is the vector of N unknown heads, $h_i = h(\mathbf{x}_i)$ is the head at the node $\mathbf{x}_i$, N is the number of nodes on the numerical grid, $\mathbf{c}$ is the $N \times N$ coefficient matrix, $\mathbf{c}^{-1}$ is the inverse of $\mathbf{c}$, and $\mathbf{b}$ is the vector containing the information regarding boundary conditions and sink/source terms. The matrix $\mathbf{c}$ depends on both the medium and fluid properties and the discretization scheme. Equation (3.287) is referred to as the space-state equation [Dettinger and Wilson,1981].

When the medium properties such as the hydraulic conductivity are treated as random space functions, so is the matrix $\mathbf{c}$. In this section, the boundary conditions and the sink/source terms are assumed to be known with certainty, the vector $\mathbf{b}$ is thus deterministic (see Dettinger and Wilson [1981] for the alternative treatment). The dependency of $\mathbf{c}$ on the vector $\mathbf{f} = (f_1, f_2, \ldots, f_N)^T$, where $f_i = \ln K_S(\mathbf{x}_i)$ is the log-transformed hydraulic conductivity (or transmissivity) at the node $\mathbf{x}_i$, can be expressed as $\mathbf{c} = \mathbf{c}(\mathbf{f})$. In the same manner, the random head vector can be expressed as $\mathbf{h} = \mathbf{h}(\mathbf{f})$. As for the (continuous, in the mean square sense) random field $f = \ln K_S(\mathbf{x})$, the random (discrete) vector $\mathbf{f}$ can be decomposed into its mean and fluctuation: $\mathbf{f} = \langle \mathbf{f} \rangle + \mathbf{f}'$ where $\langle \mathbf{f} \rangle = (\langle f_1 \rangle, \langle f_2 \rangle, \ldots, \langle f_N \rangle)^T$ and $\mathbf{f}' = (f_1', f_2', \ldots, f_N')^T$. As done by Dettinger and Wilson [1981], the random head vector can be expanded about $\langle \mathbf{f} \rangle$ by the vector form of Taylor expansion,

$$\mathbf{h} = \mathbf{h}(\langle \mathbf{f} \rangle) + \mathbf{D}_{\mathrm{fh}}\mathbf{f}' + \cdots \tag{3.288}$$

where $\mathbf{D}_{\mathrm{fh}}$ is the derivative of $\mathbf{h}$ with respect to the transpose of $\mathbf{f}$, evaluated at $\langle \mathbf{f} \rangle$. $\mathbf{D}_{\mathrm{fh}}$ is a matrix, called a *matrix derivative*. In many studies, it is often referred to as the *sensitivity matrix* of heads with respect to log-transformed hydraulic conductivities.

To first order, the mean head vector is obtained from Eq. (3.288) by taking expectation,

$$\langle \mathbf{h} \rangle = \mathbf{h}(\langle \mathbf{f} \rangle) \tag{3.289}$$

where $\mathbf{h}(\langle \mathbf{f} \rangle)$ can be solved from the second equation of Eq. (3.287) by replacing the random $\mathbf{f}$ with its mean vector. The first-order head fluctuation is given as

$$\mathbf{h}' = \mathbf{D}_{\mathrm{fh}}\mathbf{f}' \tag{3.290}$$

which is used to construct the head covariance matrix,

$$\mathbf{C_h} = \langle \mathbf{h}'\mathbf{h}'^T \rangle = \mathbf{D}_{\mathrm{fh}}\mathbf{C_f}\mathbf{D}_{\mathrm{fh}}^T \tag{3.291}$$

In Eq. (3.291), $\mathbf{C_f}$ is the $N \times N$ covariance matrix of $\mathbf{f}$, which is given as input. The cross-covariance matrix $\mathbf{C_{fh}}$ can be expressed similarly as

$$\mathbf{C_{fh}} = \langle \mathbf{f}' \mathbf{h}'^T \rangle = \mathbf{C_f} \mathbf{D}_{fh}^T \tag{3.292}$$

It is seen from Eqs. (3.291) and (3.292) that the evaluation of the second moments of heads reduces to the derivation of the sensitivity matrix $\mathbf{D_{fh}}$. Since the elements $\partial h(\mathbf{x}_l)/\partial f_k$ of $\mathbf{D_{fh}}$ are evaluated at $\langle \mathbf{f} \rangle$, it can be expressed as $\mathbf{D_{fh}}(\langle \mathbf{f} \rangle, \langle \mathbf{h} \rangle)$.

The sensitivity matrix may be evaluated with a direct method of sensitivity analysis or indirect methods such as the *adjoint state method*. In the next subsection, while the direct method is touched upon briefly, the adjoint state method will be discussed in detail.

3.8.2 Adjoint State Equations

The adjoint state method has been widely used for solving subsurface flow problems [Neuman, 1980; Sykes *et al.*, 1985; Sun and Yeh, 1992]. Below we follow the procedure outlined by Sykes *et al.* [1985] to derive the *adjoint state equations* for evaluating the sensitivity matrix.

The original (primary) flow equation can be rewritten from Eqs. (3.1)–(3.4) as

$$\frac{\partial}{\partial x_i} \left[K_S(\mathbf{x}) \frac{\partial h(\mathbf{x})}{\partial x_i} \right] = -g(\mathbf{x})$$

$$h(\mathbf{x}) = H_B(\mathbf{x}), \quad \mathbf{x} \in \Gamma_D \tag{3.293}$$

$$n_i(\mathbf{x}) K_S(\mathbf{x}) \frac{\partial h(\mathbf{x})}{\partial x_i} = -Q(\mathbf{x}), \quad \mathbf{x} \in \Gamma_N$$

where summation for repeated indices i is implied. Recall that $K_S(\mathbf{x}) = \exp[f(\mathbf{x})]$. For sensitivity analysis one may define a *performance function* or *response function* as

$$P(h, \mathbf{f}) = \int_\Omega R(h, \mathbf{f}, \mathbf{x}) \, d\mathbf{x} \tag{3.294}$$

where Ω is the domain bounded by the union of Γ_D and Γ_N, $\mathbf{f}$ is the vector of the log hydraulic conductivity values at the N nodes of a numerical grid, and R is a function yet to be specified.

The sensitivity of the performance function P to f is given as

$$\frac{\partial P(h, \mathbf{f})}{\partial f_k} = \int_\Omega \left[\frac{\partial R(h, \mathbf{f}; \mathbf{x})}{\partial f_k} + \frac{\partial R(h, \mathbf{f}; \mathbf{x})}{\partial h} \gamma \right] d\mathbf{x} \tag{3.295}$$

where $\gamma(h, f_k; \mathbf{x}) = \partial h(\mathbf{x})/\partial f_k$ is the so-called *state sensitivity*, i.e., the sensitivity of head to the parameter f_k. The state sensitivity may be evaluated directly from the following equation, obtained by differentiating Eq. (3.293) with respect to f_k,

$$\frac{\partial}{\partial x_i}\left[\frac{\partial K_s(\mathbf{x})}{\partial f_k}\frac{\partial h(\mathbf{x})}{\partial x_i} + K_S(\mathbf{x})\frac{\partial \gamma(h, f_k; \mathbf{x})}{\partial x_i}\right] = -\frac{\partial g(\mathbf{x})}{\partial f_k}$$

$$\gamma(h, f_k; \mathbf{x}) = \frac{\partial H_B(\mathbf{x})}{\partial f_k}, \quad \mathbf{x} \in \Gamma_D \tag{3.296}$$

$$n_i(\mathbf{x})\left[\frac{\partial K_S(\mathbf{x})}{\partial f_k}\frac{\partial h(\mathbf{x})}{\partial x_i} + K_S(\mathbf{x})\frac{\partial \gamma(h, f_k; \mathbf{x})}{\partial x_i}\right] = -\frac{\partial Q(\mathbf{x})}{\partial f_k}, \quad \mathbf{x} \in \Gamma_N$$

Equation (3.296) can be rewritten as

$$\frac{\partial}{\partial x_i}\left[K_S(\mathbf{x})\frac{\partial \gamma(h, f_k; \mathbf{x})}{\partial x_i}\right] = -\frac{\partial g(\mathbf{x})}{\partial f_k} - \frac{\partial}{\partial x_i}\left[\frac{\partial K_S(\mathbf{x})}{\partial f_k}\frac{\partial h(\mathbf{x})}{\partial x_i}\right]$$

$$\gamma(h, f_k; \mathbf{x}) = \frac{\partial H_B(\mathbf{x})}{\partial f_k}, \quad \mathbf{x} \in \Gamma_D \tag{3.297}$$

$$n_i(\mathbf{x})K_S(\mathbf{x})\frac{\partial \gamma(h, f_k; \mathbf{x})}{\partial x_i} = -\frac{\partial Q(\mathbf{x})}{\partial f_k} - n_i(\mathbf{x})\frac{\partial K_S(\mathbf{x})}{\partial f_k}\frac{\partial h(\mathbf{x})}{\partial x_i}, \quad \mathbf{x} \in \Gamma_N$$

This is the *direct sensitivity equation*, which has a similar structure to its original equation (3.295). Recalling that the sensitivity matrix $\mathbf{D_{fh}}$ is evaluated with $\langle \mathbf{f} \rangle$ and $\langle \mathbf{h} \rangle$, $K_S(\mathbf{x})$ and $h(\mathbf{x})$ in Eq. (3.297) are to be replaced with $\exp[\langle f(\mathbf{x})\rangle]$ and $\langle h(\mathbf{x})\rangle$. Therefore, the right-hand sides of Eq. (3.297) are known quantities, and the state sensitivity γ can be solved from this equation numerically. For each specific parameter f_k (in our case, the log hydraulic conductivity at the node k or in the element k), Eq. (3.294) needs to be solved once to evaluate the sensitivity of the performance function P to f_k. This direct sensitivity approach is time consuming if the number of parameters f_k is large.

An alternative approach is to formulate *adjoint equations* of the partial differential equations for γ. Multiplying the first equation of (3.297) by an arbitrary differentiable function $\gamma^*(\mathbf{x})$ and integrating over Ω leads to

$$\int_\Omega\left\{\gamma^*(\mathbf{x})\frac{\partial}{\partial x_i}\left[K_S(\mathbf{x})\frac{\partial \gamma(h, f_k; \mathbf{x})}{\partial x_i}\right] + \gamma^*(\mathbf{x})\frac{\partial g(\mathbf{x})}{\partial f_k}\right.$$

$$\left. + \gamma^*(\mathbf{x})\frac{\partial}{\partial x_i}\left[\frac{\partial K_S(\mathbf{x})}{\partial f_k}\frac{\partial h(\mathbf{x})}{\partial x_i}\right]\right\} d\mathbf{x} = 0 \tag{3.298}$$

Applying Green's identity to Eq. (3.298) and utilizing the boundary conditions of Eq. (3.296), Eq. (3.298) can be rewritten as [Sykes *et al.*, 1985]

$$\int_\Omega\left\{\gamma(h, f_k; \mathbf{x})\frac{\partial}{\partial x_i}\left[K_S(\mathbf{x})\frac{\partial \gamma^*(\mathbf{x})}{\partial x_i}\right] + \gamma^*(\mathbf{x})\frac{\partial g(\mathbf{x})}{\partial f_k} - \frac{\partial \gamma^*(\mathbf{x})}{\partial x_i}\frac{\partial K_S(\mathbf{x})}{\partial f_k}\frac{\partial h(\mathbf{x})}{\partial x_i}\right\} d\mathbf{x}$$

$$+ \int_\Gamma\left[\gamma(h, f_k; \mathbf{x})K_S(\mathbf{x})\frac{\partial \gamma^*(\mathbf{x})}{\partial x_i}n_i(\mathbf{x}) + \gamma^*(\mathbf{x})\frac{\partial Q(\mathbf{x})}{\partial f_k}\right] d\mathbf{x} = 0 \tag{3.299}$$

Adding Eq. (3.299) to Eq. (3.295) yields

$$
\frac{\partial P(h, \mathbf{f})}{\partial f_k} = \int_\Omega \frac{\partial R(h, \mathbf{f}; \mathbf{x})}{\partial f_k} \, d\mathbf{x}
$$

$$
+ \int_\Omega \gamma(h, f_k; \mathbf{x}) \left\{ \frac{\partial R(h, \mathbf{f}; \mathbf{x})}{\partial h} + \frac{\partial}{\partial x_i} \left[K_S(\mathbf{x}) \frac{\partial \gamma^*(\mathbf{x})}{\partial x_i} \right] \right\}
$$

$$
+ \int_\Omega \left\{ \gamma^*(\mathbf{x}) \frac{\partial g(\mathbf{x})}{\partial f_k} - \frac{\partial \gamma^*(\mathbf{x})}{\partial x_i} \frac{\partial K_S(\mathbf{x})}{\partial f_k} \frac{\partial h(\mathbf{x})}{\partial x_i} \right\} d\mathbf{x}
$$

$$
+ \int_\Gamma \left[\gamma(h, f_k; \mathbf{x}) K_S(\mathbf{x}) \frac{\partial \gamma^*(\mathbf{x})}{\partial x_i} n_i(\mathbf{x}) + \gamma^*(\mathbf{x}) \frac{\partial Q(\mathbf{x})}{\partial f_k} \right] d\mathbf{x} \qquad (3.300)
$$

To evaluate Eq. (3.300), one must supply the state sensitivity γ or eliminate its contribution there. The latter is achieved by letting the arbitrary function γ^* satisfy the following equation:

$$
\frac{\partial}{\partial x_i} \left[K_S(\mathbf{x}) \frac{\partial \gamma^*(\mathbf{x})}{\partial x_i} \right] + \frac{\partial R(h, \mathbf{f}; \mathbf{x})}{\partial h} = 0
$$

$$
\gamma^*(\mathbf{x}) = 0, \quad \mathbf{x} \in \Gamma_D \qquad (3.301)
$$

$$
n_i(\mathbf{x}) K_S(\mathbf{x}) \frac{\partial \gamma^*(\mathbf{x})}{\partial x_i} = 0, \quad \mathbf{x} \in \Gamma_N
$$

This is the so-called adjoint state equation, and γ^* is the adjoint state of γ. With Eq. (3.301), Eq. (3.300) reduces to

$$
\frac{\partial P(h, \mathbf{f})}{\partial f_k} = \int_\Omega \frac{\partial R(h, \mathbf{f}; \mathbf{x})}{\partial f_k} \, d\mathbf{x}
$$

$$
+ \int_\Omega \left\{ \gamma^*(\mathbf{x}) \frac{\partial g(\mathbf{x})}{\partial f_k} - \frac{\partial \gamma^*(\mathbf{x})}{\partial x_i} \frac{\partial K_S(\mathbf{x})}{\partial f_k} \frac{\partial h(\mathbf{x})}{\partial x_i} \right\} d\mathbf{x}
$$

$$
+ \int_{\Gamma_D} \frac{\partial H_B(\mathbf{x})}{\partial f_k} K_S(\mathbf{x}) \frac{\partial \gamma^*(\mathbf{x})}{\partial x_i} n_i(\mathbf{x}) \, d\mathbf{x} + \int_{\Gamma_N} \gamma^*(\mathbf{x}) \frac{\partial Q(\mathbf{x})}{\partial f_k} \, d\mathbf{x} \qquad (3.302)
$$

Let function R in Eq. (3.295) be

$$
R(h, \mathbf{f}; \mathbf{x}) = h(\mathbf{x}) \delta(\mathbf{x} - \mathbf{x}_l) \qquad (3.303)
$$

where $\mathbf{x}_l$ is a node on the grid, an observation location, or a point of interest. In turn, the performance function P is given as $P = h(\mathbf{x}_l)$. Hence, Eq. (3.302) becomes

$$
\frac{\partial h(\mathbf{x}_l)}{\partial f_k} = \int_\Omega \left\{ \gamma^*(\mathbf{x}) \frac{\partial g(\mathbf{x})}{\partial f_k} - \frac{\partial \gamma^*(\mathbf{x})}{\partial x_i} \frac{\partial K_S(\mathbf{x})}{\partial f_k} \frac{\partial h(\mathbf{x})}{\partial x_i} \right\} d\mathbf{x}
$$

$$
+ \int_{\Gamma_D} \frac{\partial H_B(\mathbf{x})}{\partial f_k} K_S(\mathbf{x}) \frac{\partial \gamma^*(\mathbf{x})}{\partial x_i} n_i(\mathbf{x}) \, d\mathbf{x} + \int_{\Gamma_N} \gamma^*(\mathbf{x}) \frac{\partial Q(\mathbf{x})}{\partial f_k} \, d\mathbf{x} \qquad (3.304)
$$

With Eq. (3.303), the adjoint state γ^* becomes Green's function for the Poisson operator in Eq. (3.301). For a given function $R(h, \mathbf{f}; \mathbf{x})$ such as Eq. (3.303), the adjoint state equation (3.301) has to be solved only once. Of course, Eq. (3.301) needs to be solved again if the performance function changes, for example, from $P = h(\mathbf{x}_l)$ to $P = h(\mathbf{x}_m)$.

The sensitivity matrix $\mathbf{D}_{\mathbf{fh}}(\langle \mathbf{f} \rangle, \langle \mathbf{h} \rangle)$ can be evaluated with Eq. (3.304), where the adjoint state γ^* is numerically evaluated from Eq. (3.301). Notice that both Eqs. (3.301) and (3.304) need to be evaluated at $\mathbf{f} = \langle \mathbf{f} \rangle$ and $\mathbf{h} = \langle \mathbf{h} \rangle$. With the sensitivity matrix $\mathbf{D}_{\mathbf{fh}}(\langle \mathbf{f} \rangle, \langle \mathbf{h} \rangle)$, the head covariance matrices $\mathbf{C}_{\mathbf{fh}}$ and $\mathbf{C}_{\mathbf{h}}$ are obtained with Eqs. (3.291) and (3.292).

3.9 ADOMIAN DECOMPOSITION

In this section, we discuss the application of *Adomian decomposition* in deriving statistical moment equations for flow in porous media. Adomian decomposition, or *expansion in Neumann series*, is a method of successive approximations [Adomian, 1983; Zeitoun and Braester, 1991]. The variants of this approach include the method of Volterra–Wiener series [Schetzen, 1980; Markov, 1987], that of hierarchy closure approximation [Bharucha-Reid, 1972], and Picard iterations. These methods are commonly used to solve deterministic and stochastic partial differential or integrodifferential equations. Zeitoun and Braester [1991] and others have recently applied the Adomian decomposition to flow in random porous media, as an alternative to the (log hydraulic conductivity f-based) perturbative expansion, discussed in the previous sections. Although both approaches involve writing the random head h into a formal series, the Adomian decomposition works directly with the hydraulic conductivity $K_S(\mathbf{x})$, while the f-based approach expands the $\exp[f(\mathbf{x})]$ term by Taylor series. Hence, it is believed by some authors that the Adomian decomposition is not a perturbation method and is thus superior to the f-based perturbative expansion. Below we first introduce the methodology and then discuss the issue of whether or not the approach is nonperturbative.

3.9.1 Neumann Series Expansion

The governing equation for steady-state flow can be rewritten from Eqs. (3.1)–(3.4) as

$$\nabla \cdot [K_S(\mathbf{x}) \nabla h(\mathbf{x})] = -g(\mathbf{x})$$
$$h(\mathbf{x}) = H_B(\mathbf{x}), \quad \mathbf{x} \in \Gamma_D \qquad (3.305)$$
$$K_S(\mathbf{x}) \nabla h(\mathbf{x}) \cdot \mathbf{n}(\mathbf{x}) = -Q(\mathbf{x}), \quad \mathbf{x} \in \Gamma_N$$

For simplicity, the forcing term $g(\mathbf{x})$ and the boundary terms $H_B(\mathbf{x})$ and $Q(\mathbf{x})$ are treated as deterministic quantities. Hence, the hydraulic conductivity $K_S(\mathbf{x})$ is the only source of randomness in Eq. (3.305). Let us define a new variable, $\nu(\mathbf{x}) = K_S'(\mathbf{x})/\langle K_S(\mathbf{x}) \rangle$. It is clear that the standard deviation of ν, σ_ν, is equal to the coefficient of variation C_v of K_S, defined as $C_v = \sigma_{K_S}/\langle K_S \rangle$. With $v(\mathbf{x})$ Eq. (3.305) can be written as

$$\nabla \cdot [\langle K_S(\mathbf{x}) \rangle \nabla h(\mathbf{x})] = -g(\mathbf{x}) - \nabla \cdot [\langle K_S(\mathbf{x}) \rangle v(\mathbf{x}) \nabla h(\mathbf{x})]$$

$$h(\mathbf{x}) = H_B(\mathbf{x}), \quad \mathbf{x} \in \Gamma_D \tag{3.306}$$

$$\langle K_S(\mathbf{x}) \rangle [1 + v(\mathbf{x})] \nabla h(\mathbf{x}) \cdot \mathbf{n}(\mathbf{x}) = -Q(\mathbf{x}), \quad \mathbf{x} \in \Gamma_N$$

Let us define a Green's function $G(\boldsymbol{\chi}, \mathbf{x})$ for the Poisson operator and the boundary conditions in Eq. (3.306),

$$\nabla_{\boldsymbol{\chi}} \cdot \left[\langle K_S(\boldsymbol{\chi}) \rangle \nabla_{\boldsymbol{\chi}} G(\boldsymbol{\chi}, \mathbf{x}) \right] = -\delta(\boldsymbol{\chi} - \mathbf{x})$$

$$G(\boldsymbol{\chi}, \mathbf{x}) = 0, \quad \boldsymbol{\chi} \in \Gamma_D \tag{3.307}$$

$$\langle K_S(\boldsymbol{\chi}) \rangle \nabla_{\boldsymbol{\chi}} G(\boldsymbol{\chi}, \mathbf{x}) \cdot \mathbf{n}(\boldsymbol{\chi}) = 0, \quad \boldsymbol{\chi} \in \Gamma_N$$

Note that this Green's function is deterministic. Rewriting Eq. (3.306) in terms of $\boldsymbol{\chi}$, multiplying by $G(\boldsymbol{\chi}, \mathbf{x})$ and integrating over the domain Ω leads to

$$\int_\Omega \nabla_{\boldsymbol{\chi}} \cdot \left[\langle K_S(\boldsymbol{\chi}) \rangle \nabla_{\boldsymbol{\chi}} h(\boldsymbol{\chi}) \right] G(\boldsymbol{\chi}, \mathbf{x}) \, d\boldsymbol{\chi}$$

$$= -\int_\Omega g(\boldsymbol{\chi}) G(\boldsymbol{\chi}, \mathbf{x}) \, d\boldsymbol{\chi} - \int_\Omega \nabla_{\boldsymbol{\chi}} \cdot \left[\langle K_S(\boldsymbol{\chi}) \rangle v(\boldsymbol{\chi}) \nabla_{\boldsymbol{\chi}} h(\boldsymbol{\chi}) \right] G(\boldsymbol{\chi}, \mathbf{x}) \, d\boldsymbol{\chi} \tag{3.308}$$

Applying Green's identity, twice to the left-hand side and once to the second term on the right-hand side, and using the boundary conditions in Eq. (3.307) yields

$$h(\mathbf{x}) = \int_\Omega g(\boldsymbol{\chi}) G(\boldsymbol{\chi}, \mathbf{x}) \, d\boldsymbol{\chi} - \int_\Omega \langle K_S(\boldsymbol{\chi}) \rangle v(\boldsymbol{\chi}) \nabla_{\boldsymbol{\chi}} G(\boldsymbol{\chi}, \mathbf{x}) \cdot \nabla_{\boldsymbol{\chi}} h(\boldsymbol{\chi}) \, d\boldsymbol{\chi}$$

$$- \int_{\Gamma_D} \langle K_S(\boldsymbol{\chi}) \rangle h(\boldsymbol{\chi}) \nabla_{\boldsymbol{\chi}} G(\boldsymbol{\chi}, \mathbf{x}) \cdot \mathbf{n}(\boldsymbol{\chi}) \, d\boldsymbol{\chi}$$

$$+ \int_{\Gamma_N} \langle K_S(\boldsymbol{\chi}) \rangle [1 + v(\mathbf{x})] G(\boldsymbol{\chi}, \mathbf{x}) \nabla_{\boldsymbol{\chi}} h(\boldsymbol{\chi}) \cdot \mathbf{n}(\boldsymbol{\chi}) \, d\boldsymbol{\chi} \tag{3.309}$$

With the boundary conditions in Eq. (3.306), we obtain

$$h(\mathbf{x}) = \int_\Omega g(\boldsymbol{\chi}) G(\boldsymbol{\chi}, \mathbf{x}) \, d\boldsymbol{\chi} - \int_\Omega \langle K_S(\boldsymbol{\chi}) \rangle v(\boldsymbol{\chi}) \nabla_{\boldsymbol{\chi}} G(\boldsymbol{\chi}, \mathbf{x}) \cdot \nabla_{\boldsymbol{\chi}} h(\boldsymbol{\chi}) \, d\boldsymbol{\chi}$$

$$- \int_{\Gamma_D} \langle K_S(\boldsymbol{\chi}) \rangle H_B(\boldsymbol{\chi}) \nabla_{\boldsymbol{\chi}} G(\boldsymbol{\chi}, \mathbf{x}) \cdot \mathbf{n}(\boldsymbol{\chi}) \, d\boldsymbol{\chi}$$

$$- \int_{\Gamma_N} Q(\boldsymbol{\chi}) G(\boldsymbol{\chi}, \mathbf{x}) \, d\boldsymbol{\chi} \tag{3.310}$$

Since $v(\mathbf{x})$ is a random variable, so is $h(\mathbf{x})$. Like the alternative integral expression (3.32), Eq. (3.310) is a random solution for $h(\mathbf{x})$ though exact.

The essence of the Adomian decomposition is to express the random solution of $h(\mathbf{x})$ in the following Neumann series [Adomian, 1983; Zeitoun and Braester, 1991]:

$$h(\mathbf{x}) = \sum_{i=0}^{\infty} h^i(\mathbf{x}) \tag{3.311}$$

where h^i are undefined functions. One particular way to define h^i is as follows:

$$h^0(\mathbf{x}) = \int_{\Omega} g(\chi)G(\chi, \mathbf{x})\,d\chi - \int_{\Gamma_D} \langle K_S(\chi)\rangle H_B(\chi)\nabla_\chi G(\chi, \mathbf{x}) \cdot \mathbf{n}(\chi)\,d\chi$$

$$- \int_{\Gamma_N} Q(\chi)G(\chi, \mathbf{x})\,d\chi \tag{3.312}$$

$$h^n(\mathbf{x}) = - \int_{\Omega} \langle K_S(\chi)\rangle \nu(\chi)\nabla_\chi G(\chi, \mathbf{x}) \cdot \nabla_\chi h^{n-1}(\chi)\,d\chi \tag{3.313}$$

for $n \geq 1$. Note that the first term h^0 of the Neumann series is deterministic, while the other terms are functions of the random quantity ν. It can be seen that h^n is proportional to σ_ν^n, i.e., $h^n = \mathbf{O}(\sigma_\nu^n)$.

It can be shown that the h^i terms in Eqs. (3.312) and (3.313) satisfy the following partial differential equations:

$$\left.\begin{aligned} \nabla\left[\langle K_S(\mathbf{x})\rangle\nabla h^0(\mathbf{x})\right] &= -g(\mathbf{x}) \\ h^0(\mathbf{x}) &= H_B(\mathbf{x}), \quad \mathbf{x} \in \Gamma_D \\ \langle K_S(\mathbf{x})\rangle\nabla h^0(\mathbf{x}) \cdot \mathbf{n}(\mathbf{x}) &= -Q(\mathbf{x}), \quad \mathbf{x} \in \Gamma_N \end{aligned}\right\} \tag{3.314}$$

$$\left.\begin{aligned} \nabla\left[\langle K_S(\mathbf{x})\rangle\nabla h^n(\mathbf{x})\right] &= -\nabla \cdot \left[\langle K_S(\mathbf{x})\rangle\nu(\mathbf{x})\nabla h^{n-1}(\mathbf{x})\right] \\ h^n(\mathbf{x}) &= 0, \quad \mathbf{x} \in \Gamma_D \\ \nabla h^n(\mathbf{x}) \cdot \mathbf{n}(\mathbf{x}) &= -\nu(\mathbf{x})\nabla h^{n-1}(\mathbf{x}) \cdot \mathbf{n}(\mathbf{x}), \quad \mathbf{x} \in \Gamma_N \end{aligned}\right\} \tag{3.315}$$

for $n \geq 1$. These equations may be obtained directly by substituting the Neumann series (3.311) into Eq. (3.306).

These head terms can be used to construct the statistical moments of head. It is easy to show that $\langle h\rangle = h^0$ to zeroth order in σ_ν, and $\langle h^1\rangle = 0$. To second order in σ_ν, the mean head is given as $\langle h\rangle = h^0 + \langle h^2\rangle$. As for the f-based expansion scheme, higher-order terms are increasingly more difficult to evaluate and are seldom computed, although they can, in principle, be evaluated. To first order in σ_ν^2 (to second order in σ_ν), the head covariance $C_h(\mathbf{x}, \chi) = \langle h^1(\mathbf{x})h^1(\chi)\rangle$ is given by

$$C_h(\mathbf{x}, \chi) = \int_{\Omega} \int_{\Omega} \langle K_S(\mathbf{x}')\rangle\langle K_S(\chi')\rangle C_\nu(\mathbf{x}', \chi')\nabla_{\mathbf{x}'}G(\mathbf{x}', \mathbf{x}) \cdot \nabla_{\mathbf{x}'}h^0(\mathbf{x}')$$

$$\cdot \nabla_{\chi'}G(\chi', \chi) \cdot \nabla_{\chi'}h^0(\chi')\,d\mathbf{x}'\,d\chi' \tag{3.316}$$

where $C_v(\mathbf{x}', \boldsymbol{\chi}')$ is the covariance of v. The first two moments of v are related to those of $f = \ln K_S$ via Eqs. (2.107)–(2.109) when K_S and hence v are lognormally distributed. As an alternative to the integral equation (3.316), one may write a set of partial differential equations for C_h on the basis of Eq. (3.315) with $n = 1$.

3.9.2 Discussion

The Adomian expansion, either in the integral form or partial differential equation form, is recursive in that the higher-order terms are expressed in terms of the lower ones. It has been shown [Adomian, 1983] that the Neumann expansion is unique and that the series is the solution of Eq. (3.306) if it converges. Hence, an important issue is to find out the condition for convergence of the Neumann series. Zeitoun and Braester [1991] studied conditions for the convergence of the Neumann series expansion of head under transient flow in a spatially stationary hydraulic conductivity field (i.e., $\langle K_S \rangle = $ const and $\sigma_v = $ const). The steady-state version of their head equation is exactly the same as Eq. (3.311) coupled with Eqs. (3.312) and (3.313) upon setting $\langle K_S(\mathbf{x}) \rangle$ as a constant in the latter. Their analysis implies that the condition for convergence of the (steady-state) Neumann series expansion is $\sigma_v < 1$. In addition, the speed of convergence is another issue of great importance. If the series converges rapidly as maybe in the case of $\sigma_v \ll 1$, an evaluation of a few h^i terms may suffice; otherwise, many terms need to be evaluated. The approach is nonperturbative if and only if it includes all terms of the series or at least enough terms to render its convergence under some acceptable criterion. Indeed, it has been realized by Schetzen [1980] that "the (Neumann) series representation of a physical system may converge for a limited range of the system input amplitude. The problem of convergence is the same as that encountered with the Taylor series representation of a function. This similarity should be expected, since ... the Neuman series is really a Taylor series with memory, so both the Taylor and Neumann series should have the same basic limitation."

The v-based expansion (3.311) coupled with Eqs (3.314) and (3.315) is similar to the f-based expansion (3.7) coupled with Eqs. (3.9)–(3.11) in that both expansions are given in formal series and both are recursive. However, both expansions are truncated to some low orders because high-order terms are increasingly more difficult to evaluate. It follows that like the f-based expansion, in practice the Adomian decomposition (Neumann series expansion) becomes a perturbative approach, contrary to some recent claims in the literature. As early as the 1960s, Shvidler [1964] and Matheron [1967] used the same expansion as Eq. (3.311) and derived head expressions, which are exactly the spatially stationary counterparts of Eqs. (3.312) and (3.313). It has long been recognized that their approach is perturbative and is nominally applicable to mild heterogeneities.

It is in general believed that the f-based expansion requires the condition $\sigma_f^2 < 1$, while the v-based expansion requires $\sigma_v^2 < 1$. When f is normally distributed, σ_f^2 and σ_v^2 are related by the simple relationship $\sigma_f^2 = \ln(1 + \sigma_v^2)$ (see Eq. (2.111)). Table 3.1 shows the corresponding values of σ_v^2 for some σ_f^2. It follows that the requirement of $\sigma_v^2 < 1$ in the v-based expansion is more restrictive than that of $\sigma_f^2 < 1$ in the f-based expansion as $\sigma_v^2 = 1$ corresponds to $\sigma_f^2 = 0.69$. Although the f-based expansion is nominally restricted to $\sigma_f^2 < 1$, it has been found based on Monte Carlo studies that this perturbative approach

Table 3.1. Corresponding values of σ_f^2 and σ_v^2.

σ_f^2	0.10	0.50	0.69	1.00	2.00	3.00	4.00	5.00	6.00
σ_v^2	0.11	0.65	1.00	1.72	6.39	19.09	53.60	147.41	402.43

may yield good results for large variance systems, at least, under uniform mean flow conditions. As seen from Fig. 3.2, the f-based perturbative scheme yields accurate head variances under uniform mean flow for the variance of f as large as 4. The case of $\sigma_f^2 = 4$ is equivalent to that of $\sigma_v^2 = 53.6$, which implies the coefficient of variation C_v of K_S to be about 732%. It is not clear at this point whether or not the v-based Adomian expansion scheme works for such a large variability. Although it is claimed that the Adomian decomposition is not limited by the variability of the system, no computational examples substantiate this claim [Orr and Neuman, 1994]. The maximum variability considered in the examples of Zeitoun and Braester [1991] is $C_v = 50\%$ or $\sigma_v^2 = 0.25$, which is well within the validity range of the f-based perturbative scheme. Serrano [1992] used the Adomian decomposition approach to study the problem of solute transport in a random field but also only looked at some cases of small variabilities (with C_v of velocity less than 50%).

It may be seen from Table 3.1 that when both are truncated at low orders as often (if not always) done, the f-based expansion should have some advantages over the v-based expansion. The advantages stem from the compact nature of the f-based expansion when the variabilities of v are large. Future studies are needed to investigate the speed of convergence of the v-based Adomian expansion scheme for large variances and to compare with the f-based scheme.

3.10 CLOSURE APPROXIMATIONS

As discussed in Section 1.4.1, statistical moment equations are usually unclosed because the equation for the nth moment generally requires knowledge of the $(n + 1)$th moment. Theoretically speaking, an *infinite hierarchy* of equations need to be evaluated for each moment. It is obvious that this cannot be done in practice. In the preceding sections, this closure problem is avoided by representing the flow quantities such as the hydraulic head $h(\mathbf{x})$ and the flux $\mathbf{q}(\mathbf{x})$ in formal series, through either Taylor series or Neuman series expansion. These series are truncated by evaluating the first few terms and disregarding the higher-order ones. Hence, these approaches are perturbative in nature. In this section, we discuss some closure approximations aimed to take into account of higher moments in evaluating low ones.

3.10.1 Hierarchy Closure Approximation

As discussed in Section 3.3.1, the governing equation (3.25) may be recast with stochastic and deterministic operators [e.g., Neuman and Orr, 1993; Neuman *et al.*, 1996],

$$\mathcal{L}h(\mathbf{x}) + g(\mathbf{x}) = [L + \mathcal{R}]h(\mathbf{x}) + g(\mathbf{x}) = 0 \tag{3.317}$$

where $\mathcal{L} = \nabla \cdot [K_S(\mathbf{x})\nabla]$ and $\mathcal{R} = \nabla \cdot [K'_S(\mathbf{x})\nabla]$ are stochastic operators, and $L = \langle \mathcal{L} \rangle = \nabla \cdot [\langle K_S(\mathbf{x}) \rangle \nabla]$ is a deterministic operator. One may define a random Green's function $\mathcal{G}$ and a deterministic Green's function G, respectively, as

$$\mathcal{L}\mathcal{G} + \delta = 0 \tag{3.318}$$

$$LG + \delta = 0 \tag{3.319}$$

subject to homogeneous boundary conditions. Here δ is a Dirac delta function. With these operators, the head can be expressed as Eq. (3.37), which is recalled as follows:

$$h(\mathbf{x}) = -L^{-1}g(\mathbf{x}) - L^{-1}\mathcal{R}h(\mathbf{x}) + T_{bc}(\mathbf{x}) \tag{3.320}$$

where L^{-1} is the inverse operator defined as

$$L^{-1}g(\mathbf{x}) = -\int_\Omega G(\mathbf{x}', \mathbf{x})g(\mathbf{x}')\, d\mathbf{x}' \tag{3.321}$$

so that $s = -L^{-1}g$ is the solution of $Ls + g = 0$, subject to homogeneous boundary conditions $H_B = Q = 0$, and T_{bc} is the nonhomogeneous boundary integral defined as

$$T_{bc}(\mathbf{x}) = -\int_{\Gamma_D} \langle K_S(\mathbf{x}') \rangle \nabla_{\mathbf{x}'} G(\mathbf{x}', \mathbf{x}) \cdot \mathbf{n}(\mathbf{x}') H_B(\mathbf{x}')\, d\mathbf{x}'$$

$$- \int_{\Gamma_N} G(\mathbf{x}', \mathbf{x}) \langle K_S(\mathbf{x}') \rangle [K_S(\mathbf{x}')]^{-1} Q(\mathbf{x}')\, d\mathbf{x}' \tag{3.322}$$

Applying the operator $\mathcal{R}$ to Eq. (3.320) yields

$$\mathcal{R}h(\mathbf{x}) = -\mathcal{R}L^{-1}g(\mathbf{x}) - \mathcal{R}L^{-1}\mathcal{R}h(\mathbf{x}) + \mathcal{R}T_{bc}(\mathbf{x}) \tag{3.323}$$

and substituting this back into Eq. (3.320) leads to

$$h(\mathbf{x}) = -L^{-1}g(\mathbf{x}) + L^{-1}\mathcal{R}L^{-1}g(\mathbf{x}) + L^{-1}\mathcal{R}L^{-1}\mathcal{R}h(\mathbf{x})$$

$$- L^{-1}\mathcal{R}T_{bc}(\mathbf{x}) + T_{bc}(\mathbf{x}) \tag{3.324}$$

This process may be continued indefinitely and is termed *hierarchical substitution*. After n substitutions, the head expression reads as

$$h(\mathbf{x}) = \sum_{k=0}^{n-1} (-1)^{k+1} (L^{-1}\mathcal{R})^k L^{-1}g(\mathbf{x}) + \sum_{k=0}^{n-1} (-1)^k (L^{-1}\mathcal{R})^k T_{bc}(\mathbf{x}) + (L^{-1}\mathcal{R})^n h(\mathbf{x})$$

$$\tag{3.325}$$

Taking expectation of Eq. (3.325) yields

$$\langle h(\mathbf{x}) \rangle = \sum_{k=0}^{n-1}(-1)^{k+1}\langle (L^{-1}\mathcal{R})^k \rangle L^{-1}g(\mathbf{x}) + \sum_{k=0}^{n-1}(-1)^k \langle (L^{-1}\mathcal{R})^k \rangle T_{bc}(\mathbf{x})$$
$$+ \langle (L^{-1}\mathcal{R})^n h(\mathbf{x}) \rangle \tag{3.326}$$

where the forcing term $g(\mathbf{x})$ and the boundary terms $H_B(\mathbf{x})$ and $Q(\mathbf{x})$ are assumed to be known with certainty. Since the moment $\langle (L^{-1}\mathcal{R})^n h(\mathbf{x}) \rangle$ is unknown, Eq. (3.326) is unclosed. The common practice is to invoke the closure approximation [e.g., Ghanem and Spanos, 1991],

$$\langle (L^{-1}\mathcal{R})^n h(\mathbf{x}) \rangle \approx \langle (L^{-1}\mathcal{R})^n \rangle \langle h(\mathbf{x}) \rangle \tag{3.327}$$

This is the so-called *hierarchy closure approximation* [Frisch, 1968; Adomian, 1983; also see the review by Orr and Neuman, 1994].

The explicit expression for $h(\mathbf{x})$ in Eq. (3.324) (or Eq. (3.325) with $n = 2$) is

$$h(\mathbf{x}) = \int_\Omega G(\mathbf{x}', \mathbf{x})g(\mathbf{x}')\,d\mathbf{x}'$$
$$+ \int_\Omega \int_\Omega G(\mathbf{x}', \mathbf{x})\nabla_{\mathbf{x}'} \cdot \{K_S'(\mathbf{x}')\nabla_{\mathbf{x}'}[G(\mathbf{x}'', \mathbf{x}')g(\mathbf{x}'')]\}\,d\mathbf{x}'\,d\mathbf{x}''$$
$$+ \int_\Omega \int_\Omega G(\mathbf{x}', \mathbf{x})\nabla_{\mathbf{x}'} \cdot \{K_S'(\mathbf{x}')\nabla_{\mathbf{x}'}[G(\mathbf{x}'', \mathbf{x}')\nabla_{\mathbf{x}''}$$
$$\cdot (K_S'(\mathbf{x}'')\nabla_{\mathbf{x}''}h(\mathbf{x}''))]\}\,d\mathbf{x}'\,d\mathbf{x}''$$
$$+ \int_\Omega G(\mathbf{x}', \mathbf{x})\nabla_{\mathbf{x}'} \cdot [K_S'(\mathbf{x}')\nabla_{\mathbf{x}'}T_{bc}(\mathbf{x}')]\,d\mathbf{x}' - T_{bc}(\mathbf{x}) \tag{3.328}$$

Using Green's identity and the homogeneous boundary conditions for G, Eq. (3.328) can be rewritten as

$$h(\mathbf{x}) = \int_\Omega G(\mathbf{x}', \mathbf{x})g(\mathbf{x}')\,d\mathbf{x}'$$
$$- \int_\Omega \int_\Omega \nabla_{\mathbf{x}'}^T G(\mathbf{x}', \mathbf{x})\nabla_{\mathbf{x}'}G(\mathbf{x}'', \mathbf{x}')g(\mathbf{x}'')K_S'(\mathbf{x}')\,d\mathbf{x}'\,d\mathbf{x}''$$
$$+ \int_\Omega \int_\Omega \nabla_{\mathbf{x}'}^T G(\mathbf{x}', \mathbf{x})\nabla_{\mathbf{x}'}\nabla_{\mathbf{x}''}^T G(\mathbf{x}'', \mathbf{x}')$$
$$\cdot K_S'(\mathbf{x}')K_S'(\mathbf{x}'')\nabla_{\mathbf{x}''}h(\mathbf{x}'')\,d\mathbf{x}'\,d\mathbf{x}''$$
$$+ \int_\Omega G(\mathbf{x}', \mathbf{x})\nabla_{\mathbf{x}'} \cdot [K_S'(\mathbf{x}')\nabla_{\mathbf{x}'}T_{bc}(\mathbf{x}')]\,d\mathbf{x}' - T_{bc}(\mathbf{x}) \tag{3.329}$$

It follows that the mean head $\langle h(\mathbf{x}) \rangle$ involves the joint, unknown moment $\langle K_S'(\mathbf{x}')K_S'(\mathbf{x}'')\nabla_{\mathbf{x}''}h(\mathbf{x}'') \rangle$ under the integrals. The closure approximation of Eq. (3.327) corresponding to $n = 2$ is

$$\langle (L^{-1}\mathcal{R})^2 h(\mathbf{x}) \rangle \approx \langle (L^{-1}\mathcal{R})^2 \rangle \langle h(\mathbf{x}) \rangle \tag{3.330}$$

which reads explicitly as

$$\langle K'_S(\mathbf{x}') K'_S(\mathbf{x}'') \nabla h(\mathbf{x}'') \rangle \approx \langle K'_S(\mathbf{x}') K'_S(\mathbf{x}'') \rangle \nabla \langle h(\mathbf{x}'') \rangle \tag{3.331}$$

under the integrals of Eq. (3.329). A similar approximation that involves a random Green's function $\mathcal{G}$ rather than the random head h has been made and partially evaluated by Neuman and Orr [1993]. It has been termed the *weak approximation* by the latter authors on the grounds that the approximation is made under the integrals and not at individual pairs of points. It should be noted that similar (decoupling) approximations may be made in differential equations. Since the differential and integral representations are essentially equivalent as shown in the previous sections, we expect those approximations made in the context of differential equations to be "weak" as well. The residual flux $\mathbf{r}_f(\mathbf{x}) = -\langle K'(\mathbf{x})\nabla h(\mathbf{x}) \rangle$ is obtained with Eq. (3.329) as

$$\begin{aligned}
\mathbf{r}_f(\mathbf{x}) = &\int_\Omega \int_\Omega \nabla_\mathbf{x} \nabla_{\mathbf{x}'}^T G(\mathbf{x}', \mathbf{x}) \nabla_{\mathbf{x}'} G(\mathbf{x}'', \mathbf{x}') g(\mathbf{x}'') \langle K'_S(\mathbf{x}) K'_S(\mathbf{x}') \rangle \, d\mathbf{x}' \, d\mathbf{x}'' \\
&- \int_\Omega \int_\Omega \nabla_\mathbf{x} \nabla_{\mathbf{x}'}^T G(\mathbf{x}', \mathbf{x}) \nabla_{\mathbf{x}'} \nabla_{\mathbf{x}''}^T G(\mathbf{x}'', \mathbf{x}') \\
&\quad \cdot \langle K'_S(\mathbf{x}) K'_S(\mathbf{x}') K'_S(\mathbf{x}'') \nabla_{\mathbf{x}''} h(\mathbf{x}'') \rangle \, d\mathbf{x}' \, d\mathbf{x}'' \\
&- \int_\Omega \nabla_\mathbf{x} G(\mathbf{x}', \mathbf{x}) \nabla_{\mathbf{x}'} \cdot [\langle K'_S(\mathbf{x}) K'_S(\mathbf{x}') \nabla_{\mathbf{x}'} T_{bc}(\mathbf{x}') \rangle] \, d\mathbf{x}' \\
&+ \langle K'_S(\mathbf{x}) \nabla T_{bc}(\mathbf{x}) \rangle
\end{aligned} \tag{3.332}$$

It involves the unkown moment $\langle K'_S(\mathbf{x}) K'_S(\mathbf{x}') K'_S(\mathbf{x}'') \nabla h(\mathbf{x}'') \rangle$, which also appears in $\langle h(\mathbf{x}) \rangle$ if $n = 3$ is chosen in Eq. (3.326). This fourth moment may be approximated by

$$\langle K'_S(\mathbf{x}) K'_S(\mathbf{x}') K'_S(\mathbf{x}'') \nabla h(\mathbf{x}'') \rangle \approx \langle K'_S(\mathbf{x}) K'_S(\mathbf{x}') K'_S(\mathbf{x}'') \rangle \nabla \langle h(\mathbf{x}'') \rangle \tag{3.333}$$

under the integrals. This type of approximation can be made at a higher level. However, the procedure is increasingly cumbersome as n increases. It is difficult to tell *a priori* at which level a particular approximation should be made and whether the approximation is valid.

Since these hierarchy closure approximations have not been fully tested to the writer's knowledge for porous flow problems, their validity is unknown at the present time. Heuristic arguments such as "local independence" have been made to justify the decoupling behind these closure approximations for different problems [e.g., Orr and Neuman, 1994]. The decoupling, however, has been criticized as lacking in rigor; it has been shown to be the same as truncation of a small perturbation series [e.g., Kraichnan, 1961; Frisch, 1968; Adomian, 1983]. The closure approximation of Eq. (3.331) is equivalent to neglecting the contribution of the third moment $\langle K_S'(\mathbf{x}')K_S'(\mathbf{x}'')\nabla h'(\mathbf{x}'')\rangle$ under the integrals as per the following identity:

$$\langle K_S'(\mathbf{x}')K_S'(\mathbf{x}'')\nabla h(\mathbf{x}'')\rangle = \langle K_S'(\mathbf{x}')K_S'(\mathbf{x}'')\rangle\nabla\langle h(\mathbf{x}'')\rangle$$
$$+ \langle K_S'(\mathbf{x}')K_S'(\mathbf{x}'')\nabla h'(\mathbf{x}'')\rangle \quad (3.334)$$

On similar grounds, Eq. (3.333) is equivalent to discarding the fourth moment $\langle K_S'(\mathbf{x})K_S'(\mathbf{x}')K_S'(\mathbf{x}'')\nabla h'(\mathbf{x}'')\rangle$ under the integrals. If the joint distribution of K_S and h is approximately multivariate normal, then the approximation of Eq. (3.331) is seen valid. However, neglecting the fourth moment $\langle K_S'(\mathbf{x})K_S'(\mathbf{x}')K_S'(\mathbf{x}'')\nabla h'(\mathbf{x}'')\rangle$ still seems questionable under this condition as the term neglected may be the dominant one. This motivates the following alternative closure approximation for processes that are close to normal.

3.10.2 Cumulant Neglect Hypothesis

The essence of the *cumulant neglect hypothesis* is to terminate the hierarchy of moment equations by setting the cumulant equal to zero beyond some point [e.g., Beran, 1968]. If the nth cumulant is assumed to be zero, then the nth moments can be expressed in terms of lower moments. The definition and properties of multivariate cumulants are given in Section 2.2.7. The first cumulant of U_i is equal to its mean $\langle U_i\rangle$, the second cumulant of U_i and U_j is the covariance between U_i and U_j, and higher cumulants are related to the statistical moments in a more complex manner (see, e.g., Eqs. (2.101) and (2.102)). If the random variables $U_1, U_2, \ldots, U_n$ are jointly normal, then all cumulants higher than the second are identically zero. The cumulant neglect hypothesis is also called *the quasi-normality approximation*. As the latter name implies, the approximation would be exact if the distribution under consideration were normal.

Let us take the fourth moment $\langle K_S'(\mathbf{x})K_S'(\mathbf{x}')K_S'(\mathbf{x}'')\nabla h'(\mathbf{x}'')\rangle$ as an example. On the basis of Eq. (2.102), the fourth cumulant $\langle\langle K_S'(\mathbf{x})K_S'(\mathbf{x}')K_S'(\mathbf{x}'')\nabla h'(\mathbf{x}'')\rangle\rangle$ is given as

$$\langle\langle K_S'(\mathbf{x})K_S'(\mathbf{x}')K_S'(\mathbf{x}'')\nabla h'(\mathbf{x}'')\rangle\rangle - \langle K_S'(\mathbf{x})K_S'(\mathbf{x}')K_S'(\mathbf{x}'')\nabla h'(\mathbf{x}'')\rangle$$
$$- \langle K_S'(\mathbf{x})K_S'(\mathbf{x}')\rangle\langle K_S'(\mathbf{x}'')\nabla h(\mathbf{x}'')\rangle$$
$$- \langle K_S'(\mathbf{x})K_S'(\mathbf{x}'')\rangle\langle K_S'(\mathbf{x}')\nabla h(\mathbf{x}'')\rangle$$
$$- \langle K_S'(\mathbf{x}')K_S'(\mathbf{x}'')\rangle\langle K_S'(\mathbf{x})\nabla h(\mathbf{x}'')\rangle \quad (3.335)$$

Setting the cumulant $\langle\langle K_S'(\mathbf{x})K_S'(\mathbf{x}')K_S'(\mathbf{x}'')\nabla h'(\mathbf{x}'')\rangle\rangle = 0$ is equivalent to expressing the fourth moment as

$$\langle K_S'(\mathbf{x})K_S'(\mathbf{x}')K_S'(\mathbf{x}'')\nabla h'(\mathbf{x}'')\rangle = \langle K_S'(\mathbf{x})K_S'(\mathbf{x}')\rangle\langle K_S'(\mathbf{x}'')\nabla h'(\mathbf{x}'')\rangle$$
$$+ \langle K_S'(\mathbf{x})K_S'(\mathbf{x}'')\rangle\langle K_S'(\mathbf{x}')\nabla h'(\mathbf{x}'')\rangle$$
$$+ \langle K_S'(\mathbf{x}')K_S'(\mathbf{x}'')\rangle\langle K_S'(\mathbf{x})\nabla h'(\mathbf{x}'')\rangle \quad (3.336)$$

Hence, we have the following closure approximation,

$$
\begin{aligned}
\langle K'_S(\mathbf{x})K'_S(\mathbf{x}')K'_S(\mathbf{x}'')\nabla h(\mathbf{x}'')\rangle \approx\ & \langle K'_S(\mathbf{x})K'_S(\mathbf{x}')K'_S(\mathbf{x}'')\rangle \nabla \langle h(\mathbf{x}'')\rangle \\
& + \langle K'_S(\mathbf{x})K'_S(\mathbf{x}')\rangle \langle K'_S(\mathbf{x}'')\nabla h'(\mathbf{x}'')\rangle \\
& + \langle K'_S(\mathbf{x})K'_S(\mathbf{x}'')\rangle \langle K'_S(\mathbf{x}')\nabla h'(\mathbf{x}'')\rangle \\
& + \langle K'_S(\mathbf{x}')K'_S(\mathbf{x}'')\rangle \langle K'_S(\mathbf{x})\nabla h'(\mathbf{x}'')\rangle
\end{aligned}
\tag{3.337}
$$

in which the fourth moment is expressed in terms of the first, second, and third moments. Note that the moments $\langle K'_S(\mathbf{x})K'_S(\boldsymbol{\chi})K'_S(\mathbf{y})\rangle$ and $\langle K'_S(\mathbf{x})K'_S(\boldsymbol{\chi})\rangle$ are the statistics of the input variable, which can be assumed known. An expression for the second moment $\langle K'_S(\boldsymbol{\chi})\nabla h'(\mathbf{x})\rangle$ can be given in analogy to $\mathbf{r}_f(\mathbf{x}) = -\langle K'_S(\mathbf{x})\nabla h'(\mathbf{x})\rangle$.

Note that the statistical moments resulting from neglect of the fourth cumulant are different from those that would result from a normal distribution. If the hydraulic conductivity field $K_S(\mathbf{x})$ were multivariate normal, the third moment $\langle K'_S(\mathbf{x})K'_S(\boldsymbol{\chi})K'_S(\mathbf{y})\rangle$ in Eq. (3.337) would be zero. It is seen that the closure approximation in Eq. (3.337) on the basis of neglect of the fourth cumulant is very different but perhaps more reasonable than the third hierarchy approximation of Eq. (3.333). However, an approximation with neglect of the third cumulant $\langle\langle K'_S(\mathbf{x}')K'_S(\mathbf{x}'')\nabla h'(\mathbf{x}'')\rangle\rangle$ would be equivalent to the second hierarchy approximation of Eq. (3.331).

The head covariance $C_h(\mathbf{x}, \boldsymbol{\chi}) = \langle h(\mathbf{x})h(\boldsymbol{\chi})\rangle - \langle h(\mathbf{x})\rangle\langle h(\boldsymbol{\chi})\rangle$ can be expressed with the aid of Eq. (3.329). It can be shown that the head covariance involves the sixth moment $\langle K'_S(\mathbf{x}')K'_S(\mathbf{x}'')K'_S(\boldsymbol{\chi}')K'_S(\boldsymbol{\chi}'')\nabla^T h(\mathbf{x}'')\ \nabla h(\boldsymbol{\chi}'')\rangle$ and other fifth and fourth ones. Setting the fourth and higher cumulants equal to zero renders closure to these moments, which are now expressed in terms of first, second, and third moments. If a lower approximation were invoked by setting the third cumulant equal to zero, these higher moments would be expressed in terms of the first two moments.

As for the hierarchy closure approximations, it is not known *a priori* at which order the cumulant neglect hypothesis should be invoked and whether or not it is valid. Since the equations given above have never been evaluated for porous flow problems, we shall not be able to comment on the range of validity regarding the cumulant neglect hypothesis for the problems under consideration. However, the cumulant neglect hypothesis has been investigated extensively in turbulence [e.g., Beran, 1968; Leslie, 1973]. Although this type of approximation may sometimes yield unreasonable results of negative values of the power spectrum, the cumulant neglect hypothesis should be a good approximation for processes that are nearly normal. As the cumulant neglect approximation is made under the integral (or in the context of differential equations) and not locally at individual points, it should be weaker (milder) than imposing quasi-normality on the dependent and independent variables. This type of closure approximation may deserve further studies for groundwater problems.

3.11 MONTE CARLO SIMULATION METHOD

The various methods discussed in the preceding sections may be best categorized as *moment equation methods*. In these methods, equations governing the statistical moments of flow quantities are first derived from the (original) stochastic differential equations and are then solved numerically or analytically. Monte Carlo simulation is an alternative and perhaps the most straightforward method for solving stochastic equations. This widely used approach is conceptually simple and is based on the idea of approximating stochastic processes by a large number of equally probable realizations.

3.11.1 A Historical Review

The principle behind the *Monte Carlo method* is statistical sampling, which dates back to the late eighteenth century. However, because of the labor and time required the method was not applied to any significant extent until the advent of electronic computers. The development of a computerized statistical sampling method was initiated at Los Alamos Laboratory in 1946 by Stanislaw Ulam, a Los Alamos scientist, and John von Neumann, a consultant to Los Alamos [Eckhardt, 1987]. The slightly racy but entirely appropriate name of Monte Carlo method was given to this statistical technique by another Los Alamos scientist, Nicholas C. Metropolis [Metropolis, 1987]. The problem motivated Los Alamos scientists to develop the Monte Carlo method using neutron diffusion and multiplication. Since then, the method has been applied successfully to a vast number of scientific problems in various fields.

The early applications of the Monte Carlo method to flow in porous media include Warren and Price [1961] and Shvidler [1964]. Warren and Price [1961] performed an extensive set of Monte Carlo simulations studying flow through heterogeneous porous media and examined the influence of probability distributions for hydraulic conductivity on the results for flow in one and three dimensions. Independently, Shvidler [1964], the original Russian text of which was published in 1963, applied the Monte Carlo method to some problems of flow in porous media. However, the Monte Carlo method had largely escaped the attention of the porous flow community until the work of Freeze [1975]. He used the Monte Carlo method to study both steady-state and transient flow in a one-dimensional bounded domain, where the hydraulic conductivity is assumed to be statistically independent in adjacent blocks. On the basis of large head variances found from his simulations, Freeze [1975] questioned the utility of deterministic modeling in groundwater hydrology. Although his results were obtained under a number of oversimplifying assumptions, Freeze's [1975] work is arguably the beginning of the stochastic era for flow in porous media, during which more sophisticated Monte Carlo approaches and formal stochastic theories have been developed.

3.11.2 Methodology

The Monte Carlo approach involves three steps. The first one is to generate multiple realizations of the geological formation of interest on the basis of given statistical moments and distributions of formation properties. A given realization is deterministic and provides a complete representation of the formation properties, but it is selected by a probabilistic procedure analogous to coin-flipping and is only one possible representative of the unknown formation. The second step is to solve, for each realization, the deterministic governing equations by numerical methods such as finite differences and finite elements. The third step is to average over the solutions of many realizations to obtain the statistical moments or distributions of the dependent variables. Below we briefly discuss the procedure of Monte Carlo simulations.

Random field generation The generation of numerical samples of random functions or stochastic processes (e.g., hydraulic conductivity and porosity) plays a crucial role in Monte Carlo simulations. The purpose of this step is to create realizations of stochastic processes with known statistical properties. For the steady-state saturated flow problem under study in this chapter, hydraulic conductivity is treated as a random space function or random field. The generated realizations of the hydraulic conductivity field are required to honor the given statistical moments such as the mean, variance, and covariance. In particular, care must be taken in honoring the spatial structure of a random field as reflected in its covariance, since this quantity is difficult to reproduce compared to the mean and variance. How well the generated realizations honor the statistics of the random field of interest directly impacts the quality of the Monte Carlo simulations. The realizations are said to be conditional if they honor available measurements in the field.

There are several techniques available for generating random fields, including, but not limited to, the *turning bands method*, *spectral methods*, *matrix decomposition*, and *Gaussian sequential simulation*. Let us briefly discuss these techniques.

1. Turning bands method. The turning bands method was first proposed by Matheron [1973] and further developed and used by various researchers [e.g., Journel and Huijbregts, 1978; Mantoglou and Wilson, 1982; Mantoglou, 1987; Tompson *et al.*, 1989]. The principle advantage of the method is that it reduces the generation of a two- or three-dimensional random field to that of one-dimensional random space functions. The reduction in dimensionality is made possible by the fact that the transformation from two- and three-dimensional covariance function into an equivalent one-dimensional covariance function can be uniquely defined [Matheron, 1973; Mantoglou and Wilson, 1982].

After the equivalent one-dimensional covariance is determined from the multi-dimensional covariance, a one-dimensional multivariate process Z is generated along each line with an appropriate autoregressive or moving average technique, or a spectral method. Theoretically, it requires an infinite number of lines to attain a multidimensional random field. In practice, only a finite number of lines are usually used. The sample at each point $\mathbf{x}$ in the field is calculated as the average contributions projected from all lines onto this point,

$$Y(\mathbf{x}) = \frac{1}{\sqrt{N}} \sum_{j=1}^{N} Z(\mathbf{x}_j, j) \qquad (3.338)$$

where j is the line index and $\mathbf{x}_j$ is located on line j such that $\mathbf{x}$ is orthogonal to $\mathbf{x}_j$ with respect to line j. For generating three-dimensional random fields, Tompson *et al.* [1989] suggested using a large number (say, $N = 100$) of randomly distributed lines. In two dimensions, Orr [1993] found that at least 32 evenly spaced lines are needed in order to avoid a distortion effect associated with the appearance of line-like patterns in both the generated fields and their spatial covariances. These artificial patterns usually diminish with the increase in the number of lines and, of course, with the price of increased computational effort.

2. Spectral methods. Spectral methods take advantage of the spectral representation of stochastic processes, which, as discussed in Section 2.4.4, express a spatially correlated random field in terms of an uncorrelated (white noise) process in the spectral domain. Among the family of spectral methods, those using fast Fourier transforms (FFTs) are most efficient and widely used. The spectral methods are applicable to one-, two-, and three-dimensional fields but are limited to second-order stationary random fields. Interested readers are referred to Gutjahr [1989], Christakos [1992], Robin *et al.* [1993] and Harter [1994] for a detailed description of these methods.

3. Matrix decomposition. Matrix decomposition is an elegant and direct method for gen-erating correlated random fields in that the covariance matrix is used directly. The method was developed independently by Clifton and Neuman [1982] and Elishakoff [1983].

For n points on a grid where samples are to be generated, one can obtain the covariance matrix $\mathbf{C}$ of size $n \times n$ from the given covariance function $C(\mathbf{r})$ (where $\mathbf{r}$ is the separation vector between two points). As discussed in Chapter 2, this covariance matrix $\mathbf{C}$ must be symmetric and positive definite. Hence, this matrix can be decomposed into a lower triangular matrix $\mathbf{L}$ and an upper triangular matrix $\mathbf{U}$,

$$\mathbf{C} = \mathbf{LU} \tag{3.339}$$

A realization of the random field is obtained by multiplying the lower matrix $\mathbf{L}$ with a vector $\boldsymbol{\alpha} = (\alpha_1, \ldots, \alpha_n)^T$ of n independent, normally distributed random numbers with zero mean and unit variance,

$$\mathbf{Y} = \mathbf{L}\boldsymbol{\alpha} \tag{3.340}$$

The mean of the generated vector $\mathbf{Y}$ is zero and its covariance is $\langle \mathbf{Y}\mathbf{Y}^T \rangle = \mathbf{L}(\boldsymbol{\alpha}\boldsymbol{\alpha}^T)\mathbf{U} = \mathbf{LIU} = \mathbf{C}$ where $\mathbf{I}$ is the identity matrix. A nonzero (uniform or spatially varying) mean can be simply added to $\mathbf{Y}$ to obtain nonzero mean realizations.

It is seen that once $\mathbf{C}$ is decomposed as in Eq. (3.339), a new realization $\mathbf{Y}$ is readily generated by Eq. (3.340) for a new vector $\boldsymbol{\alpha}$, which can be created by any good random number generator. The additional cost to generate new realizations is small. Note that the method is independent of the dimensionality of the random field except for the size of the matrix. However, this straightforward method for generating correlated random fields has a large memory requirement. For example, the generation of a realization on a two-dimensional grid of 50×50 points involves a covariance matrix of 2500×2500 components, which requires about 50 megabytes (MB) of memory; a 100×100 grid requires about 800 MB of memory.

4. Gaussian sequential simulation. Sequential simulation is probably the most gen-eral algorithm for generating both Gaussian continuous and categorical random fields

[Deutsch and Journel, 1998]. The sequential simulation technique goes back to the definition of a conditional density function, given by Eq. (2.78). With this definition, one has

$$p_U(u_1, \ldots, u_n) = p_U(u_1, \ldots, u_{n-1}) p_U(u_n | u_1, \ldots, u_{n-1})$$
$$\cdots$$
$$= p_U(u_1) p_U(u_2 | u_1) \cdots p_U(u_n | u_1, \ldots, u_{n-1}) \qquad (3.341)$$

It follows that the n-variate probability density function $p_U(u_1, \ldots, u_n)$ can be expressed as the product of $n - 1$ univariate conditional density functions and the unconditional density function of u_1. When conditional realizations are desirable which honor measurements, say, $\{u^m\}$, Eq. (3.341) can be modified as

$$p_U(u_1, \ldots, u_n | \{u^m\}) = p_U(u_1 | \{u^m\}) p_U(u_2 | u_1; \{u^m\}) \cdots$$
$$p_U(u_n | u_1, \ldots, u_{n-1}; \{u^m\}) \qquad (3.342)$$

In the case that u_i depends on the position $\mathbf{x}_i$ in continuous space, $u(\mathbf{x})$ becomes a random field. The procedure for generating a realization is, in principle, straightforward: (1) generate a value at the first location $\mathbf{x}_1$, according to unconditional $p_U(u_1)$ or conditional $p_U(u_1 | \{u^m\})$; (2) at the next point $\mathbf{x}_2$, find the conditional density given the already generated data point(s) and, if for a conditional realization, the measurements; (3) draw a value from the conditional distribution; and (4) repeat steps 2 and 3 until the whole field is generated. Other realizations are generated with the same procedure. Although Eq. (3.341) provides a general way to generate random fields of any distribution, the conditional density functions are difficult to obtain except for some special distributions such as Gaussian. A Gaussian sequential simulator was first presented by Gomez-Hernandez [1991], where the conditional moments are computed by the system of equations (2.221)–(2.223).

As a collaborative effort, Orr [1993] and Harter [1994] performed a detailed study to evaluate these techniques. The key findings of their study are summarized as follows:

- In general, two conditions must be fulfilled for a random field generator to give statistical results that converge in mean square to the desired moments of the random field. First, the random number generator must be able to generate independent random numbers with given mean and variance in the limit as the sample size becomes large. This condition can be tested separately and should be tested prior to Monte Carlo simulations. Second, the numerical implementation of the random field generator must be free of deterministic artificial patterns caused by the generating algorithm.
- For the above four random field generators, their performance study indicated that local sample statistics, such as sample means, variances, and covariances for each or selected points in the field, deviate significantly from the input statistics unless the number of realizations is large (e.g., larger than 1000 for the cases that they tested).
- The spatial averages of the local sample statistics, however, converge more rapidly in the mean square sense to the input ensemble statistics as the number of realizations (M) increases and as the size of the field increases.

- Of the four random field generators tested, the matrix-decomposition-based generator showed the least artificial bias and honored best the input statistics.

The random field generators discussed above generate multivariate normal (Gaussian) fields, which are uniquely characterized by their first two moments, i.e., mean and covariance. When the random variable under consideration is not univariate normal, it can always be transformed through a normal score transform [Deutsch and Journel, 1998]. However, univariate normality of the transformed variable, say Y, does not guarantee the spatially distributed values $Y(\mathbf{x})$ to be multivariate normal. Although the bivariate normality of a data set may be examined, there are seldom enough field data to check the multivariate normality beyond this. Therefore, in practice, a stationary random field is usually approximated by a multivariate normal field if the sample statistics from the available data do not violate the bivariate normal properties. The most common example is the log transformation of the hydraulic conductivity field. Although field data support the univariate normality of Y, the multivariate normality of $Y(\mathbf{x})$ is essentially an unproven hypothesis. The log-transformed hydraulic conductivity field $Y(\mathbf{x})$ is first generated and then transformed back to the $K(\mathbf{x}) = \exp[Y(\mathbf{x})]$ field when the flow equations are solved numerically.

The Gaussian (continuous) random field generators are, however, inadequate to reproduce categorical features such as hydrofacies, hydrostratigraphic units, petroleum lithologies, and mineral classifications. Such categorical random fields may be created with a *two-step generation* approach [Deutsch and Journel, 1998], in which the geometry of these features is first represented and attributes within each distinct unit are then generated. *Indicator*-based algorithms and *transition probability* approaches are particularly useful for generating such random fields [Deutsch and Journel, 1998; Carle and Fogg, 1996, 1997].

As found in the preceding sections, even in a stationary hydraulic conductivity field the flow field can be nonstationary in the presence of finite boundaries and nonuniform flow configurations. In such a situation, spatial averaging of local sample statistics may not be meaningful. Hence, a large number of realizations may be needed to preserve the input (ensemble) statistics at each points and to obtain meaningful statistics for the nonstationary flow quantities such as hydraulic head and flux. Plots of the local sample statistics at some representative points versus the number of realizations M may be used to examine the convergence of the generated random fields.

The quality of Monte Carlo simulation results strongly depends on how well the realizations are generated. In particular, when Monte Carlo simulations are used to validate or invalidate approximate theories or to assess the validity range of such theories, one must make sure that the realizations reproduce the input ensemble statistics of the random field, which are the common ground for the theories and the Monte Carlo simulations.

While the matrix decomposition method generates high-quality realizations, it has a large requirement of computer memory and requires a big initial computational time in decomposing the $n \times n$ covariance matrix (where n is the number of points to be generated). For small random fields (e.g., less than 6000 points) with a large number of realizations, the matrix decomposition method is an efficient way to generate good quality realizations, which are crucial for the purpose of comparing with stochastic theories. As the size of the field increases, this method becomes cumbersome. However, the problem of high memory demand may be alleviated with the advent of fast and cheap computers with large memory.

Numerical solution of each realization For a given (continuous or categorical) realization, the governing equations become deterministic and can thus be solved with conventional numerical techniques such as finite differences, finite elements, and mixed finite elements. Here we shall not go into details of these numerical techniques as they can be easily found in standard textbooks.

Any new code or existing simulator that is capable of handing spatial (and, for transient flow, temporal) variabilities may be used for solving these realizations. However, care must be taken to ensure the quality of numerical solutions by comparing with analytical solutions (for some idealized cases) and other numerical solutions and by studying the sensitivities of spatial (and temporal) discretizations. Note that the spatial discretization (Δ) should be inversely related to the level of spatial variability, e.g., the standard deviation of log hydraulic conductivity σ_f. Ababou *et al.* [1989] suggested the following relationship between λ_f/Δ (the number of cells within each integral scale λ_f) and σ_f: $\lambda_f/\Delta \gg 1 + \sigma_f$. Several studies indicated that at least five cells within each integral scale should be used in order to minimize the filtering (local averaging) effect of spatial discretization [e.g., Ababou *et al.*, 1989; Van Lent and Kitanidis, 1996; Salandin and Fiorotto, 1998]. Since a larger number of cells may be required to resolve the more rapidly varying hydraulic conductivity field as σ_f increases, the maximum level of spatial variability that the Monte Carlo method could handle is limited, in practice, by the available computer power and by the ability of numerical simulators to resolve high variabilities.

Statistical postprocessing Once the realizations are solved, the resulting solutions of flow quantities such as hydraulic head and flux can be averaged to obtain statistical moments or distributions of these flow quantities. For example, the sample mean and variance of hydraulic head at any point $\mathbf{x}$ can be computed, respectively, by

$$\langle h(\mathbf{x}) \rangle_s = \frac{1}{M} \sum_{m=1}^{M} h_m(\mathbf{x}) \tag{3.343}$$

$$\text{var}[h(\mathbf{x})]_s = \frac{1}{M} \sum_{m=1}^{M} [h_m(\mathbf{x})]^2 - [\langle h(\mathbf{x}) \rangle_s]^2 \tag{3.344}$$

where M is the total number of realizations and $h_m(\mathbf{x})$ is the head at $\mathbf{x}$ from the mth realization. Similarly, the covariance of head at the points $\mathbf{x}$ and $\boldsymbol{\chi}$ may be computed by

$$\text{cov}[h(\mathbf{x}), h(\boldsymbol{\chi})]_s = \frac{1}{M} \sum_{m=1}^{M} h_m(\mathbf{x}) h_m(\boldsymbol{\chi}) - \langle h(\mathbf{x}) \rangle_s \langle h(\boldsymbol{\chi}) \rangle_s \tag{3.345}$$

As for the generated random hydraulic conductivity field, some plots of the local hydraulic head moments at representative locations versus M may be used to identify the minimum number of realizations needed to achieve convergence in the statistical moments of the hydraulic head field. The number M needed could be different for different dependent variables and for the independent and dependent variables. The sample moments for flux can be computed in the same way. As mentioned before, the head moments are strongly nonstationary (location dependent) and are thus not ergodic due to the effect of finite boundaries. However, the flux moments are less affected by the presence of boundaries and become approximately stationary a few integral scales away from the boundaries. As such, many Monte Carlo simulations designed to compare with stationary flow and transport theories treat the flow fields as stationary by only retaining results from the inner core region that is at least three integral scales from the closest boundary and averaging the local sample moments spatially [e.g., Bellin *et al.*, 1992; Chin and Wang, 1992; Chin, 1997; Zhang and Lin, 1998]. Higher moments and one- or multiple-point histograms can be also computed from the Monte Carlo results, although these statistical quantities may require an even larger number of realizations.

3.11.3 Discussion

The Monte Carlo approach can handle complex geometry and boundary conditions and requires fewer assumptions than does the moment equation approach. Most importantly, the Monte Carlo approach can, in principle, deal with extremely large variability in the independent variables so long as the number of realizations is large. However, this leads to the main disadvantage of the Monte Carlo approach, which is that it must solve many realizations of a given formation. This requires considerable computation and a careful examination of the results. In general, two types of error are associated with Monte Carlo simulation results: numerical and statistical. The former depends on the numerical method and the particular solver used as well as the spatial and temporal discretizations. The larger the spatial variability, the finer the required spatial discretization is likely to be. Hence, the level of spatial variability that the Monte Carlo method can handle is limited by the available computer power and the ability of the numerical techniques. Statistical errors arise from the method used to generate realizations and the number of realizations.

On the other hand, moment equations are derived under the assumption of small perturbations or with some kind of closure approximation, either of which may introduce error. Compared to the Monte Carlo method, the moment equation approach has some important distinctions. For example, the coefficients of the moment equations are relatively smooth because they are averaged quantities. Thus the moment equations can be solved on relatively coarse grids. In addition, the moment equations are available in analytical form, even though they are usually solved numerically in applications. This holds the potential for increased physical understanding of the mechanisms of uncertainty through qualitative analysis. Last but not least, the results from the moment equation approach are less ambiguous than those of the Monte Carlo approach. The ambiguity in the Monte Carlo simulation results may stem from its multiple sources of potential error, which are difficult to either ascertain or control [Neuman, 1997]. It is because of this ambiguity that different sets of Monte Carlo simulations may give significantly different results even for the same physical problem. Therefore, care must be taken to ensure the quality of Monte Carlo simulations, in particular, when the Monte Carlo method is used to validate or invalidate approximate theories.

It is worthwhile to note that the numerical grid used in the moment equation approach can be much coarser than that in the Monte Carlo approach for the same problem because in the former the coefficients as averaged quantities are much smoother. For a small- or medium-sized domain, the moment equation approach not only reveals more physical insight but also is more computationally efficient than the Monte Carlo approach. For a large domain, the number of times needed to solve the moment (cross-covariance) equations may be more than the number of realizations in the Monte Carlo approach; but the number of cells (or nodes) used in the Monte Carlo approach is always much larger than that in the moment equation approach. This excludes any simple comparison on the computational issue. A carefully designed study is needed to make a firm conclusion.

3.12 SOME REMARKS

3.12.1 Application to Gas Flow

In this chapter, a variety of stochastic methods are discussed for Eqs. (3.1)–(3.4), which are for incompressible, steady-state single-phase fluid flow. The most common example of such flow is saturated flow in porous media. These stochastic techniques are also applicable to compressible fluid flow such as gas flow.

Steady-state gas flow in porous media is governed by the following equations [e.g., Aziz and Settari, 1979]:

$$\nabla \cdot [\rho(P)\mathbf{q}(\mathbf{x})] = g_\rho(\mathbf{x})$$

$$q_i(\mathbf{x}) = -\frac{k(\mathbf{x})}{\mu(P)} \frac{\partial}{\partial x_i}[P(\mathbf{x}) + \rho g x_3]$$

$$P(\mathbf{x}) = P_B(\mathbf{x}), \quad \mathbf{x} \in \Gamma_D$$

$$\rho(P)\mathbf{q}(\mathbf{x}) \cdot \mathbf{n}(\mathbf{x}) = Q_\rho(\mathbf{x}), \quad \mathbf{x} \in \Gamma_N$$

(3.346)

where $\mathbf{q}(\mathbf{x})$ is the specific discharge (flux) vector, $g_\rho(\mathbf{x})$ is the mass source/sink term (due to recharge, pumping, or injection; positive for source and negative for sink), $P(\mathbf{x})$ is the gas pressure, ρ is the gas density, g is the gravitational acceleration factor, x_3 is the elevation, $k(\mathbf{x})$ is the (random) intrinsic or absolute permeability, μ is the gas viscosity, $P_B(\mathbf{x})$ is the prescribed pressure on Dirichlet boundary segments Γ_D, and $Q_\rho(\mathbf{x})$ is the prescribed mass flux across Neumann boundary segments Γ_N. Equation (3.346) is supplemented by the *equation of state*, or, the gas law,

$$\rho = \frac{PM}{ZRT}$$

(3.347)

where M is the molecular weight of the gas, R is the gas constant, T is the absolute temperature, and Z is the compressibility factor. The compressibility factor Z is a function of both P and T, i.e., $Z = Z(P, T)$. Under isothermal conditions where T is a constant, we may write Z as $Z(P)$ and ρ as $\rho(P)$.

Equation (3.346) is nonlinear in that the coefficients ρ and μ depend on the dependent variable P. In the case of random $k(\mathbf{x})$, this equation becomes a nonlinear stochastic partial differential equation. While Chapter 5 deals specifically with *nonlinear stochastic equations*, here we show an alternative procedure which directly utilizes the stochastic methods already discussed for linear equations.

It is well known in the literature [e.g., Aziz and Settari, 1979] that Eq. (3.346) can be made linear through the so-called *Kirchhoff integral transform*,

$$\psi(\mathbf{x}) = 2 \int_{P_o}^{P} \frac{p}{\mu(p)Z(p)} \, dp \tag{3.348}$$

where ψ is a *pseudo-pressure*. Substituting Eq. (3.348) into Eq. (3.346) and neglecting the effect of gravitation on gas flow yields

$$\nabla \cdot [k(\mathbf{x})\nabla\psi(\mathbf{x})] = -\frac{2RT}{M} g_\rho(\mathbf{x})$$

$$\psi(\mathbf{x}) = \psi_B(\mathbf{x}), \quad \mathbf{x} \in \Gamma_D \tag{3.349}$$

$$k(\mathbf{x})\nabla\psi(\mathbf{x}) \cdot \mathbf{n}(\mathbf{x}) = -\frac{2RT}{M} Q_\rho(\mathbf{x}), \quad \mathbf{x} \in \Gamma_N$$

where ψ_B is the Kirchhoff transform of P_B on the Dirichlet boundary. It is clear that Eq. (3.349) is linear. With an approximation Eq. (3.346) can be also made linear in terms of P^2 [e.g., Aziz and Settari, 1979]. The approximate equation is exactly the same as Eq. (3.349) if letting $\psi = P^2$ there. The linearized equation through the approximate procedure may not be satisfactory when the gradients of P are significant. For an ideal gas under isothermal condition, the two procedures are, however, equivalent.

The linear equation (3.349) is analogous to Eqs. (3.25)–(3.27) for groundwater flow, which are restated from Eqs. (3.1)–(3.4). Similarly, Darcy's law may be written in terms of the mass flux $\rho\mathbf{q}$ and the pseudo-pressure ψ,

$$\rho(P)\mathbf{q}(\mathbf{x}) = -\frac{M}{2RT} k(\mathbf{x})\nabla\psi(\mathbf{x}) \tag{3.350}$$

Therefore, the stochastic methods introduced earlier are applicable to Eqs. (3.349) and (3.350) and the results for the hydraulic head $h(\mathbf{x})$ and the flux $\mathbf{q}(\mathbf{x})$ can be directly translated to the pseudo-pressure $\psi(\mathbf{x})$ and the mass flux $\rho(P)\mathbf{q}(\mathbf{x})$. With this linearization procedure, Tartakovsky [1999] has recently studied gas flow using the Green's-function-based moment equation method. He also gave explicit expressions for the effective permeability relating $\langle\rho\mathbf{q}\rangle$ and $\nabla\langle\psi\rangle$ on the basis of literature results regarding effective conductivities of uniform mean groundwater flow (see Section 3.5.1). However, it is not so straightforward to convert the moments of $\psi(\mathbf{x})$ and $\rho(P)\mathbf{q}(\mathbf{x})$ to those of $P(\mathbf{x})$ and $\mathbf{q}(\mathbf{x})$, if the latter moments are of primary interest.

3.12.2 Conditional Moment Equations

In this chapter, the stochastic methods are discussed in the context of unconditional statistical moments. In these unconditional stochastic methods, actual measurements are not utilized in obtaining the moments of dependent variables, although they may be used to infer the statistical moments of the independent variables. The moments are said to be *conditional* when they are conditioned on the actual measurements of the independent or dependent variables from the formation under study. The conditional moments are often desirable because they provide site-specific estimates and reduced estimation uncertainties [e.g., Dagan, 1982a, 1985; Rubin and Dagan, 1987, 1988; Graham and McLaughlin, 1989b; Rubin, 1991; Neuman and Orr, 1993; Neuman, 1993; Zhang and Neuman, 1995a,b; Harter and Yeh, 1996b; Guadagnini and Neuman, 1999a,b].

The measurement conditioning may enter into the stochastic solution at different stages. For example, one may first derive the unconditional statistical moments of flow on the basis of unconditional moments of log hydraulic conductivity $\langle f(\mathbf{x}) \rangle$ and $C_f(\mathbf{x}, \boldsymbol{\chi})$ and then make the flow moments conditioned on the measurements of f and/or hydraulic head $h(\mathbf{x})$ by the scheme of Gaussian conditioning. The latter scheme has been discussed in Section 2.3.3 for a single stochastic process (see Eqs. (2.221)–(2.223)), and it is generalized here for the joint $f(\mathbf{x})$ and $h(\mathbf{x})$ fields,

$$\langle h(\mathbf{x}) \rangle^c = \langle h(\mathbf{x}) \rangle + \sum_{i=1}^{M_h} a_i(\mathbf{x})[h(\mathbf{x}_i) - \langle h(\mathbf{x}_i) \rangle] + \sum_{j=1}^{M_f} b_j(\mathbf{x})[f(\mathbf{x}_j) - \langle f(\mathbf{x}_j) \rangle] \quad (3.351)$$

$$C_h^c(\mathbf{x}, \boldsymbol{\chi}) = C_h(\mathbf{x}, \boldsymbol{\chi}) - \sum_{i=1}^{M_h} a_i(\mathbf{x}) C_h(\mathbf{x}_i, \boldsymbol{\chi}) - \sum_{j=1}^{M_f} b_j(\mathbf{x}) C_{fh}(\mathbf{x}_j, \boldsymbol{\chi}) \quad (3.352)$$

where M_h and M_f are the number of measurements for h and f, respectively; $\langle f(\mathbf{x}) \rangle$ and $C_f(\mathbf{x}, \boldsymbol{\chi})$ are the input unconditional moments of f; $\langle h(\mathbf{x}) \rangle$, $C_h(\mathbf{x}, \boldsymbol{\chi})$, and $C_{fh}(\mathbf{x}, \boldsymbol{\chi})$ are the unconditional moments obtained by solving unconditional moment equations or from unconditional Monte Carlo simulations; and the coefficients $a_i(\mathbf{x})$ and $b_j(\mathbf{x})$ are solutions of the following two linear systems of equations:

$$\sum_{i=1}^{M_h} a_i(\mathbf{x}) C_h(\mathbf{x}_i, \mathbf{x}_k) + \sum_{j=1}^{M_f} b_j(\mathbf{x}) C_{fh}(\mathbf{x}_j, \mathbf{x}_k) = C_h(\mathbf{x}, \mathbf{x}_k) \quad (3.353)$$

$$\sum_{i=1}^{M_h} a_i(\mathbf{x}) C_{fh}(\mathbf{x}_l, \mathbf{x}_i) + \sum_{j=1}^{M_f} b_j(\mathbf{x}) C_f(\mathbf{x}_j, \mathbf{x}_l) = C_{fh}(\mathbf{x}_l, \mathbf{x}) \quad (3.354)$$

for $1 \le k \le M_h$ and $1 \le l \le M_f$. Similar equations can be written for the moments of $f(\mathbf{x})$ as well as those of flux $q_i(\mathbf{x})$ by conditioning on the measurements of f and h. Equations (3.351)–(3.354) can result from either Gaussian conditioning [e.g., Rubin and Dagan, 1987] (see also Section 2.3.3) or cokriging [e.g., Hoeksema and Kitanidis, 1984, 1985b; de Marsily, 1986]. For Gaussian conditioning, which requires the random fields f and h to be jointly multivariate normal, this set of equations yields the conditional mean and covariance. On the other hand, cokriging gives unbiased, minimum variance estimates but not the conditional moments, while it does not assume normality. Therefore, it is expected that Eqs. (3.351) (3.354) yield exact conditional moments when the underlying distributions are normal, whereas they may provide approximate conditional moments when the normality requirement is not met.

An alternative procedure is to first derive equations governing the conditional moments of flow from the original stochastic equations and then solve these *conditional moment equations* numerically or analytically. Below we briefly discuss the methodology. As for unconditional moment equation methods, the random field $f(\mathbf{x})$ is decomposed into the mean (conditional mean, in this case) $\langle f(\mathbf{x}) \rangle^c$ and the zero-mean fluctuation $f'(\mathbf{x})$, where the superscript c stands for conditioning. The conditioning of $f(\mathbf{x})$ is made on available measurements of f and/or h. The conditional moments $\langle f(\mathbf{x}) \rangle^c \equiv \langle f(\mathbf{x})|c(\mathbf{x}) \rangle$ and $C_f^c(\mathbf{x}, \chi)$ can be obtained from a set of equations similar to Eqs. (3.351)–(3.354).

The essence of a conditional moment equation method is to derive equations for conditional flow moments such as $\langle h(\mathbf{x}) \rangle^c \equiv \langle h(\mathbf{x})|c(\mathbf{x}) \rangle$ and $C_h^c(\mathbf{x}, \chi)$ in terms of the conditional moments of $f(\mathbf{x})$. The general procedure is similar to that for unconditional moment equations, except for some issues specific to conditional moments. For example, taking conditional expectation of Eq. (3.5) yields

$$\left\langle \frac{\partial^2 h(\mathbf{x})}{\partial x_i^2} \right\rangle^c + \left\langle \frac{\partial f(\mathbf{x})}{\partial x_i} \frac{\partial h(\mathbf{x})}{\partial x_i} \right\rangle^c = -g(\mathbf{x}) \left\langle e^{-f(\mathbf{x})} \right\rangle^c \tag{3.355}$$

where the boundary conditions are left out for the time being and $g(\mathbf{x})$ is assumed deterministic. One key step in the derivation of unconditional moment equations is to commute between the unconditional expectation and differentiation, namely

$$\left\langle \frac{\partial h(\mathbf{x})}{\partial x_i} \right\rangle = \frac{\partial \langle h(\mathbf{x}) \rangle}{\partial x_i} \tag{3.356}$$

However, the conditional counterpart of Eq. (3.356) may not be true. A discussion is given in Section 2.4.3 on the conditions that may warrant the following approximation:

$$\left\langle \frac{\partial h(\mathbf{x})}{\partial x_i} \middle| c(\mathbf{x}) \right\rangle \approx \frac{\partial \langle h(\mathbf{x})|c(\mathbf{x}) \rangle}{\partial x_i} \tag{3.357}$$

It seems that many researchers have utilized Eq. (3.357) without realizing its approximate nature. While detailed studies are needed to determine the conditions for Eq. (3.357) and its effects on the resulting conditional moments, it may be reasonable to expect Eq. (3.357) to be a good approximation if the conditioning measurements are sparse or vary slowly with respect to x_i.

The other significant feature of conditional moments is that they are spatially nonstationary even if the unconditional moments are assumed stationary and the effect of boundary conditions is neglected. With this recognition and the untilization of Eq. (3.357), the unconditional moment equation methods discussed in this chapter are directly applicable to the derivation of conditional moment equations. The resulting equations for the conditional head moments $\langle h(\mathbf{x})\rangle^c$ and $C_h^c(\mathbf{x}, \chi)$ will be the same as Eqs. (3.73)–(3.76) if using the partial-differential-equation-based perturbative approach or as Eqs. (3.82)–(3.87) if using the Green's-function-based perturbative approach, upon replacing the unconditional moments of f by their conditional counterparts $\langle f(\mathbf{x})\rangle^c$ and $C_f^c(\mathbf{x}, \chi)$. These conditional moment equations can be solved in the same way as their (nonstationary) unconditional counterparts.

3.12.3 Field Theoretic Methods

The methods of field theory have recently found applications in studying flow in heterogeneous porous media. In particular, the methods of *renormalization*, *renormalization group*, and *Feynman diagrams* are receiving more and more attention in the literature of flow in porous media. In this subsection, we briefly discuss these field theoretic methods and refer the reader to appropriate references for more details.

Renormalization and renormalization group Renormalization generally means the procedure for redefining fundamental processes into larger scales. The renormalization procedure was developed originally for the purpose of removing divergences in quantum field theory. A huge body of literature on renormalization exists in quantum field theory, statistical physics, and other fields. Early applications of the renormalization procedure to flow in porous media include King [1989], who used this technique for upscaling permeability values on numerical grids. As illustrated in Fig. 3.18, the permeability values at the two-dimensional, fine grid of 2^6 (or, more generally, 2^{Nd}, where d is the space dimensionality) meshes are processed to obtain the rescaled (or renormalized) values at the coarser one with 2^4 (or, $2^{(N-1)d}$) meshes. This procedure is repeated until a grid of the desired size is reached. It is seen that renormalization is a recursive algorithm. The permeability values at the finer grid are implicitly accounted for through the renormalized values at the coarse grid. However, there is no universal theoretical formula in two and three dimensions for calculating the renormalized (upscaled, block, or equivalent) permeabilities from finer scales. In the two-dimensional case, the renormalized permeability at each block needs to be calculated from the four sub-blocks at the finer level. This calculation can be done either numerically or analytically. A numerical approach for calculating the renormalized (block) permeability values can be computationally very expensive if a large number of blocks are involved. Using successive star–triangle transformations, King [1989] derived

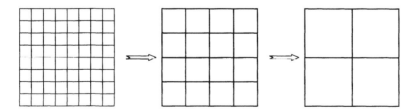

Figure 3.18. Illustration of the renormalization procedure.

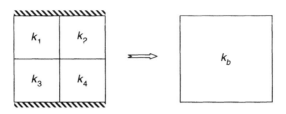

Figure 3.19. Calculation of the renormalized block permeability.

an approximate expression for the renormalized (block) permeability,

$$k_b = 4(k_1 + k_3)(k_2 + k_4)[k_2k_4(k_1 + k_3) + k_1k_3(k_2 + k_4)]$$
$$\cdot \{[k_2k_4(k_1 + k_3) + k_1k_3(k_2 + k_4)][k_1 + k_2 + K_3 + k_4]$$
$$+ 3(k_1 + k_2)(k_3 + k_4)(k_1 + k_3)(k_2 + k_4)\}^{-1} \qquad (3.358)$$

where k_i ($i = 1, 2, 3$, and 4) are the permeability values of the four sub-blocks shown in Fig. 3.19. Note that the four sub-blocks are numbered in a particular way and that Eq. (3.358) is derived with the specific boundary conditions: no-flow along the 1–2 and 3–4 sides, and constant pressure along the 1–3 side and the 2–4 side. This treatment of boundary conditions is not rigorous for the internal blocks of the flow domain. Renard and de Marsily [1997] reviewed some other approximate renormalization formulae. It may be worthwhile to mention that although it is customary to group four (or eight) sub-blocks in 2-D (or 3-D), other ways of grouping may be used for renormalization.

A number of studies demonstrated that renormalization is a promising and powerful approach for upscaling. It can also be easily implemented. However, there may be large systematic errors, for instance, in the presence of a large anisotropy [King, 1989]. Moreover, there are no general error estimation procedures associated with the renormalization procedure, although King [1996] has recently made some progress in addressing this problem.

One shall distinguish renormalization from *renormalization group* as the latter is one way for achieving renormalization. In particular, the renormalization group theory was initially devised for improving perturbation theory by exploiting the nonuniqueness in the renormalization procedure and was subsequently used for establishing connections between physics at different scale levels. The renormalization group method has been found useful in studying multiple space- and time-scale problems in a variety of fields such as statistical thermodynamics, quantum electrodynamics, and turbulence. In particular, the work of Wilson [1975] in developing the renormalization group theory for critical phenomena led to his being awarded the Nobel Prize in Physics in 1982. Applications of renormalization group perturbation expansions to subsurface flow and transport problems include Zhang [1995, 1997] and Jaekel and Vereecken [1997].

Feynman diagrams As seen earlier in this chapter, formal perturbation expansions of the stochastic flow equation are complex and may be represented by partial differential or integrodifferential equations. Furthermore, they can be written in terms of stochastic and deterministic operators, which, as shown in Section 3.10, may facilitate developing closure approximations for closing the formal perturbation series.

Feynman [1948] was the first to represent formal perturbation series graphically by diagrams, which are the so-called *Feynman diagrams*. Since then, diagrammatic methods have been used in many scientific disciplines. The first application of Feynman diagrams to flow in porous media was probably due to King [1987], who represented an infinite series of the stochastic pressure equation in random porous media graphically. From this graphical representation, he evaluated the effective permeability and the first two moments of pressure. More recent works include Christakos *et al.* [1995], Zhang [1995, 1997], Jaekel and Vereecken [1997], and Noetinger and Gautier [1998].

The method of Feynman diagrams provides a way to visualize complicated multifold integrals that appear in the perturbation series expansion of the stochastic flow equation. As it allows a more attractive graphical interpretation of the perturbation series, it may facilitate the development of strategies for partial summations that include an infinite number of selected terms from the series. One of the strategies is to include the diagrams that have a more restrictive topological structure than the full diagrammatic expansion set. Unlike the conventional perturbation series methods that truncate the series at some finite orders, this strategy retains infinite-order correlations. However, the validity of a particular strategy cannot be revealed merely on the basis of the graphical representation but should be examined with other physical or numerical means as for any other perturbation methods.

3.13 EXERCISES

1. The moment partial differential equations (3.12)–(3.16) are derived without any assumption on the statistical nature of the log-transformed hydraulic conductivity $f(\mathbf{x})$, the forcing term $g(\mathbf{x})$, the boundary head term $H_B(\mathbf{x})$, and the boundary flux term $Q(\mathbf{x})$.

   (a) Rewrite these equations under the condition that f, q, H_B, and Q are mutually uncorrelated.
   (b) Show that these partial differential equations are equivalent to the integrodifferential equations (3.62)–(3.66).

2. The seepage velocity $\mathbf{u}(\mathbf{x})$ is related to the flux $\mathbf{q}(\mathbf{x})$ via $\mathbf{u} = \mathbf{q}/\phi$. In most stochastic theories, the porosity ϕ is treated deterministic. However, when the spatial variability of $\phi(\mathbf{x})$ is significant, it may have to be treated as a random space function.

   (a) When $\phi(\mathbf{x})$ is a random space function, write general expressions for the relationship between the moments of $\mathbf{u}$ and those of $\mathbf{q}$ and ϕ.
   (b) Simplify these expressions for the following conditions: (i) log hydraulic conductivity f and ϕ are uncorrelated; and (ii) they are linearly correlated.

3. When $f(\mathbf{x})$ is second-order stationary, it can be decomposed as $f(\mathbf{x}) = \langle f \rangle + \sigma_f f_*'(\mathbf{x})$ where $f_*'(\mathbf{x})$ has a zero mean and a unit standard deviation. If the head term $h(\mathbf{x})$ is

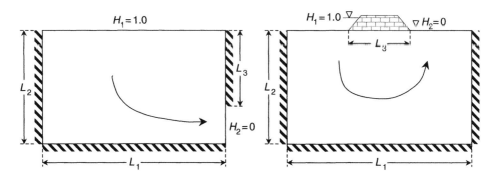

Figure 3.20. Sketch of two-dimensional nonuniform flows due to boundary conditions.

expanded as $h(\mathbf{x}) = h_*^0 + \sigma_f h_*^1 + \sigma_f^2 h_*^2 + \cdots$, derive equations for $h_*^i(\mathbf{x})$ (where $i = 0, 1$, and 2) and compare them with Eqs. (3.73)–(3.76).

4. Write a numerical code with the method of finite differences or finite elements to solve the one-dimensional version of the moment partial differential equations (3.73)–(3.76). The log hydraulic conductivity $f(\mathbf{x})$ may be assumed to be second-order stationary with a constant mean and an exponential or Gaussian covariance. Run the cases described in Fig. 1.6 (except for the white noise case) and compare results with Fig. 1.6.

5. The statistical moments of flow quantities such as the hydraulic head h and the flux q_i strongly depend on the boundary configuration. Consider the two situations illustrated in Fig. 3.20. It is very tedious, if possible, to obtain analytical solutions for the flow moments in such cases. For flow with a complex configuration, numerical approaches are indispensable.

   (a) Compute the first two moments of h and q_i for the above two flow configurations. Assume f to be second-order stationary with $\langle f \rangle = -2$ and $\sigma_f^2 = 1$; assume its covariance to to be an isotropic, exponential function with the integral scale $\lambda_f = 1$; let the parameters indicated in the figure take the following values: $L_1 = L_2 = 10$ [L], $L_3 = 3$, and $H_1 = 1$ and $H_2 = 0$. (The computation may be done with the numerical moment equation model STO-SAT, which may be obtained from the writer, or any other available moment equation codes.)
   (b) Plot the mean head and the head standard deviation fields and make a qualitative comparison with the Monte Carlo results of Smith and Freeze [1979, Figures 7 and 13]. Use some streamlines to depict the mean flow behaviors in each case.

4

TRANSIENT SATURATED FLOW

4.1 INTRODUCTION

In this chapter, we consider transient, single-phase fluid flow satisfying the following continuity equation and Darcy's law [Bear, 1972]:

$$S_s(\mathbf{x}) \frac{\partial h(\mathbf{x}, t)}{\partial t} + \nabla \cdot \mathbf{q}(\mathbf{x}, t) = g(\mathbf{x}, t) \tag{4.1}$$

$$q_i(\mathbf{x}, t) = -K_S(\mathbf{x}) \frac{\partial h(\mathbf{x}, t)}{\partial x_i} \tag{4.2}$$

subject to initial and boundary conditions

$$h(\mathbf{x}, 0) = H_o(\mathbf{x}), \quad \mathbf{x} \in \Omega \tag{4.3}$$

$$h(\mathbf{x}, t) = H_B(\mathbf{x}, t), \quad \mathbf{x} \in \Gamma_D \tag{4.4}$$

$$\mathbf{q}(\mathbf{x}, t) \cdot \mathbf{n}(\mathbf{x}) = Q(\mathbf{x}, t), \quad \mathbf{x} \in \Gamma_N \tag{4.5}$$

where $\mathbf{q}$ is the specific discharge (flux), $S_s(\mathbf{x})$ is the specific storage, $g(\mathbf{x}, t)$ is the source/sink function (due to recharge, pumping, or injection; positive for source and negative for sink), $h(\mathbf{x}, t) = P(\mathbf{x})/(\rho g) + z$ is the hydraulic head (where P is pressure, ρ fluid density, g gravitational acceleration factor, and z elevation), $K_S(\mathbf{x}) = k(\mathbf{x})\rho g/\mu$ is the hydraulic conductivity (where k is intrinsic or absolute permeability, and μ fluid viscosity), $i = 1, \ldots, d$ (where d is the number of space dimensions), $H_o(\mathbf{x})$ is the initial head, $H_B(\mathbf{x}, t)$ is the prescribed head on Dirichlet boundary segments Γ_D, Q is the prescribed flux across Neumann boundary segments Γ_N, and $\mathbf{n}(\mathbf{x}) = (n_1, \ldots, n_d)^T$ is an outward unit vector normal to the boundary. As usual, K_S is assumed to be isotropic locally and is treated as a random space function. $H_o(\mathbf{x})$, $H_B(\mathbf{x}, t)$, Q, and $g(\mathbf{x}, t)$ are assumed to be deterministic (i.e., known with certainty). In addition, $S_s(\mathbf{x})$ is assumed to be a deterministic function, as done by Indelman [1996] and Tartakovsky and Neuman [1998a,b].

When the hydraulic conductivity $K_S(\mathbf{x})$ or its log-transformed value $f(\mathbf{x}) = \ln K_S(\mathbf{x})$ is treated as a random spatial function, Eqs. (4.1)–(4.5) become stochastic partial differential equations. The stochastic equations may be solved by the method of moment equations, the Monte Carlo method, or the probability density function (PDF) method. The various moment equation methods and the Monte Carlo method have been discussed in detail in Chapter 3 for steady-state flow. As mentioned in Chapter 1, any discussion of the PDF method is largely omitted in this book, since this method has not been applied to subsurface flow problems to any significant extent. In this chapter, we introduce several methods for formulating and solving moment equations.

4.2 MOMENT PARTIAL DIFFERENTIAL EQUATIONS

In this section, we discuss how to derive and solve partial differential equations governing the statistical moments of flow quantities with perturbative expansions. The approach is similar to that for steady-state flow introduced in Section 3.2.

Substituting Eq. (4.2) into Eq. (4.1) and utilizing the transformation $f(\mathbf{x}) = \ln K_S(\mathbf{x})$ yields

$$\frac{\partial^2 h(\mathbf{x}, t)}{\partial x_i^2} + \frac{\partial f(\mathbf{x})}{\partial x_i} \frac{\partial h(\mathbf{x}, t)}{\partial x_i} = S_s(\mathbf{x}) e^{-f(\mathbf{x})} \frac{\partial h(\mathbf{x}, t)}{\partial t} - g(\mathbf{x}, t) e^{-f(\mathbf{x})} \qquad (4.6)$$

Summation for repeated indices i is implied. The log-transformed hydraulic conductivity f is decomposed as $f(\mathbf{x}) = \langle f(\mathbf{x}) \rangle + f'(\mathbf{x})$ where $\langle f(\mathbf{x}) \rangle$ is the mean log hydraulic conductivity and $f'(\mathbf{x})$ is the zero-mean fluctuation. Hence the head h and other flow quantities are also random. Since the randomness of h depends on that of f, we may expand $h(\mathbf{x})$ as $h^{(0)} + h^{(1)} + h^{(2)} + \cdots$, where $h^{(n)}$ is a term of nth order in σ_f, i.e., $h^{(n)} = \mathbf{O}(\sigma_f^n)$, and σ_f is the standard deviation of f. Therefore, we have

$$\frac{\partial^2}{\partial x_i^2} [h^{(0)} + h^{(1)} + h^{(2)} + \cdots] + \frac{\partial}{\partial x_i} [\langle f(\mathbf{x}) \rangle + f'(\mathbf{x})] \frac{\partial}{\partial x_i} [h^{(0)} + h^{(1)} + h^{(2)} + \cdots]$$

$$= \frac{S_s(\mathbf{x})}{K_G(\mathbf{x})} \left[1 - f' + \frac{1}{2} f'^2 + \cdots \right] \frac{\partial}{\partial t} [h^{(0)} + h^{(1)} + h^{(2)} + \cdots]$$

$$- \frac{g(\mathbf{x}, t)}{K_G(\mathbf{x})} \left[1 - f' + \frac{1}{2} f'^2 + \cdots \right] \qquad (4.7)$$

with initial and boundary conditions

$$h^{(0)}(\mathbf{x}, 0) + h^{(1)}(\mathbf{x}, 0) + h^{(2)}(\mathbf{x}, 0) + \cdots = H_o(\mathbf{x}), \qquad \mathbf{x} \in \Omega \qquad (4.8)$$

$$h^{(0)}(\mathbf{x}, t) + h^{(1)}(\mathbf{x}, t) + h^{(2)}(\mathbf{x}, t) + \cdots = H_B(\mathbf{x}, t), \qquad \mathbf{x} \in \Gamma_D \qquad (4.9)$$

$$n_i(\mathbf{x}) \frac{\partial}{\partial x_i} [h^{(0)}(\mathbf{x}, t) + h^{(1)}(\mathbf{x}, t) + h^{(2)}(\mathbf{x}, t) + \cdots] = -Q(\mathbf{x}, t), \qquad \mathbf{x} \in \Gamma_N \qquad (4.10)$$

In the above, $K_G(\mathbf{x}) = \exp[\langle f(\mathbf{x})\rangle]$. Collecting terms at each separate order, we have

$$
\left.\begin{aligned}
&\frac{\partial^2 h^{(0)}(\mathbf{x}, t)}{\partial x_i^2} + \frac{\partial \langle f(\mathbf{x})\rangle}{\partial x_i} \frac{\partial h^{(0)}(\mathbf{x}, t)}{\partial x_i} - \frac{S_s(\mathbf{x})}{K_G(\mathbf{x})} \frac{\partial h^{(0)}(\mathbf{x}, t)}{\partial t} = -\frac{g(\mathbf{x}, t)}{K_G(\mathbf{x})} \\
&\qquad h^{(0)}(\mathbf{x}, 0) = H_o(\mathbf{x}), \quad \mathbf{x} \in \Omega \\
&\qquad h^{(0)}(\mathbf{x}, t) = H_B(\mathbf{x}, t), \quad \mathbf{x} \in \Gamma_D \\
&\qquad n_i(\mathbf{x}) \frac{\partial h^{(0)}(\mathbf{x}, t)}{\partial x_i} = -Q(\mathbf{x}, t), \quad \mathbf{x} \in \Gamma_N
\end{aligned}\right\} \tag{4.11}
$$

$$
\left.\begin{aligned}
&\frac{\partial^2 h^{(n)}(\mathbf{x}, t)}{\partial x_i^2} + \frac{\partial \langle f(\mathbf{x})\rangle}{\partial x_i} \frac{\partial h^{(n)}(\mathbf{x}, t)}{\partial x_i} \\
&\quad = -\frac{\partial h^{(n-1)}(\mathbf{x}, t)}{\partial x_i} \frac{\partial f'(\mathbf{x})}{\partial x_i} - \frac{(-1)^n}{n!} \frac{g(\mathbf{x})}{K_G(\mathbf{x})} [f'(\mathbf{x})]^n \\
&\qquad + \sum_{m=0}^n \frac{S_s(\mathbf{x})}{K_G(\mathbf{x})} \frac{(-1)^{n-m}}{(n-m)!} [f'(\mathbf{x})]^{n-m} \frac{\partial h^{(m)}(\mathbf{x}, t)}{\partial t} \\
&\qquad\qquad h^{(n)}(\mathbf{x}, 0) = 0, \quad \mathbf{x} \in \Omega \\
&\qquad\qquad h^{(n)}(\mathbf{x}, t) = 0, \quad \mathbf{x} \in \Gamma_D \\
&\qquad\qquad n_i(\mathbf{x}) \frac{\partial h^{(n)}(\mathbf{x}, t)}{\partial x_i} = 0, \quad \mathbf{x} \in \Gamma_N
\end{aligned}\right\} \tag{4.12}
$$

where $n \geq 1$. In the above, no summation for repeated indices n is implied. It can be shown that $\langle h^{(0)}\rangle = h^{(0)}$, and $\langle h^{(1)}\rangle = 0$. Hence, the mean head is $\langle h\rangle = h^{(0)}$ to zeroth or first order in σ_f and $\langle h\rangle = h^{(0)} + \langle h^{(2)}\rangle$ to second order. The head fluctuation is $h' = h^{(1)}$ to first order. Therefore, the head covariance is $C_h(\mathbf{x}, t; \boldsymbol{\chi}, \tau) = \langle h^{(1)}(\mathbf{x}, t)h^{(1)}(\boldsymbol{\chi}, \tau)\rangle$ to first order in σ_f^2. If the log hydraulic conductivity were stationary such that $\langle f\rangle$ is constant, then the second terms in Eqs. (4.11) and (4.12) would disappear. Here we let it vary. Therefore, we may include any spatial trend in permeability caused by, for example, downward or upward sand coarsening, or periodic sand–shale patterns. Other types of medium nonstationarities have been discussed in Section 2.3.3.

The head covariance $C_h(\mathbf{x}, t; \boldsymbol{\chi}, \tau)$ is obtained by multiplying Eq. (4.12) for $n = 1$ with $h^{(1)}$ at another space–time point $(\boldsymbol{\chi}, \tau)$ and then taking the ensemble average,

$$
\begin{aligned}
&\frac{\partial^2 C_h(\mathbf{x}, t; \boldsymbol{\chi}, \tau)}{\partial x_i^2} + \frac{\partial \langle f(\mathbf{x})\rangle}{\partial x_i} \frac{\partial C_h(\mathbf{x}, t; \boldsymbol{\chi}, \tau)}{\partial x_i} - \frac{S_s(\mathbf{x})}{K_G(\mathbf{x})} \frac{\partial C_h(\mathbf{x}, t; \boldsymbol{\chi}, \tau)}{\partial t} \\
&\quad = J_i(\mathbf{x}, t) \frac{\partial C_{fh}(\mathbf{x}; \boldsymbol{\chi}, \tau)}{\partial x_i} + \left[\frac{g(\mathbf{x}, t)}{K_G(\mathbf{x})} - \frac{S_s(\mathbf{x})}{K_G(\mathbf{x})} J_t(\mathbf{x}, t)\right] C_{fh}(\mathbf{x}; \boldsymbol{\chi}, \tau)
\end{aligned}
$$

$$C_h(\mathbf{x}, 0; \boldsymbol{\chi}, \tau) = 0, \quad \mathbf{x} \in \Omega$$

$$C_h(\mathbf{x}, t; \boldsymbol{\chi}, \tau) = 0, \quad \mathbf{x} \in \Gamma_D \tag{4.13}$$

$$n_i(\mathbf{x}) \frac{\partial C_h(\mathbf{x}, t; \boldsymbol{\chi}, \tau)}{\partial x_i} = 0, \quad \mathbf{x} \in \Gamma_N$$

where $J_i = -\partial h^{(0)}(\mathbf{x}, t)/\partial x_i$ is the negative of the spatial gradient of the (zeroth-order) mean hydraulic head, $J_t = \partial h^{(0)}(\mathbf{x}, t)/\partial t$ is the temporal gradient of the mean head, and the cross-covariance $C_{fh}(\mathbf{x}; \boldsymbol{\chi}, \tau)$ is obtained similarly as

$$\frac{\partial^2 C_{fh}(\mathbf{x}; \boldsymbol{\chi}, \tau)}{\partial \chi_i^2} + \frac{\partial \langle f(\boldsymbol{\chi}) \rangle}{\partial \chi_i} \frac{\partial C_{fh}(\mathbf{x}; \boldsymbol{\chi}, \tau)}{\partial \chi_i} - \frac{S_s(\boldsymbol{\chi})}{K_G(\boldsymbol{\chi})} \frac{\partial C_{fh}(\mathbf{x}; \boldsymbol{\chi}, \tau)}{\partial \tau}$$

$$= J_i(\boldsymbol{\chi}, \tau) \frac{\partial C_f(\mathbf{x}; \boldsymbol{\chi})}{\partial \chi_i} + \left[\frac{g(\boldsymbol{\chi}, \tau)}{K_G(\boldsymbol{\chi})} - \frac{S_s(\boldsymbol{\chi})}{K_G(\boldsymbol{\chi})} J_t(\boldsymbol{\chi}, \tau) \right] C_f(\mathbf{x}, \boldsymbol{\chi})$$

$$C_{fh}(\mathbf{x}; \boldsymbol{\chi}, 0) = 0, \quad \boldsymbol{\chi} \in \Omega \tag{4.14}$$

$$C_{fh}(\mathbf{x}; \boldsymbol{\chi}, \tau) = 0, \quad \boldsymbol{\chi} \in \Gamma_D$$

$$n_i(\boldsymbol{\chi}) \frac{\partial C_{fh}(\mathbf{x}; \boldsymbol{\chi}, \tau)}{\partial \chi_i} = 0, \quad \boldsymbol{\chi} \in \Gamma_N$$

All terms on the right-hand side of Eq. (4.14) are known with the solution of $h^{(0)}$ from Eq. (4.11), hence the equation governing C_{fh} is deterministic and fully solvable. With C_{fh}, C_h is solvable from Eq. (4.13).

The flux in Eq. (4.2) can be rewritten as

$$\mathbf{q} = -K_G \left[1 + f' + \frac{f'^2}{2} + \cdots \right] \nabla \left[h^{(0)} + h^{(1)} + h^{(2)} + \cdots \right] \tag{4.15}$$

Collecting terms at each separate order, we have

$$\mathbf{q}^{(0)}(\mathbf{x}, t) = -K_G(\mathbf{x}) \nabla h^{(0)}(\mathbf{x}, t) \tag{4.16}$$

$$\mathbf{q}^{(1)}(\mathbf{x}, t) = -K_G(\mathbf{x}) \left[f'(\mathbf{x}) \nabla h^{(0)}(\mathbf{x}, t) + \nabla h^{(1)}(\mathbf{x}, t) \right] \tag{4.17}$$

$$\mathbf{q}^{(2)}(\mathbf{x}, t) = -K_G(\mathbf{x}) \left[f'(\mathbf{x}) \nabla h^{(1)}(\mathbf{x}, t) + \tfrac{1}{2} f'^2(\mathbf{x}) \nabla h^{(0)}(\mathbf{x}, t) + \nabla h^{(2)}(\mathbf{x}, t) \right] \tag{4.18}$$

It can be shown that the mean flux is $\langle \mathbf{q} \rangle = \mathbf{q}^{(0)}$ to zeroth or first order in σ_f, $\langle \mathbf{q} \rangle = \mathbf{q}^{(0)} + \langle \mathbf{q}^{(2)} \rangle$ to second order, and the flux fluctuation is $\mathbf{q}' = \mathbf{q}^{(1)}$ to first order. Therefore, the flux covariance is given as

$$
C_{q_{ij}}(\mathbf{x}, t; \boldsymbol{\chi}, \tau) = K_G(\mathbf{x}) K_G(\boldsymbol{\chi}) \bigg[J_i(\mathbf{x}, t) J_j(\boldsymbol{\chi}, \tau) C_f(\mathbf{x}, \boldsymbol{\chi})
$$

$$
- J_i(\mathbf{x}, t) \frac{\partial}{\partial \chi_j} C_{fh}(\mathbf{x}; \boldsymbol{\chi}, \tau) - J_j(\boldsymbol{\chi}, \tau) \frac{\partial}{\partial x_i} C_{fh}(\boldsymbol{\chi}; \mathbf{x}, t)
$$

$$
+ \frac{\partial^2}{\partial x_i \partial \chi_j} C_h(\mathbf{x}, t; \boldsymbol{\chi}, \tau) \bigg] \tag{4.19}
$$

The mean of a flow quantity (e.g., head or flux) estimates the spatial distribution of this quantity and its standard deviation is a measure of uncertainty associated with this estimation. The first two moments can be used to approximate confidence intervals for the flow quantities at each point in the domain at each time. The second-order correction terms for the mean head and flux are given as

$$
\left.\begin{aligned}
& \frac{\partial^2 \langle h^{(2)}(\mathbf{x}, t) \rangle}{\partial x_i^2} + \frac{\partial \langle f(\mathbf{x}) \rangle}{\partial x_i} \frac{\partial \langle h^{(2)}(\mathbf{x}, t) \rangle}{\partial x_i} - \frac{S_s(\mathbf{x})}{K_G(\mathbf{x})} \frac{\partial \langle h^{(2)}(\mathbf{x}, t) \rangle}{\partial t} \\[2mm]
& = -\frac{\partial}{\partial x_i} \frac{\partial}{\partial \chi_i} C_{fh}(\mathbf{x}; \boldsymbol{\chi}, t) \bigg|_{\mathbf{x}=\boldsymbol{\chi}} + \frac{1}{2} \left[-\frac{g(\mathbf{x}, t)}{K_G(\mathbf{x})} + \frac{S_s(\mathbf{x})}{K_G(\mathbf{x})} J_t(\mathbf{x}, t) \right] \sigma_f^2(\mathbf{x}) \\[2mm]
& \quad - \frac{S_s(\mathbf{x})}{K_G(\mathbf{x})} \frac{\partial}{\partial t} C_{fh}(\mathbf{x}; \mathbf{x}, t) \\[3mm]
& \qquad \langle h^{(2)}(\mathbf{x}, 0) \rangle = 0, \quad \mathbf{x} \in \Omega \\[2mm]
& \qquad \langle h^{(2)}(\mathbf{x}, t) \rangle = 0, \quad \mathbf{x} \in \Gamma_D \\[2mm]
& \qquad n_i(\mathbf{x}) \frac{\partial \langle h^{(2)}(\mathbf{x}, t) \rangle}{\partial x_i} = 0, \quad \mathbf{x} \in \Gamma_N
\end{aligned}\right\} \tag{4.20}
$$

$$
\langle q_i^{(2)}(\mathbf{x}, t) \rangle = K_G(\mathbf{x}) \left[-\frac{\partial}{\partial x_i} \langle h^{(2)}(\mathbf{x}, t) \rangle - \frac{\partial}{\partial x_i} C_{fh}(\boldsymbol{\chi}; \mathbf{x}, t) \bigg|_{\boldsymbol{\chi}=\mathbf{x}} + \frac{J_i(\mathbf{x}, t)}{2} \sigma_f^2(\mathbf{x}) \right] \tag{4.21}
$$

Higher-order terms may be given similarly.

4.2.1 Numerical Solution Strategies

It is very difficult to derive analytical solutions for the statistical moments without further simplifying assumptions such as medium stationarity, unbounded domain, and slowly varying, uniform mean gradient. Here we solve these moment equations numerically. Hence, we do not invoke any additional assumption except for that inherent in the perturbative approach. We approximate the spatial derivatives by the central-difference scheme and the temporal derivatives by the implicit (backward) method. As found for steady-state flow in Section 3.4, the solution is facilitated by recognizing that $h^{(0)}$, $\langle h^{(2)} \rangle$, C_{Yh}, and C_h are governed by the same type of equations but with different forcing terms. Once $h^{(0)}(\mathbf{x}, t)$ is solved from Eq. (4.11), $C_{Yh}(\mathbf{x}; \boldsymbol{\chi}, \tau)$ at $\tau = t$ can be solved from Eq. (4.14). Then $\langle h^{(2)}(\mathbf{x}, t) \rangle$ in Eq. (4.20) is solvable. The details for solving the steady-state version of these equations have already been given in Section 3.4.1 and are applicable to the transient equations at each time step. It is, however, worthwhile to outline the solution procedure for $C_h(\mathbf{x}, t; \boldsymbol{\chi}, \tau)$. We may approximate the temporal derivative as

$$\frac{\partial C_h(\mathbf{x}, t; \boldsymbol{\chi}, \tau)}{\partial t} \approx \frac{C_h(\mathbf{x}, t; \boldsymbol{\chi}, \tau) - C_h(\mathbf{x}, t - \Delta t; \boldsymbol{\chi}, \tau)}{\Delta t} \tag{4.22}$$

Here the time derivative only operates on the first time argument. For simplicity, we let τ take the value of t. $C_h(\mathbf{x}, t; \boldsymbol{\chi}, \tau = t)$ is the solution that we are after, while $C_h(\mathbf{x}, t - \Delta t; \boldsymbol{\chi}, \tau = t)$ is also unknown. Therefore, we need to first find the solution of $C_h(\mathbf{x}, t - \Delta t; \boldsymbol{\chi}, \tau = t)$, which depends on $C_h(\mathbf{x}, t - 2\Delta t; \boldsymbol{\chi}, \tau = t)$. Apparently, in order to obtain $C_h(\mathbf{x}, t; \boldsymbol{\chi}, \tau = t)$ we need to solve for $C_h(\mathbf{x}, t'; \boldsymbol{\chi}, \tau = t)$ from $t' = 0$ to $t' = t$. For a large t, this may cause a significant computational burden. However, at the large time when the flow does not change rapidly, $C_h(\mathbf{x}, t - \Delta t; \boldsymbol{\chi}, \tau = t)$ may be well approximated by $C_h(\mathbf{x}, t - \Delta t; \boldsymbol{\chi}, t - \Delta t)$ which is the solution at the previous time step. This may greatly speed up the computation. Under what conditions this approximation is warranted deserves further investigation.

The finite difference scheme for transient flow moment equations has been implemented in STO-SAT, the code already mentioned in Section 3.4.1 for steady-state flow. The numerical examples shown below are computed with this code. Again, the code is available from the writer upon request.

4.2.2 Illustrative Examples

We first look at a simple case of flow in a rectangular domain of size $L_1 = 200$ [L] by $L_2 = 100$ [L] (where L in [ ] is an arbitrary length unit). A sketch of such a domain is given in Fig. 3.1. In this case, $S_s = 10^{-5}$ [1/L], $\langle f \rangle = -2.0$ (i.e., $K_G = 0.135$ [L/T]), $\sigma_f^2 = 0.23$ and $\lambda = 18$ [L]; the domain is initially at static state with $H_o = 50$ [L]; the boundary condition of $H = 100$ [L] at $x_1 = 0$ and $H = 0$ [L] at $x_1 = 200$ is imposed at $t > 0$; and the other two sides ($x_2 = 0$ and $x_2 = 100$) are impervious. The steady-state version of this case has been discussed in Section 3.4.5; it has also been studied by Smith and Freeze [1979] with the method of Monte Carlo simulations. Figure 4.1 shows the mean head (Fig. 4.1a), head standard deviation (Fig. 4.1c), mean flux (Fig. 4.1b), and flux standard deviation (Fig. 4.1d) along the horizontal centerline $x_2 = 100$ [L] at different times $t = 0.01$ [T], 0.06 [T], 0.15 [T], and 1.50 [T]. The solid curves in these figures correspond to the steady-state case, which can be computed by letting $t = \infty$ or setting $S_s = 0$. At $t = 0.01$ [T] when the flow has just started, the center of the domain is largely unaffected with constant mean head, zero head standard deviation, zero mean flow, and

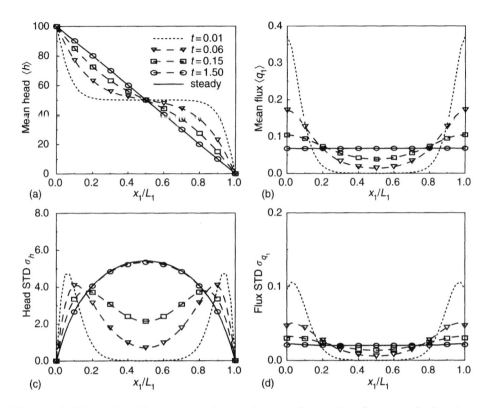

Figure 4.1. The head and flux moments along the horizontal centerline of a rectangular domain as functions of time. (Adapted from Zhang [1999a]. Copyright 1999 by Elsevier Science B.V.)

zero flux standard deviation at the domain center, while the flow is significant at the two ends. With time, the flow propagates through the entire domain such that the head standard deviation, the mean flux and the flux standard deviation increase at the domain center. At $t = 1.50$ [T], the flow quantities are identical to those at steady state except for the head standard deviation, which takes a little longer to completely reach its steady state. It is seen that the transient effect is the strongest in the head standard deviation. In Section 3.4.5, we have compared the steady-state head standard deviation with that from Smith and Freeze [1979] and found a good agreement (see Fig. 3.2).

In the above case, $\Delta x_1 = \Delta x_2 = 5$ [L], and Δt is specified as follows: $\Delta t = 0.0025$ [T] for two time steps, $\Delta t = 0.005$ [T] for one, $\Delta t = 0.05$ [T] for one, $\Delta t = 0.09$ [T] for one, $\Delta t = 0.15$ [T] for three, and $\Delta t = 0.225$ [T] for the last four time steps. Figure 4.2 shows the first two moments of head and flux at time $t = 0.06$ [T] as functions of different Δx_i and Δt. When Δx_i is changed from 5 to 10 while Δt remains the same, the statistical moments of head and flux do not change significantly. This implies that the spatial discretization is fine enough. When each of the above time steps is reduced by half while Δx_i is kept to be 5, the impacts on the statistical moments are relatively large. Even though the temporal

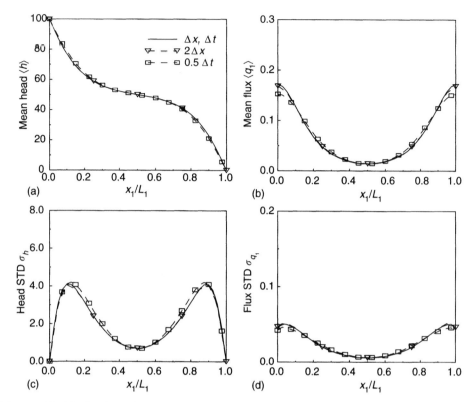

Figure 4.2. The head and flux moments along the horizontal centerline of a rectangular domain at $t = 0.06$ [T] as functions of spatial and temporal discretizations. (Adapted from Zhang [1999a]. Copyright 1999 by Elsevier Science B.V.)

discretizations used for the case shown in Fig. 4.1 are probably not fine enough, the results at large times reduce to those at steady state. This indicates that the choice of time steps has less impact on the results at late times.

We now look at two cases in the presence of wells. In the first example, a producing well of a fixed pressure ($h = 0$) is placed in the center of a square domain of size $L_i = 100$ [L] and with closed boundaries. The medium properties are the same as described in the previous subsection. The initial pressure head is $H_o = 100$ [L], the spatial discretization is $\Delta x_i = 5.0$ [L], and the time steps are as follows: $\Delta t = 0.0025$ for the first two time steps, $\Delta t = 0.005$, 0.05, and 0.09 each for one, $\Delta t = 0.15$ for three, and $\Delta t = 0.225$ for four time steps. Figure 4.3 shows the statistical moments of head (pressure) and flux along the horizontal centerline of the square domain as functions of time. In this case, the statistical moments are exactly the same along the transverse centerline. It is seen from Fig. 4.3a that the zone of pressure drop propagates with time and so does the pressure standard deviation. The mean flux and the corresponding standard deviation are the highest in the immediate vicinity of the producing well (point) and their magnitudes decrease with time. The mean

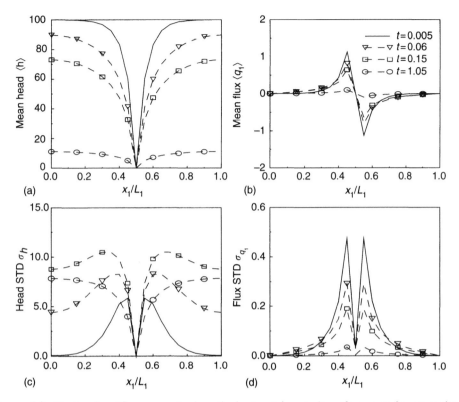

Figure 4.3. The head and flux moments along the horizontal centerline of a square domain in the presence of a pumping well. (Adapted from Zhang [1999a]. Copyright 1999 by Elsevier Science B.V.)

flow rate and the associated uncertainty at the well may be computed by integrating the flux moments around it.

Since the moment equations are solved numerically, we can handle any number of wells in any arbitrary configuration. The pressure or flow rate at the pumping/injection wells may vary with time. Figure 4.4 shows an example of three wells in the same square domain as used in Fig. 4.3. The two wells of a fixed pressure ($h = 0$) are located at (25, 75) and (75, 75), and the one with a specified pumping rate of 16.67 [L^3/T] is at (50, 25). In this example, the thickness of the domain is assumed to be 1 [L], the spatial discretization is $\Delta x_i = 2.5$ [L], and the time steps are the same as for the case shown in Fig. 4.3. The moments shown in Fig. 4.4 correspond to $t = 0.15$ [T]. The upper two plots show the first- and second-order mean heads, which are very similar. This means that the second-order correction term is not large in this case. The lower two plots show the head standard deviation and the mean flux vector. It is seen that the head standard deviation is the largest at the well of the specified rate, while it is the smallest (zero) at the wells of the fixed pressure (head). The mean head can be used to estimate the spatial distribution of head and the head standard deviation is a measure of uncertainty associated with the estimation. The mean

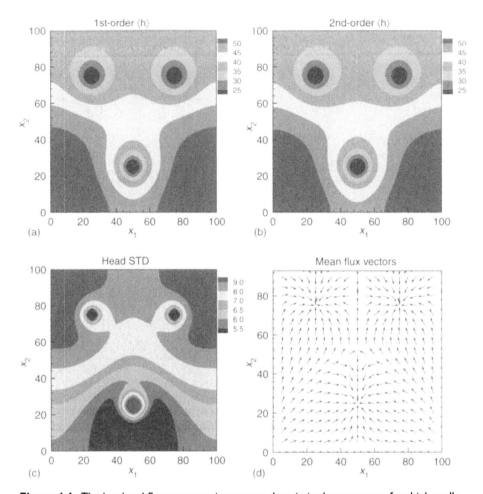

Figure 4.4. The head and flux moments in a square domain in the presence of multiple wells.

head and head standard deviation can be used to construct confidence intervals for head (pressure) at each point in the domain. This is also true for other flow quantities.

4.3 MOMENT INTEGRODIFFERENTIAL EQUATIONS

As an alternative to the partial differential equations, we may formulate the moment equations in terms of integrodifferential representations. This is commonly done with the aid of Green's function. As for steady-state flow, there are two ways to derive moment integrodifferential equations for transient flow. One is to first derive exact but unclosed moment equations and then close the equations by closure approximations or perturbation schemes [e.g., Tartakovsky and Neuman, 1998a,b]; the other is to invoke the perturbation

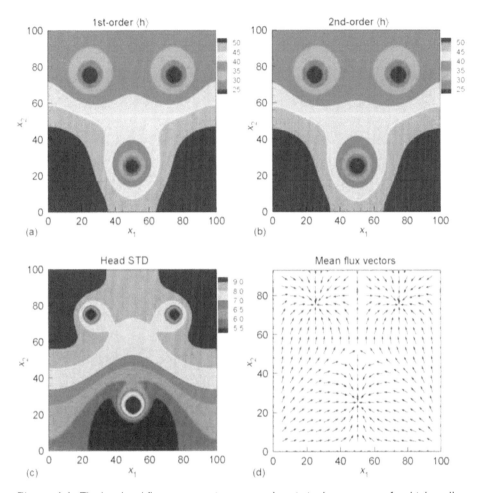

Figure 4.4. The head and flux moments in a square domain in the presence of multiple wells.

schemes from the onset and then formulate the moment integrodifferential equations [e.g., Dagan, 1982c].

With the aid of a random Green's function, Tartakovsky and Neuman [1998a,b] derived integrodifferential equations governing the expected values and covariances of hydraulic head and flux. The methodology is the same as that utilized in Section 3.2 for steady-state flow. The resulting equations are exact in that they are essentially the restatement of the original stochastic equations without any approximation. However, they are unclosed and thus unsolvable without high-resolution Monte Carlo simulations or additional approximations. In the work of Tartakovsky and Neuman [1998a,b], these integrodifferential equations are made workable by using perturbative expansions. As shown for steady-state flow, when such a perturbation scheme is utilized this approach yields the same equations as does the second procedure where the perturbation expansions are invoked from the onset. Below we choose the alternative yet more straightforward procedure to derive the flow moment equations.

4.3.1 Green's Function Representation

We begin by rewriting Eqs. (4.1)–(4.5) as

$$\nabla \cdot [K_S(\mathbf{x})\nabla h(\mathbf{x}, t)] = S_s(\mathbf{x})\frac{\partial h(\mathbf{x}, t)}{\partial t} - g(\mathbf{x}, t)$$

$$h(\mathbf{x}, 0) = H_o(\mathbf{x}), \quad \mathbf{x} \in \Omega$$

$$h(\mathbf{x}, t) = H_B(\mathbf{x}, t), \quad \mathbf{x} \in \Gamma_D$$

$$K_S(\mathbf{x})\nabla h(\mathbf{x}, t) \cdot \mathbf{n}(\mathbf{x}) = -Q(\mathbf{x}, t), \quad \mathbf{x} \in \Gamma_N$$

(4.23)

Substituting the asymptotic expansions of $K_S(\mathbf{x}) = \exp[f(\mathbf{x})]$ and $h(\mathbf{x}, t)$ into Eq. (4.23) and separating terms at each order yields

$$\left.\begin{array}{l}\nabla \cdot \left[K_G(\mathbf{x})\nabla h^{(0)}(\mathbf{x}, t)\right] = S_s(\mathbf{x})\dfrac{\partial h^{(0)}(\mathbf{x}, t)}{\partial t} - g(\mathbf{x}, t) \\[8pt] h^{(0)}(\mathbf{x}, 0) = H_o(\mathbf{x}), \quad \mathbf{x} \in \Omega \\[6pt] h^{(0)}(\mathbf{x}, t) = H_B(\mathbf{x}, t), \quad \mathbf{x} \in \Gamma_D \\[6pt] K_G(\mathbf{x})\nabla h^{(0)}(\mathbf{x}, t) \cdot \mathbf{n}(\mathbf{x}) = -Q(\mathbf{x}, t), \quad \mathbf{x} \in \Gamma_N \end{array}\right\}$$

(4.24)

$$\left.\begin{array}{l}\nabla \cdot \left\{K_G(\mathbf{x})\left[\nabla h^{(1)}(\mathbf{x}, t) + f'(\mathbf{x})\nabla h^{(0)}(\mathbf{x}, t)\right]\right\} = S_s(\mathbf{x})\dfrac{\partial h^{(1)}(\mathbf{x}, t)}{\partial t} \\[8pt] h^{(1)}(\mathbf{x}, 0) = 0, \quad \mathbf{x} \in \Omega \\[6pt] h^{(1)}(\mathbf{x}, t) = 0, \quad \mathbf{x} \in \Gamma_D \\[6pt] \left\{K_G(\mathbf{x})[\nabla h^{(1)}(\mathbf{x}, t) + f'(\mathbf{x})\nabla h^{(0)}(\mathbf{x}, t)]\right\} \cdot \mathbf{n}(\mathbf{x}) = 0, \quad \mathbf{x} \in \Gamma_N \end{array}\right\}$$

(4.25)

$$\nabla \cdot \left\{ K_G(\mathbf{x}) \left[\nabla h^{(2)}(\mathbf{x}, t) + f'(\mathbf{x}) \nabla h^{(1)}(\mathbf{x}, t) + \frac{f'^2(\mathbf{x})}{2} \nabla h^{(0)}(\mathbf{x}, t) \right] \right\}$$

$$= S_s(\mathbf{x}) \frac{\partial h^{(2)}(\mathbf{x}, t)}{\partial t}$$

$$h^{(2)}(\mathbf{x}, 0) = 0, \quad \mathbf{x} \in \Omega$$

$$h^{(2)}(\mathbf{x}, t) = 0, \quad \mathbf{x} \in \Gamma_D \qquad\qquad (4.26)$$

$$\left\{ K_G(\mathbf{x}) \left[\nabla h^{(2)}(\mathbf{x}, t) + f'(\mathbf{x}) \nabla h^{(1)}(\mathbf{x}, t) \right. \right.$$

$$\left. \left. + \frac{f'^2(\mathbf{x})}{2} \nabla h^{(0)}(\mathbf{x}, t) \right] \right\} \cdot \mathbf{n}(\mathbf{x}) = 0, \quad \mathbf{x} \in \Gamma_N$$

Equations for higher-order terms can be given similarly. Taking expectation of Eqs. (4.24) and (4.26) yields the zeroth-order mean head and the second-order mean head correction. Multiplying Eq. (4.25) by $h^{(1)}(\boldsymbol{\chi}, \tau)$ and taking expectation yields the equation for the first-order head covariance $C_h(\mathbf{x}, t; \boldsymbol{\chi}, \tau) = \langle h^{(1)}(\mathbf{x}, t) h^{(1)}(\boldsymbol{\chi}, \tau) \rangle$,

$$\nabla \cdot \left\{ K_G(\mathbf{x}) \left[\nabla C_h(\mathbf{x}, t; \boldsymbol{\chi}, \tau) + C_{fh}(\mathbf{x}; \boldsymbol{\chi}, \tau) \nabla h^{(0)}(\mathbf{x}, t) \right] \right\}$$

$$= S_s(\mathbf{x}) \frac{\partial C_h(\mathbf{x}, t; \boldsymbol{\chi}, \tau)}{\partial t}$$

$$C_h(\mathbf{x}, 0; \boldsymbol{\chi}, \tau) = 0, \quad \mathbf{x} \in \Omega \qquad\qquad (4.27)$$

$$C_h(\mathbf{x}, t; \boldsymbol{\chi}, \tau) = 0, \quad \mathbf{x} \in \Gamma_D$$

$$\left\{ K_G(\mathbf{x}) \left[\nabla C_h(\mathbf{x}, t; \boldsymbol{\chi}, \tau) + C_{fh}(\mathbf{x}; \boldsymbol{\chi}, \tau) \nabla h^{(0)}(\mathbf{x}, t) \right] \right\} \cdot \mathbf{n}(\mathbf{x}) = 0, \quad \mathbf{x} \in \Gamma_N$$

where $C_{fh}(\mathbf{x}; \boldsymbol{\chi}, \tau)$ is the cross-covariance between f and h. A similar equation may be written for C_{fh}. Alternatively, it can be represented in an integral form.

Define the zeroth-order mean Green's function as

$$\nabla_{\mathbf{x}'} \cdot \left[K_G(\mathbf{x}') \nabla_{\mathbf{x}'} G^{(0)}(\mathbf{x}', \mathbf{x}; t - t') \right] + S_s(\mathbf{x}') \frac{\partial G^{(0)}(\mathbf{x}', \mathbf{x}; t - t')}{\partial t'}$$

$$= -\delta(\mathbf{x} - \mathbf{x}') \delta(t - t')$$

$$G^{(0)}(\mathbf{x}', \mathbf{x}; t) = 0, \quad \mathbf{x}' \in \Omega \qquad\qquad (4.28)$$

$$G^{(0)}(\mathbf{x}', \mathbf{x}; t - t') = 0, \quad \mathbf{x}' \in \Gamma_D$$

$$\nabla_{\mathbf{x}'} G^{(0)}(\mathbf{x}', \mathbf{x}; t - t') \cdot \mathbf{n}(\mathbf{x}') = 0, \quad \mathbf{x}' \in \Gamma_N$$

Rewriting Eq. (4.25) in terms of $(\mathbf{x}', t')$, multiplying by the mean Green's function, integrating over domain Ω and from 0 to t, and applying Green's identity, yields

$$h^{(1)}(\mathbf{x}, t) = -\int_0^t \int_\Omega K_G(\mathbf{x}') f'(\mathbf{x}') \nabla_{\mathbf{x}'}^T h^{(0)}(\mathbf{x}', t') \nabla_{\mathbf{x}'} G^{(0)}(\mathbf{x}', \mathbf{x}; t - t') \, d\mathbf{x}' \, dt' \qquad (4.29)$$

Hence, the cross-covariance can be expressed as

$$
C_{fh}(\mathbf{x}; \boldsymbol{\chi}, \tau) = -\int_0^\tau \int_\Omega K_G(\boldsymbol{\chi}')C_f(\mathbf{x}, \boldsymbol{\chi}')\nabla_{\boldsymbol{\chi}'}^T h^{(0)}(\boldsymbol{\chi}', \tau')
$$
$$
\cdot \nabla_{\boldsymbol{\chi}'} G^{(0)}(\boldsymbol{\chi}', \boldsymbol{\chi}; \tau - \tau')\,d\boldsymbol{\chi}'\,d\tau' \tag{4.30}
$$

In passing, this expression may be obtained directly from Eq. (4.14) coupled with Eq. (4.11). Similarly, one may derive an integral expression for C_h.

To the writer's knowledge, the integrodifferential covariance equations (4.27) and (4.30) or their variants have not actually been solved for the case of transient flow. However, the finite element scheme introduced in Section 3.4.2 should be directly applicable to the transient equations. As the integrodifferential equations are equivalent to the moment partial differential equations (4.13) and (4.14), the first two flow moments should behave as those illustrated in Section 4.2.2.

4.3.2 Transient Effective Hydraulic Conductivity

Most previous studies of stochastic transient flow have concerned themselves with the solution of transient effective hydraulic conductivity $\mathbf{K}^{ef}(\mathbf{x}, t)$. Similar to its steady-state counterpart defined in Section 3.5.1, the transient effective hydraulic conductivity is defined as the tensor that relates the expected values of flux and hydraulic head gradient via

$$
\langle \mathbf{q}(\mathbf{x}, t) \rangle = -\mathbf{K}^{ef}(\mathbf{x}, t)\nabla\langle h(\mathbf{x}, t) \rangle \tag{4.31}
$$

It is seen from Eqs. (4.16) and (4.17) that the zeroth- and first-order effective hydraulic conductivity is

$$
\mathbf{K}^{ef}(\mathbf{x}, t) = K_G(\mathbf{x})\mathbf{I} \tag{4.32}
$$

where $\mathbf{I}$ is the identity tensor. This means that to first order in σ_f, the effective hydraulic conductivity does not change with time and is equal to the geometric mean hydraulic conductivity. To second order in σ_f, it is given by

$$
\langle \mathbf{q}^{(0)}(\mathbf{x}, t) \rangle + \langle \mathbf{q}^{(2)}(\mathbf{x}, t) \rangle = -\mathbf{K}^{ef}(\mathbf{x}, t)[\nabla\langle h^{(0)}(\mathbf{x}, t) \rangle + \nabla\langle h^{(2)}(\mathbf{x}, t) \rangle] \tag{4.33}
$$

It follows that one needs to evaluate both the second-order correction terms $\langle \mathbf{q}^{(2)}(\mathbf{x}, t) \rangle$ and $\langle h^{(2)}(\mathbf{x}, t) \rangle$. In bounded domains, $\langle h^{(2)}(\mathbf{x}, t) \rangle$ is neither zero nor uniform in space. However, for simplicity a common treatment is adopted by assuming that $\nabla\langle h^{(2)}(\mathbf{x}, t) \rangle$ is negligible compared to $\nabla\langle h^{(0)}(\mathbf{x}, t) \rangle$ [Dagan, 1982c; Tartakovsky and Neuman, 1998c]. Together with

Eqs. (4.16) and (4.21), the second-order $\mathbf{K}^{ef}(\mathbf{x}, t)$ is usually approximated as

$$\mathbf{K}^{ef}(\mathbf{x}, t)\mathbf{J}(\mathbf{x}, t) = K_G(\mathbf{x})\left\{\left[1 + \frac{\sigma_f^2(\mathbf{x})}{2}\right]\mathbf{I}\mathbf{J}(\mathbf{x}, t) - \boldsymbol{\alpha}(\mathbf{x}, t)\right\} \tag{4.34}$$

where $\mathbf{J}(\mathbf{x}, t) = -\nabla\langle h^{(0)}(\mathbf{x}, t)\rangle$ is the zeroth- or first-order mean head gradient, and

$$\boldsymbol{\alpha}(\mathbf{x}, t) = \nabla_{\mathbf{x}} C_{fh}(\boldsymbol{\chi}; \mathbf{x}, t)\big|_{\boldsymbol{\chi}=\mathbf{x}}$$

$$= \int_0^t \int_\Omega K_G(\mathbf{x}')C_f(\mathbf{x}, \mathbf{x}')\,\nabla_{\mathbf{x}}\nabla_{\mathbf{x}'}^T G^{(0)}(\mathbf{x}', \mathbf{x}; t - t')\mathbf{J}(\mathbf{x}', t')\,d\mathbf{x}'\,dt' \tag{4.35}$$

Hence, the evaluation of the second-order $\mathbf{K}^{ef}$ reduces to that of Eq. (4.35). On the basis of Eq. (4.35), the mean flow is nonlocal and non-Darcian in that the mean head gradient $\mathbf{J}(\mathbf{x}', t')$ resides under the integrals. Furthermore, the effective hydraulic conductivity $\mathbf{K}^{ef}$ depends not only on the medium properties (e.g., K_G and C_f) but also on the flow configuration (e.g., $G^{(0)}$ and $\mathbf{J}$). The transient effective hydraulic conductivity is, however, local in Laplace space [Indelman, 1996; Tartakovsky and Neuman, 1998c]. For transient flow in unbounded domains of stationary f fields, Indelman [1996] derived expressions for the transient effective hydraulic conductivity with the Fourier and Laplace transform. The mean flow is local in the Fourier–Laplace space.

When the mean flow varies slowly in space and time such that $\mathbf{J}(\mathbf{x}', t') \approx \mathbf{J}(\mathbf{x}, t)$, Eq. (4.35) takes the following local form:

$$\boldsymbol{\alpha}(\mathbf{x}, t) \approx \boldsymbol{\beta}(\mathbf{x}, t)\mathbf{J}(\mathbf{x}, t) \tag{4.36}$$

where

$$\boldsymbol{\beta}(\mathbf{x}, t) = \int_0^t \int_\Omega K_G(\mathbf{x}')C_f(\mathbf{x}, \mathbf{x}')\nabla_{\mathbf{x}}\nabla_{\mathbf{x}'}^T G^{(0)}(\mathbf{x}', \mathbf{x}; t - t')\,d\mathbf{x}'\,dt' \tag{4.37}$$

Thus, the transient effective hydraulic conductivity is given as

$$\mathbf{K}^{ef}(\mathbf{x}, t) = K_G(\mathbf{x})\left\{\left[1 + \frac{\sigma_f^2(\mathbf{x})}{2}\right]\mathbf{I} - \boldsymbol{\beta}(\mathbf{x}, t)\right\} \tag{4.38}$$

Dagan [1982c] evaluated Eq. (4.37) in the case of a stationary f field, isotropic C_f, and unbounded domains. In unbounded domains, the mean Green's function reads as [Dagan, 1982c; Carslaw and Jaeger, 1959]

$$G^{(0)}(\mathbf{x}', \mathbf{x}; t - t') = \frac{1}{2^d[\pi c(t - t')]^{d/2}}\exp\left[-\frac{r^2}{4c(t - t')}\right] \tag{4.39}$$

where d is the number of space dimensions, $c = K_G/S_s$, and $r^2 = \sum_{i=1}^d (x_i - x_i')^2$. With this Green's function and for a given covariance function C_f, $\boldsymbol{\beta}$ in Eq. (4.37) may be integrated analytically. Dagan [1982c] found that the effective hydraulic conductivity is stationary in space but varying in time and takes the following form:

$$K^{ef}(t) = K_G \left\{ 1 + \sigma_f^2 \left[\frac{1}{2} - \frac{1}{d} + \frac{1}{d} b_d(t) \right] \right\} \tag{4.40}$$

where $b_d(t)$ is a function dependent on the space dimensionality d. He also found that b_d is equal to unity for $t = 0$ and tends to zero as $t \to \infty$. Hence, the transient $K^{ef}(t)$ changes in the range

$$K^{ef}(0) = K_G \left(1 + \frac{\sigma_f^2}{2} \right) \tag{4.41}$$

$$K^{ef}(\infty) = K_G \left[1 + \sigma_f^2 \left(\frac{1}{2} - \frac{1}{d} \right) \right] \tag{4.42}$$

While Eq. (4.41) may be regarded as a first-order approximation of the arithmetic mean conductivity $K_A = K_G \exp(\sigma_f^2/2)$, Eq. (4.42) is identical to the steady-state effective conductivity of Eq. (3.158) under uniform mean flow. It is thus seen that for space–time slowly varying mean flow in unbounded domains, the transient effective hydraulic conductivity may have the following bounds:

$$K^{ef}(\infty) \le K^{ef}(t) \le K_A \tag{4.43}$$

Tartakovsky and Neuman [1998c] evaluated the transient effective hydraulic conductivity $\mathbf{K}^{ef}(\mathbf{x}, t)$ for spatially uniform, temporally slowly varying mean flow in bounded domains. In the case of box-shaped domains, the mean Green's function $G^{(0)}$ can be constructed with Eq. (4.39) by the method of images [Carslaw and Jaeger, 1959; Tartakovsky and Neuman, 1998c]. Tartakovsky and Neuman [1998c] found that $\beta(\mathbf{x}, t)$ is both space and time dependent and may be decomposed into the steady state and the transient components, both of which need to be evaluated by numerical quadratures.

4.4 SOME REMARKS

4.4.1 Adjoint State Equations

In the previous sections, equations governing the first two statistical moments of flow quantities are derived with a perturbative technique, and then the moment equations are solved by numerical schemes such as finite differences and finite elements. There exists an alternative numerical approach, in which the original governing equations are first discretized on specified grids using finite differences or finite elements, and then the resulting (discretized) equation is used to derive the statistical moments of the flow quantities by the adjoint state method. This approach has been discussed in detail in Section 3.8 for steady-state flow. The extension of this method to transient flow should be straightforward and is left as an exercise. Sun and Yeh [1992] formulated adjoint state equations on the basis of Eq. (4.12) with $n = 1$.

4.4.2 Laplace-Transformed Moment Equations

In the preceding sections, partial differential or integrodifferential moment equations are formulated in real space–time domains. Alternatively, one may first transform the original stochastic equations into Fourier and Laplace domains and then derive corresponding moment equations. This is exactly the approach that Indelman [1996] employed to study transient flow in unbounded domains of stationary media. For flow in bounded domains and/or for the case of nonstationary media, the Fourier transformation is no longer advantageous, while the Laplace transform may be still appropriate. The Laplace transform of Eq. (4.23) yields

$$\nabla \cdot [K_S(\mathbf{x})\nabla \tilde{h}(\mathbf{x}, s)] = s S_s(\mathbf{x})\tilde{h}(\mathbf{x}, s) - \tilde{g}(\mathbf{x}, s) - H_o(\mathbf{x})$$

$$\tilde{h}(\mathbf{x}, s) = \tilde{H}_B(\mathbf{x}, s), \quad \mathbf{x} \in \Gamma_D \tag{4.44}$$

$$K_S(\mathbf{x})\nabla \tilde{h}(\mathbf{x}, s) \cdot \mathbf{n}(\mathbf{x}) = -\tilde{Q}(\mathbf{x}, s), \quad \mathbf{x} \in \Gamma_N$$

where s is the Laplace variable, and $\tilde{u}$ is the Laplace transform of u with $u = h$, g, H_B, or Q. The definition and properties of the Laplace transform can be found in Section 2.4.5.

As $K_S(\mathbf{x})$ is a random space function, Eq. (4.44) is a stochastic partial differential equation. The moment equation methods introduced in Chapter 3 for steady-state flow may be used to formulate moment equations in Laplace space. For example, Zeitoun and Braester [1991] used the method of Adomian decomposition to derive the moment equations in the Laplace space. After solving these moment equations either analytically or numerically, the flow moments are inverted into the real-time domain with the inverse Laplace transform.

4.4.3 Quasi-Steady State Flow

There are situations where the specific storage S_s is so small that it may be neglected. In such situations, the governing flow equation (4.23) can be approximated as

$$\nabla \cdot [K_S(\mathbf{x})\nabla h(\mathbf{x}, t)] = -g(\mathbf{x}, t)$$

$$h(\mathbf{x}, t) = H_B(\mathbf{x}, t), \quad \mathbf{x} \in \Gamma_D \tag{4.45}$$

$$K_S(\mathbf{x})\nabla h(\mathbf{x}, t) \cdot \mathbf{n}(\mathbf{x}) = -Q(\mathbf{x}, t), \quad \mathbf{x} \in \Gamma_N$$

Thus, the transients in the system are caused by the time-varying sink/source term and/or the boundary terms. Any change in these terms affects the whole domain instantaneously. At each time, the flow is just like at steady state. As such, it is called *quasi-steady-state* flow.

The various moment equation methods introduced in Chapter 3 are directly applicable to quasi-steady-state flow. The resulting moment equations would be the same as their steady-state counterparts except for the additional dependency on time. Zhang and Neuman [1996] and Dagan *et al.* [1996] studied the special case of quasi-steady-state uniform mean flow in a two-dimensional, unbounded domain. In this case, the mean head gradient $J_i(t) = -\partial h^{(0)}/\partial x_i$ is spatially uniform (thus stationary in space) but time dependent.

The time dependency of J_i is caused by the time-varying boundaries set far away and may be reflected as changes in both the magnitude and the direction of $\mathbf{J} = (J_1, J_2)^T$. Independently, the above-mentioned two studies obtained analytical expressions for the flow moments and investigated the impacts of flow transients on solute transport on the basis of these quasi-steady-state moments. Under such a quasi-steady-state flow condition and with an isotropic, exponential covariance $C_f(\mathbf{r})$, the cross-covariance $C_{fh}(\mathbf{r}; \tau)$ and the head variogram $\Gamma_h(\mathbf{r}; t, \tau) = C_h(0; t, \tau) - C_h(\mathbf{r}; t, \tau)$ are given as [Zhang and Neuman, 1996]

$$C_{fh}(\mathbf{r}; \tau) = \lambda_f \sigma_f^2 \sum_{i=1}^{2} \frac{J_i(\tau) r_i'}{r'^2} [(1 + r') \exp(-r') - 1] \tag{4.46}$$

$$\Gamma_h(\mathbf{r}; t, \tau)$$

$$= \lambda_f^2 \sigma_f^2 \sum_{i=1}^{2} \sum_{j=1}^{2} J_i(t) J_j(\tau) \left\{ \left(\frac{r_i' r_j'}{r'^2} - \frac{\delta_{ij}}{2} \right) \cdot \left[\frac{1}{2} + \frac{(3 + 3r' + r'^2) - 3}{r'^2} \exp(-r') \right] \right.$$

$$\left. + \delta_{ij} [-\mathrm{Ei}(-r') + \ln r' + \exp(-r') - 0.4228] \right\} \tag{4.47}$$

where $\mathbf{r}' = \mathbf{r}/\lambda_f$, $\mathbf{r} = \mathbf{x} - \boldsymbol{\chi}$, $r_i' = r_i/\lambda_f$, $r_i = x_i - \chi_i$, and Ei is the exponential integral function. For other flow moments, the reader is referred to Zhang and Neuman [1996] and Dagan *et al.* [1996].

4.5 EXERCISES

1. Formulate partial differential equations for the first two (mean and covariance) moments of hydraulic head for the case that the specific storage $S_s(\mathbf{x})$ is also a random space function and can be decomposed as $\langle S_s(\mathbf{x}) \rangle + S_s'(\mathbf{x})$. Compare the resulting equations with Eq. (4.11) and Eqs. (4.13) and (4.14).

2. Use the adjoint state method introduced in Section 3.8 to derive moment equations for transient saturated flow in random porous media. The essence of the approach is that the flow equations are first discretized on a given grid by finite differences or finite elements and the resulting equations are then expanded by Taylor series to formulate approximate expressions for the flow statistical moments. The sensitivity matrices in these expressions are obtained with the method of adjoint state.

3. Taking the Laplace transform of Eqs. (4.34) and (4.35) leads to the relationship for the effective hydraulic conductivity in Laplace space. Show that the Laplace-transformed effective hydraulic conductivity is local.

4. Formulate partial differential equations for the first two (mean and covariance) moments for the Laplace-transformed head $\tilde{h}(\mathbf{x}, s)$ in Eq. (4.44).

(a) Compare them with the steady-state moment equations (3.73)–(3.76).

(b) Outline the solution procedures for the Laplace-transformed moment equations.

(c) Take the Laplace transform of the partial differential moment equations (4.11) and (4.13), (4.14), and compare the resulting equations with those just obtained with the different procedure.

5. Consider the case of a single pumping well of specified rate Q_w in an unbounded domain of a stationary, random hydraulic conductivity field. With the method of Green's function, derive analytical or semi-analytical expressions for the mean head, the cross-covariance of head and log-transformed hydraulic conductivity, and the head covariance for the following two cases: (a) the covariance of f is approximated by the white noise covariance $C_f(\mathbf{r}) = S_o \delta(\mathbf{r})$; (b) $C_f(\mathbf{r})$ takes the Gaussian form of Eq. (2.154).

5

UNSATURATED FLOW

5.1 INTRODUCTION

In this chapter, we consider the situation where water and air coexist. Such a zone in the subsurface is usually called the *vadose zone*, and flow in the vadose zone is called *unsaturated flow*. In recent years, consideration of the vadose zone has received increasing attention among scientists, engineers, and regulators. In the context of groundwater pollution, the vadose zone acts as a buffer and a conveyor belt between the land surface, where most contaminants originate, and groundwater, which is a resource protected under a number of environmental regulations. Unsaturated flow processes play an important role in determining the pathways of a contaminant plume before it reaches the aquifer, particularly in semiarid and arid regions where the vadose zone may be several tens to several thousands of meters thick. As mentioned earlier, geological media are heterogeneous and this heterogeneity significantly impacts fluid flow and solute transport in the subsurface. In an unsaturated system, the problem is further complicated by the fact that the flow equations are *nonlinear* because the unsaturated hydraulic conductivity depends on the pressure head.

5.1.1 Governing Equations

Let us first consider steady-state flow in unsaturated media, which satisfies the following continuity equation and Darcy's law [Bear, 1972]:

$$\nabla \cdot \mathbf{q}(\mathbf{x}) = g(\mathbf{x}) \tag{5.1}$$

$$q_i(\mathbf{x}) = -K[\psi(\mathbf{x}), \cdot]\frac{\partial \Phi(\mathbf{x})}{\partial x_i} \tag{5.2}$$

subject to boundary conditions

$$\psi(\mathbf{x}) = H_B(\mathbf{x}), \quad \mathbf{x} \in \Gamma_D \tag{5.3}$$

$$\mathbf{q}(\mathbf{x}) \cdot \mathbf{n}(\mathbf{x}) = Q(\mathbf{x}), \quad \mathbf{x} \in \Gamma_N \tag{5.4}$$

where $\mathbf{q}$ is the specific discharge (flux), g is the fluid sink/source function, $\Phi(\mathbf{x}) = \psi(\mathbf{x}) + x_1$ is the total head, $\psi = P(\mathbf{x})/(\rho g)$ is the pressure head which is negative under unsaturated conditions, $\iota = 1, \dots, d$ (where d is the number of space dimensions), $H_B(\mathbf{x})$ is the prescribed (pressure) head on Dirichlet boundary segments Γ_D, Q is the prescribed flux across Neumann boundary segments Γ_N, $\mathbf{n}(\mathbf{x}) = (n_1, \dots, n_d)^T$ is an outward unit vector normal to the boundary, and $K[\psi(\mathbf{x}), \cdot]$ is the unsaturated hydraulic conductivity which depends also on soil properties at $\mathbf{x}$. For convenience, it will be written as $K(\mathbf{x})$ in the sequel. The elevation x_1 is directed vertically upward. In these coordinates, the recharge has a negative sign. The seepage velocity at $\mathbf{x}$ is related to the specific flux q_i by

$$u_i(\mathbf{x}) = \frac{q_i(\mathbf{x})}{\theta_e(\mathbf{x})} \tag{5.5}$$

where $\theta_e \equiv \theta_e[\psi(\mathbf{x}), \cdot]$ is the effective volumetric water content, which depends on ψ and soil properties.

The governing equation for pressure head ψ as given in Eqs. (5.1) and (5.2) is nonlinear in that the coefficient (i.e., the unsaturated hydraulic conductivity $K(\mathbf{x})$) depends on the dependent variable ψ. This nonlinearity makes the problem of unsaturated flow more difficult to solve than that of saturated flow.

5.1.2 Constitutive Relations

It is clear that some model is needed to describe the *constitutive relationships* of K versus ψ and θ_e versus ψ. No universal models are available for the constitutive relationships. Instead, several empirical models are commonly used, including the Gardner–Russo model [Gardner, 1958; Russo, 1988], the Brooks–Corey model [Brooks and Corey, 1964], and the van Genuchten–Mualem model [van Genuchten, 1980]. Among these three models, the Gardner–Russo constitutive relation is the simplest and is often used for deriving analytical solutions and for stochastic modeling. However, it is well known that the more complex van Genuchten model and the Brooks–Corey model usually fit measured data better. These three models are given below.

The Gardner–Russo model reads as

$$K(\mathbf{x}) = K_S(\mathbf{x}) \exp[\alpha(\mathbf{x}) \psi(\mathbf{x})] \tag{5.6}$$

$$\theta_e(\mathbf{x}) = (\theta_s - \theta_r) \{\exp[0.5\alpha(\mathbf{x}) \psi(\mathbf{x})][1 - 0.5\alpha(\mathbf{x}) \psi(\mathbf{x})]\}^{2/(m+2)} \tag{5.7}$$

where α is the soil parameter related to pore size distribution, m is a parameter related to tortuosity, θ_r is the residual (irreducible) water content, and θ_s is the saturated water content.

The Brooks–Corey model reads as

$$K(\mathbf{x}) = K_S(\mathbf{x})[-\alpha(\mathbf{x})\psi(\mathbf{x})]^{-[2+3\beta(\mathbf{x})]}, \quad |\psi(\mathbf{x})| > 1/\alpha(\mathbf{x})$$
$$K(\mathbf{x}) = K_S(\mathbf{x}), \quad |\psi(\mathbf{x})| \le 1/\alpha(\mathbf{x}) \tag{5.8}$$

$$\theta_e(\mathbf{x}) = (\theta_s - \theta_r)[-\alpha(\mathbf{x})\psi(\mathbf{x})]^{-\beta(\mathbf{x})}, \quad |\psi(\mathbf{x})| > 1/\alpha(\mathbf{x})$$
$$\theta_e(\mathbf{x}) = \theta_s - \theta_r, \quad |\psi(\mathbf{x})| \le 1/\alpha(\mathbf{x}) \tag{5.9}$$

where α is the inverse of the absolute value of air entry pressure, and β is a parameter related to the pore size distribution.

The van Genuchten–Mualem model reads as

$$K(\mathbf{x}) = K_S(\mathbf{x})\sqrt{S(\mathbf{x})}[1 - (1 - S^{1/m})^m]^2 \tag{5.10}$$

$$\theta_e(\mathbf{x}) = (\theta_s - \theta_r)\left[1 + |\alpha(\mathbf{x})\psi(\mathbf{x})|^n\right]^{-m} \tag{5.11}$$

where $S = \theta_e/(\theta_s - \theta_r)$ is the effective saturation, α, n, and m are fitting parameters, and $m = 1 - 1/n$.

5.1.3 Spatial Variabilities

While the spatial variability of saturated hydraulic conductivity K_S is well documented in the literature, there are relatively few studies that investigate the statistical properties of rock or soil unsaturated parameters such as α, m, and β. An excellent review on the variability of Gardner–Russo model parameter α has recently been given by Tartakovsky *et al.* [1999]. The values of α were found to strongly depend on soil texture and vegetation: ranging from $0.05\,\mathrm{cm}^{-1}$ for clay to $0.71\,\mathrm{cm}^{-1}$ for gravelly loam fine sand [White and Sully, 1987]; 0.15–$1.34\,\mathrm{cm}^{-1}$ for grassland; 0.36–$0.37\,\mathrm{cm}^{-1}$ for woodland; and 0.28–$0.89\,\mathrm{cm}^{-1}$ for arable land [Ragab and Cooper, 1993a,b].

According to the above-mentioned review, the variance of log-transformed α, $\ln\alpha$, can be either larger or smaller than that of $f = \ln K_S$. Unlu *et al.* [1990b] reported the $\ln\alpha$ variances in the range 0.045–0.112, corresponding to 21.5%–34.4% for the coefficient of variation α, $C_{v_\alpha} = \sigma_\alpha/\langle\alpha\rangle$. They reported the variances for f in the range 0.391–0.96, which are equivalent to 69.2%–127% for the coefficient of variation of K_S, $C_{v_{K_S}}$. Russo *et al.* [1997] found the $\ln\alpha$ variance to be 0.425 (i.e., $C_{v_\alpha} = 72.8\%$), compared to 1.242 for the variance of f (i.e., $C_{v_{K_S}} = 156.9\%$). White and Sully [1992] and Russo and Bouton [1992] found the variances of $\ln\alpha$ and f to be of similar order, whereas Ragab and Cooper [1993a,b] observed the variance of $\ln\alpha$ to exceed that of f.

Since the expected value of α is usually small (say, $\langle\alpha\rangle = 0.1\,\mathrm{cm}^{-1}$), a moderate variability of $\sigma_{\ln\alpha}^2 = 0.22$ (i.e., $C_{v_\alpha} = 50\%$) is equivalent to a small variance value of $\sigma_\alpha^2 = 0.0025$. Thus, it would be misleading to just look at the variance of α as done in some previous studies. The better measures for the variability of α (and any other quantities) are the variance of the log-transformed α and the coefficient of variation of α, both of which are unitless.

Since the variabilities of θ_s and θ_r are likely to be small compared to that of the effective water content θ_e [Russo and Bouton, 1992], θ_r and θ_s are assumed to be constant spatially. Adding their variabilities to the moment equations to be developed is straightforward by expressing $\theta_r = \langle\theta_r\rangle + \theta_r'$ and $\theta_s = \langle\theta_s\rangle + \theta_s'$ and then following the same procedures outlined below. The soil parameters $\alpha(\mathbf{x})$ and $\beta(\mathbf{x})$, and the log-transformed saturated hydraulic conductivity $f(\mathbf{x}) = \ln K_S(\mathbf{x})$ are treated as random space functions with known means and covariances.

5.2 SPATIALLY NONSTATIONARY FLOW

The vadose zone is bounded by the land surface at the top and by the water table at the bottom. These boundaries make flow in the vadose zone complicated in nature. In particular, the statistical moments of the flow quantities are strongly dependent on location in the vadose zone even though the medium properties may be spatially stationary. In this section, we develop and solve moment equations for the full regime of unsaturated flow in the vadose zone.

5.2.1 Moment Partial Differential Equations

In this section, the random space functions $\alpha(\mathbf{x})$ and $f(\mathbf{x})$ are assumed to be second-order stationary such that their expected values are constant and their covariances depend on the relative distance of two points. This requirement is relaxed in Section 5.5.

Substituting Eq. (5.2) into Eq. (5.1) and utilizing $Y(\mathbf{x}) = \ln K(\mathbf{x})$ yields

$$\frac{\partial^2 \psi(\mathbf{x})}{\partial x_i^2} + \frac{\partial Y(\mathbf{x})}{\partial x_i}\left[\frac{\partial \psi(\mathbf{x})}{\partial x_i} + \delta_{i1}\right] = -g(\mathbf{x})e^{-Y(\mathbf{x})}$$

$$\psi(\mathbf{x}) = H_B(\mathbf{x}), \quad \mathbf{x} \in \Gamma_D \tag{5.12}$$

$$n_i(\mathbf{x})\left[\frac{\partial \psi(\mathbf{x})}{\partial x_i} + \delta_{i1}\right]\exp[Y(\mathbf{x})] = -Q(\mathbf{x}), \quad \mathbf{x} \in \Gamma_N$$

Summation for repeated indices i is implied. Since the variability of ψ depends on those of soil properties and the variability of Y depends on those of ψ and the soil properties, we may express these quantities as formal series in the following form: $\psi(\mathbf{x}) = \psi^{(0)} + \psi^{(1)} + \psi^{(2)} + \cdots$ and $Y(\mathbf{x}) = Y^{(0)} + Y^{(1)} + Y^{(2)} + \cdots$. Then, we have

$$\frac{\partial^2}{\partial x_i^2}[\psi^{(0)} + \psi^{(1)} + \psi^{(2)} + \cdots] + \frac{\partial}{\partial x_i}[Y^{(0)} + Y^{(1)} + Y^{(2)} + \cdots]$$

$$\cdot\left\{\frac{\partial}{\partial x_i}[\psi^{(0)} + \psi^{(1)} + \psi^{(2)} + \cdots] + \delta_{i1}\right\}$$

$$= -\frac{g(\mathbf{x})}{K_m(\mathbf{x})}\left\{1 - Y^{(1)} - Y^{(2)} + \frac{1}{2}[Y^{(1)}]^2 + \cdots\right\} \tag{5.13}$$

$$\psi^{(0)}(\mathbf{x}) + \psi^{(1)}(\mathbf{x}) + \psi^{(2)}(\mathbf{x}) + \cdots = H_B(\mathbf{x}), \quad \mathbf{x} \in \Gamma_D$$

$$n_i(\mathbf{x})\left\{\frac{\partial}{\partial x_i}[\psi^{(0)}(\mathbf{x}) + \psi^{(1)}(\mathbf{x}) + \psi^{(2)}(\mathbf{x}) + \cdots] + \delta_{i1}\right\}$$

$$\cdot\left\{1 + Y^{(1)} + Y^{(2)} + \frac{1}{2}[Y^{(1)}]^2 + \cdots\right\} = -\frac{Q(\mathbf{x})}{K_m(\mathbf{x})}, \quad \mathbf{x} \in \Gamma_N$$

In the above, $K_m(\mathbf{x}) = \exp[Y^{(0)}(\mathbf{x})]$. Collecting terms at separate order, we have

$$
\left.
\begin{aligned}
\frac{\partial^2 \psi^{(0)}(\mathbf{x})}{\partial x_i^2} + \frac{\partial Y^{(0)}(\mathbf{x})}{\partial x_i}\left[\frac{\partial \psi^{(0)}(\mathbf{x})}{\partial x_i} + \delta_{i1}\right] = -\frac{g(\mathbf{x})}{K_m(\mathbf{x})} \\
\psi^{(0)}(\mathbf{x}) = H_B(\mathbf{x}), \quad \mathbf{x} \in \Gamma_D \\
n_i(\mathbf{x})\left[\frac{\partial \psi^{(0)}(\mathbf{x})}{\partial x_i} + \delta_{i1}\right] = -\frac{Q(\mathbf{x})}{K_m(\mathbf{x})}, \quad \mathbf{x} \in \Gamma_N
\end{aligned}
\right\}
\tag{5.14}
$$

$$
\left.
\begin{aligned}
\frac{\partial^2 \psi^{(1)}(\mathbf{x})}{\partial x_i^2} + \frac{\partial Y^{(0)}(\mathbf{x})}{\partial x_i}\frac{\partial \psi^{(1)}(\mathbf{x})}{\partial x_i} = -J_i(\mathbf{x})\frac{\partial Y^{(1)}(\mathbf{x})}{\partial x_i} + \frac{g(\mathbf{x})}{K_m(\mathbf{x})}Y^{(1)}(\mathbf{x}) \\
\psi^{(1)}(\mathbf{x}) = 0, \quad \mathbf{x} \in \Gamma_D \\
n_i(\mathbf{x})\left[\frac{\partial \psi^{(1)}(\mathbf{x})}{\partial x_i} + J_i(\mathbf{x})Y^{(1)}(\mathbf{x})\right] = 0, \quad \mathbf{x} \in \Gamma_N
\end{aligned}
\right\}
\tag{5.15}
$$

where $J_i(\mathbf{x}) = \partial \psi^{(0)}/\partial x_i + \delta_{i1}$ is the gradient of the mean total head. Equations governing higher-order terms can be written similarly.

A constitutive relation between K and ψ must be specified in order to close Eqs. (5.14) and (5.15). Although the Brooks–Corey model [Brooks and Corey, 1964] may have certain mathematical advantages over the Gardner–Russo model in low-order stochastic analyses [Zhang et al., 1998], we use the latter for simplicity (see Zhang and Winter [1998] for the treatment based on the Brooks–Corey model). In the Gardner–Russo model, we further assume θ_s, θ_r, and m to be deterministic constants.

With $f(\mathbf{x}) = \langle f \rangle + f'(\mathbf{x})$ and $\alpha(\mathbf{x}) = \langle \alpha \rangle + \alpha'(\mathbf{x})$ and on the basis of Eq. (5.6), the log-transformed hydraulic conductivity $Y(\mathbf{x})$ can be written as

$$
\begin{aligned}
Y(\mathbf{x}) &= f(\mathbf{x}) + \alpha(\mathbf{x})\psi(\mathbf{x}) \\
&= \langle f \rangle + f'(\mathbf{x}) + [\langle \alpha \rangle + \alpha'(\mathbf{x})][\psi^{(0)} + \psi^{(1)} + \psi^{(2)} + \cdots]
\end{aligned}
\tag{5.16}
$$

Hence, we have

$$
Y^{(0)}(\mathbf{x}) = \langle f \rangle + \langle \alpha \rangle \psi^{(0)}(\mathbf{x})
\tag{5.17}
$$

$$
Y^{(1)}(\mathbf{x}) = f'(\mathbf{x}) + \langle \alpha \rangle \psi^{(1)}(\mathbf{x}) + \psi^{(0)}(\mathbf{x})\alpha'(\mathbf{x})
\tag{5.18}
$$

Substituting Eq. (5.17) into Eq. (5.14) yields

$$
\left.
\begin{aligned}
\frac{\partial^2 \psi^{(0)}(\mathbf{x})}{\partial x_i^2} + a_i(\mathbf{x})\frac{\partial \psi^{(0)}(\mathbf{x})}{\partial x_i} = -\frac{g(\mathbf{x})}{K_m(\mathbf{x})} \\
\psi^{(0)}(\mathbf{x}) = H_B(\mathbf{x}), \quad \mathbf{x} \in \Gamma_D \\
n_i(\mathbf{x})\frac{\partial \psi^{(0)}(\mathbf{x})}{\partial x_i} = -\frac{Q(\mathbf{x})}{K_m(\mathbf{x})} - \delta_{i1}n_i(\mathbf{x}), \quad \mathbf{x} \in \Gamma_N
\end{aligned}
\right\}
\tag{5.19}
$$

where $K_m(\mathbf{x}) = \exp[\langle f \rangle]\exp[\langle \alpha \rangle \psi^{(0)}]$ and $a_i(\mathbf{x}) = \langle \alpha \rangle J_i(\mathbf{x})$. This equation is nonlinear because the coefficients $K_m(\mathbf{x})$ and $a_i(\mathbf{x})$ are functions of the dependent variable $\psi^{(0)}$.

Substituting Eq. (5.18) into Eq. (5.15) yields

$$
\frac{\partial^2 \psi^{(1)}(\mathbf{x})}{\partial x_i^2} + b_i(\mathbf{x}) \frac{\partial \psi^{(1)}(\mathbf{x})}{\partial x_i} + c(\mathbf{x})\psi^{(1)}(\mathbf{x})
$$
$$
= -J_i(\mathbf{x}) \frac{\partial f'(\mathbf{x})}{\partial x_i} - J_i(\mathbf{x})\psi^{(0)}(\mathbf{x}) \frac{\partial \alpha'(\mathbf{x})}{\partial x_i} + d_1(\mathbf{x})f'(\mathbf{x}) + d_2(\mathbf{x})\alpha'(\mathbf{x}) \qquad (5.20)
$$
$$
\psi^{(1)}(\mathbf{x}) = 0, \quad \mathbf{x} \in \Gamma_D
$$
$$
n_i(\mathbf{x}) \frac{\partial \psi^{(1)}(\mathbf{x})}{\partial x_i} + d_3(\mathbf{x})\psi^{(1)}(\mathbf{x}) = d_4(\mathbf{x})f'(\mathbf{x}) + d_5(\mathbf{x})\alpha'(\mathbf{x}), \quad \mathbf{x} \in \Gamma_N
$$

where $b_i(\mathbf{x}) = [2J_i(\mathbf{x}) - \delta_{i1}]\langle \alpha \rangle$, $c(\mathbf{x}) = -\langle \alpha \rangle g(\mathbf{x})/K_m(\mathbf{x})$, $d_1(\mathbf{x}) = g(\mathbf{x})/K_m(\mathbf{x})$, $d_2(\mathbf{x}) = [g(\mathbf{x})/K_m(\mathbf{x})]\psi^{(0)}(\mathbf{x}) - J_i(\mathbf{x})\partial\psi^{(0)}/\partial x_i$, $d_3(\mathbf{x}) = \langle \alpha \rangle n_i(\mathbf{x})J_i(\mathbf{x})$, $d_4(\mathbf{x}) = -n_i(\mathbf{x})J_i(\mathbf{x})$, and $d_5(\mathbf{x}) = -n_i(\mathbf{x})J_i(\mathbf{x})\psi^{(0)}$.

Since the fluid sink/source g, and the boundary terms $H_B(\mathbf{x})$ and $Q(\mathbf{x})$ are assumed to be known with certainty and the mean soil parameters $\langle f \rangle$ and $\langle \alpha \rangle$ are given, there is no random quantity in Eq. (5.19) for $\psi^{(0)}$, i.e., $\langle \psi^{(0)} \rangle = \psi^{(0)}$. It is clear from Eq. (5.20) that $\langle \psi^{(1)} \rangle = 0$. Hence, the mean head is $\langle \psi \rangle = \psi^{(0)}$ to zeroth or first order in σ_Y and $\langle \psi \rangle = \psi^{(0)} + \langle \psi^{(2)} \rangle$ to second order. The head fluctuation is $\psi' = \psi^{(1)}$ to first order. Therefore, the head covariance is $C_\psi(\mathbf{x}, \boldsymbol{\chi}) = \langle \psi^{(1)}(\mathbf{x})\psi^{(1)}(\boldsymbol{\chi}) \rangle$ to first order in σ_Y^2 (or second order in σ_Y). Multiplying Eq. (5.20) by $\psi^{(1)}(\boldsymbol{\chi})$ and taking the ensemble mean yields

$$
\frac{\partial^2 C_\psi(\mathbf{x}, \boldsymbol{\chi})}{\partial x_i^2} + b_i(\mathbf{x}) \frac{\partial C_\psi(\mathbf{x}, \boldsymbol{\chi})}{\partial x_i} + c(\mathbf{x})C_\psi(\mathbf{x}, \boldsymbol{\chi})
$$
$$
= -J_i(\mathbf{x}) \left[\frac{\partial C_{f\psi}(\mathbf{x}, \boldsymbol{\chi})}{\partial x_i} + \psi^{(0)}(\mathbf{x}) \frac{\partial C_{\alpha\psi}(\mathbf{x}, \boldsymbol{\chi})}{\partial x_i} \right] + d_1(\mathbf{x})C_{f\psi}(\mathbf{x}, \boldsymbol{\chi})
$$
$$
+ d_2(\mathbf{x})C_{\alpha\psi}(\mathbf{x}, \boldsymbol{\chi}) \qquad (5.21)
$$
$$
C_\psi(\mathbf{x}, \boldsymbol{\chi}) = 0, \quad \mathbf{x} \in \Gamma_D
$$
$$
n_i(\mathbf{x}) \frac{\partial C_\psi(\mathbf{x}, \boldsymbol{\chi})}{\partial x_i} + d_3(\mathbf{x})C_\psi(\mathbf{x}, \boldsymbol{\chi})
$$
$$
= d_4(\mathbf{x})C_{f\psi}(\mathbf{x}, \boldsymbol{\chi}) + d_5(\mathbf{x})C_{\alpha\psi}(\mathbf{x}, \boldsymbol{\chi}), \quad \mathbf{x} \in \Gamma_N
$$

where $C_{f\psi}$ and $C_{\alpha\psi}$ are the solutions of the following equations:

$$
\left.
\begin{aligned}
&\frac{\partial^2 C_{f\psi}(\mathbf{x}, \boldsymbol{\chi})}{\partial \chi_i^2} + b_i(\boldsymbol{\chi}) \frac{\partial C_{f\psi}(\mathbf{x}, \boldsymbol{\chi})}{\partial \chi_i} + c(\boldsymbol{\chi})C_{f\psi}(\mathbf{x}, \boldsymbol{\chi}) \\
&= -J_i(\boldsymbol{\chi}) \left[\frac{\partial C_f(\mathbf{x}, \boldsymbol{\chi})}{\partial \chi_i} + \psi^{(0)}(\boldsymbol{\chi}) \frac{\partial C_{f\alpha}(\mathbf{x}, \boldsymbol{\chi})}{\partial \chi_i} \right] \\
&\quad + d_1(\boldsymbol{\chi})C_f(\mathbf{x}, \boldsymbol{\chi}) + d_2(\boldsymbol{\chi})C_{f\alpha}(\mathbf{x}, \boldsymbol{\chi}) \\
&\qquad C_{f\psi}(\mathbf{x}, \boldsymbol{\chi}) = 0, \quad \boldsymbol{\chi} \in \Gamma_D \\
&n_i(\boldsymbol{\chi}) \frac{\partial C_{f\psi}(\mathbf{x}, \boldsymbol{\chi})}{\partial \chi_i} + d_3(\boldsymbol{\chi})C_{f\psi}(\mathbf{x}, \boldsymbol{\chi}) \\
&= d_4(\boldsymbol{\chi})C_f(\mathbf{x}, \boldsymbol{\chi}) + d_5(\boldsymbol{\chi})C_{f\alpha}(\mathbf{x}, \boldsymbol{\chi}), \quad \boldsymbol{\chi} \in \Gamma_N
\end{aligned}
\right\} \qquad (5.22)
$$

$$\left.\begin{array}{l} \dfrac{\partial^2 C_{\alpha\psi}(\mathbf{x}, \boldsymbol{\chi})}{\partial \chi_i^2} + b_i(\boldsymbol{\chi})\dfrac{\partial C_{\alpha\psi}(\mathbf{x}, \boldsymbol{\chi})}{\partial \chi_i} + c(\boldsymbol{\chi})C_{\alpha\psi}(\mathbf{x}, \boldsymbol{\chi}) \\[3mm] = -J_i(\boldsymbol{\chi})\left[\dfrac{\partial C_{\alpha f}(\mathbf{x}, \boldsymbol{\chi})}{\partial \chi_i} + \psi^{(0)}(\boldsymbol{\chi})\dfrac{\partial C_{\alpha}(\mathbf{x}, \boldsymbol{\chi})}{\partial \chi_i}\right] \\[3mm] \quad + d_1(\boldsymbol{\chi})C_{\alpha f}(\mathbf{x}, \boldsymbol{\chi}) + d_2(\boldsymbol{\chi})C_{\alpha}(\mathbf{x}, \boldsymbol{\chi}) \\[3mm] \qquad\qquad C_{\alpha\psi}(\mathbf{x}, \boldsymbol{\chi}) = 0, \quad \boldsymbol{\chi} \in \Gamma_D \\[3mm] n_i(\boldsymbol{\chi})\dfrac{\partial C_{\alpha\psi}(\mathbf{x}, \boldsymbol{\chi})}{\partial \chi_i} + d_3(\boldsymbol{\chi})C_{\alpha\psi}(\mathbf{x}, \boldsymbol{\chi}) \\[3mm] = d_4(\boldsymbol{\chi})C_{\alpha f}(\mathbf{x}, \boldsymbol{\chi}) + d_5(\boldsymbol{\chi})C_{\alpha}(\mathbf{x}, \boldsymbol{\chi}), \quad \boldsymbol{\chi} \in \Gamma_N \end{array}\right\} \quad (5.23)$$

In the above, C_f, $C_{f\alpha}$, and C_{α} are the input covariances of the soil properties. Note that although the equation governing the first moment $\psi^{(0)}$ is nonlinear, Eqs. (5.22) and (5.23) are linear and solvable with the solution of $\psi^{(0)}(\mathbf{x})$. With $C_{f\psi}$ and $C_{\alpha\psi}$, C_ψ can be solved from Eq. (5.21).

The flux in Eq. (5.2) can be rewritten as

$$q_i(\mathbf{x}) = -K_m(\mathbf{x})\left\{1 + Y^{(1)} + Y^{(2)} + \tfrac{1}{2}[Y^{(1)}]^2 + \cdots\right\}$$
$$\cdot \left\{\frac{\partial}{\partial x_i}[\psi^{(0)} + \psi^{(1)} + \psi^{(2)} + \cdots] + \delta_{i1}\right\} \quad (5.24)$$

Collecting terms at separate order, we have

$$q_i^{(0)}(\mathbf{x}) = -K_m(\mathbf{x})\left[\frac{\partial \psi^{(0)}(\mathbf{x})}{\partial x_i} + \delta_{i1}\right] \quad (5.25)$$

$$q_i^{(1)}(\mathbf{x}) = -K_m(\mathbf{x})\left[J_i(\mathbf{x})Y^{(1)}(\mathbf{x}) + \frac{\partial \psi^{(1)}(\mathbf{x})}{\partial x_i}\right] \quad (5.26)$$

$$q_i^{(2)}(\mathbf{x}) = -K_m(\mathbf{x})\left\{\frac{1}{2}J_i(\mathbf{x})\left[Y^{(1)}(\mathbf{x})^2 + Y^{(2)}(\mathbf{x})\right]\right.$$
$$\left. + \frac{\partial \psi^{(1)}(\mathbf{x})}{\partial x_i}Y^{(1)}(\mathbf{x}) + \frac{\partial \psi^{(2)}(\mathbf{x})}{\partial x_i}\right\} \quad (5.27)$$

It can be shown that the mean flux is $\langle \mathbf{q} \rangle = \mathbf{q}^{(0)}$ to zeroth or first order in σ_Y, $\langle \mathbf{q} \rangle = \mathbf{q}^{(0)} + \langle \mathbf{q}^{(2)} \rangle$ to second order, and the flux fluctuation is $\mathbf{q}' = \mathbf{q}^{(1)}$ to first order. Therefore, to second order in σ_Y, the flux covariances are given as

$$C_{q_i q_j}(\mathbf{x}, \boldsymbol{\chi}) = K_m(\mathbf{x})K_m(\boldsymbol{\chi})\left[J_i(\mathbf{x})J_j(\boldsymbol{\chi})C_Y(\mathbf{x}, \boldsymbol{\chi}) + J_i(\mathbf{x})\frac{\partial C_{Y\psi}(\mathbf{x}, \boldsymbol{\chi})}{\partial \chi_i}\right.$$
$$\left. + J_j(\boldsymbol{\chi})\frac{\partial C_{Y\psi}(\boldsymbol{\chi}, \mathbf{x})}{\partial x_i} + \frac{\partial^2 C_\psi(\mathbf{x}, \boldsymbol{\chi})}{\partial x_i \partial \chi_j}\right] \quad (5.28)$$

where C_Y and $C_{Y\psi}$ are given as

$$C_Y(\mathbf{x}, \boldsymbol{\chi}) - C_f(\mathbf{x}, \boldsymbol{\chi}) + \langle u \rangle^2 C_\psi(\mathbf{x}, \boldsymbol{\chi}) + \psi^{(0)}(\mathbf{x})\psi^{(0)}(\boldsymbol{\chi})C_\alpha(\mathbf{x}, \boldsymbol{\chi})$$
$$+ \langle\alpha\rangle C_{f\psi}(\mathbf{x}, \boldsymbol{\chi}) + \langle\alpha\rangle C_{f\psi}(\boldsymbol{\chi}, \mathbf{x}) + \psi^{(0)}(\boldsymbol{\chi})C_{f\alpha}(\mathbf{x}, \boldsymbol{\chi})$$
$$+ \psi^{(0)}(\mathbf{x})C_{f\alpha}(\boldsymbol{\chi}, \mathbf{x}) + \psi^{(0)}(\mathbf{x})\langle\alpha\rangle C_{\alpha\psi}(\mathbf{x}, \boldsymbol{\chi})$$
$$+ \psi^{(0)}(\boldsymbol{\chi})\langle\alpha\rangle C_{\alpha\psi}(\boldsymbol{\chi}, \mathbf{x}) \tag{5.29}$$

$$C_{Y\psi}(\mathbf{x}, \boldsymbol{\chi}) = C_{f\psi}(\mathbf{x}, \boldsymbol{\chi}) + \langle\alpha\rangle C_\psi(\mathbf{x}, \boldsymbol{\chi}) + \psi^{(0)}(\mathbf{x})C_{\alpha\psi}(\mathbf{x}, \boldsymbol{\chi}) \tag{5.30}$$

With the moments involving head, one may evaluate these flux moments readily. For the case of $m = 0$, Eq. (5.7) can be rewritten as

$$\theta_e(\mathbf{x}) = (\theta_s - \theta_r)\exp\{0.5[\langle\alpha\rangle + \alpha'(\mathbf{x})][\psi^{(0)} + \psi^{(1)} + \psi^{(2)} + \cdots]\}$$
$$\cdot \{1 - 0.5[\langle\alpha\rangle + \alpha'(\mathbf{x})][\psi^{(0)} + \psi^{(1)} + \psi^{(2)} + \cdots]\}$$
$$= (\theta_s - \theta_r)\exp[0.5\langle\alpha\rangle\psi^{(0)}]$$
$$\cdot \exp\{0.5[\langle\alpha\rangle\psi^{(1)} + \psi^{(0)}\alpha'(\mathbf{x}) + \alpha'\psi^{(1)} + \langle\alpha\rangle\psi^{(2)} + \cdots]\}$$
$$\cdot \{1 - 0.5[\langle\alpha\rangle\psi^{(0)} + \langle\alpha\rangle\psi^{(1)} + \psi^{(0)}\alpha'(\mathbf{x}) + \alpha'\psi^{(1)} + \langle\alpha\rangle\psi^{(2)} + \cdots]\}$$
$$= (\theta_s - \theta_r)\exp[0.5\langle\alpha\rangle\psi^{(0)}]\{1 + 0.5\langle\alpha\rangle\psi^{(1)} + 0.5\psi^{(0)}\alpha'(\mathbf{x}) + 0.5\alpha'\psi^{(1)}$$
$$+ 0.5\langle\alpha\rangle\psi^{(2)} + 0.125\langle\alpha\rangle^2[\psi^{(1)}]^2 + 0.125[\psi^{(0)}]^2[\alpha']^2 \ldots]\}$$
$$\cdot \{1 - 0.5[\langle\alpha\rangle\psi^{(0)} + \langle\alpha\rangle\psi^{(1)} + \psi^{(0)}\alpha'(\mathbf{x}) + \alpha'\psi^{(1)} + \langle\alpha\rangle\psi^{(2)} + \cdots]\}$$
$$\tag{5.31}$$

The treatment for $m \neq 0$ can be found in Zhang *et al.* [1998] and Harter and Zhang [1999]. Collecting terms at separate order, we have

$$\theta_e^{(0)}(\mathbf{x}) = (\theta_s - \theta_r)\exp[0.5\langle\alpha\rangle\psi^{(0)}][1 - 0.5\langle\alpha\rangle\psi^{(0)}] \tag{5.32}$$

$$\theta_e^{(1)}(\mathbf{x}) = -(\theta_s - \theta_r)\exp[0.5\langle\alpha\rangle\psi^{(0)}]\{0.25\langle\alpha\rangle^2\psi^{(0)}\psi^{(1)} + 0.25\langle\alpha\rangle[\psi^{(0)}]^2\alpha'\} \tag{5.33}$$

$$\theta_e^{(2)}(\mathbf{x}) = -(\theta_s - \theta_r)\exp[0.5\langle\alpha\rangle\psi^{(0)}]\{0.75\langle\alpha\rangle\psi^{(0)}\alpha'\psi^{(1)} + 0.25\langle\alpha\rangle^2\psi^{(0)}\langle\alpha\rangle\psi^{(2)}$$
$$+ [0.125 + 0.0625\langle\alpha\rangle\psi^{(0)}][\langle\alpha\rangle^2[\psi^{(1)}]^2 + [\psi^{(0)}]^2[\alpha']^2]\} \tag{5.34}$$

The covariance of θ_e is obtained with Eq. (5.33) as

$$C_{\theta_e}(\mathbf{x}, \boldsymbol{\chi}) = (\theta_s - \theta_r)^2\exp[0.5\langle\alpha\rangle\psi^{(0)}(\mathbf{x})]\exp[0.5\langle\alpha\rangle\psi^{(0)}(\boldsymbol{\chi})]$$
$$\cdot [p_3(\mathbf{x})p_3(\boldsymbol{\chi})C_\psi(\mathbf{x}, \boldsymbol{\chi}) + p_3(\mathbf{x})p_4(\boldsymbol{\chi})C_{\alpha\psi}(\boldsymbol{\chi}, \mathbf{x})$$
$$+ p_3(\boldsymbol{\chi})p_4(\mathbf{x})C_{\alpha\psi}(\mathbf{x}, \boldsymbol{\chi}) + p_4(\mathbf{x})p_4(\boldsymbol{\chi})C_\alpha(\mathbf{x}; \boldsymbol{\chi})] \tag{5.35}$$

where $p_3(\mathbf{x}) = 0.25\langle\alpha\rangle^2\psi^{(0)}(\mathbf{x})$ and $p_4(\mathbf{x}) = 0.25\langle\alpha\rangle[\psi^{(0)}(\mathbf{x})]^2$.

We may rewrite Eq. (5.5) as

$$
\begin{aligned}
u_i(\mathbf{x}) &= \frac{q_i^{(0)}(\mathbf{x}) + q_i^{(1)}(\mathbf{x}) + q_i^{(2)}(\mathbf{x}) + \cdots}{\theta_e^{(0)}(\mathbf{x}) + \theta_e^{(1)}(\mathbf{x}) + \theta_e^{(2)}(\mathbf{x}) + \cdots} \\
&= \frac{q_i^{(0)}(\mathbf{x}) + q_i^{(1)}(\mathbf{x}) + q_i^{(2)}(\mathbf{x}) + \cdots}{\theta_e^{(0)}(\mathbf{x})} \left[1 - \frac{\theta_e^{(1)}(\mathbf{x}) + \theta_e^{(2)}(\mathbf{x})}{\theta_e^{(0)}(\mathbf{x})} + \cdots \right]
\end{aligned}
\tag{5.36}
$$

To first order, we have

$$
u_i^{(0)}(\mathbf{x}) = \frac{q_i^{(0)}(\mathbf{x})}{\theta_e^{(0)}(\mathbf{x})}
\tag{5.37}
$$

$$
u_i^{(1)}(\mathbf{x}) = \frac{q_i^{(1)}(\mathbf{x})}{\theta_e^{(0)}(\mathbf{x})} - \frac{q_i^{(0)}(\mathbf{x})\theta_e^{(1)}(\mathbf{x})}{[\theta_e^{(0)}(\mathbf{x})]^2}
\tag{5.38}
$$

$$
\begin{aligned}
C_{u_{ij}}(\mathbf{x}, \boldsymbol{\chi}) = \frac{1}{\theta_e^{(0)}(\mathbf{x})\theta_e^{(0)}(\boldsymbol{\chi})} &[C_{q_{ij}}(\mathbf{x}, \boldsymbol{\chi}) - u_j^{(0)}(\boldsymbol{\chi})C_{q_i\theta_e}(\mathbf{x}, \boldsymbol{\chi}) \\
&- u_i^{(0)}(\mathbf{x})C_{q_j\theta_e}(\boldsymbol{\chi}, \mathbf{x}) + u_i^{(0)}(\mathbf{x})u_j^{(0)}(\boldsymbol{\chi})C_{\theta_e}(\mathbf{x}, \boldsymbol{\chi})]
\end{aligned}
\tag{5.39}
$$

It is seen from Eq. (5.32) that to the first-order mean $\theta_e^{(0)}(\mathbf{x}) \equiv \theta_e[\psi^{(0)}(\mathbf{x})]$. Hence, even if the mean specific flux is constant, the mean velocity is generally a function of mean head and spatial location. If the variability in θ_e were neglected as done in most previous stochastic theories, only the first term in the right-hand side of Eq. (5.39) would exist.

Multiplying $q_i^{(1)}(\mathbf{x})$ in Eq. (5.26) with $\theta_e^{(1)}(\boldsymbol{\chi})$ in Eq. (5.33) and taking the ensemble mean yields the cross-covariance $C_{q_i\theta_e}$ required to evaluate Eq. (5.39),

$$
\begin{aligned}
C_{q_i\theta_e}(\mathbf{x}, \boldsymbol{\chi}) = (\theta_s - \theta_r)K_m(\mathbf{x})\exp[0.5\langle\alpha\rangle\psi^{(0)}(\boldsymbol{\chi})] \\
\cdot \bigg[J_i(\mathbf{x})p_3(\boldsymbol{\chi})C_{Y\psi}(\mathbf{x}, \boldsymbol{\chi}) + J_i(\mathbf{x})p_4(\boldsymbol{\chi})C_{Y\alpha}(\mathbf{x}, \boldsymbol{\chi}) \\
+ p_3(\boldsymbol{\chi})\frac{\partial C_\psi(\mathbf{x}, \boldsymbol{\chi})}{\partial x_i} + p_4(\boldsymbol{\chi})\frac{\partial C_{\alpha\psi}(\boldsymbol{\chi}, \mathbf{x})}{\partial x_i} \bigg]
\end{aligned}
\tag{5.40}
$$

Numerical solution Note that C_ψ, $C_{f\psi}$, and $C_{\alpha\psi}$ in Eqs. (5.21)–(5.23) are governed by the same type of equations but with different forcing terms. This facilitates the solution. Once $\psi^{(0)}(\mathbf{x})$ is solved from Eq. (5.19), $C_{f\psi}$ and $C_{\alpha\psi}$ can be solved from Eqs. (5.22) and (5.23). Then C_ψ can be obtained from Eq. (5.21). However, it is very difficult to obtain analytical or semi-analytical solutions for the statistical moments without further simplifying assumptions such as the (local) stationarity of flow quantities, unbounded domain, and slowly varying, uniform mean gradient. In this section, we solve these moment equations numerically. Hence, we do not have to invoke any additional assumption except for the one inherent in the perturbative expansions, i.e., the smallness in the variabilities of soil properties.

As mentioned before, the mean pressure head equation (5.19) is nonlinear. It requires some special considerations to solve this nonlinear equation for steady-state unsaturated flow. A Picard or Newton–Raphson iteration scheme may or may not lead to convergence, depending on the initial guess. There are two ways to circumvent this difficulty. The first one is to use a transient time matching approach, which solves a transient version of the mean pressure head equation (see Eq. (5.100) in Section 5.5.1) until reaching steady state. The second one is specifically for unidirectional mean flow, which occurs under some particular boundary conditions and with stationary media. Without loss of generality, we use steady-state, vertical mean flow as an example, which requires $\langle q_1 \rangle = \text{const}$ and $\langle q_2 \rangle = \langle q_3 \rangle = 0$. In this case, the mean pressure head can be solved from the following nonlinear equation by an iterative scheme:

$$\langle q_1 \rangle = -K_m(\mathbf{x}) \left[\frac{\partial \psi^{(0)}(\mathbf{x})}{\partial x_i} + \delta_{i1} \right] \tag{5.41}$$

where $K_m(\mathbf{x}) = \exp(\langle f \rangle) \exp[\langle \alpha \rangle \psi^{(0)}(\mathbf{x})]$.

With the solution for $\psi^{(0)}(\mathbf{x})$, the covariances C_ψ, $C_{f\psi}$, and $C_{\alpha\psi}$ can be solved from the linear moment equations (5.21)–(5.23). As in the case of saturated flow, the moment equations may be solved by conventional numerical techniques such as finite differences and finite elements. As in Section 3.4.1, we approximate the spatial derivatives by the central-difference scheme,

$$\frac{\partial C_{pq}}{\partial x_1} \approx \frac{C_{pq}(x_1 + \Delta x_1, x_2; \chi_1, \chi_2) - C_{pq}(x_1 - \Delta x_1, x_2; \chi_1, \chi_2)}{2\Delta x_1} \tag{5.42}$$

$$\frac{\partial^2 C_{pq}}{\partial x_1^2} \approx \frac{1}{\Delta x_1^2} [C_{pq}(x_1 + \Delta x_1, x_2; \chi_1, \chi_2) - 2C_{pq}(x_1, x_2; \chi_1, \chi_2)$$
$$+ C_{pq}(x_1 - \Delta x_1, x_2; \chi_1, \chi_2)] \tag{5.43}$$

where pq stands for $f\psi$, $\alpha\psi$, or ψ. With these discretizations, the covariance equations become a set of linear algebraic equations similar to Eq. (3.81), which may be solved by Gauss–Jordan elimination, lower–upper (LU) decomposition, successive over-relaxation, or the conjugate gradient method. Recently, Zhang and Winter [1998] and Zhang [1999b] solved these equations by LU decomposition with forward and back substitutions. For a specific grid, $C_\psi(\mathbf{x}, \chi)$ in Eq. (5.21) needs to be solved for each selected reference point χ. However, as $C_{f\psi}(\mathbf{x}, \chi)$ and $C_{\alpha\psi}(\mathbf{x}, \chi)$ are solved on the grid for χ, they must be solved as many times as the number of nodes on the grid $\mathbf{x}$ in order to obtain the derivatives $\partial C_{f\psi}(\mathbf{x}, \chi)/\partial x_i$ and $\partial C_{\alpha\psi}(\mathbf{x}, \chi)/\partial x_i$ required in Eq. (5.21).

Illustrative examples In this section, we illustrate the numerical moment equation model through some one- and two-dimensional examples of unsaturated flow in a hypothetical soil with different boundary conditions. In this soil, the log saturated hydraulic conductivity f and the pore size distribution parameter α are assumed to be second-order stationary, and their spatial covariances take the following exponential form:

$$C_p(\mathbf{x} - \chi) = \sigma_p^2 \exp \left[-\frac{|\mathbf{x} - \chi|}{\lambda_p} \right] \tag{5.44}$$

where $p = f$ or α, σ_p^2 is the variance of p, and λ_p is the correlation scale of p. For ease of illustration, f and α are assumed to be uncorrelated, i.e., $C_{f\alpha} = 0$. It should be pointed out, the numerical moment equation approach is able to easily handle medium nonstationarity, other forms of covariances, statistical anisotropy, and correlation between input variables. The input parameters are given as $\langle f \rangle = 0$, $\sigma_f^2 = 1$, and $\lambda_f = 10$ cm; $\langle \alpha \rangle = 0.04$ cm^{-1}, $\sigma_\alpha^2 = 0.0001$ (cm^{-1})^2, and $\lambda_\alpha = 10$ cm; and $\theta_s - \theta_r = 0.3$. In terms of the coefficient of variation, the variability is 131% for K_S or 25% for α.

We consider vertical infiltration in a one-dimensional domain of the above-mentioned soils under different conditions. In the first case, the boundary conditions are specified as follows: a specified recharge ($q_n = 0.001$) at the top ($x_1 = 600$ cm) and the water table ($\psi = 0$) at the bottom ($x_1 = 0$). Here, $q_n = |q|/K_G$ is the absolute value of recharge normalized with respect to the geometric mean saturated hydraulic conductivity. The steady-state (divergence-free) condition requires the flux q_1 to be constant throughout the one-dimensional domain. Hence, Eq. (5.41) is used for the solution of the mean pressure head. Figure 5.1 shows the expected values and standard deviations of the pressure head ψ and the effective water content θ_e versus the normalized distance from the water table

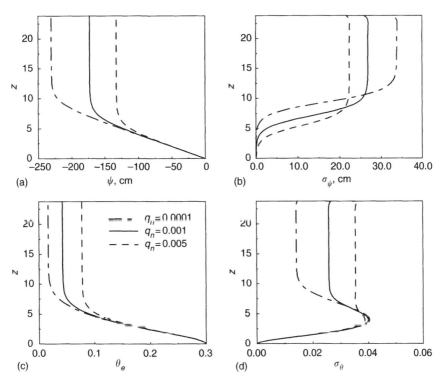

Figure 5.1. The expected values and standard deviations of the pressure head ψ and the effective water content θ_e versus the normalized distance z as functions of normalized recharge q_n (1-D case).

$(z = x_1\langle\alpha\rangle)$ as functions of q_n. It is seen from Fig. 5.1a that the (first-order) mean pressure head decreases with the distance from the water table and then approaches a constant after some critical distance for each recharge value. That is to say, the mean head is nonstationary when the distance from the water table is less than $z \approx 7.5$, 10, and 12.5 for the respective cases of recharge $q_n = 0.005$, 0.001, and 0.0001. Thus, we see that the larger is the recharge (or the smaller is the geometric mean of saturated hydraulic conductivity), the larger is the mean pressure head, and the smaller is the normalized distance at which stationarity is approached. Figure 5.1c shows the (first-order) mean water content $\langle\theta_e\rangle$ versus the normalized distance z from the water table. As does the mean pressure head, the mean water content approaches its stationary limit when the normalized distance from the water table becomes large (and exceeds the height of the *capillary fringe*). Again, the distance for it to become stationary increases as the normalized recharge q_n decreases. It is of interest to mention that the other flow quantities such as the pressure head, water content, and velocity are nonstationary although the flux is constant in the one-dimensional case. A similar case but in a semi-unbounded domain has been investigated by Indelman *et al.* [1993].

Figures 5.1b and d show the standard deviations of the pressure head and the effective water content. It is seen that both the stationary head standard deviation σ_ψ and the normalized distance at which stationarity is approached increase as the normalized recharge decreases. In addition, it seems that the distance for the head standard deviation to become constant is slightly larger than that for the mean head at a certain recharge value. The behavior of the water content standard deviation σ_{θ_e} is a little more complicated. First, it is not a monotonic function of the depth. For cases shown in this figure, the standard deviation initially increases with the distance z from the water table, but after reaching its peak it decreases rapidly towards the stationary constant. Though not shown, the detailed behavior of the water content standard deviation is a strong function of σ_f^2, σ_α^2, and their combination. Second, the water content standard deviation increases as q_n increases, contrary to that of the pressure head.

In the area where the mean pressure head and the mean effective water content are constant, flow is said to be *mean gravity-dominated*. There, the mean (total) head gradient is $J_1 \equiv 1$. In the mean gravity-dominated region, the mean pressure head takes a finite value depending on the normalized recharge, whereas it is zero at the water table. Therefore, there is a transition from the water table to the mean gravity-dominated region. The smaller is the q_n, the larger is the transition. For a certain soil type (i.e., with fixed average soil properties), it likely requires a larger distance to accomplish a larger transition. The stationary limits and the critical vertical distance at which stationarity is attained also depend on the soil type [Zhang and Winter, 1998]: the coarser the soil texture is, the smaller the distance.

Figure 5.2 shows the confidence intervals for the pressure head and the effective water content as functions of the normalized distance z for the case of $q_n = 0.001$. These curves are obtained by adding and subtracting one standard deviation from the mean quantity. They correspond to the 68% confidence intervals for the flow quantities; this is in an approximate sense because the flow quantities may not be normally distributed. Nevertheless, these intervals provide a way to quantify uncertainties due to incomplete information about the medium heterogeneity. It is seen from Figs. 5.2a and c that the confidence intervals strongly

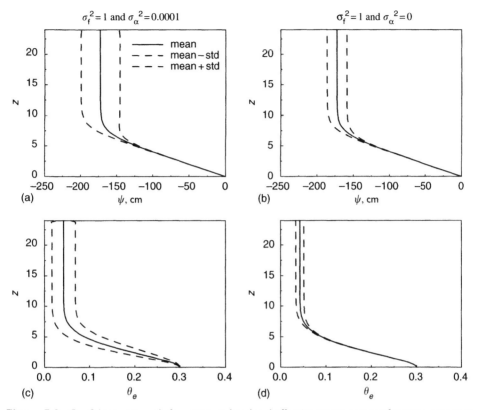

Figure 5.2. Confidence intervals for pressure head and effective water content for two scenarios: (a) and (c) $\sigma_f^2 = 1$ and $\sigma_\alpha^2 = 0.0001$; (b) and (d) $\sigma_f^2 = 1$ and $\sigma_\alpha^2 = 0$ (1-D case).

depend on the distance from the water table. Near the water table, the intervals are narrow with varying widths. At larger distances, the intervals approach their stationary limits. It is of interest to note that at locations near the water table the confidence intervals are much tighter for the head than for the water content. This is caused by the specific boundary conditions. If one specifies water content instead of head at the boundaries, one will see opposite behaviors.

We next look at the effect of the variability of the soil parameter α on the variabilities of flow quantities. When the variance of α is reduced to zero from the previous case of $\sigma_\alpha^2 = 0.0001$, the variabilities of ψ and θ_e are reduced drastically (Figs. 5.2b and d). This result may be surprising because the variability of α, which appears to be very small if in terms of σ_α^2, is indeed not very large in terms of the coefficient of variation $C_{v_\alpha} = 25\%$, compared to $C_{v_{K_S}} = 131\%$. A comparison between Figs. 5.2b and d and Figs. 5.2a and c reveals that the variabilities of unsaturated flow quantities are much more sensitive to the variability of α than that of the log saturated hydraulic conductivity f. This observation is consistent with the finding of Section 1.5.2 based on analytical solutions. This substantiates

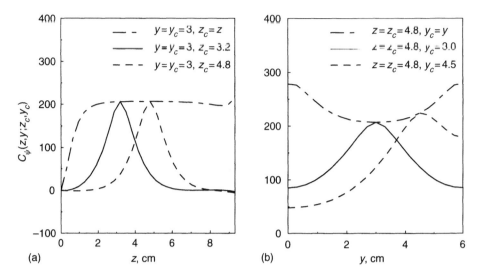

Figure 5.3. Pressure head covariances along the vertical (a) and horizontal (b) centerlines (2-D case).

the earlier assertion that the variability of the pore size distribution parameter α should not be neglected even though its variance may appear to be a small value.

We finally consider unsaturated flow in a two-dimensional domain. In this case, the soil properties are the same as specified earlier and the normalized recharge is $q_n = 0.001$. The vertical extent of the domain is 240 cm or $24\lambda_f$, and the horizontal (lateral) size is 150 cm or $15\lambda_f$. The domain is bounded with the specified pressure head of $H_B = -172.694$ cm at the bottom, the specified flux ($q_n = 0.001$) boundary at the top and no-flow boundaries at the two sides. This represents an upper portion of a vadose zone. In this specific setting, the mean pressure head is uniform throughout the domain and thus the flow is mean gravity-dominated. However, the presence of the boundaries may render the flow nonstationary.

Figures 5.3a and b show the head covariance $C_\psi(z, y; z_c, y_c) = C_\psi(x_1, x_2; \chi_1, \chi_2)$ along the vertical centerline $y = 3$ and the horizontal centerline $z = 4.8$, respectively, where $z = \langle\alpha\rangle x_1$ and $z_c = \langle\alpha\rangle\chi_1$ are the normalized distances from the water table and $y = \langle\alpha\rangle x_2$ and $y_c = \langle\alpha\rangle\chi_2$ are the normalized horizontal coordinates. When $z = z_c$ and $y = y_c$, the head covariance becomes the head variance. It is seen that as in the one-dimensional case, the head variance approaches its asymptotic limit and the head covariance becomes more symmetric with the increase of the distance from the water table. Note that near the top boundary of specified flux, the head variance deviates from the stationary limit. This may be a numerical artifact caused by inconsistent finite difference schemes at the top boundary and in the domain interior. Along the horizontal direction, the head variance is the largest at the lateral boundaries and decreases towards the center of the domain. It is seen that the head variance becomes independent of the actual location when the horizontal distance from the boundary is larger than a few integral scales as for saturated flow. The constant value of the head variance attained away from the lateral boundaries is, however, strongly dependent on the lateral extent of the domain.

Though not shown, the statistical moments of other flow quantities behave similarly. In general, the unsaturated flow quantities such as pressure head, water content, unsaturated hydraulic conductivity, flux, and velocity are nonstationary near the water table and approach stationarity as the vertical distance from the water table increases. The stationary limits and the critical vertical distance at which stationarity is attained depend on the soil type and the recharge rate. The flow quantities are also nonstationary in the horizontal direction near the lateral boundaries.

5.2.2 Adjoint State Moment Equations

In the preceding section, equations governing the first two statistical moments of flow quantities are derived with a perturbative technique, and then the moment equations are solved by numerical schemes such as finite differences and finite elements. In this section, we will introduce an alternative numerical approach, in which the original governing equation is first discretized on specified grids using finite differences or finite elements, and then the resulting (discretized) equation is used to derive the statistical moments of the flow quantities by the adjoint state method discussed in Section 3.8. The adjoint state method has been widely used for saturated flow problems [Neuman, 1980; Dettinger and Wilson, 1981; Hoeksema and Kitanidis, 1984; Sykes *et al.*, 1985; Sun and Yeh, 1992] while its applications to unsaturated flow are relatively limited [Yeh and Zhang, 1996; Zhang and Yeh, 1997; Li and Yeh, 1998; Hughson and Yeh, 2000].

Finite difference or finite element discretizations of the original (stochastic) partial differential equations (5.1)–(5.4) may lead to the following equation in a matrix form:

$$\mathbf{c}\boldsymbol{\psi} = \mathbf{b}, \qquad \boldsymbol{\psi} = \mathbf{c}^{-1}\mathbf{b} \tag{5.45}$$

where $\boldsymbol{\psi} = (\psi_1, \psi_2, \ldots, \psi_N)^T$ is the vector of N unknown pressure heads, $\psi_i = \psi(\mathbf{x}_i)$ is the pressure head at the node $\mathbf{x}_i$, N is the number of nodes on the numerical grid, $\mathbf{c}$ is the $N \times N$ coefficient matrix, $\mathbf{c}^{-1}$ is the inverse of $\mathbf{c}$, and $\mathbf{b}$ is the vector containing the information regarding boundary conditions and sink/source terms. The matrix $\mathbf{c}$ depends on both the medium and fluid properties and the discretization scheme.

When the medium properties such as the log saturated hydraulic conductivity $f(\mathbf{x}) = \ln K_S(\mathbf{x})$ and the pore size distribution parameter $\alpha(\mathbf{x})$ are treated as random space functions, so should the matrix $\mathbf{c}$. The dependency of $\mathbf{c}$ on the vectors $\mathbf{f} = (f_1, f_2, \ldots, f_N)^T$ and $\boldsymbol{\alpha} = (\alpha_1, \alpha_2, \ldots, \alpha_N)^T$, where $f_i = f(\mathbf{x}_i)$ and $\alpha_i = \alpha(\mathbf{x}_i)$, can be expressed as $\mathbf{c} = \mathbf{c}(\mathbf{f}, \boldsymbol{\alpha})$. In the same manner, the random head vector can be expressed as $\boldsymbol{\psi} = \boldsymbol{\psi}(\mathbf{f}, \boldsymbol{\alpha})$. One may decompose the random vectors into their means and fluctuations: $\mathbf{f} = \langle \mathbf{f} \rangle + \mathbf{f}'$ and $\boldsymbol{\alpha} = \langle \boldsymbol{\alpha} \rangle + \boldsymbol{\alpha}'$ where $\langle \mathbf{f} \rangle = (\langle f_1 \rangle, \langle f_2 \rangle, \ldots, \langle f_N \rangle)^T$, $\mathbf{f}' = (f_1', f_2', \ldots, f_N')^T$, $\langle \boldsymbol{\alpha} \rangle = (\langle \alpha_1 \rangle, \langle \alpha_2 \rangle, \ldots, \langle \alpha_N \rangle)^T$, and $\boldsymbol{\alpha}' = (\alpha_1', \alpha_2', \ldots, \alpha_N')^T$. As done in Section 3.8, the random pressure head vector can be expanded about $\langle \mathbf{f} \rangle$ and $\langle \boldsymbol{\alpha} \rangle$ by the vector form of

Taylor expansion,

$$\boldsymbol{\psi} = \boldsymbol{\psi}(\langle \mathbf{f} \rangle, \langle \boldsymbol{\alpha} \rangle) + \mathbf{D}_{\mathbf{f}\boldsymbol{\psi}}\mathbf{f}' + \mathbf{D}_{\boldsymbol{\alpha}\boldsymbol{\psi}}\boldsymbol{\alpha}' + \cdots \tag{5.46}$$

where $\mathbf{D}_{\mathbf{f}\boldsymbol{\psi}}$ and $\mathbf{D}_{\boldsymbol{\alpha}\boldsymbol{\psi}}$ are the respective derivatives of $\boldsymbol{\psi}$ with respect to the transpose of $\mathbf{f}$ and $\boldsymbol{\alpha}$, evaluated at $\langle \mathbf{f} \rangle$ and $\langle \boldsymbol{\alpha} \rangle$. $\mathbf{D}_{\mathbf{f}\boldsymbol{\psi}}$ and $\mathbf{D}_{\boldsymbol{\alpha}\boldsymbol{\psi}}$ are called matrix derivatives or sensitivity matrices.

To first order, the mean pressure head vector is obtained from Eq. (5.46) by taking expectation,

$$\langle \boldsymbol{\psi} \rangle = \boldsymbol{\psi}(\langle \mathbf{f} \rangle, \langle \boldsymbol{\alpha} \rangle) \tag{5.47}$$

The first-order pressure head fluctuation is given as

$$\boldsymbol{\psi}' = \mathbf{D}_{\mathbf{f}\boldsymbol{\psi}}\mathbf{f}' + \mathbf{D}_{\boldsymbol{\alpha}\boldsymbol{\psi}}\boldsymbol{\alpha}' \tag{5.48}$$

which is used to construct the pressure head covariance matrix,

$$\begin{aligned}\mathbf{C}_{\boldsymbol{\psi}} &= \langle \boldsymbol{\psi}'\boldsymbol{\psi}'^T \rangle \\ &= \mathbf{D}_{\mathbf{f}\boldsymbol{\psi}}\mathbf{C}_{\mathbf{f}}\mathbf{D}_{\mathbf{f}\boldsymbol{\psi}}^T + \mathbf{D}_{\mathbf{f}\boldsymbol{\psi}}\mathbf{C}_{\mathbf{f}\boldsymbol{\alpha}}\mathbf{D}_{\boldsymbol{\alpha}\boldsymbol{\psi}}^T + \mathbf{D}_{\boldsymbol{\alpha}\boldsymbol{\psi}}\mathbf{C}_{\boldsymbol{\alpha}\mathbf{f}}\mathbf{D}_{\mathbf{f}\boldsymbol{\psi}}^T + \mathbf{D}_{\boldsymbol{\alpha}\boldsymbol{\psi}}\mathbf{C}_{\boldsymbol{\alpha}}\mathbf{D}_{\boldsymbol{\alpha}\boldsymbol{\psi}}^T \end{aligned} \tag{5.49}$$

where $\mathbf{C}_{\mathbf{f}}$ and $\mathbf{C}_{\boldsymbol{\alpha}}$ are the input covariance matrices of $\mathbf{f}$ and $\boldsymbol{\alpha}$, and $\mathbf{C}_{\mathbf{f}\boldsymbol{\alpha}}$ is the input cross-covariance matrix between $\mathbf{f}$ and $\boldsymbol{\alpha}$. The cross-covariance matrices can be expressed similarly as

$$\mathbf{C}_{\mathbf{f}\boldsymbol{\psi}} = \langle \mathbf{f}'\boldsymbol{\psi}'^T \rangle = \mathbf{C}_{\mathbf{f}}\mathbf{D}_{\mathbf{f}\boldsymbol{\psi}}^T + \mathbf{C}_{\mathbf{f}\boldsymbol{\alpha}}\mathbf{D}_{\boldsymbol{\alpha}\boldsymbol{\psi}}^T \tag{5.50}$$

$$\mathbf{C}_{\boldsymbol{\alpha}\boldsymbol{\psi}} = \langle \boldsymbol{\alpha}'\boldsymbol{\psi}'^T \rangle = \mathbf{C}_{\boldsymbol{\alpha}\mathbf{f}}\mathbf{D}_{\mathbf{f}\boldsymbol{\psi}}^T + \mathbf{C}_{\boldsymbol{\alpha}}\mathbf{D}_{\boldsymbol{\alpha}\boldsymbol{\psi}}^T \tag{5.51}$$

It is seen from Eqs. (5.49)–(5.51) that the evaluation of the second moments of pressure heads reduces to the derivation of the sensitivity matrices $\mathbf{D}_{\mathbf{f}\boldsymbol{\psi}}$ and $\mathbf{D}_{\boldsymbol{\alpha}\boldsymbol{\psi}}$.

The original (primary) flow equation can be rewritten from Eqs. (5.1)–(5.4) as

$$\frac{\partial}{\partial x_i}\left[K(\mathbf{x})\frac{\partial \Phi(\mathbf{x})}{\partial x_i} \right] = -g(\mathbf{x})$$
$$\psi(\mathbf{x}) = H_B(\mathbf{x}), \quad \mathbf{x} \in \Gamma_D \tag{5.52}$$
$$n_i(\mathbf{x})K(\mathbf{x})\frac{\partial \Phi(\mathbf{x})}{\partial x_i} = -Q(\mathbf{x}), \quad \mathbf{x} \in \Gamma_N$$

where summation for repeated indices i is implied and $\Phi = \psi + x_1$ is the total head. Equation (5.52) is usually called the *Richards equation*. Recall that $K(\mathbf{x}) = K_S(\mathbf{x})\exp[\alpha(\mathbf{x})\psi(\mathbf{x})]$ for the Gardner–Russo constitutive model. A performance function may be defined as

$$P(\Phi, \boldsymbol{\beta}) = \int_\Omega R(\Phi, \boldsymbol{\beta}; \mathbf{x})\, d\mathbf{x} \tag{5.53}$$

where Ω is the domain bounded by the union of Γ_D and Γ_N, $\boldsymbol{\beta}$ is the vector of system parameters such as the log saturated hydraulic conductivity f and the pore size distribution parameter α values, and R is a function yet to be specified.

The sensitivity of the performance function P to a specific parameter β_k is given as

$$\frac{\partial P(\Phi, \boldsymbol{\beta})}{\partial \beta_k} = \int_\Omega \left[\frac{\partial R(\Phi, \boldsymbol{\beta}; \mathbf{x})}{\partial \beta_k} + \frac{\partial R(\Phi, \boldsymbol{\beta}; \mathbf{x})}{\partial \Phi} \gamma \right] d\mathbf{x} \tag{5.54}$$

where $\gamma(\Phi, \beta_k; \mathbf{x}) = \partial \Phi(\mathbf{x})/\partial \beta_k$ is the state sensitivity, i.e., the sensitivity of total head Φ to the parameter β_k. The state sensitivity may be evaluated directly from the following equation, obtained by differentiating Eq. (5.52) with respect to β_k:

$$\frac{\partial}{\partial x_i} \left[K(\mathbf{x}) \frac{\partial \gamma(\Phi, \beta_k; \mathbf{x})}{\partial x_i} \right] = -\frac{\partial g(\mathbf{x})}{\partial \beta_k} - \frac{\partial}{\partial x_i} \left[\frac{\partial K(\mathbf{x})}{\partial \beta_k} \frac{\partial \Phi(\mathbf{x})}{\partial x_i} \right]$$

$$\gamma(\Phi, \beta_k; \mathbf{x}) = \frac{\partial H_B(\mathbf{x})}{\partial \beta_k}, \quad \mathbf{x} \in \Gamma_D \tag{5.55}$$

$$n_i(\mathbf{x}) K(\mathbf{x}) \frac{\partial \gamma(\Phi, \beta_k; \mathbf{x})}{\partial x_i} = -\frac{\partial Q(\mathbf{x})}{\partial \beta_k} - n_i(\mathbf{x}) \frac{\partial K(\mathbf{x})}{\partial \beta_k} \frac{\partial \Phi(\mathbf{x})}{\partial x_i}, \quad \mathbf{x} \in \Gamma_N$$

This is the direct sensitivity equation, which has a similar structure to its original equation (5.52). Recalling that the sensitivity matrices $\mathbf{D}_{f\psi}$ and $\mathbf{D}_{\alpha\psi}$ are evaluated with $\langle \mathbf{f} \rangle$, $\langle \boldsymbol{\alpha} \rangle$, and $\langle \boldsymbol{\psi} \rangle$, $K(\mathbf{x})$ and $\Phi(\mathbf{x})$ in Eq. (5.55) are to be replaced with $\exp[\langle f \rangle + \langle \alpha \rangle \langle \psi(\mathbf{x}) \rangle]$ and $\langle \psi(\mathbf{x}) \rangle + x_1$, respectively. Therefore, the right-hand sides of Eq. (5.55) are known quantities, and the state sensitivity γ can be solved from this equation numerically. However, as discussed in Section 3.8, this direct sensitivity approach is time consuming if the number of parameters β_k is large.

An alternative approach is to formulate adjoint equations of the partial differential equations for γ. Multiplying the first equation of (5.55) by an arbitrary differentiable function $\gamma^*(\mathbf{x})$ and integrating over Ω leads to

$$\int_\Omega \left\{ \gamma^*(\mathbf{x}) \frac{\partial}{\partial x_i} \left[K(\mathbf{x}) \frac{\partial \gamma(\Phi, \beta_k; \mathbf{x})}{\partial x_i} \right] + \gamma^*(\mathbf{x}) \frac{\partial g(\mathbf{x})}{\partial \beta_\lambda} \right.$$

$$\left. + \gamma^*(\mathbf{x}) \frac{\partial}{\partial x_i} \left[\frac{\partial K(\mathbf{x})}{\partial \beta_k} \frac{\partial \Phi(\mathbf{x})}{\partial x_i} \right] \right\} d\mathbf{x} = 0 \tag{5.56}$$

Applying Green's identity to Eq. (5.56) and utilizing the boundary conditions of Eq. (5.54), Eq. (5.56) can be rewritten as

$$\int_\Omega \left\{ \gamma(\Phi, \beta_k; \mathbf{x}) \frac{\partial}{\partial x_i} \left[K(\mathbf{x}) \frac{\partial \gamma^*(\mathbf{x})}{\partial x_i} \right] + \gamma^*(\mathbf{x}) \frac{\partial g(\mathbf{x})}{\partial \beta_k} - \frac{\partial \gamma^*(\mathbf{x})}{\partial x_i} \frac{\partial K(\mathbf{x})}{\partial \beta_k} \frac{\partial \Phi(\mathbf{x})}{\partial x_i} \right\} d\mathbf{x}$$

$$+ \int_\Gamma \left[\gamma(\Phi, \beta_k; \mathbf{x}) K(\mathbf{x}) \frac{\partial \gamma^*(\mathbf{x})}{\partial x_i} n_i(\mathbf{x}) + \gamma^*(\mathbf{x}) \frac{\partial Q(\mathbf{x})}{\partial \beta_k} \right] d\mathbf{x} = 0 \tag{5.57}$$

Adding Eq. (5.57) to Eq. (5.54) yields

$$\frac{\partial \Gamma(\Phi, \boldsymbol{\beta})}{\partial \beta_k} = \int_{\Omega} \frac{\partial R(\Phi, \boldsymbol{\beta}; \mathbf{x})}{\partial \beta_k} \, d\mathbf{x}$$

$$+ \int_{\Omega} \gamma(\Phi, \beta_k; \mathbf{x}) \left\{ \frac{\partial R(\Phi, \boldsymbol{\beta}; \mathbf{x})}{\partial \Phi} + \frac{\partial}{\partial x_i} \left[K(\mathbf{x}) \frac{\partial \gamma^*(\mathbf{x})}{\partial x_i} \right] \right\}$$

$$+ \int_{\Omega} \left\{ \gamma^*(\mathbf{x}) \frac{\partial g(\mathbf{x})}{\partial \beta_k} - \frac{\partial \gamma^*(\mathbf{x})}{\partial x_i} \frac{\partial K(\mathbf{x})}{\partial \beta_k} \frac{\partial \Phi(\mathbf{x})}{\partial x_i} \right\} d\mathbf{x}$$

$$+ \int_{\Gamma} \left[\gamma(\Phi, \beta_k; \mathbf{x}) K(\mathbf{x}) \frac{\partial \gamma^*(\mathbf{x})}{\partial x_i} n_i(\mathbf{x}) + \gamma^*(\mathbf{x}) \frac{\partial Q(\mathbf{x})}{\partial \beta_k} \right] d\mathbf{x} \qquad (5.58)$$

To eliminate the contribution of the unknown state sensitivities γ, one may let the arbitrary function γ^* satisfy the following equation:

$$\frac{\partial}{\partial x_i} \left[K(\mathbf{x}) \frac{\partial \gamma^*(\mathbf{x})}{\partial x_i} \right] + \frac{\partial R(\Phi, \boldsymbol{\beta}; \mathbf{x})}{\partial \Phi} = 0$$

$$\gamma^*(\mathbf{x}) = 0, \quad \mathbf{x} \in \Gamma_D \qquad (5.59)$$

$$n_i(\mathbf{x}) K(\mathbf{x}) \frac{\partial \gamma^*(\mathbf{x})}{\partial x_i} = 0, \quad \mathbf{x} \in \Gamma_N$$

This is the adjoint state equation, and γ^* is the adjoint state of γ. With Eq. (5.59), Eq. (5.58) reduces to

$$\frac{\partial P(\Phi, \boldsymbol{\beta})}{\partial \beta_k} = \int_{\Omega} \frac{\partial R(\Phi, \boldsymbol{\beta}; \mathbf{x})}{\partial \beta_k} d\mathbf{x} + \int_{\Omega} \left\{ \gamma^*(\mathbf{x}) \frac{\partial g(\mathbf{x})}{\partial \beta_k} - \frac{\partial \gamma^*(\mathbf{x})}{\partial x_i} \frac{\partial K(\mathbf{x})}{\partial \beta_k} \frac{\partial \Phi(\mathbf{x})}{\partial x_i} \right\} d\mathbf{x}$$

$$+ \int_{\Gamma_D} \frac{\partial H_B(\mathbf{x})}{\partial \beta_k} K(\mathbf{x}) \frac{\partial \gamma^*(\mathbf{x})}{\partial x_i} n_i(\mathbf{x}) d\mathbf{x} + \int_{\Gamma_N} \gamma^*(\mathbf{x}) \frac{\partial Q(\mathbf{x})}{\partial \beta_k} d\mathbf{x} \qquad (5.60)$$

Let function R in Eq. (5.53) be

$$R(\Phi, \boldsymbol{\beta}; \mathbf{x}) = \Phi(\mathbf{x}) \delta(\mathbf{x} - \mathbf{x}_l) \qquad (5.61)$$

where $\mathbf{x}_l$ is a node on the grid, an observation location, or a point of interest. In turn, the performance function P is given as $P = \Phi(\mathbf{x}_l) = \psi(\mathbf{x}) + x_1$. Hence, Eq. (5.60) becomes

$$\frac{\partial \psi(\mathbf{x}_l)}{\partial \beta_k} = \int_{\Omega} \left\{ \gamma^*(\mathbf{x}) \frac{\partial g(\mathbf{x})}{\partial \beta_k} - \frac{\partial \gamma^*(\mathbf{x})}{\partial x_i} \frac{\partial K(\mathbf{x})}{\partial \beta_k} \frac{\partial \Phi(\mathbf{x})}{\partial x_i} \right\} d\mathbf{x}$$

$$+ \int_{\Gamma_D} \frac{\partial H_B(\mathbf{x})}{\partial \beta_k} K(\mathbf{x}) \frac{\partial \gamma^*(\mathbf{x})}{\partial x_i} n_i(\mathbf{x}) \, d\mathbf{x} + \int_{\Gamma_N} \gamma^*(\mathbf{x}) \frac{\partial Q(\mathbf{x})}{\partial \beta_k} d\mathbf{x} \qquad (5.62)$$

When β_k denotes the log saturated hydraulic conductivity f or the pore size distribution parameter α, the sink/source term $g(\mathbf{x})$ and the boundary terms $H_B(\mathbf{x})$ and $Q(\mathbf{x})$ are not sensitive to β_k at all as they are known with certainty in our case. Therefore, Eq. (5.62) reduces to

$$\frac{\partial \psi(\mathbf{x}_l)}{\partial \beta_k} = -\int_\Omega \frac{\partial \gamma^*(\mathbf{x})}{\partial x_i} \frac{\partial K(\mathbf{x})}{\partial \beta_k} \frac{\partial \Phi(\mathbf{x})}{\partial x_i} d\mathbf{x} \qquad (5.63)$$

With Eq. (5.61), Eq. (5.59) becomes

$$\frac{\partial}{\partial x_i} \left[K(\mathbf{x}) \frac{\partial \gamma^*(\mathbf{x})}{\partial x_i} \right] = -\delta(\mathbf{x} - \mathbf{x}_l)$$

$$\gamma^*(\mathbf{x}) = 0, \quad \mathbf{x} \in \Gamma_D \qquad (5.64)$$

$$n_i(\mathbf{x}) K(\mathbf{x}) \frac{\partial \gamma^*(\mathbf{x})}{\partial x_i} = 0, \quad \mathbf{x} \in \Gamma_N$$

It is thus seen that the adjoint state γ^* is Green's function for the Poisson operator, evaluated at $\langle \mathbf{f} \rangle$ and $\langle \boldsymbol{\alpha} \rangle$. The adjoint state equation (5.59) needs only to be solved once for a given function $R(\Phi, \mathbf{f}; \mathbf{x})$ independent of the parameter β_k. When the performance function changes, for example, from $P = \psi(\mathbf{x}_l)$ to $P = \psi(\mathbf{x}_m)$, Eq. (5.59) is to be solved again.

The sensitivity matrices $\mathbf{D}_{\mathbf{f}\psi}(\langle \mathbf{f} \rangle, \langle \boldsymbol{\psi} \rangle)$ and $\mathbf{D}_{\boldsymbol{\alpha}\psi}(\langle \mathbf{f} \rangle, \langle \boldsymbol{\psi} \rangle)$ can be evaluated with Eq. (5.62) or (5.63), where the adjoint state γ^* is numerically evaluated from Eq. (5.64). Notice that both Eqs. (5.64) and (5.62) or (5.63) need to be evaluated at $\mathbf{f} = \langle \mathbf{f} \rangle$, $\boldsymbol{\alpha} = \langle \boldsymbol{\alpha} \rangle$, and $\boldsymbol{\psi} = \langle \boldsymbol{\psi} \rangle$. With the sensitivity matrices, the pressure head covariance matrices $\mathbf{C}_{\mathbf{f}\psi}$, $\mathbf{C}_{\boldsymbol{\alpha}\psi}$, and $\mathbf{C}_\psi$ are obtained with Eqs. (5.49)–(5.51).

5.3 GRAVITY-DOMINATED FLOW

Most previous stochastic studies of unsaturated flow have concerned themselves with the special case of mean gravity-dominated flow [e.g., Yeh *et al.*, 1985a,b; Russo, 1993, 1995b; Harter and Yeh, 1996a,b; Yang *et al.*, 1996; Zhang *et al.*, 1998; Harter and Zhang, 1999]. Together with the treatment of unbounded domains, the unsaturated flow then becomes stationary so that analytical or semi-analytical solutions may be derived. Under mean gravity-dominated flow conditions, the mean pressure head is constant throughout the domain such that the mean total head gradient is $J_i = \partial \langle \psi(\mathbf{x}) \rangle / \partial x_i + \delta_{i1} = \delta_{i1}$. It has not been fully recognized in the past that the mean total head gradient must be $J_i = \delta_{i1}$ in order to satisfy the stationary unsaturated flow condition, which is a necessary requirement in many previous theories. As shown by Zhang and Winter [1998, p. 1094], a value of $J_1 \neq 1$ would violate the stationary flow condition in the vadose zone.

Under mean gravity-dominated flow conditions, the zeroth-order mean pressure head gradient must be $\partial \psi^{(0)}(\mathbf{x})/\partial x_i = 0$ and the fluid sink/source term g must be absent. Hence, the solution of the mean pressure head equation (5.19) is trivial, being a constant in space. The solution of $\psi^{(0)}$ is entirely determined by the normalized recharge rate. Under the same conditions, the first-order pressure head fluctuation equation (5.20) is simplified as

$$\frac{\partial^2 \psi^{(1)}(\mathbf{x})}{\partial x_i^2} + \langle \alpha \rangle \frac{\partial \psi^{(1)}(\mathbf{x})}{\partial x_1} = -\frac{\partial f'(\mathbf{x})}{\partial x_i} - \psi^{(0)} \frac{\partial \alpha'(\mathbf{x})}{\partial x_1}$$

$$\psi^{(1)}(\mathbf{x}) = 0, \quad \mathbf{x} \in \Gamma_D \tag{5.65}$$

$$n_i(\mathbf{x}) \frac{\partial \psi^{(1)}(\mathbf{x})}{\partial x_i} + n_1(\mathbf{x}) \langle \alpha \rangle \psi^{(1)}(\mathbf{x}) = -n_1(\mathbf{x}) f'(\mathbf{x}) - n_1(\mathbf{x}) \psi^{(0)} \alpha'(\mathbf{x}), \quad \mathbf{x} \in \Gamma_N$$

Equation (5.65) can be used to construct covariance equations like Eqs. (5.21)–(5.23) by the partial differential moment equation approach or to derive integrodifferential covariance equations with the aid of Green's function. The spectral methods introduced in Section 3.6 are another viable choice.

5.3.1 Spectral Analysis

In bounded domains, the pressure head and other flow quantities are nonstationary so that the nonstationary spectral method discussed in Section 3.6.2 is more appropriate than the stationary spectral method. It should be noted that the nonstationary spectral method is also applicable to the general nonstationary unsaturated flow discussed in Section 5.2, as long as the input soil properties $f(\mathbf{x})$ and $\alpha(\mathbf{x})$ are stationary. As in the many studies mentioned earlier, in this section we restrict our attention to mean gravity-dominated flow in unbounded domains. Under such conditions, the flow is stationary and thus admits the classical spectral representation.

As discussed in Section 3.6.1 for saturated flow, the stationary fluctuations $f'(\mathbf{x})$, $\alpha'(\mathbf{x})$, and $\psi'(\mathbf{x}) = \psi^{(1)}$ can be expressed by the following Fourier–Stieltjes integral representations:

$$f'(\mathbf{x}) = \int \exp(\iota \mathbf{k} \cdot \mathbf{x}) \, dZ_f(\mathbf{k}) \tag{5.66}$$

$$\alpha'(\mathbf{x}) = \int \exp(\iota \mathbf{k} \cdot \mathbf{x}) \, dZ_\alpha(\mathbf{k}) \tag{5.67}$$

$$\psi'(\mathbf{x}) = \int \exp(\iota \mathbf{k} \cdot \mathbf{x}) \, dZ_\psi(\mathbf{k}) \tag{5.68}$$

where $\mathbf{k} = (k_1, \ldots, k_d)^T$ is the wave number space vector (where d is the space dimensionality), $\iota = \sqrt{-1}$, and $dZ_f(\mathbf{k})$, $dZ_\alpha(\mathbf{k})$, and $dZ_\psi(\mathbf{k})$ are the complex Fourier increments of the fluctuations at $\mathbf{k}$. The integration is d-fold from $-\infty$ to ∞. The properties of the Fourier–Stieltjes integral are discussed in Section 2.4.4. Of particular importance is the orthogonality property,

$$\langle dZ_U(\mathbf{k}) \, dZ_U^*(\mathbf{k}') \rangle = \delta(\mathbf{k} - \mathbf{k}') S_U(\mathbf{k}) \, d\mathbf{k} \, d\mathbf{k}' \tag{5.69}$$

where $U = f$, α, or ψ, dZ_U^* is the complex conjugate of dZ_U, and S_U is the spectrum or spectral density of U.

Substitution of Eqs. (5.66)–(5.68) into the first equation of Eq. (5.65) yields

$$dZ_\psi(\mathbf{k}) = \frac{\imath k_1[dZ_f(\mathbf{k}) + \psi^{(0)}dZ_\alpha(\mathbf{k})]}{k^2 - i\langle\alpha\rangle k_1} \tag{5.70}$$

where $k^2 = \sum_{i=1}^d k_i^2$. Therefore, the spectral density $S_\psi(\mathbf{k})$ of pressure head is obtained by multiplying Eq. (5.70) by its complex conjugate, taking expectation and utilizing the orthogonality property of Eq. (5.69),

$$S_\psi(\mathbf{k}) = \frac{k_1^2\{S_f(\mathbf{k}) + \psi^{(0)}S_{f\alpha}(\mathbf{k}) + \psi^{(0)}S_{\alpha f}(\mathbf{k}) + [\psi^{(0)}]^2 S_\alpha(\mathbf{k})\}}{k^4 + \langle\alpha\rangle^2 k_1^2} \tag{5.71}$$

where $S_{f\alpha}$ is the *cross-spectral density* of f and α, given via

$$\langle dZ_f(\mathbf{k})\, dZ_\alpha^*(\mathbf{k}')\rangle = \delta(\mathbf{k} - \mathbf{k}')S_{f\alpha}(\mathbf{k})\, d\mathbf{k}\, d\mathbf{k}' \tag{5.72}$$

Premultiplying the complex conjugate of Eq. (5.70) by dZ_f or dZ_α, taking expectation and utilizing the orthogonality property leads to

$$S_{f\psi}(\mathbf{k}) = \frac{-\imath k_1[S_f(\mathbf{k}) + \psi^{(0)}S_{f\alpha}(\mathbf{k})]}{k^2 + i\langle\alpha\rangle k_1} \tag{5.73}$$

$$S_{\alpha\psi}(\mathbf{k}) = \frac{-\imath k_1[S_{\alpha f}(\mathbf{k}) + \psi^{(0)}S_\alpha(\mathbf{k})]}{k^2 + i\langle\alpha\rangle k_1} \tag{5.74}$$

Similarly, we may obtain the Fourier–Stieltjes integral representations for $Y'(\mathbf{x})$, $q_i^{(1)}(\mathbf{x})$, $\theta_e^{(1)}(\mathbf{x})$, and $u_i^{(1)}(\mathbf{x})$ on the basis of Eqs. (5.18), (5.26), (5.33), and (5.38), respectively. Then, various spectra or cross-spectra regarding these variables can be derived.

5.3.2 Covariance Evaluation

The covariances or cross-covariances may be obtained through the inverse Fourier transform,

$$C_{pq}(\mathbf{r}) = \int \exp(\imath\mathbf{k} \cdot \mathbf{r})S_{pq}(\mathbf{k})\, d\mathbf{k} \tag{5.75}$$

When the input covariances $C_f(\mathbf{r})$, $C_\alpha(\mathbf{r})$, and $C_{f\alpha}(\mathbf{r})$, or their Fourier transforms $S_f(\mathbf{k})$, $S_\alpha(\mathbf{k})$, and $S_{f\alpha}(\mathbf{k})$, are known, the covariances of flow quantities may be evaluated with Eq. (5.75) coupled with expressions such as Eq. (5.71) and Eqs. (5.73) and (5.74). These covariances may be evaluated analytically or numerically.

Analytical solutions For the mean gravity-dominated flow in one-dimensional unbounded domains, analytical solutions of the auto- and cross-covariances and variances of head are available in the literature [Yeh *et al.*, 1985a,b; Zhang *et al.*, 1998] (scc also Section 1.5.2). In multiple dimensions, it is, however, more difficult to derive fully analytical solutions. Yeh *et al.* [1985a,b] derived analytical solutions for the head variances for the cases that the covariances of f and α have the same correlation structure with the same correlation length, and f and α are either uncorrelated or fully correlated. The resulting head covariances are, however, not fully analytical in that they could not be evaluated without the aid of numerical quadratures. On the basis of these head covariances, Russo [1993, 1995b] derived semi-analytical expressions for the flux covariances, which were used to study solute spreading in unsaturated media. The flux covariances also need to be evaluated numerically. It should be noted that although the head and flux covariance expressions were originally given for any constant mean total pressure gradient J_i in the above-mentioned studies, they are actually only valid for the case of $J_1 = 1$ and $J_2 = J_3 = 0$ because of the requirement of the mean gravity-dominated and stationary flow in the derivations of those studies.

Fast Fourier transformation Since most of the flow covariance expressions need to be evaluated numerically, an alternative approach is to directly invert the auto- or cross-spectral expressions $S_p(\mathbf{k})$ or $S_{pq}(\mathbf{k})$ by inverse fast Fourier transform (inverse FFT). FFT provides an efficient way to numerically obtain the flow covariances such as $C_\psi(\mathbf{r})$, $C_{f\psi}(\mathbf{r})$, and $C_{\alpha\psi}(\mathbf{r})$ from the spectral expressions of $S_\psi(\mathbf{k})$, $S_{f\psi}(\mathbf{k})$, and $S_{\alpha\psi}(\mathbf{k})$ [Harter, 1994; Harter and Zhang, 1999]. In addition, other flow covariances such as those of water content θ_e, flux q_i, and velocity u_i can be obtained similarly. Using this method, Harter and Zhang [1999] were able to perform a detailed study of the effects of water content variability on solute transport in mean gravity-dominated and stationary flow. For the details of FFT algorithms, the reader is referred to Press *et al.* [1992].

5.4 KIRCHHOFF TRANSFORMATION

In both the nonstationary perturbative approaches of Section 5.2 and the stationary spectral method of Section 5.3, the coefficient $K(\mathbf{x}) = K(\psi, f, \alpha)$ in the Richards equation is expanded by Taylor expansions. Then equations governing each term in the series $\psi = \psi^{(0)} + \psi^{(1)} + \psi^{(2)} + \cdots$ are obtained by collecting terms of separate order. This procedure leads to linear equations for the pressure head terms $\psi^{(i)}$ when $i \geq 1$ and is thus often called *linearization*. Recently, Tartakovsky *et al.* [1999] developed an alternative approach, which attempts to avoid or delay the linearization procedure. Their approach is based on the following Kirchhoff transformation:

$$\Psi(\mathbf{x}) = \int_{-\infty}^{\psi(\mathbf{x})} k_r(\tau) \, d\tau \qquad (5.76)$$

where $k_r(\psi) = K[\psi(\mathbf{x}), \cdot]/K_S(\mathbf{x})$ is the *relative permeability*. For the Gardner–Russo model, $k_r(\psi) = \exp[\alpha(\mathbf{x})\psi(\mathbf{x})]$.

In order to fully take advantage of the Kirchhoff transformation, Tartakovsky *et al.* [1999] assumed the pore size distribution parameter α of the Gardner–Russo model to be a random constant. With this assumption, α no longer varies spatially and the integral of Eq. (5.76) can be integrated out explicitly,

$$\Psi(\mathbf{x}) = \frac{1}{\alpha} k_r(\psi) = \frac{1}{\alpha} \exp[\alpha\psi(\mathbf{x})] \tag{5.77}$$

Substituting Eqs. (5.76) and (5.77) into the nonlinear Richards equation (5.52) yields the following linear partial differential equations [Tartakovsky *et al.*, 1999]:

$$\frac{\partial}{\partial x_i} \left[K_S(\mathbf{x}) \frac{\partial \Psi(\mathbf{x})}{\partial x_i} \right] + \alpha \frac{\partial}{\partial x_1} [K_S(\mathbf{x})\Psi(\mathbf{x})] = -g(\mathbf{x})$$

$$\Psi(\mathbf{x}) = \Psi_B(\mathbf{x}), \quad \mathbf{x} \in \Gamma_D \tag{5.78}$$

$$n_i(\mathbf{x})K_S(\mathbf{x})\frac{\partial \Psi(\mathbf{x})}{\partial x_i} + \alpha n_1(\mathbf{x})K_S(\mathbf{x})\Psi(\mathbf{x}) = -Q(\mathbf{x}), \quad \mathbf{x} \in \Gamma_N$$

where $\Psi_B(\mathbf{x}) = (1/\alpha) \exp[\alpha H_B(\mathbf{x})]$.

As the problem of solving for the pressure head $\psi(\mathbf{x})$ is translated into that for the transformed head $\Psi(\mathbf{x})$, the governing equation becomes linear. Hence, the various moment equation methods discussed in Chapter 3 for linear, saturated flow are applicable to Eq. (5.78). As usual, $K_S(\mathbf{x})$, $f(\mathbf{x}) = \ln K_S$, and α may be decomposed as $K_S(\mathbf{x}) = \langle K_S \rangle + K_S'(\mathbf{x})$, $f(\mathbf{x}) = \langle f \rangle + f'(\mathbf{x})$, and $\alpha = \langle \alpha \rangle + \alpha'$, while $\Psi(\mathbf{x})$ may be expanded into the following formal series:

$$\Psi(\mathbf{x}) = \Psi^{(0)}(\mathbf{x}) + \Psi^{(1)}(\mathbf{x}) + \Psi^{(2)}(\mathbf{x}) + \tag{5.79}$$

Below we briefly outline the application of the f-based perturbation approach on Eq. (5.78). With $f = \ln K_S$, Eq. (5.78) reads as

$$\frac{\partial^2 \Psi(\mathbf{x})}{\partial x_i^2} + \frac{\partial f(\mathbf{x})}{\partial x_i} \frac{\partial \Psi(\mathbf{x})}{\partial x_i} + \alpha \frac{\partial \Psi(\mathbf{x})}{\partial x_1} + \alpha \frac{\partial f(\mathbf{x})}{\partial x_1} \Psi(\mathbf{x}) = -g(\mathbf{x}) \exp[-f(\mathbf{x})]$$

$$\Psi(\mathbf{x}) = \Psi_B(\mathbf{x}), \quad \mathbf{x} \in \Gamma_D \tag{5.80}$$

$$n_i(\mathbf{x})\frac{\partial \Psi(\mathbf{x})}{\partial x_i} + \alpha n_1(\mathbf{x})\Psi(\mathbf{x}) = -Q(\mathbf{x}) \exp[-f(\mathbf{x})], \quad \mathbf{x} \in \Gamma_N$$

Substitution of the expansions for f and Ψ into Eq. (5.80) yields

$$\frac{\partial^2}{\partial x_i^2}\left[\Psi^{(0)}(\mathbf{x}) + \Psi^{(1)}(\mathbf{x}) + \Psi^{(2)}(\mathbf{x}) + \cdots\right]$$

$$+ \frac{\partial f'(\mathbf{x})}{\partial x_i}\frac{\partial}{\partial x_i}\left[\Psi^{(0)}(\mathbf{x}) + \Psi^{(1)}(\mathbf{x}) + \Psi^{(2)}(\mathbf{x}) + \cdots\right]$$

$$+ (\langle\alpha\rangle + \alpha')\frac{\partial}{\partial x_1}\left[\Psi^{(0)}(\mathbf{x}) + \Psi^{(1)}(\mathbf{x}) + \Psi^{(2)}(\mathbf{x}) + \cdots\right]$$

$$+ (\langle\alpha\rangle + \alpha')\frac{\partial f'(\mathbf{x})}{\partial x_1}\left[\Psi^{(0)}(\mathbf{x}) + \Psi^{(1)}(\mathbf{x}) + \Psi^{(2)}(\mathbf{x}) + \cdots\right]$$

$$= -\frac{g(\mathbf{x})}{K_G}\left[1 - f'(\mathbf{x}) + \frac{1}{2}f'^2(\mathbf{x}) + \cdots\right] \qquad (5.81)$$

$$\Psi^{(0)}(\mathbf{x}) + \Psi^{(1)}(\mathbf{x}) + \Psi^{(2)}(\mathbf{x}) + \cdots$$

$$= \Psi_B^{(0)}(\mathbf{x}) + \Psi_B^{(1)}(\mathbf{x}) + \Psi_B^{(2)}(\mathbf{x}) + \cdots, \quad \mathbf{x} \in \Gamma_D$$

$$n_i(\mathbf{x})\frac{\partial}{\partial x_i}\left[\Psi^{(0)}(\mathbf{x}) + \Psi^{(1)}(\mathbf{x}) + \Psi^{(2)}(\mathbf{x}) + \cdots\right]$$

$$+ (\langle\alpha\rangle + \alpha')n_1(\mathbf{x})\left[\Psi^{(0)}(\mathbf{x}) + \Psi^{(1)}(\mathbf{x}) + \Psi^{(2)}(\mathbf{x}) + \cdots\right]$$

$$= -\frac{Q(\mathbf{x})}{K_G}\left[1 - f'(\mathbf{x}) + \frac{1}{2}f'^2(\mathbf{x}) + \cdots\right], \quad \mathbf{x} \in \Gamma_N$$

where we have assumed $\langle f\rangle$ to be constant and the $g(\mathbf{x})$, $H_B(\mathbf{x})$, and $Q(\mathbf{x})$ terms to be known with certainty. Collecting terms of separate orders yields

$$\left.\begin{array}{c}\dfrac{\partial^2\Psi^{(0)}(\mathbf{x})}{\partial x_i^2} + \langle\alpha\rangle\dfrac{\partial\Psi^{(0)}(\mathbf{x})}{\partial x_1} = -\dfrac{g(\mathbf{x})}{K_G} \\[2mm] \Psi^{(0)}(\mathbf{x}) = \Psi_B^{(0)}(\mathbf{x}), \quad \mathbf{x} \in \Gamma_D \\[2mm] n_i(\mathbf{x})\dfrac{\partial\Psi^{(0)}(\mathbf{x})}{\partial x_i} + \langle\alpha\rangle n_1(\mathbf{x})\Psi^{(0)}(\mathbf{x}) = -\dfrac{Q(\mathbf{x})}{K_G}, \quad \mathbf{x} \in \Gamma_N\end{array}\right\} \quad (5.82)$$

$$\left.\begin{array}{c}\dfrac{\partial^2\Psi^{(1)}(\mathbf{x})}{\partial x_i^2} + \dfrac{\partial f'(\mathbf{x})}{\partial x_i}\dfrac{\partial\Psi^{(0)}(\mathbf{x})}{\partial x_i} + \langle\alpha\rangle\dfrac{\partial\Psi^{(1)}(\mathbf{x})}{\partial x_1} + \alpha'\dfrac{\partial\Psi^{(0)}(\mathbf{x})}{\partial x_1} \\[2mm] + \langle\alpha\rangle\dfrac{\partial f'(\mathbf{x})}{\partial x_1}\Psi^{(0)}(\mathbf{x}) = \dfrac{g(\mathbf{x})}{K_G}f'(\mathbf{x}) \\[2mm] \Psi^{(1)}(\mathbf{x}) = \Psi_B^{(1)}(\mathbf{x}), \quad \mathbf{x} \in \Gamma_D \\[2mm] n_i(\mathbf{x})\dfrac{\partial\Psi^{(1)}(\mathbf{x})}{\partial x_i} + \langle\alpha\rangle n_1(\mathbf{x})\Psi^{(1)}(\mathbf{x}) + \alpha' n_1(\mathbf{x})\Psi^{(0)}(\mathbf{x}) \\[2mm] = \dfrac{Q(\mathbf{x})}{K_G}f'(\mathbf{x}), \quad \mathbf{x} \in \Gamma_N\end{array}\right\} \quad (5.83)$$

where the fluctuations $f'(\mathbf{x})$ and α' are treated as terms of the same order. The actual expressions for $\Psi_B^{(0)}(\mathbf{x})$ and $\Psi_B^{(1)}(\mathbf{x})$ can be obtained from the definition of $\Psi_B(\mathbf{x})$ as

$$\Psi_B^{(0)}(\mathbf{x}) = \frac{1}{\langle\alpha\rangle}\exp\left[\langle\alpha\rangle H_B(\mathbf{x})\right] \tag{5.84}$$

$$\Psi_B^{(1)}(\mathbf{x}) = \frac{1}{\langle\alpha\rangle}\exp\left[\langle\alpha\rangle H_B(\mathbf{x})\right]\left[H_B(\mathbf{x}) - \frac{1}{\langle\alpha\rangle}\right]\alpha' \tag{5.85}$$

Equations for higher terms of Ψ can be given similarly. It is verifiable that the mean transformed head is $\langle\Psi(\mathbf{x})\rangle = \Psi^{(0)}$ to first order in the variability of f and α and $\langle\Psi(\mathbf{x})\rangle = \Psi^{(0)} + \langle\Psi^{(2)}\rangle$ to second order. The first-order covariance of Ψ is $C_\Psi(\mathbf{x}, \boldsymbol{\chi}) = \langle\Psi^{(1)}(\mathbf{x})\Psi^{(1)}(\boldsymbol{\chi})\rangle$.

The covariance equation is obtained by multiplying $\Psi^{(1)}(\boldsymbol{\chi})$ to Eq. (5.83), utilizing Eq. (5.85), and taking expectation,

$$\frac{\partial^2 C_\Psi(\mathbf{x}, \boldsymbol{\chi})}{\partial x_i^2} + \langle\alpha\rangle\frac{\partial C_\Psi(\mathbf{x}, \boldsymbol{\chi})}{\partial x_1} = \frac{g(\mathbf{x})}{K_G}C_{f\Psi}(\mathbf{x}, \boldsymbol{\chi}) - \left[\langle\alpha\rangle\Psi^{(0)}(\mathbf{x}) + \frac{\partial\Psi^{(0)}(\mathbf{x})}{\partial x_i}\right]$$
$$\cdot\frac{\partial C_{f\Psi}(\mathbf{x}, \boldsymbol{\chi})}{\partial x_i} - \frac{\partial\Psi^{(0)}(\mathbf{x})}{\partial x_1}C_{\alpha\Psi}(\boldsymbol{\chi})$$

$$C_\Psi(\mathbf{x}, \boldsymbol{\chi}) = \frac{1}{\langle\alpha\rangle}\exp\left[\langle\alpha\rangle H_B(\mathbf{x})\right]\left[H_B(\mathbf{x}) - \frac{1}{\langle\alpha\rangle}\right]C_{\alpha\Psi}(\boldsymbol{\chi}), \quad \mathbf{x} \in \Gamma_D \tag{5.86}$$

$$n_i(\mathbf{x})\frac{\partial C_\Psi(\mathbf{x}, \boldsymbol{\chi})}{\partial x_i} + \langle\alpha\rangle n_1(\mathbf{x})C_\Psi(\mathbf{x}, \boldsymbol{\chi})$$
$$= \frac{Q(\mathbf{x})}{K_G}C_{f\Psi}(\mathbf{x}, \boldsymbol{\chi}) - n_1(\mathbf{x})\Psi^{(0)}(\mathbf{x})C_{\alpha\Psi}(\boldsymbol{\chi}), \quad \mathbf{x} \in \Gamma_N$$

where the cross-covariances $C_{f\Psi}(\mathbf{x}, \boldsymbol{\chi})$ and $C_{\alpha\Psi}(\boldsymbol{\chi})$ are governed by the following equations:

$$\left.\begin{array}{c}\dfrac{\partial^2 C_{f\Psi}(\mathbf{x}, \boldsymbol{\chi})}{\partial \chi_i^2} + \langle\alpha\rangle\dfrac{\partial C_{f\Psi}(\mathbf{x}, \boldsymbol{\chi})}{\partial \chi_1} \\[2mm] = \dfrac{g(\boldsymbol{\chi})}{K_G}C_f(\mathbf{x}, \boldsymbol{\chi}) - \left[\langle\alpha\rangle\Psi^{(0)}(\boldsymbol{\chi}) + \dfrac{\partial\Psi^{(0)}(\boldsymbol{\chi})}{\partial \chi_i}\right]\dfrac{\partial C_f(\mathbf{x}, \boldsymbol{\chi})}{\partial \chi_1} \\[2mm] C_{f\Psi}(\mathbf{x}, \boldsymbol{\chi}) = 0, \quad \boldsymbol{\chi} \in \Gamma_D \\[2mm] n_i(\boldsymbol{\chi})\dfrac{\partial C_{f\Psi}(\mathbf{x}, \boldsymbol{\chi})}{\partial \chi_i} + \langle\alpha\rangle n_1(\boldsymbol{\chi})C_{f\Psi}(\mathbf{x}, \boldsymbol{\chi}) = \dfrac{Q(\boldsymbol{\chi})}{K_G}C_f(\mathbf{x}, \boldsymbol{\chi}), \quad \boldsymbol{\chi} \in \Gamma_N\end{array}\right\} \tag{5.87}$$

$$\left.\begin{array}{c}\dfrac{\partial^2 C_{\alpha\Psi}(\boldsymbol{\chi})}{\partial \chi_i^2} + \langle\alpha\rangle\dfrac{\partial C_{\alpha\Psi}(\boldsymbol{\chi})}{\partial \chi_1} = -\dfrac{\partial\Psi^{(0)}(\boldsymbol{\chi})}{\partial \chi_1}\sigma_\alpha^2 \\[2mm] C_{\alpha\Psi}(\boldsymbol{\chi}) = \dfrac{1}{\langle\alpha\rangle}\exp\left[\langle\alpha\rangle H_B(\boldsymbol{\chi})\right]\left[H_B(\boldsymbol{\chi}) - \dfrac{1}{\langle\alpha\rangle}\right]\sigma_\alpha^2, \quad \boldsymbol{\chi} \in \Gamma_D \\[2mm] n_i(\boldsymbol{\chi})\dfrac{\partial C_{\alpha\Psi}(\boldsymbol{\chi})}{\partial \chi_i} + \langle\alpha\rangle n_1(\boldsymbol{\chi})C_{\alpha\Psi}(\mathbf{x}, \boldsymbol{\chi}) = -n_1(\boldsymbol{\chi})\Psi^{(0)}(\boldsymbol{\chi})\sigma_\alpha^2, \quad \boldsymbol{\chi} \in \Gamma_N\end{array}\right\} \tag{5.88}$$

In deriving Eqs. (5.87) and (5.88), we have taken advantage of the property that the random constant α and the random space function $f(\mathbf{x})$ are uncorrelated, i.e., $C_{\alpha f}(\chi) = 0$. With the statistical moments $\langle\alpha\rangle$, σ_α^2, $\langle f\rangle$, and $C_f(\mathbf{x}, \chi)$, the first two moments of Ψ can be solved from Eqs. (5.84) and (5.86)–(5.88). As for the moment equations for saturated flow in Chapter 3, these moment equations may have to be solved numerically except for some special cases.

An alternative procedure has been developed in the work of Tartakovsky *et al.* [1999], where the Green's function approach is used to first derive exact but unclosed equations for the statistical moments of Ψ and then a perturbative scheme is devised to obtain solvable equations for the moments. For the case of one-dimensional vertical flow with the variance of α taken to be zero, Tartakovsky *et al.* [1999] derived analytical solutions for $\langle\Psi^{(0)}(\mathbf{x})\rangle$, $\langle\Psi^{(2)}(\mathbf{x})\rangle$, and $C_\Psi(\mathbf{x}, \chi)$. More recently, Lu [2000] solved both the unconditional and conditional versions of the Ψ moment equations in two dimensions by the method of finite elements. In the work of Lu [2000], the variability of α is also neglected.

The statistical moments of Ψ are only intermediate quantities, which need to be related to those of the pressure head ψ. The statistical moments of ψ may be obtained from those of Ψ on the basis of Eq. (5.77). Tartakovsky *et al.* [1999] expanded Eq. (5.77) as

$$\alpha\Psi(\mathbf{x}) = k_r(\psi) = \exp\{(\langle\alpha\rangle + \alpha')[\psi^{(0)}(\mathbf{x}) + \psi^{(1)}(\mathbf{x}) + \psi^{(2)}(\mathbf{x}) + \cdots]\}$$
$$= \exp[\langle\alpha\rangle\psi^{(0)}(\mathbf{x})]\{1 + \langle\alpha\rangle\psi^{(1)}(\mathbf{x}) + \alpha'\psi^{(0)}(\mathbf{x}) + \langle\alpha\rangle\psi^{(2)}(\mathbf{x})$$
$$+ \alpha'\psi^{(1)}(\mathbf{x}) + \tfrac{1}{2}[\langle\alpha\rangle\psi^{(1)}(\mathbf{x}) + \alpha'\psi^{(0)}(\mathbf{x})]^2 + \cdots\} \tag{5.89}$$

Substituting Eq. (5.79) into Eq. (5.89) and collecting terms of the same order leads to equations relating $\psi^{(i)}$ to $\Psi^{(i)}$. Together with a similar expansion based on $\alpha^2\Psi^2 = k_r^2$, Tartakovsky *et al.* [1999] obtained explicit equations for the moments of ψ in the absence of α variability,

$$\langle\psi^{(0)}(\mathbf{x})\rangle = \frac{1}{\alpha}\ln[\alpha\langle\Psi^{(0)}(\mathbf{x})\rangle] \tag{5.90}$$

$$\sigma_\psi^2(\mathbf{x}) = \frac{1}{\alpha^2}\frac{\sigma_\Psi^2(\mathbf{x})}{\langle\Psi^{(0)}(\mathbf{x})\rangle^2} \tag{5.91}$$

$$\langle\psi^{(2)}(\mathbf{x})\rangle = \frac{1}{\alpha}\ln\left[\frac{\langle\Psi^{(2)}(\mathbf{x})\rangle}{\langle\Psi^{(0)}(\mathbf{x})\rangle} - \frac{1}{2}\frac{\sigma_\Psi^2(\mathbf{x})}{\langle\Psi^{(0)}(\mathbf{x})\rangle^2}\right] \tag{5.92}$$

It should be noted that the expansion of $k_r(\psi)$ in Eq. (5.89) is exactly the same as Eq. (5.16) utilized in Eq. (5.13). It seems that the direct perturbation approach of Section 5.2 and the Kirchhoff transform approach use the same expansion and linearization, though at different stages of their derivations. This may suggest that linearization in the relative permeability $k_r(\psi)$ cannot be avoided even though it may be delayed in the Kirchhoff transform approach. At present, it seems that the direct approach is more general in that it does not require any restrictive assumption on the correlation structure of α and it is also applicable to constitutive relations other than the Gardner–Russo model. It is a challenge to extend the

Kirchhoff transform approach to transient flow (even for the Gardner–Russo model), which, however, does not cause any additional difficulty for the direct perturbation approach.

5.5 TRANSIENT FLOW IN NONSTATIONARY MEDIA

In this section, we consider transient unsaturated flow in statistically nonhomogeneous porous media. Transient unsaturated flow is described by the Richards equation,

$$C[h, \cdot] \frac{\partial \psi(\mathbf{x}, t)}{\partial t} + \nabla \cdot \mathbf{q}(\mathbf{x}, t) = g(\mathbf{x}, t) \tag{5.93}$$

$$q_i(\mathbf{x}, t) = -K[h, \cdot] \frac{\partial}{\partial x_i} [\psi(\mathbf{x}, t) + x_1] \tag{5.94}$$

subject to initial and boundary conditions

$$\psi(\mathbf{x}, 0) = H_o(\mathbf{x}), \quad \mathbf{x} \in \Omega \tag{5.95}$$

$$\psi(\mathbf{x}, t) = H_B(\mathbf{x}, t), \quad \mathbf{x} \in \Gamma_D \tag{5.96}$$

$$\mathbf{q}(\mathbf{x}, t) \cdot \mathbf{n}(\mathbf{x}) = Q(\mathbf{x}, t), \quad \mathbf{x} \in \Gamma_N \tag{5.97}$$

$$\frac{\partial}{\partial x_1} [\psi(\mathbf{x}, t) + x_1] = 1, \quad \mathbf{x} \in \Gamma_G \tag{5.98}$$

where $\mathbf{q}$ is the specific discharge (flux), $\psi(\mathbf{x}, t) + x_1$ is the total head, h is the pressure head, $i = 1, \ldots, d$ (where d is the number of space dimensions), $g(\mathbf{x}, t)$ is the fluid source/sink term, $H_o(\mathbf{x})$ is the initial pressure head distribution in the domain Ω, $H_B(\mathbf{x}, t)$ is the prescribed head on Dirichlet boundary segments Γ_D at time t, $Q(\mathbf{x}, t)$ is the prescribed flux across Neumann boundary segments Γ_N at time t, $\mathbf{n}(\mathbf{x}) = (n_1, \ldots, n_d)^T$ is an outward unit vector normal to the boundary, Γ_G is the boundary segment where flow is gravity-dominated so that the (total) pressure gradient is unity, $C[\psi, \cdot] \equiv d\theta_e/d\psi$ is the *specific moisture capacity*, and $K[\psi, \cdot]$ is the unsaturated hydraulic conductivity (assumed to be isotropic locally). Both C and K are functions of pressure head and soil properties at $\mathbf{x}$. For convenience, they will be written as $C(\mathbf{x}, t)$ and $K(\mathbf{x}, t)$ in the sequel.

As for steady-state flow in Eqs. (5.1)–(5.4), Eqs. (5.93)–(5.98) are not closed without some constitutive relationships of $K(\psi)$ and $C(\psi)$. For simplicity, we use the Gardner–Russo model of Eqs. (5.6) and (5.7). For the case of $m = 0$, $C(\psi)$ takes the following functional form:

$$C(\mathbf{x}, t) = -\frac{\theta_s - \theta_r}{4} \alpha^2(\mathbf{x}) \psi(\mathbf{x}, t) \exp[0.5\alpha(\mathbf{x}) \psi(\mathbf{x}, t)] \tag{5.99}$$

The extension to the case of $m \neq 0$ can be made by following the treatment of Zhang *et al.* [1998] for steady-state flow. We further assume θ_s and θ_r to be deterministic constants. The pore size distribution parameter $\alpha(\mathbf{x})$, and the log-transformed saturated hydraulic conductivity $f(\mathbf{x}) = \ln K_s(\mathbf{x})$ are treated as general random space functions, which can be either stationary or nonstationary in space. In turn, the dependent variables like pressure and flux are also (nonstationary) random space functions and the Richards equation becomes a set of stochastic partial differential equations.

As for steady-state unsaturated flow, we aim to derive equations governing the first two moments of the dependent variables based on the statistical moments of input parameters. In this section, both α and f are taken to be nonstationary such that their expected values may vary in space and their covariances may depend on the actual locations of two points. Hence, flow in such situations as distinct geological layers, zones, and facies can be handled. For more discussion on different scenarios of nonstationary media, the reader is referred to Section 2.3.3. Without loss of generality, $H_o(\mathbf{x})$, $H_B(\mathbf{x}, t)$, $Q(\mathbf{x}, t)$, and $g(\mathbf{x}, t)$ are assumed to be deterministic (i.e., known with certainty).

5.5.1 Moment Partial Differential Equations

As usual, the moment equation approach starts with decomposing the input random variables as $f(\mathbf{x}) = \langle f(\mathbf{x}) \rangle + f'(\mathbf{x})$ and $\alpha(\mathbf{x}) = \langle \alpha(\mathbf{x}) \rangle + \alpha'(\mathbf{x})$, where $\langle \ \rangle$ stands for the ensemble average (expectation) and the primed quantities are the zero-mean fluctuations. Since the variability of $\psi(\mathbf{x}, t)$ depends on those of the medium properties α and f and the variabilities of $Y = \ln K(\mathbf{x}, t)$ and $C(\mathbf{x}, t)$ depend on those of ψ and the medium properties, one may express these quantities as an infinite series in the following form: $\psi(\mathbf{x}, t) = \psi^{(0)} + \psi^{(1)} + \psi^{(2)} + \cdots$, $Y(\mathbf{x}, t) = Y^{(0)} + Y^{(1)} + Y^{(2)} + \cdots$, and $C(\mathbf{x}, t) = C^{(0)} + C^{(1)} + C^{(2)} + \cdots$. In these series, the order of each term is with respect to the variability of the medium properties. After substituting these expressions into the Richards equation and collecting terms at separate order, and after some mathematical manipulations, one arrives at equations governing $\psi^{(n)}$ ($n = 1, 2, \ldots$) [Zhang, 1999]. It can be shown that $\langle \psi^{(0)} \rangle = \psi^{(0)}$ and $\langle \psi^{(1)} \rangle = 0$. Hence, the mean head is $\langle \psi \rangle = \psi^{(0)}$ to zeroth or first order in terms of the variability of medium properties, and $\langle \psi \rangle = \psi^{(0)} + \langle \psi^{(2)} \rangle$ to second order. The head fluctuation is $\psi' = \psi^{(1)}$ to first order. Therefore, to first order the head covariance is $C_\psi(\mathbf{x}, t; \boldsymbol{\chi}, \tau) = \langle \psi^{(1)}(\mathbf{x}, t) \psi^{(1)}(\boldsymbol{\chi}, \tau) \rangle$.

The resulting equation for the first-order mean head reads as

$$\frac{\partial^2 \langle \psi(\mathbf{x}, t) \rangle}{\partial x_i^2} + a_i(\mathbf{x}, t) \frac{\partial \langle \psi(\mathbf{x}, t) \rangle}{\partial x_i} + c_2(\mathbf{x}, t) \langle \psi(\mathbf{x}, t) \rangle$$

$$= e(\mathbf{x}, t) \frac{\partial \langle \psi(\mathbf{x}, t) \rangle}{\partial t} - \frac{g(\mathbf{x}, t)}{K_m(\mathbf{x}, t)} + d_6(\mathbf{x}, t)$$

$$\langle \psi(\mathbf{x}, 0) \rangle = H_o(\mathbf{x}), \quad \mathbf{x} \in \Omega$$

$$\langle \psi(\mathbf{x}, t) \rangle = H_B(\mathbf{x}, t), \quad \mathbf{x} \in \Gamma_D \tag{5.100}$$

$$n_i(\mathbf{x}) \frac{\partial \langle \psi(\mathbf{x}, t) \rangle}{\partial x_i} = -\frac{Q(\mathbf{x}, t)}{K_m(\mathbf{x}, t)} - \delta_{i1} n_i(\mathbf{x}), \quad \mathbf{x} \in \Gamma_N$$

$$\frac{\partial \langle \psi(\mathbf{x}, t) \rangle}{\partial x_1} = 0, \quad \mathbf{x} \in \Gamma_G$$

where $K_m(\mathbf{x}, t) = \exp[\langle f \rangle] \exp[\langle \alpha \rangle \langle \psi(\mathbf{x}, t) \rangle]$, $a_i(\mathbf{x}, t) = \langle \alpha \rangle J_i(\mathbf{x}, t)$, $c_2(\mathbf{x}, t) = J_i(\mathbf{x}, t) \partial \langle \alpha(\mathbf{x}) \rangle / \partial x_i$, $d_6(\mathbf{x}, t) = -J_i(\mathbf{x}, t) \partial \langle f(\mathbf{x}) \rangle / \partial x_i$, $J_i(\mathbf{x}, t) = \partial \langle \psi(\mathbf{x}, t) \rangle / \partial x_i + \delta_{i1}$, and $e(\mathbf{x}, t) = C^{(0)}(\mathbf{x}, t) / K_m(\mathbf{x}, t)$ with $C^{(0)} = -0.25(\theta_s - \theta_r) \langle \alpha \rangle^2 \langle \psi(\mathbf{x}, t) \rangle \cdot \exp[0.5 \langle \alpha \rangle \langle \psi(\mathbf{x}, t) \rangle]$. This equation is nonlinear because the coefficients $K_m(\mathbf{x}, t)$, $a_i(\mathbf{x}, t)$,

$c_2(\mathbf{x})$, $d_6(\mathbf{x}, t)$, and $e(\mathbf{x}, t)$ all are functions of the dependent variable $\langle \psi \rangle$. The equation for the first-order head covariance is given as

$$\frac{\partial^2 C_\psi(\mathbf{x}, t; \boldsymbol{\chi}, \tau)}{\partial x_i^2} + b_i(\mathbf{x}, t)\frac{\partial C_\psi(\mathbf{x}, t; \boldsymbol{\chi}, \tau)}{\partial x_i} + c(\mathbf{x}, t)C_\psi(\mathbf{x}, t; \boldsymbol{\chi}, \tau)$$

$$= e(\mathbf{x}, t)\frac{\partial C_\psi(\mathbf{x}, t; \boldsymbol{\chi}, \tau)}{\partial t} - J_i(\mathbf{x}, t)\frac{\partial C_{f\psi}(\mathbf{x}; \boldsymbol{\chi}, \tau)}{\partial x_i}$$

$$- J_i(\mathbf{x}, t)\langle\psi(\mathbf{x}, t)\rangle\frac{\partial C_{\alpha\psi}(\mathbf{x}; \boldsymbol{\chi}, \tau)}{\partial x_i} + d_1(\mathbf{x}, t)C_{f\psi}(\mathbf{x}; \boldsymbol{\chi}, \tau)$$

$$+ d_2(\mathbf{x}, t)C_{\alpha\psi}(\mathbf{x}; \boldsymbol{\chi}, \tau)$$

$$C_\psi(\mathbf{x}, 0; \boldsymbol{\chi}, \tau) = 0, \quad \mathbf{x} \in \Omega \tag{5.101}$$

$$C_\psi(\mathbf{x}, t; \boldsymbol{\chi}, \tau) = 0, \quad \mathbf{x} \in \Gamma_D$$

$$n_i(\mathbf{x})\frac{\partial C_\psi(\mathbf{x}, t; \boldsymbol{\chi}, \tau)}{\partial x_i} + d_3(\mathbf{x}, t)C_\psi(\mathbf{x}, t; \boldsymbol{\chi}, \tau)$$

$$= d_4(\mathbf{x}, t)C_{f\psi}(\mathbf{x}; \boldsymbol{\chi}, \tau) + d_5(\mathbf{x}, t)C_{\alpha\psi}(\mathbf{x}; \boldsymbol{\chi}, \tau), \quad \mathbf{x} \in \Gamma_N$$

$$\frac{\partial C_\psi(\mathbf{x}, t; \boldsymbol{\chi}, \tau)}{\partial x_1} = 0, \quad \mathbf{x} \in \Gamma_G$$

where $b_i(\mathbf{x}, t) = [2J_i(\mathbf{x}, t) - \delta_{i1}]\langle\alpha\rangle + \partial\langle f(\mathbf{x})\rangle/\partial x_i + \langle\psi(\mathbf{x}, t)\rangle\partial\langle\alpha(\mathbf{x})\rangle/\partial x_i$, $c(\mathbf{x}, t) = -\langle\alpha\rangle d_1(\mathbf{x}, t) - J_t(\mathbf{x}, t)p_1(\mathbf{x}, t)/K_m(\mathbf{x}, t) + c_2(\mathbf{x}, t)$ with $J_t(\mathbf{x}, t) = \partial\langle\psi(\mathbf{x}, t)\rangle/\partial t$, $d_1(\mathbf{x}, t) = [g(\mathbf{x}, t) - J_t(\mathbf{x}, t)C^{(0)}(\mathbf{x}, t)]/K_m(\mathbf{x}, t)$, $d_2(\mathbf{x}, t) = d_1(\mathbf{x}, t)\langle\psi(\mathbf{x}, t)\rangle - J_i(\mathbf{x}, t)[J_i(\mathbf{x}, t) - \delta_{i1}] + J_t(\mathbf{x}, t)p_2(\mathbf{x}, t)/K_m(\mathbf{x}, t)$, $p_1(\mathbf{x}, t) = -0.25(\theta_s - \theta_r)\exp[0.5\langle\alpha\rangle\langle\psi(\mathbf{x}, t)\rangle][\langle\alpha\rangle^2 + 0.5\langle\alpha\rangle^3\langle\psi(\mathbf{x}, t)\rangle]$, $p_2(\mathbf{x}, t) = -0.5(\theta_s - \theta_r)\exp[0.5\langle\alpha\rangle\langle\psi(\mathbf{x}, t)\rangle][\langle\alpha\rangle\langle\psi(\mathbf{x}, t)\rangle + 0.25\langle\alpha\rangle^2[\langle\psi(\mathbf{x}, t)\rangle]^2]$, $d_3(\mathbf{x}, t) = \langle\alpha\rangle n_i(\mathbf{x}) J_i(\mathbf{x}, t)$, $d_4(\mathbf{x}, t) = -n_i(\mathbf{x})J_i(\mathbf{x}, t)$, and $d_5(\mathbf{x}, t) = -n_i(\mathbf{x})J_i(\mathbf{x}, t)\langle\psi(\mathbf{x}, t)\rangle$. The cross-covariances $C_{f\psi}$ and $C_{\alpha\psi}$ are the solutions of the following equations:

$$\left.\begin{aligned}
&\frac{\partial^2 C_{f\psi}(\mathbf{x}; \boldsymbol{\chi}, \tau)}{\partial \chi_i^2} + b_i(\boldsymbol{\chi}, \tau)\frac{\partial C_{f\psi}(\mathbf{x}; \boldsymbol{\chi}, \tau)}{\partial \chi_i} + c(\boldsymbol{\chi}, \tau)C_{f\psi}(\mathbf{x}; \boldsymbol{\chi}, \tau) \\
&= e(\boldsymbol{\chi}, \tau)\frac{\partial C_{f\psi}(\mathbf{x}; \boldsymbol{\chi}, \tau)}{\partial \tau} - J_i(\boldsymbol{\chi}, \tau)\frac{\partial C_f(\mathbf{x}; \boldsymbol{\chi})}{\partial \chi_i} \\
&\quad - J_i(\boldsymbol{\chi}, \tau)\langle\psi(\boldsymbol{\chi}, \tau)\rangle\frac{\partial C_{f\alpha}(\mathbf{x}; \boldsymbol{\chi})}{\partial \chi_i} + d_1(\boldsymbol{\chi}, \tau)C_f(\mathbf{x}; \boldsymbol{\chi}) \\
&\quad + d_2(\boldsymbol{\chi}, \tau)C_{f\alpha}(\mathbf{x}; \boldsymbol{\chi}) \\
&\qquad\qquad C_{f\psi}(\mathbf{x}; \boldsymbol{\chi}, 0) = 0, \quad \boldsymbol{\chi} \in \Omega \\
&\qquad\qquad C_{f\psi}(\mathbf{x}; \boldsymbol{\chi}, \tau) = 0, \quad \boldsymbol{\chi} \in \Gamma_D \\
&n_i(\boldsymbol{\chi})\frac{\partial C_{f\psi}(\mathbf{x}; \boldsymbol{\chi}, \tau)}{\partial \chi_i} + d_3(\boldsymbol{\chi}, \tau)C_{f\psi}(\mathbf{x}; \boldsymbol{\chi}, \tau) \\
&= d_4(\boldsymbol{\chi}, \tau)C_f(\mathbf{x}; \boldsymbol{\chi}) + d_5(\boldsymbol{\chi}, \tau)C_{f\alpha}(\mathbf{x}; \boldsymbol{\chi}), \quad \boldsymbol{\chi} \in \Gamma_N \\
&\qquad\qquad \frac{\partial C_{f\psi}(\mathbf{x}; \boldsymbol{\chi}, \tau)}{\partial \chi_1} = 0, \quad \boldsymbol{\chi} \in \Gamma_G
\end{aligned}\right\} \tag{5.102}$$

$$
\left.\begin{array}{l}
\dfrac{\partial^2 C_{\alpha\psi}(\mathbf{x};\boldsymbol{\chi},\tau)}{\partial \chi_i^2} + b_i(\boldsymbol{\chi},\tau)\dfrac{\partial C_{\alpha\psi}(\mathbf{x};\boldsymbol{\chi},\tau)}{\partial \chi_i} + c(\boldsymbol{\chi},\tau)C_{\alpha\psi}(\mathbf{x};\boldsymbol{\chi},\tau) \\[2ex]
= e(\boldsymbol{\chi},\tau)\dfrac{\partial C_{\alpha\psi}(\mathbf{x};\boldsymbol{\chi},\tau)}{\partial \tau} - J_i(\boldsymbol{\chi},\tau)\dfrac{\partial C_{\alpha f}(\mathbf{x};\boldsymbol{\chi})}{\partial \chi_i} \\[2ex]
\quad - J_i(\boldsymbol{\chi},\tau)\langle\psi(\boldsymbol{\chi},\tau)\rangle\dfrac{\partial C_{\alpha}(\mathbf{x};\boldsymbol{\chi})}{\partial \chi_i} + d_1(\boldsymbol{\chi},\tau)C_{\alpha f}(\mathbf{x};\boldsymbol{\chi}) \\[2ex]
\quad + d_2(\boldsymbol{\chi},\tau)C_{\alpha}(\mathbf{x};\boldsymbol{\chi}) \\[2ex]
\qquad\qquad C_{\alpha\psi}(\mathbf{x};\boldsymbol{\chi},0) = 0, \quad \boldsymbol{\chi}\in\Omega \\[1ex]
\qquad\qquad C_{\alpha\psi}(\mathbf{x};\boldsymbol{\chi},\tau) = 0, \quad \boldsymbol{\chi}\in\Gamma_D \\[2ex]
n_i(\boldsymbol{\chi})\dfrac{\partial C_{\alpha\psi}(\mathbf{x};\boldsymbol{\chi},\tau)}{\partial \chi_i} + d_3(\boldsymbol{\chi},\tau)C_{\alpha\psi}(\mathbf{x};\boldsymbol{\chi},\tau) \\[2ex]
= d_4(\boldsymbol{\chi},\tau)C_{\alpha f}(\mathbf{x};\boldsymbol{\chi}) + d_5(\boldsymbol{\chi},\tau)C_{\alpha}(\mathbf{x};\boldsymbol{\chi}), \quad \boldsymbol{\chi}\in\Gamma_N \\[2ex]
\qquad\qquad \dfrac{\partial C_{\alpha\psi}(\mathbf{x};\boldsymbol{\chi},\tau)}{\partial \chi_1} = 0, \quad \boldsymbol{\chi}\in\Gamma_G
\end{array}\right\} \tag{5.103}
$$

In the above, C_f, $C_{f\alpha}$, and C_α are the input covariances of the medium properties. In the case of stationary f and α, moment equations similar to Eqs. (5.100)–(5.103) have been given and solved by Zhang [1999b]. Note that although the equation governing the first moment $\langle\psi\rangle$ is nonlinear, Eqs. (5.102) and (5.103) are linear and solvable with the solution of $\langle\psi(\mathbf{x},t)\rangle$. With $C_{f\psi}$ and $C_{\alpha\psi}$, C_ψ can be solved from Eq. (5.101).

With these head moments, the statistical moments of water content, unsaturated conductivity, flux, and velocity can be derived with the same procedure utilized for steady-state flow in Section 5.2.1. These moments look like those for steady-state flow in Eqs. (5.24)–(5.40) except for the additional dependency on time. In the case of stationary soil properties, these moments have been given by Zhang [1999b].

This moment-equation-based stochastic model of transient flow is applicable to the entire domain of a bounded vadose zone in the presence of sinks/sources and nonstationary medium features. Due to the mathematical complexity of the equations, in general they need to be solved numerically. The numerical moment equation approach, however, has the flexibility in handling nonstationary medium features as well as different boundary conditions, internal sink/source terms, input covariance structures and soil constitutive relationships. The results from this stochastic model are the first two moments of flow quantities. The first two moments of a flow quantity may be used to approximate the confidence intervals for the quantity, which are a measure of uncertainty caused by incomplete knowledge of medium heterogeneity.

Numerical solution Under steady-state conditions and in the absence of medium non-stationarity, the moment equations derived in this section reduce to Eqs. (5.19) and (5.21)–(5.23). As pointed out earlier, the solution is facilitated by recognizing that the moments C_ψ, $C_{f\psi}$, and $C_{\alpha\psi}$ are governed by the same type of equations but with different forcing terms.

The nonlinear mean pressure head equation (5.100) may be solved by a Picard or Newton–Raphson iteration scheme. With the mean pressure head, $C_{f\psi}$ and $C_{\alpha\psi}$ can be solved from Eqs. (5.102) and (5.103). Then C_ψ can be obtained from Eq. (5.101). We may approximate the spatial derivatives by the central-difference scheme and the temporal derivatives by the implicit (backward) method. At each time step, the solution of these covariance equations is similar to that of the steady-state equations discussed already in Section 5.2.1. It is, however, worthwhile to outline the solution procedure for $C_\psi(\mathbf{x}, t; \boldsymbol{\chi}, \tau)$. We may approximate the temporal derivative as

$$\frac{\partial C_\psi(\mathbf{x}, t; \boldsymbol{\chi}, \tau)}{\partial t} \approx \frac{C_\psi(\mathbf{x}, t; \boldsymbol{\chi}, \tau) - C_\psi(\mathbf{x}, t - \Delta t; \boldsymbol{\chi}, \tau)}{\Delta t} \tag{5.104}$$

where the time derivative only operates on the first time argument. For simplicity, we let τ take the value of t. $C_\psi(\mathbf{x}, t; \boldsymbol{\chi}, \tau = t)$ is the solution that we are after while $C_\psi(\mathbf{x}, t - \Delta t; \boldsymbol{\chi}, \tau = t)$ is also unknown. Therefore, we need to first find the solution of $C_\psi(\mathbf{x}, t - \Delta t; \boldsymbol{\chi}, \tau = t)$, which depends on $C_\psi(\mathbf{x}, t - 2\Delta t; \boldsymbol{\chi}, \tau = t)$. In order to obtain $C_\psi(\mathbf{x}, t; \boldsymbol{\chi}, \tau = t)$ we need to solve for $C_\psi(\mathbf{x}, t'; \boldsymbol{\chi}, \tau)$ from $t' = 0$ to $t' = t$. For a large t, this may cause a significant computational burden. However, at the large time when the flow does not change rapidly, $C_\psi(\mathbf{x}, t - \Delta t; \boldsymbol{\chi}, \tau = t)$ may be well approximated by $C_\psi(\mathbf{x}, t - \Delta t; \boldsymbol{\chi}, t - \Delta t)$ which is the solution at the previous time step. This may greatly speed up the computation. Under what conditions this approximation is warranted deserves further investigations. Since $C_{f\psi}$ and $C_{\alpha\psi}$ involve only one time argument, their solution procedure is straightforward.

The finite difference scheme for unsaturated flow moment equations has been implemented in a computer code called STO-UNSAT, which is capable of handling both steady-state and transient cases in spatially stationary and nonstationary media. The code is available from the writer upon request.

Illustrative examples In this section, we illustrate the moments of transient unsaturated flow through some one- and two-dimensional examples of unsaturated flow in hypothetical soils with different boundary conditions. The first few examples involve soils of stationary properties, while the other ones deal with nonstationary media. The properties of the stationary soil are the same as discussed in Section 5.2.1 for steady-state flow.

We first consider vertical infiltration in a one-dimensional domain of the above-mentioned soils under different conditions. In this case, the boundary conditions are specified heads: $\psi = -172.694$ cm at the top and $\psi = 0$ at the bottom. The head of -172.694 cm at the top would correspond to a recharge value of 0.001 [cm/T] in a homogeneous soil of saturated hydraulic conductivity $K_S = \exp(\langle f \rangle) = 1$ [cm/T] under gravity-dominated flow condition (where T is an arbitrary time unit). It is assumed that the head distribution is initially known with certainty; and at time $t = 0$ the above-mentioned boundary conditions start to be effective. Figure 5.4 shows the first two moments of pressure head ψ and effective water content θ_e as functions of dimensionless distance $z = x_1\langle \alpha \rangle$ at different times. The solid curves in these plots correspond to the steady-state case, which can be computed by letting $t \to \infty$ or independently using the steady-state approach of

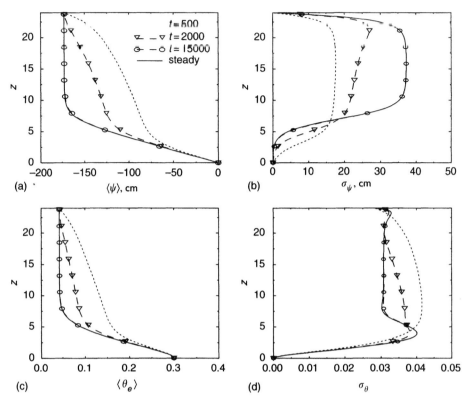

Figure 5.4. The expected values and standard deviations of the pressure head ψ and the effective water content θ_e versus the normalized distance z as functions of normalized recharge q_n. (Adapted from Zhang [1999b]. Copyright 1999 by the American Geophysical Union.)

Section 5.2. It is seen from Fig. 5.4a that at steady state, the mean (expected) pressure head decreases with the distance from the water table. After some critical distance, it approaches the constant specified at the top. The steady-state head standard deviation increases with the distance from the water table, then becomes stablized for some distances, and finally decreases toward zero at the top boundary of specified head (Fig. 5.4b). The distance from the water table for the head standard deviation to become constant is slightly larger than that for the mean head, as observed in Fig. 5.1 for specified flux condition at the top. It is seen that with time, the mean head and head standard deviation approach their respective steady-state limits. Although the mean and standard deviation profiles strongly depend on the initial condition (not shown here), at large times they always approach their respective steady-state profiles, which are dependent on the boundary conditions but are independent of the initial condition, as one should expect. We see similar behaviors for the two moments of water content (Figs. 5.4c and d), except for the water content standard deviation at the top boundary. Unlike the head standard deviation, the water content standard deviation is not zero at the top boundary. This is so because the uncertainty in the water content comes

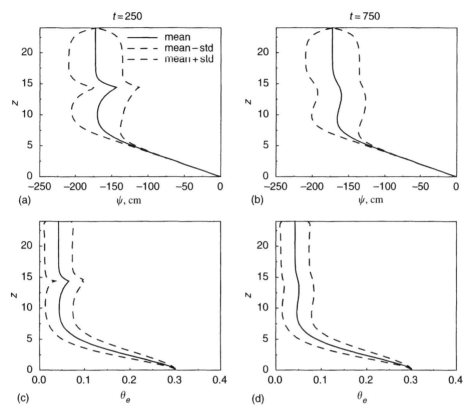

Figure 5.5. Confidence intervals for pressure head and effective water content at two different times in the presence of an internal source. (Adapted from Zhang [1999b]. Copyright 1999 by the American Geophysical Union.)

from that in the soil parameter α even though the head is specified with certainty there. Mathematically speaking, although the first three terms on the right-hand side of Eq. (5.35) are zero the last one is not, which makes $\sigma_{\theta_e}^2(\mathbf{x}, t) = C_{\theta_e}(\mathbf{x}, t; \mathbf{x}, t)$ nonzero.

The second case is the same as the first one except that there is an internal source of specified flow rate. This may correspond to the situation of a field injection. The soil column is run to steady state with the above-mentioned boundary conditions, then an injection of strength $g = 0.001\,[\mathrm{T}^{-1}]$ starts at this time (let it be $t = 0$) and lasts for $250\,[\mathrm{T}]$. At the end of the fluid injection, the mean head and water content profiles are shown in Figs. 5.5a and c. It is seen that the impact of the injection is the increase of the pressure head and water content at the injection point and its vicinity, as expected. Also presented are the 68% confidence intervals for the flow quantities. It is seen that the impacts of the injection on the confidence intervals dissipate with time after the injection stops (Figs. 5.5b and d).

Figure 5.6 shows the impact of deterministic yet time-varying recharge. Here we specify a time-varying flux condition at the top and the water table condition at the bottom. Two

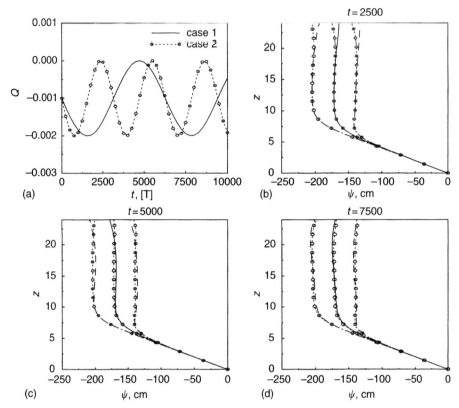

Figure 5.6. Two scenarios of time-varying recharge (Q) and the corresponding confidence intervals for pressure head at three different times. (Adapted from Zhang [1999b]. Copyright 1999 by the American Geophysical Union.)

scenarios of time-varying recharges are depicted in Fig. 5.6a, which vary with the same magnitude but different frequencies. The head confidence intervals at three different times are shown in the rest of this figure. In the case of a constant recharge, the head and saturation profiles are constant when away from the boundaries (see Fig. 5.2). It is seen that only near the land surface do the pressure head and saturation confidence intervals for the cases of time-varying recharges significantly deviate from their respective behaviors in the case of a constant recharge. That is, the effects of the transient (periodic) recharge are mainly limited to the upper portion of the domain. By comparing the two cases of different frequencies, we find that the effects of periodic recharge decrease with an increase in variation frequency. In other words, the temporal variations of high frequency tend to be damped out. This observation is similar to that made by Zhang and Neuman [1996] for solute transport in saturated systems. However, the observations based on this figure come with the condition of a given magnitude. For a given frequency, a larger magnitude of variation in recharge produces a more profound effect on flow (though not shown here). Physically speaking,

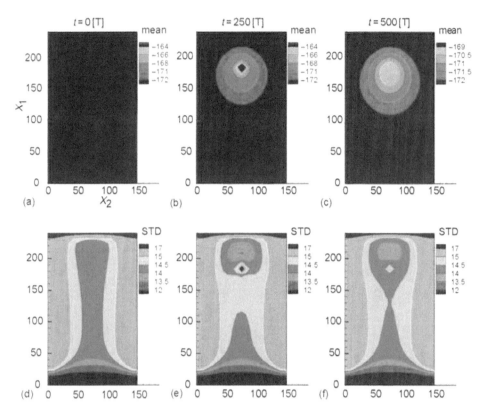

Figure 5.7. Contours of mean pressure head and head standard deviation at three different times in the presence of an internal source for the case of stationary media.

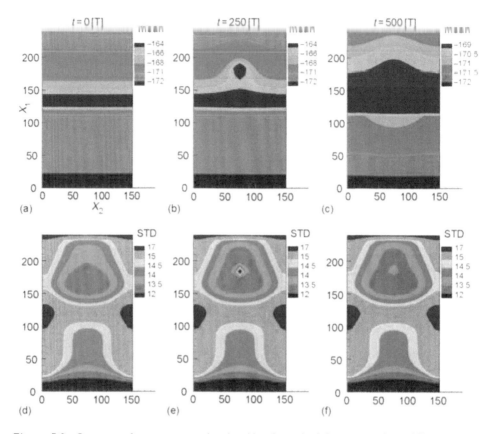

Figure 5.8. Contours of mean pressure head and head standard deviation at three different times in the presence of an internal source for the case of an embedded layer of coarser soils.

the transient effects of time-varying recharge on both the mean behavior of flow and the associated uncertainty may be negligible away from the top boundary if the extent of the vadose zone is large, the magnitude of variation is small, and/or the frequency is high. Otherwise, the transient effects can be significant for the entire domain of a vadose zone.

We now look at some two-dimensional examples. Like in the case shown in Fig. 5.3, here the domain is of size 240 cm by 150 cm and represents an upper portion of an unsaturated zone. The boundary conditions are also the same as in the case shown in Fig. 5.3. At steady state ($t = 0$), the flow is mean gravity-dominated such that the mean pressure head is constant through the whole domain (Fig. 5.7a). However, the head standard deviation is not uniform in such a bounded domain (Fig. 5.7d). The head standard deviation is zero at the bottom boundary of constant head and increases in the vertical direction with distance from there. In the horizontal direction and away from the bottom boundary, the head standard deviation is the largest at the two lateral (no-flow) boundaries and decreases toward the center of the domain. Profiles of the head variance along the vertical and horizontal centerlines are depicted in Fig. 5.3. At $t = 0^+$, a source of strength $g = 0.001$ [T^{-1}] is introduced at

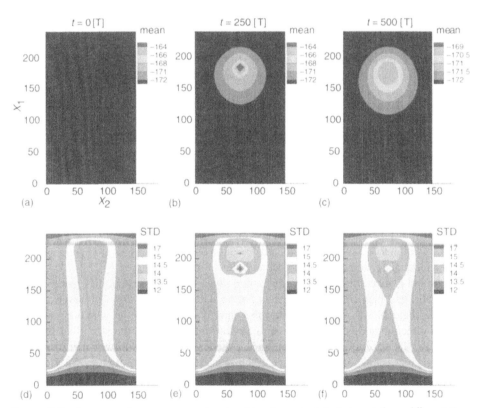

Figure 5.7. Contours of mean pressure head and head standard deviation at three different times in the presence of an internal source for the case of stationary media.

$(x_1, x_2) = (192, 75)$ in the upper portion of the domain and lasts for 250 [T]. At the end of the injection when $t = 250$ [T], the mean pressure head and head standard deviation contours are shown in Figs. 5.7b and e, respectively. As expected, the mean head contours are asymmetric in the vertical direction about the point of injection due to gravitational effect, while they are symmetric in the horizontal direction. It is seen from Fig. 5.7e that the head standard deviation increases in the immediate vicinity of the injection. The head standard deviation slightly increases at the downstream locations of the injection point and slightly decreases at the upstream points. With time, the moisture plume continues to spread horizontally and vertically, as reflected in both the mean head (Fig. 5.7c) and the head standard deviation (Fig. 5.7f). Confidence intervals such as those shown in Figs. 5.2 and 5.5 can be drawn with these contours. The uncertainty in the pressure head includes two parts: one due to the incomplete knowledge of the head field before injection, and the other due to the redistribution of the injected fluid. However, in this case both of them are caused by the incomplete knowledge of the spatial variability of soil properties. In many field experiments, the field conditions cannot be known without uncertainty. It is thus important to consider the effects of the initial uncertainty in predicting or interpreting experimental results.

Figure 5.8 investigates the effects of a thin layer of slightly different soils embedded in the otherwise spatially stationary domain. This layer is of thickness 16 cm and width 150 cm with its center at $x_1 = 128$ cm. This layer has coarser soils than does the rest of the domain. In this layer, the mean pore size distribution parameter is $\langle \alpha \rangle = 0.045$ cm^{-1}, compared to 0.04 cm^{-1} outside of this layer. Otherwise, this case is the same as that shown in Fig. 5.7. Comparisons between Figs. 5.8 and 5.7 reveal that this nonstationary medium feature has a significant impact on the expected values of the pressure head field and the associated prediction uncertainty. First, the steady-state mean flow is no longer uniform in space and is thus nonstationary (Fig. 5.8a). The head standard deviation contours of Fig. 5.8d are also very different from their counterparts in Fig. 5.7d for the stationary medium. At the end of injection ($t = 250$ [T]), the moisture plume moves mainly in the lateral direction above the thin layer (see Fig. 5.8b), while it moves preferentially downward for the case of stationary medium as shown in Fig. 5.7b. It is seen that this embedded layer of slightly coarser soils acts like a *capillary barrier*, which induces lateral moisture movement but inhibits vertical migration when the layer is relatively dry. At $t = 500$ [T], the moisture starts to move vertically past the thin layer, while it continues to migrate horizontally. The fluid injection renders a higher pressure head standard deviation and hence a larger prediction uncertainty in the immediate vicinity of the injection point.

At first glance, it is surprising that a thin layer with such a small difference in the soil properties has such significant impacts on the statistics of the flow. However, this result is understandable if one realizes that this thin layer is across the whole domain in the lateral direction and is thus no longer a random feature. When the size of such a feature is smaller, its effect is also less profound as shown in Fig. 5.9. In this figure, the width of the thin layer is only 16 cm thick and 15 cm wide with its center at (128,75). This feature is small compared to the size of the domain but still large relative to the point source and the moisture plume, which explains why its impact on the flow moments is still relatively significant. This may suggest that whether medium features can be treated statistically without explicitly (and deterministically) considering them should depend on the scale of

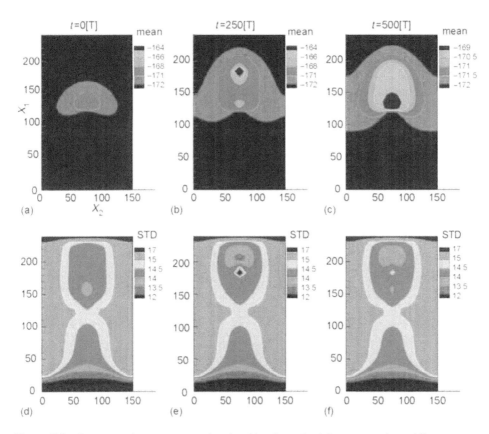

Figure 5.9. Contours of mean pressure head and head standard deviation at three different times in the presence of an internal source for the case of an embedded zone of coarser soils.

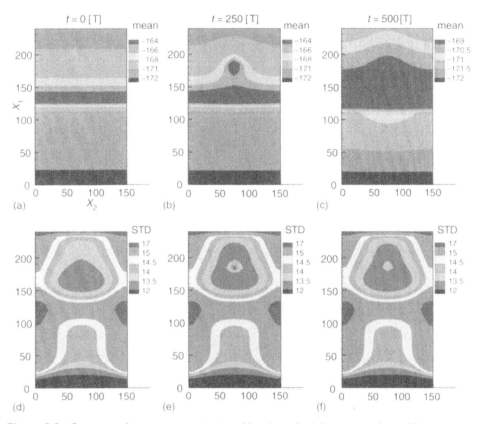

Figure 5.8. Contours of mean pressure head and head standard deviation at three different times in the presence of an internal source for the case of an embedded layer of coarser soils.

interest. If we are interested in flow behaviors at a pore scale, then minute features of porous media are important and flow is described by the Navier–Stokes equation. When our interest is primarily at the lab scale, the detailed flow behaviors at the pore scale are then less important and may be averaged out so that Darcy's law is appropriate. At the field scale, random features much smaller than this scale may be treated stochastically without explicitly accounting for them. However, if there are persistent features of significant size compared to the domain of interest or to characteristic lengths of the flow, a nonstationary (statistically nonhomogeneous) description, such as the one utilized in this section, may be needed.

In summary, flow in the vadose zone is usually nonstationary no matter whether the medium is stationary or nonstationary. Flow nonstationarity may significantly impact solute spreading in the vadose zone [Sun and Zhang, 2000]. In the absence of medium nonstationary, the vadose zone may be divided into three flow regimes, as illustrated in Fig. 5.10: the recharge zone, the gravity-dominated zone, and the capillary fringe. The first one is near the land surface and is greatly influenced by time-varying net recharge after accounting for precipitation and evapotranspiration. This zone is nonstationary in

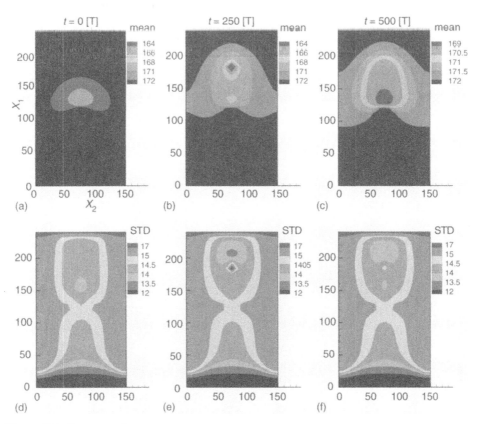

Figure 5.9. Contours of mean pressure head and head standard deviation at three different times in the presence of an internal source for the case of an embedded zone of coarser soils.

space–time and its size depends on the magnitude and frequency of the variation in the net recharge. The zone of capillary fringe is caused by the presence of the water table at the bottom of the vadose zone. Although the variability in this zone is usually small when the water table does not vary significantly in time, the pressure head in this zone is strongly nonstationary in space. The intermediate zone is less affected by the two boundaries and may be considered stationary in space and/or time. The size of this zone is dependent on the size of the vadose zone, the value of the recharge and the texture of the soil: the larger the vadose zone, the bigger the normalized recharge (with respect to the geometric mean saturated hydraulic conductivity) or the coarser the soil texture, the larger is the mean gravity-dominated zone.

5.5.2 Discussion

There are a number of works in the literature which study transient unsaturated flow in random porous media. Mantoglou and Gelhar [1987] extended the stationary spectral

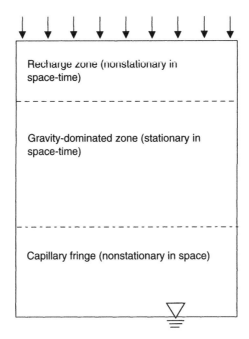

Figure 5.10. Schematical flow regimes of the vadose zone.

perturbation approach [Yeh *et al.* 1985a,b] (see also Section 5.3.1) to transient unsaturated flow. However, owing to the nature of the stationary spectral approach it was necessary to make some restrictive assumptions such as spatial and temporal stationarity of local flow characteristics (e.g., slowly varying mean gradient), stratified soil formation, and infinite flow domain. Unlu *et al.* [1990a] studied one-dimensional transient redistribution of unsaturated flow through Monte Carlo simulations and found a good agreement with the spectral perturbation approach except near the boundaries. Mantoglou [1992] developed a more general stochastic framework for transient unsaturated flow using the distributed parameter estimation theory of McLaughlin and Wood [1988a,b]. The theoretical framework consists of a set of partial differential equations for mean head and head covariances. These equations are nonlinearly coupled and have a high dimensionality of $2d + 1$ (where d is the number of space dimensions). Due to the nonlinear coupling and dimensionality problems, Mantoglou [1992] was not able to implement the theoretical equations except for some simplified cases. The covariance equations of Mantoglou [1992] are different from those developed in Section 5.5.1 in that the latter are linear and decoupled from the nonlinear mean equation and can thus be solved sequentially and efficiently. Liedl [1994] proposed a slightly different perturbation model for transient unsaturated flow. Similar to Mantoglou [1992], the results of the model are a set of partial differential equations governing the statistical moments of saturation. Liedl [1994] implemented the model in one dimension.

Recently, Foussereau *et al.* [2000] derived moment equations for transient unsaturated flow subject to random recharge but in the absence of soil parameter α variabilities. Their equations are somewhat similar to those given in Section 5.5.1. With a number of restrictive assumptions such as mean gravity-dominated flow at late time and unbounded domains, the flow moment equations were approximated in one dimension with Laplace–Fourier transforms and numerical inversions.

The adjoint state method is another viable approach for deriving moment equations for transient unsaturated flow. Protopapas and Bras [1990] and Li and Yeh [1998] studied transient unsaturated flow in heterogeneous porous media using a vector state-space approach coupled with the adjoint state method. This approach was originally proposed by Dettinger and Wilson [1981] for saturated flow in heterogeneous porous media. The details of the approach can be found in Section 3.8 and its application to steady-state unsaturated flow is discussed in Section 5.2.2. The essence of the approach is to first discretize the governing equations by a numerical scheme such as finite differences or finite elements and then expand random variables contained in the system by Taylor series to derive approximate statistical moments of flow quantities. Protopapas and Bras [1990] studied flow and solute transport in one-dimensional soil columns. Li and Yeh [1998] investigated the behavior of head variances for transient unsaturated flow in two dimensions.

5.6 EXERCISES

1. Show that for unsaturated flow, even if the mean flow is unidirectional and uniform, the mean head gradient is, in general, not uniform, in space. The only exception is the case of mean gravity-dominated flow (i.e., the mean total head gradient $J_i = \delta_{i1}$).

2. Represent the steady-state pressure head fluctuation of Eq. (5.65) with the aid of Green's function. Then derive integral expressions for the covariances $C_{f\psi}(\mathbf{x}, \boldsymbol{\chi})$, $C_{\alpha\psi}(\mathbf{x}, \boldsymbol{\chi})$, and $C_{\psi}(\mathbf{x}, \boldsymbol{\chi})$.

3. In a bounded unsaturated zone, the pressure head fluctuation $\psi'(\mathbf{x}) = \psi^{(1)}(\mathbf{x})$ of Eq. (5.65) is generally nonstationary even under the simplified condition of mean gravity-dominated flow. Express $\psi'(\mathbf{x})$ with the nonstationary spectral representation discussed in Section 3.6.2. Then derive integral expressions for the covariances $C_{f\psi}(\mathbf{x}, \boldsymbol{\chi})$, $C_{\alpha\psi}(\mathbf{x}, \boldsymbol{\chi})$, and $C_{\psi}(\mathbf{x}, \boldsymbol{\chi})$, and compare them with those derived on the basis of Green's function.

4. In this chapter, the pore size distribution parameter $\alpha(\mathbf{x})$ has been directly decomposed into its mean and zero-mean fluctuation: $\alpha(\mathbf{x}) = \langle\alpha\rangle + \alpha'(\mathbf{x})$. An alternative is to work with log-transformed α, $a(\mathbf{x}) = \ln \alpha(\mathbf{x})$. By letting $a(\mathbf{x}) = \langle a\rangle + a'(\mathbf{x})$, one has $\alpha(\mathbf{x}) = \exp[a(\mathbf{x})] = \exp(\langle a\rangle)\exp[a'(\mathbf{x})]$. The term $\exp[a'(\mathbf{x})]$ may be expanded by Taylor expansion. With this treatment in Eq. (5.16), derive alternative equations for the first two moments of ψ and θ_e.

5. The adjoint state moment equation method is discussed in Section 5.2.2 for steady-state unsaturated flow. Express the sensitivity matrices of the water content and the specific flux with respect to $\mathbf{f}$ and $\boldsymbol{\alpha}$ in terms of the pressure head sensitivity matrices, and then derive corresponding expressions for the water content and flux covariance matrices.

6. With the perturbation approach discussed in Section 5.2.1, derive the equation for the second-order pressure head term $\psi^{(2)}(\mathbf{x})$. Taking expectation of it results in the equation

for the second-order mean correction. Analyze this mean equation, determine whether it is linear or not, and find out what quantities are needed to solve it.

7. Transient flow in saturated formations is governed by Eqs. (4.1)–(4.5) while its counterpart in unsaturated media is governed by Eqs. (5.93)–(5.98). Flow in an unsaturated/saturated system may be described with the following equations:

$$[S_s H(\psi) + H(-\psi)C[\psi, \cdot]] \frac{\partial \psi(\mathbf{x}, t)}{\partial t} + \nabla \cdot \mathbf{q}(\mathbf{x}, t) = g(\mathbf{x}, t) \qquad (5.105)$$

$$q_i(\mathbf{x}, t) = -K[\psi(\mathbf{x}), \cdot] \frac{\partial}{\partial x_i}[\psi(\mathbf{x}, t) + x_1] \qquad (5.106)$$

subject to appropriate initial and boundary conditions, where $H(\psi)$ is the Heaviside function, being zero when $\psi < 0$ and one when $\psi > 0$. Other symbols are the same as those for saturated and unsaturated flows. Derive partial differential equations for the first two moments of ψ, and compare them with both the respective saturated and unsaturated moment equations.

6

TWO-PHASE FLOW

6.1 INTRODUCTION

In this chapter, we study a special case of *multiphase flow*, *two-phase flow*, in randomly heterogeneous porous media. The problem of multiphase flow in porous media is of great importance for a number of applications such as oil/gas recovery from hydrocarbon reservoirs, subsurface remediation of organic contaminants, and carbon dioxide sequestration in geological formations. Some commonly encountered cases of two-phase flows are oil–water and gas–oil systems. The unsaturated flow of gas and water studied in Chapter 5 is a special case of two-phase flow.

We consider a simultaneous flow of two *immiscible, incompressible* fluids in porous media. For each fluid, a continuity equation is written as [Bear, 1972]

$$\phi(\mathbf{x})\frac{\partial S_i(\mathbf{x}, t)}{\partial t} + \nabla \cdot \mathbf{q}_i(\mathbf{x}, t) = 0 \tag{6.1}$$

where ϕ denotes the porosity, S_i is the saturation of fluid i, $i = 1$ or 2 standing for fluid 1 or 2, and $\mathbf{q}_i$ is the flux of fluid i. The flux $\mathbf{q}_i$ is given by the *two-phase extension* of the single-phase Darcy's law,

$$\mathbf{q}_i = -\lambda_i[\nabla P_i(\mathbf{x}, t) + \rho_i \mathbf{g}] \tag{6.2}$$

where P_i is the pressure of fluid i, ρ_i is its density, $\mathbf{g}$ is the gravitation vector, and λ_i is the phase mobility of fluid i,

$$\lambda_i = k(\mathbf{x})\frac{k_{ri}(S_i)}{\mu_i} \tag{6.3}$$

Here $k(\mathbf{x})$ is the absolute (intrinsic) permeability, k_{ri} is the (phase) relative permeability of fluid i, and μ_i is the (phase) viscosity of fluid i.

In the literature of hydrology, the *two-pressure approach* is commonly used, in which the governing equations are written in terms of the pressures in each of the two phases. This can be done through substituting Eq. (6.2) into Eq. (6.1):

$$\phi(\mathbf{x})\frac{\partial S_w(\mathbf{x}, t)}{\partial t} - \nabla \cdot \{\lambda_w(\mathbf{x}, t)[\nabla P_w(\mathbf{x}, t) + \rho_w \mathbf{g}]\} = \mathbf{0} \tag{6.4}$$

$$\phi(\mathbf{x})\frac{\partial S_n(\mathbf{x}, t)}{\partial t} - \nabla \cdot \{\lambda_n(\mathbf{x}, t)[\nabla P_n(\mathbf{x}, t) + \rho_n \mathbf{g}]\} = \mathbf{0} \tag{6.5}$$

subject to appropriate boundary and initial conditions, where subscripts w and n stand for *wetting* and *nonwetting* fluid phases, respectively. This is a system of two equations of the four unknowns S_w, S_n, P_w, and P_n, and is usually supplemented by the following two relations:

$$S_w + S_n = 1 \tag{6.6}$$

$$P_n - P_w = P_c(S_w) \tag{6.7}$$

where P_c is called the *capillary pressure*, whose functional form is to be given externally (through physical measurements, microscopic numerical simulations, or theoretical studies). It is clear that the two-phase flow equations are nonlinear and must, in general, be solved numerically in an iterative manner.

In the literature of petroleum engineering, the two-phase flow problem is often expressed with the approach of *fractional flow*, in which the problem is treated as a *total fluid flow* of a mixed fluid with the individual phases as fractions of the total flow. This approach leads to two equations: the pressure equation and the saturation equation. The pressure equation is derived by first summing Eq. (6.1) for $i = w$ and n,

$$\nabla \cdot \mathbf{q}(\mathbf{x}, t) = 0 \tag{6.8}$$

where $\mathbf{q} = \mathbf{q}_w + \mathbf{q}_n$ is the *total flux*. Then a new variable, called the *global pressure P*, is introduced to express the total flux in terms of the gradient of the global pressure instead of that of the individual phase pressures. A global pressure may be defined as [e.g., Ewing, 1997; Binning and Celia, 1999],

$$P = \frac{1}{2}(P_w + P_n) + \int_{S_c}^{S_w} \left(\frac{1}{2} - f_w\right)\frac{dP_c}{dS_w} dS_w \tag{6.9}$$

where S_c is the saturation with $P_c(S_c) = 0$, and f_w is the *fractional flow function* defined as

$$f_w = \frac{\lambda_w}{\lambda_w + \lambda_n} = \frac{k_{rw}/\mu_w}{(k_{rw}/\mu_w) + (k_{rn}/\mu_n)} \tag{6.10}$$

Finally, the total flow may be expressed as

$$\mathbf{q} = -\lambda_T[\nabla P + \mathbf{G}(S_w)] \tag{6.11}$$

where $\lambda_T = \lambda_w + \lambda_n$ is the *total mobility* and $\mathbf{G}$ accounts for the gravitational effects,

$$\mathbf{G} = [f_w \rho_w + (1 - f_w)\rho_n]\mathbf{g} \tag{6.12}$$

The saturation equation is obtained by manipulating Eq. (6.1) with $i = w$. The phase flux $\mathbf{q}_w$ is first related to the total flux via [e.g., Binning and Celia, 1999]

$$\mathbf{q}_w = f_w \left\{ \mathbf{q} + \lambda_n \left[\frac{dP_c}{dS_w} \nabla S_w + (\rho_w - \rho_n)\mathbf{g} \right] \right\} \tag{6.13}$$

Substitution of Eq. (6.13) into Eq. (6.1) yields

$$\phi \frac{\partial S_w}{\partial t} + \mathbf{w}(S_w) \cdot \nabla S_w - \nabla \cdot [\mathbf{d}(S_w) \cdot \nabla S_w] = 0 \tag{6.14}$$

where $\mathbf{w}(S_w)$ is the *apparent flux* defined as

$$\mathbf{w}(S_w) = \frac{d}{dS_w}[f_w \mathbf{q} + f_w \lambda_n (\rho_w - \rho_n)\mathbf{g}] \tag{6.15}$$

and $\mathbf{d}(S_w)$ is the *capillary diffusion term* defined as

$$\mathbf{d}(S_w) = -f_w \lambda_n \frac{dP_c}{dS_w} \tag{6.16}$$

It is seen that Eq. (6.14) is a nonlinear *advection–diffusion equation*. When the capillary term $\nabla \cdot [\mathbf{d}(S_w) \cdot \nabla S_w]$ is small compared to the advection term $\mathbf{w}(S_w) \cdot \nabla S_w$, the equation is hyperbolic and advection dominated. This is the case if the capillary pressure $P_c(S_w)$ is negligible, or if its derivative dP_c/dS_w is negligible. In homogeneous media, analytical solutions to the hyperbolic version of Eq. (6.14) are available [e.g., Marle, 1981].

When the absolute permeability k and the parameters characterizing the constitutive relationships $k_{ri}(S_i)$ and $P_c(S_w)$ are treated as random space functions, the governing equations become stochastic partial differential equations. In this chapter, we discuss some methods for solving these nonlinear stochastic equations. The choice of stochastic methods depends on the characteristics of the problem under consideration.

For the problem of *water infiltration* in the vadose zone, the common treatment is to regard the movement of gas instantaneously. Thus, the pressure of gas, P_n, may be considered uniform throughout the domain of interest. By setting the atmospheric pressure as the reference pressure such that $P_n \equiv 0$, we may recover the unsaturated flow equations (5.93) and (5.94) from Eqs. (6.4) and (6.5). For water infiltration, the mechanism of capillarity is at least as important as advection. However, capillarity may be neglected for studying water–oil displacement in oil reservoirs [e.g., Marle, 1981]. Hence, the displacement of oil by water, usually called *Buckley–Leverett displacement*, becomes advection dominated. In the next section, we formulate moment partial differential equations with the usual (Eulerian) approach and discuss some difficulties in solving these moment equations for the Buckley–Leverett displacement in random porous media. In Section 6.3, we use a Lagrangian approach for solving the Buckley–Leverett displacement problem by taking advantage of the advection-dominated nature of the problem. In Section 6.4, on the basis of the two-pressure equations we use Eulerian approaches to develop moment equations and discuss published solutions for some simplified situations.

6.2 BUCKLEY–LEVERETT DISPLACEMENT

In the absence of capillary and gravitational effects, the flux of the wetting phase fluid (water), $\mathbf{q}_w$, can be written as

$$\mathbf{q}_w = \mathbf{q} f_w \tag{6.17}$$

where f_w is the fractional flow function of Eq. (6.10). In turn, Eq. (6.14) becomes

$$\frac{\partial S_w(\mathbf{x}, t)}{\partial t} + f'_w(S_w)\mathbf{u}(\mathbf{x}, t) \cdot \nabla S_w(\mathbf{x}, t) = 0 \tag{6.18}$$

where $\mathbf{u} = \mathbf{q}(\mathbf{x}, t)/\phi(\mathbf{x})$ is the total (seepage) velocity vector, and $f'_w = df_w/dS_w$. Equation (6.18) is "quasi-linear" in that it is linear with respect to the derivatives of the unknown but not to the function itself. For convenience, we rewrite it as

$$\frac{\partial S(\mathbf{x}, t)}{\partial t} + \mathbf{v}(S, \mathbf{x}, t) \cdot \nabla S(\mathbf{x}, t) = 0 \tag{6.19}$$

where $\mathbf{v}(S, \mathbf{x}, t) \equiv \mathbf{v}[S(\mathbf{x}, t), \mathbf{x}, t] \equiv \mathbf{u}(df_w/dS_w)$ is the "apparent" velocity vector and $S = S_w$. Since $\mathbf{u}$ is a random space function, so are $\mathbf{v}$, S, k_{rp}, and f_w. Decompose $\mathbf{v}$ and S as

$$\mathbf{v}(S, \mathbf{x}, t) = \langle \mathbf{v}(S, \mathbf{x}, t) \rangle + \mathbf{v}'(S, \mathbf{x}, t) \tag{6.20}$$

$$S(\mathbf{x}, t) = \langle S(\mathbf{x}, t) \rangle + S'(\mathbf{x}, t) \tag{6.21}$$

where $\langle \, \rangle$ stands for ensemble mean (expected value) and the primed quantities represent the zero-mean fluctuations (perturbations). Below we attempt to derive equations governing the statistical moments of $S(\mathbf{x}, t)$.

6.2.1 Moment Partial Differential Equations

Taking the ensemble mean of Eq. (6.19) yields

$$\frac{\partial \langle S(\mathbf{x}, t) \rangle}{\partial t} + \langle \mathbf{v}^T(S, \mathbf{x}, t) \rangle \nabla \langle S(\mathbf{x}, t) \rangle + \langle \mathbf{v}'^T(S, \mathbf{x}, t) \nabla S'(\mathbf{x}, t) \rangle = 0 \tag{6.22}$$

which can be rewritten as

$$\frac{\partial \langle S(\mathbf{x}, t) \rangle}{\partial t} + \langle \mathbf{v}^T(S, \mathbf{x}, t) \rangle \nabla \langle S(\mathbf{x}, t) \rangle + \nabla \cdot \mathbf{Q}(\mathbf{x}, t) = 0 \tag{6.23}$$

Here $\mathbf{Q}(\mathbf{x}, t) = \langle \mathbf{v}'(S, \tilde{\mathbf{x}}, \tilde{t}) S'(\mathbf{x}, t) \rangle$ is the *dispersive flux*, and $\mathbf{v}'(S, \tilde{\mathbf{x}}, \tilde{t})$ is evaluated at $\tilde{\mathbf{x}} = \mathbf{x}$ and $\tilde{t} = t$ but is immune to operations with respect to $\mathbf{x}$ and t. Subtracting Eq. (6.22) from Eq. (6.19) gives

$$\frac{\partial S'(\mathbf{x}, t)}{\partial t} + \langle \mathbf{v}^T(S, \mathbf{x}, t) \rangle \nabla S'(\mathbf{x}, t) + \mathbf{v}'^T(S, \mathbf{x}, t) \nabla S'(\mathbf{x}, t) - \langle \mathbf{v}'^T(S, \mathbf{x}, t) \nabla S'(\mathbf{x}, t) \rangle$$

$$= -\mathbf{v}'^T(S, x, t) \nabla \langle S(\mathbf{x}, t) \rangle \tag{6.24}$$

Multiplying Eq. (6.24) by $\mathbf{v}'(S, \mathbf{y}, \theta)$ at different space–time $(\mathbf{y}, \theta)$ and taking the ensemble mean yields

$$\frac{\partial \langle \mathbf{v}'(S, \mathbf{y}, \theta) S'(\mathbf{x}, t) \rangle}{\partial t} + \nabla_{\mathbf{x}}^T \langle \mathbf{v}'(S, \mathbf{y}, \theta) S'(\mathbf{x}, t) \rangle \langle \mathbf{v}(S, \mathbf{x}, t) \rangle$$

$$+ \langle \mathbf{v}'(S, \mathbf{y}, \theta) \mathbf{v}'^T(S, \mathbf{x}, t) \nabla S'(\mathbf{x}, t) \rangle = -\langle \mathbf{v}'(S, \mathbf{y}, \theta) \mathbf{v}'^T(S, \mathbf{x}, t) \rangle \nabla \langle S(\mathbf{x}, t) \rangle \tag{6.25}$$

Here, we see that the equation for the second moment involves the third moment $\langle \mathbf{v}' \mathbf{v}'^T \nabla S' \rangle$. It is anticipated that the equation for the third moment will involve higher moments. In general, the equation for the nth moment involves $(n + 1)$th moments. This is the closure problem that has been already discussed in Chapters 1 and 3. Therefore, a closure approximation is needed. The commonly used one is to neglect the third moment in Eq. (6.25). Hence, one has

$$\frac{\partial \langle \mathbf{v}'(S, \mathbf{y}, \theta) S'(\mathbf{x}, t) \rangle}{\partial t} + \langle \mathbf{v}^T(S, \mathbf{x}, t) \rangle \nabla_{\mathbf{x}} \langle \mathbf{v}'(S, \mathbf{y}, \theta) S'(\mathbf{x}, t) \rangle$$

$$= -\langle \mathbf{v}'(S, \mathbf{y}, \theta) \mathbf{v}'^T(S, \mathbf{x}, t) \rangle \nabla \langle S(\mathbf{x}, t) \rangle \tag{6.26}$$

Letting $\mathbf{y} = \tilde{\mathbf{x}}$ and $\theta = \tilde{t}$, one has

$$\frac{\partial \mathbf{Q}(\mathbf{x}, t)}{\partial t} + \langle \mathbf{v}^T(S, \mathbf{x}, t) \rangle \nabla \mathbf{Q}(\mathbf{x}, t) = -\langle \mathbf{v}'(S, \tilde{\mathbf{x}}, \tilde{t}) \mathbf{v}'^T(S, \mathbf{x}, t) \rangle \nabla \langle S(\mathbf{x}, t) \rangle \tag{6.27}$$

Hence, in principle, $\langle S(\mathbf{x}, t) \rangle$ can be obtained by solving the coupled equations (6.23) and (6.27). However, these two equations are nonlinear like the original equation (6.19) and can only be solved in an iterative manner. To facilitate discussion, we associate them with an iteration index i,

$$\frac{\partial \langle S(\mathbf{x}, t) \rangle^{(i)}}{\partial t} + \langle \mathbf{v}^T(\mathbf{x}, t) \rangle^{(i-1)} \nabla \langle S(\mathbf{x}, t) \rangle^{(i)} + \nabla \cdot \mathbf{Q}^{(i)}(\mathbf{x}, t) = 0 \tag{6.28}$$

$$\frac{\partial \mathbf{Q}^{(i)}(\mathbf{x}, t)}{\partial t} + \langle \mathbf{v}^T(\mathbf{x}, t) \rangle^{(i-1)} \nabla \mathbf{Q}^{(i)}(\mathbf{x}, t) = -\langle \mathbf{v}'(\tilde{\mathbf{x}}, \tilde{t}) \mathbf{v}'^T(\mathbf{x}, t) \rangle^{(i-1)} \nabla \langle S(\mathbf{x}, t) \rangle^i \tag{6.29}$$

At each iteration, these equations become linear. From now on, the dependence of $\mathbf{v}$ on S is suppressed for brevity. Equations (6.28) and (6.29) are similar to averaged concentration equations for solute transport in random velocity fields [e.g., Neuman, 1993;

Zhang and Neuman, 1995a]. The solution of $\mathbf{Q}^{(i)}(\mathbf{x}, t)$ can be expressed as

$$Q^{(i)}(\mathbf{x}, t) = -\int_0^t \int G(\mathbf{x}, t; \mathbf{x}', t')\langle \mathbf{v}'(\tilde{\mathbf{x}}, \tilde{t})\mathbf{v}'^T(\mathbf{x}', t')\rangle^{(i-1)} \nabla_{\mathbf{x}'}\langle S(\mathbf{x}', t')\rangle^{(i)} d\mathbf{x}' dt' \quad (6.30)$$

where $G(\mathbf{x}, t; \mathbf{x}', t')$ is the Green's function for Eq. (6.29). It can be verified that $G(\mathbf{x}, t; \mathbf{x}', t') = \delta[\mathbf{x}' - \langle \chi(t')\rangle]$ where δ is the Dirac delta function and $\langle \chi(t')\rangle = \mathbf{x} - \int_{t'}^t \langle \mathbf{v}[\chi(t'')]\rangle^{(i-1)} dt''$ is the mean upstream position at time t', $0 \le t' \le t$, of an indivisible particle which, at later time t, is found at the downstream location $\mathbf{x}$ [Neuman, 1993; Zhang and Neuman, 1995a]. Hence

$$\mathbf{Q}^{(i)}(\mathbf{x}, t) = -\int_0^t \langle \mathbf{v}'(\tilde{\mathbf{x}}, \tilde{t})\mathbf{v}'^T[\langle \chi(t')\rangle]\rangle^{(i-1)} \nabla_{\chi}\langle S[\langle \chi(t')\rangle]\rangle^{(i)} dt' \quad (6.31)$$

Under *pseudo-Fickian* regime where $\langle S(\mathbf{x}, t)\rangle$ has spread sufficiently so that its gradient varies slowly across the mean travel distance [Zhang and Neuman, 1995a], one may approximate Eq. (6.31) by

$$\mathbf{Q}^{(i)}(\mathbf{x}, t) = -\mathbf{D}^{(i-1)}(\mathbf{x}, t)\nabla\langle S(\mathbf{x}, t)\rangle^{(i)} \quad (6.32)$$

where $\mathbf{D}^{(i-1)}(\mathbf{x}, t)$ is in analogy to the macrodispersion term in solute transport in heterogeneous media,

$$\mathbf{D}^{(i-1)}(\mathbf{x}, t) = \int_0^t \langle \mathbf{v}'(\tilde{\mathbf{x}}, \tilde{t})\mathbf{v}'^T[\langle \chi(t')\rangle]\rangle^{(i-1)} dt' \quad (6.33)$$

Therefore, the mean saturation at the ith iteration is given by

$$\frac{\partial\langle S(\mathbf{x}, t)\rangle^{(i)}}{\partial t} + \langle \mathbf{v}^T(\mathbf{x}, t)\rangle^{(i-1)}\nabla\langle S(\mathbf{x}, t)\rangle^{(i)} - \nabla \cdot [\mathbf{D}^{(i-1)}(\mathbf{x}, t)\nabla\langle S(\mathbf{x}, t)\rangle^{(i)}] = 0 \quad (6.34)$$

subject to initial and boundary conditions. It will be shown later that, in order to evaluate $\langle \mathbf{v}'\mathbf{v}'^T\rangle$, the second moments $\langle S'(\mathbf{y}, \theta)S'(\mathbf{x}, t)\rangle$ and $\langle \mathbf{u}'(\mathbf{y}, \theta)S'(\mathbf{x}, t)\rangle$ are required. Similar to Eqs. (6.26) and (6.29), they may be given as

$$\frac{\partial\langle S'(\mathbf{y}, \theta)S'(\mathbf{x}, t)\rangle^{(i)}}{\partial t} + \langle \mathbf{v}^T(\mathbf{x}, t)\rangle^{(i-1)}\nabla_{\mathbf{x}}\langle S'(\mathbf{y}, \theta)S'(\mathbf{x}, t)\rangle^{(i)}$$
$$= -\langle S'(\mathbf{y}, \theta)\mathbf{v}'^T(\mathbf{x}, t)\rangle^{(t-1)}\nabla\langle S(\mathbf{x}, t)\rangle^{(i)} \quad (6.35)$$

$$\frac{\partial\langle \mathbf{u}'(\mathbf{y}, \theta)S'(\mathbf{x}, t)\rangle^{(i)}}{\partial t} + \langle \mathbf{v}^T(\mathbf{x}, t)\rangle^{(i-1)}\nabla_{\mathbf{x}}\langle \mathbf{u}'(\mathbf{y}, \theta)S'(\mathbf{x}, t)\rangle^{(i)}$$
$$= -\langle \mathbf{u}'(\mathbf{y}, \theta)\mathbf{v}'^T(\mathbf{x}, t)\rangle^{(i-1)}\nabla\langle S(\mathbf{x}, t)\rangle^{(i)} \quad (6.36)$$

In analogy to Eqs. (6.29)–(6.33), we have

$$\langle S'(\mathbf{y}, \theta)S'(\mathbf{x}, t)\rangle^{(i)} = -\mathbf{E}^{(i-1)^T}(\mathbf{x}, t)\nabla\langle S(\mathbf{x}, t)\rangle^{(i)} \tag{6.37}$$

$$\langle \mathbf{u}'(\mathbf{y}, \theta)S'(\mathbf{x}, t)\rangle^{(i)} = -\mathbf{F}^{(i-1)}(\mathbf{x}, t)\nabla\langle S(\mathbf{x}, t)\rangle^{(i)} \tag{6.38}$$

where $\mathbf{E}^{(i-1)}(\mathbf{x}, t)$ and $\mathbf{F}^{(i-1)}(\mathbf{x}, t)$ are given as

$$\mathbf{E}^{(i-1)^T}(\mathbf{x}, t) = \int_0^t \langle S'(\mathbf{y}, \theta)\mathbf{v}'^T[\langle \chi(t')\rangle]\rangle^{(i-1)} dt' \tag{6.39}$$

$$\mathbf{F}^{(i-1)}(\mathbf{x}, t) = \int_0^t \langle \mathbf{u}'(\mathbf{y}, \theta)\mathbf{v}'^T[\langle \chi(t')\rangle]\rangle^{(i-1)} dt' \tag{6.40}$$

6.2.2 Statistical Moments of Apparent Velocity

We may rewrite Eq. (6.10) as

$$f_w = \frac{mk_{rw}}{mk_{rw} + k_{rw}} \tag{6.41}$$

where $m = \mu_n/\mu_w$. The explicit expressions of k_{rw} and k_{rn} as functions of S must be known in order to derive the statistical moments of the apparent velocity, $\langle \mathbf{v}\rangle$ and $\langle \mathbf{v}'\mathbf{v}'^T\rangle$. Recognizing the dependency of f_w on S we may write, in general,

$$\begin{aligned}
\mathbf{v}(\mathbf{x}, t) &= \mathbf{u}(\mathbf{x}, t)f_w'(\langle S\rangle + S') \\
&= [\langle \mathbf{u}(\mathbf{x}, t)\rangle + \mathbf{u}'(\mathbf{x}, t)]\{f_w'[\langle S(\mathbf{x}, t)\rangle] + S'(\mathbf{x}, t)f_w''[\langle S(\mathbf{x}, t)\rangle] + \cdots\}
\end{aligned} \tag{6.42}$$

where $f_w'[\langle S\rangle] = df_w(S)/dS$ evaluated at $\langle S\rangle$ and $f_w''(S) = df_w'/dS$. Note that S' stands for the fluctuation of S. To first order, we have

$$\langle \mathbf{v}(\mathbf{x}, t)\rangle = \langle \mathbf{u}(\mathbf{x}, t)\rangle f_w'[\langle S(\mathbf{x}, t)\rangle] \tag{6.43}$$

$$\mathbf{v}'(\mathbf{x}, t) = a(\mathbf{x}, t)\mathbf{u}'(\mathbf{x}, t) + \mathbf{b}(\mathbf{x}, t)S'(\mathbf{x}, t) \tag{6.44}$$

$$\begin{aligned}
\langle \mathbf{v}'(\mathbf{y}, \theta)\mathbf{v}'^T(\mathbf{x}, t)\rangle = {}& a(\mathbf{x}, t)a(\mathbf{y}, \theta)\langle \mathbf{u}'(\mathbf{y}, \theta)\mathbf{u}'^T(\mathbf{x}, t)\rangle + a(\mathbf{y}, \theta)\langle \mathbf{u}'(\mathbf{y}, \theta)S'(\mathbf{x}, t)\rangle\mathbf{b}^T(\mathbf{x}, t) \\
& + \mathbf{b}(\mathbf{y}, \theta)a(\mathbf{x}, t)\langle \mathbf{u}'^T(\mathbf{x}, t)S'(\mathbf{y}, \theta)\rangle + \mathbf{b}(\mathbf{y}, \theta)\mathbf{b}^T(\mathbf{x}, t)\langle S'(\mathbf{y}, \theta)S'(\mathbf{x}, t)\rangle
\end{aligned} \tag{6.45}$$

where

$$a(\mathbf{x}, t) = f_w'[\langle S(\mathbf{x}, t)\rangle] \tag{6.46}$$

$$\mathbf{b}(\mathbf{x}, t) = \langle \mathbf{u}(\mathbf{x}, t)\rangle f_w''[\langle S(\mathbf{x}, t)\rangle] \tag{6.47}$$

Similarly, we have

$$\langle S'(\mathbf{y}, \theta)\mathbf{v}'^T(\mathbf{x}, t)\rangle = a(\mathbf{x}, t)\langle \mathbf{u}'^T(\mathbf{x}, t)S'(\mathbf{y}, \theta)\rangle + \mathbf{b}^T(\mathbf{x}, t)\langle S'(\mathbf{y}, \theta)S'(\mathbf{x}, t)\rangle \qquad (6.48)$$

$$\langle \mathbf{u}'(\mathbf{y}, \theta)\mathbf{v}'^T(\mathbf{x}, t)\rangle = a(\mathbf{x}, t)\langle \mathbf{u}'(\mathbf{y}, \theta)\mathbf{u}'^T(\mathbf{x}, t)\rangle + \langle \mathbf{u}'(\mathbf{y}, \theta)S'(\mathbf{x}, t)\rangle \mathbf{b}^T(\mathbf{x}, t) \qquad (6.49)$$

6.2.3 Solutions

It is seen from Eqs. (6.8) and (6.11) that the total velocity $\mathbf{u}(\mathbf{x}, t)$ is, in general, a function of S. Equations governing the statistical moments of the total velocity ($\langle \mathbf{u} \rangle$ and $\langle \mathbf{u}'\mathbf{u}'^T \rangle$) may be derived with the approaches discussed in Chapters 3 and 4 for single-phase flow. They are, however, coupled with the saturation moment equations. The first two moments of saturation are summarized as follows:

$$\frac{\partial \langle S(\mathbf{x}, t)\rangle^{(i)}}{\partial t} + \langle \mathbf{v}^T(\mathbf{x}, t)\rangle^{(i-1)}\nabla\langle S(\mathbf{x}, t)\rangle^{(i)} - \nabla \cdot \left[\mathbf{D}^{(i-1)}(\mathbf{x}, t)\nabla\langle S(\mathbf{x}, t)\rangle^{(i)}\right] = 0 \quad (6.50)$$

$$\langle S'(\mathbf{y}, \theta)S'(\mathbf{x}, t)\rangle^{(i)} = -\mathbf{E}^{(i-1)}(\mathbf{x}, t)\nabla\langle S(\mathbf{x}, t)\rangle^{(i)} \qquad (6.51)$$

under appropriate initial and boundary conditions. The terms in the above two equations are given as

$$\langle \mathbf{v}(\mathbf{x}, t)\rangle^{(i-1)} = \langle \mathbf{u}(\mathbf{x}, t)\rangle f_w'[\langle S(\mathbf{x}, t)\rangle^{(i-1)}] \qquad (6.52)$$

$$\mathbf{D}^{(i-1)}(\mathbf{x}, t) = \int_0^t \langle \mathbf{v}'(\tilde{\mathbf{x}}, \tilde{t})\mathbf{v}'^T[\langle \chi(t')\rangle]\rangle^{(i-1)} dt' \qquad (6.53)$$

$$\langle \mathbf{v}'(\mathbf{y}, \theta)\mathbf{v}'^T(\mathbf{x}, t)\rangle^{(i-1)} = a^{(i-1)}(\mathbf{x}, t)a^{(i-1)}(\mathbf{y}, \theta)\langle \mathbf{u}'(\mathbf{y}, \theta)\mathbf{u}'^T(\mathbf{x}, t)\rangle$$
$$+ a^{(i-1)}(\mathbf{y}, \theta)\langle \mathbf{u}'(\mathbf{y}, \theta)S'(\mathbf{x}, t)\rangle^{(i-1)}[\mathbf{b}^{(i-1)}(\mathbf{x}, t)]^T$$
$$+ \mathbf{b}^{(i-1)}(\mathbf{y}, \theta)a^{(i-1)}(\mathbf{x}, t)\langle \mathbf{u}'^T(\mathbf{x}, t)S'(\mathbf{y}, \theta)\rangle^{(i-1)}$$
$$+ \mathbf{b}^{(i-1)}(\mathbf{y}, \theta)[\mathbf{b}^{(i-1)}(\mathbf{x}, t)]^T\langle S'(\mathbf{y}, \theta)S'(\mathbf{x}, t)\rangle^{(i-1)} \qquad (6.54)$$

$$\langle S'(\mathbf{y}, \theta)S'(\mathbf{x}, t)\rangle^{(i-1)} = -\mathbf{E}^{(i-2)^T}(\mathbf{x}, t)\nabla\langle S(\mathbf{x}, t)\rangle^{(i-1)} \qquad (6.55)$$

$$\langle \mathbf{u}'(\mathbf{y}, \theta)S'(\mathbf{x}, t)\rangle^{(i-1)} = -\mathbf{F}^{(i-2)}(\mathbf{x}, t)\nabla\langle S(\mathbf{x}, t)\rangle^{(i-1)} \qquad (6.56)$$

$$\mathbf{E}^{(i-2)^T}(\mathbf{x}, t) = \int_0^t \left\{ a^{(i-2)}[\langle \chi(t')\rangle]\langle \mathbf{u}'^T[\langle \chi(t')\rangle]S'(\mathbf{y}, \theta)\rangle^{(i-2)} \right.$$
$$\left. + [\mathbf{h}^{(i-2)}(\langle \chi(t')\rangle)]^T\langle S'(\mathbf{y}, \theta)S'[\langle \chi(t')\rangle]\rangle^{(i-2)} \right\} dt' \qquad (6.57)$$

$$\mathbf{F}^{(i-2)}(\mathbf{x}, t) = \int_0^t \left\{ a^{(i-2)}[\langle \chi(t')\rangle]\langle \mathbf{u}'(\mathbf{y}, \theta)\mathbf{u}'^T[\langle \chi(t')\rangle]\rangle \right.$$
$$\left. + \langle \mathbf{u}'(\mathbf{y}, \theta)S'[\langle \chi(t')\rangle]\rangle^{(i-2)}[\mathbf{b}^{(i-2)}(\langle \chi(t')\rangle)]^T \right\} dt' \qquad (6.58)$$

It is seen that the mean saturation equation (6.50) is a *parabolic* second-order partial differential equation, while the original stochastic equation (6.19) is a *hyperbolic* first-order equation. The term $\mathbf{D}(\mathbf{x}, t)$ is called *macrodispersion*, accounting for *heterogeneity fingering*. The first two moments of saturation given in Eqs. (6.50)–(6.58) are nonlinearly coupled and thus must be solved iteratively. To solve such a large number of coupled equations is not an easy task. To the writer's knowledge, no attempt has been made to solve these general moment equations of saturation. Recently, Langlo and Espedal [1994] derived and solved simplified equations under some more restrictive conditions. Below we demonstrate how these general equations can be simplified under those conditions and discuss some specific results obtained by them.

Simplifications The first simplification commonly made is that the dependency of the total velocity $\mathbf{u}$ on saturation S may be negligible. This is a good approximation for moderate mobility ratios. It has been argued [e.g., Langlo and Espedal, 1994] that the total velocity is fairly weakly dependent on changes in the saturation, even for a high mobility ratio. Further, when the boundary conditions (including injecting/pumping rates) do not change with time, the total velocity does not vary temporally, although it may vary spatially. Hence, the total flow is governed by

$$\nabla \cdot \mathbf{q}(\mathbf{x}) = 0, \qquad \mathbf{q}(\mathbf{x}) = -\lambda_T(\mathbf{x})\nabla P(\mathbf{x}) \tag{6.59}$$

The statistical moments of the total velocity can be obtained independently of the saturation moments, with the various approaches introduced in Chapter 3.

Another simplification is that the second derivative of f_w is approximately zero, i.e., $f_w''(S) \approx 0$ [Langlo and Espedal, 1994]. With this approximation, the coefficient $\mathbf{b}(\mathbf{x}, t)$ is zero. Thus, the moment equations of saturation can be simplified significantly as

$$\frac{\partial \langle S(\mathbf{x}, t)\rangle}{\partial t} + \langle \mathbf{v}^T(\mathbf{x}, t)\rangle\nabla\langle S(\mathbf{x}, t)\rangle - \nabla \cdot [\mathbf{D}(\mathbf{x}, t)\nabla\langle S(\mathbf{x}, t)\rangle] = 0 \tag{6.60}$$

subject to appropriate initial and boundary conditions. The coefficients are given as

$$\langle \mathbf{v}(\mathbf{x}, t)\rangle = \langle \mathbf{u}(\mathbf{x})\rangle f_w'[\langle S(\mathbf{x}, t)\rangle] \tag{6.61}$$

$$\mathbf{D}(\mathbf{x}, t) = \int_0^t \langle \mathbf{v}'(\tilde{\mathbf{x}}, \tilde{t})\mathbf{v}'^T[\langle \boldsymbol{\chi}(t')\rangle]\rangle\, dt' \tag{6.62}$$

where

$$\langle \mathbf{v}'(\mathbf{y}, \theta)\mathbf{v}'^T(\mathbf{x}, t)\rangle = a(\mathbf{x}, t)a(\mathbf{y}, \theta)\langle \mathbf{u}'(\mathbf{y})\mathbf{u}'^T(\mathbf{x})\rangle \tag{6.63}$$

where $a(\mathbf{x}, t) = f'[\langle S(\mathbf{x}, t)\rangle]$. Since both $\langle \mathbf{v}(\mathbf{x}, t)\rangle$ and $\mathbf{D}(\mathbf{x}, t)$ are functions of $\langle S(\mathbf{x}, t)\rangle$, the mean equation (6.60) is nonlinear and must be solved in an iterative manner. However, the second moment (saturation covariance) is no longer coupled with the first moment in this case. With the solution for $\langle S(\mathbf{x}, t)\rangle$ one may compute the covariance directly from

$$\langle S'(\mathbf{y}, \theta)S'(\mathbf{x}, t)\rangle = -\mathbf{E}(\mathbf{x}, t)\nabla\langle S(\mathbf{x}, t)\rangle \tag{6.64}$$

where

$$\mathbf{E}(\mathbf{x}, t) = \int_0^t a[\langle\boldsymbol{\chi}(t')\rangle]\langle\mathbf{u}'[\langle\boldsymbol{\chi}(t')\rangle]\rangle S'(\mathbf{y}, \theta)\rangle \, dt' \tag{6.65}$$

$$\langle\mathbf{u}'(\mathbf{y})S'(\mathbf{x}, t)\rangle = -\mathbf{F}(\mathbf{x}, t)\nabla\langle S(\mathbf{x}, t)\rangle \tag{6.66}$$

$$\mathbf{F}(\mathbf{x}, t) = \int_0^t a[\langle\boldsymbol{\chi}(t')\rangle]\langle\mathbf{u}'(\mathbf{y})\mathbf{u}'^T[\langle\boldsymbol{\chi}(t')\rangle]\rangle \, dt' \tag{6.67}$$

In addition, Langlo and Espedal [1994] assumed $f_w'(S)$ to be slowly varying in time. As such, the macrodispersion coefficient may be approximated as

$$\mathbf{D}(\mathbf{x}, t) \approx \left\{f'[\langle S(\mathbf{x}, t)\rangle]\right\}^2 \int_0^t \langle\mathbf{u}'(\mathbf{x})\mathbf{u}'^T[\langle\boldsymbol{\chi}(t')\rangle]\rangle \, dt' \tag{6.68}$$

It is seen that the saturation-dependent two-phase macrodispersion is a quadratic function of $f'[\langle S\rangle]$. For the case of uniform mean flow such that $\langle\mathbf{u}(\mathbf{x})\rangle = \langle\mathbf{u}\rangle$, Langlo and Espedal [1994] further approximated the Lagrangian particle position $\langle\boldsymbol{\chi}(t')\rangle$ with

$$\langle\boldsymbol{\chi}(t')\rangle = \mathbf{x} - \int_{t'}^t \langle\mathbf{v}[\boldsymbol{\chi}(t'')]\rangle \, dt'' \approx \mathbf{x} - f'[\langle S(\mathbf{x}, t)\rangle]\langle\mathbf{u}\rangle(t - t') \tag{6.69}$$

Therefore, the integral in Eq. (6.68) can be evaluated with the velocity covariance matrix $\langle\mathbf{u}'(\mathbf{x})\mathbf{u}'^T(\boldsymbol{\chi})\rangle$. This integral has been evaluated by many researchers for studying solute transport in random (single-phase) flow fields. Under the condition of uniform mean flow in unbounded domains, the velocity covariances are given in Section 3.5.1 and the integral may be integrated out analytically [e.g., Dagan, 1989; Langlo and Espedal, 1994]. For solute transport in (single-phase) flow, the following integral, called (macroscopic) dispersivities, is usually considered:

$$\alpha_{ij}(t) = \frac{1}{\langle u\rangle}\int_0^t \langle u_i'(\mathbf{x})u_j'[\boldsymbol{\chi}(t')]\rangle \, dt' \tag{6.70}$$

where $\langle u\rangle$ is the magnitude of the mean total velocity assumed to be aligned with x_1. Langlo and Espedal [1994] related the (two-phase) macrodispersion coefficients D_{ij} to the (single-phase) dispersivities α_{ij} as

$$D_{ij}(\mathbf{x}, t) = \delta_{ij}\left\{f'[\langle S(\mathbf{x}, t)\rangle]\right\}^2 \langle u\rangle\alpha_{ii}(t) \tag{6.71}$$

The dispersivities α_{ij} are generally time dependent. However, they tend asymptotically to their longitudinal and transverse constants with time.

For nonuniform (total) flows, results would be more complicated. The numerical moment equation approaches introduced in Section 3.4 are needed to obtain the total velocity covariances and the integral in Eq. (6.68) is to be evaluated numerically.

6.3 LAGRANGIAN APPROACH

In the absence of capillarity and gravity, the simplified Buckley–Leverett equation is of hyperbolic type and may be solved by the *method of characteristics*. In this section, we utilize this property of the equation in formulating moment equations of the saturation field. The Buckley–Leverett equation is recalled from Eq. (6.18) as

$$\frac{\partial S(\mathbf{x}, t)}{\partial t} + f'_w(S)\mathbf{u}(\mathbf{x}, t) \cdot \nabla S(\mathbf{x}, t) = 0 \tag{6.72}$$

where $S = S_w$ and $\mathbf{u}(\mathbf{x}, t) = \mathbf{q}(\mathbf{x}, t)/\phi(x)$ is the total velocity. We assume that the total flux is a function of space and varies little with time so that $\mathbf{q}(\mathbf{x}, t) = \mathbf{q}(\mathbf{x})$. Since either $\mathbf{q}$ or ϕ, or both, is a random space function, so are $\mathbf{u}$, S, k_{rp}, and f_w. Decompose $\mathbf{u}$ and S into their respective means and fluctuations:

$$\mathbf{u}(\mathbf{x}) = \langle \mathbf{u}(\mathbf{x}) \rangle + \mathbf{u}'(\mathbf{x}) \tag{6.73}$$

$$S(\mathbf{x}, t) = \langle S(\mathbf{x}, t) \rangle + S'(\mathbf{x}, t) \tag{6.74}$$

The main purpose of this section is to estimate the saturation profiles and the associated uncertainty for Buckley–Leverett displacement in heterogeneous media based on some statistical moments of medium properties such as permeability and porosity. We first show how this may be accomplished in a straightforward manner with a Lagrangian approach for one-dimensional flow, and then extend this approach to two dimensions.

6.3.1 One-Dimensional Flow

In this subsection, we consider a one-dimensional (1-D) version of the equation, which represents displacement along a streamline or a result of averaging over many streamlines. With the transformation

$$\frac{d\tau}{dx} = \frac{1}{u(x)} \tag{6.75}$$

we can rewrite Eq. (6.72) as

$$\frac{\partial S(\tau, t)}{\partial t} + f'_w(S)\frac{\partial S(\tau, t)}{\partial \tau} = 0 \tag{6.76}$$

where $\tau(x; x_o)$ is the *travel time* (time-of-flight) of a particle from x_o to x in the (total) velocity field $u(x)$. Since u is a random variable, so is τ.

Consider the case that the domain is initially occupied by oil at saturation $S_n = 1 - S_{wr}$ such that $S(x, 0) = S_{wr}$, and a continuous injection of water at rate q is placed at $x = 0$ such that $S(0, t) = 1 - S_{nr}$, where S_{wr} and S_{nr} are the respective residual (irreducible) saturations of water and oil. Equation (6.76) has the so-called *shock wave* solution

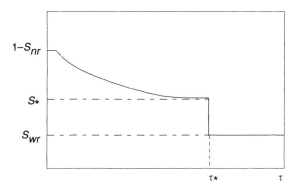

Figure 6.1. Shock wave saturation profile along a streamline.

[e.g., Marle, 1981; Dagan and Cvetkovic, 1996]

$$S(\tau, t) = s(\tau/t) H[f'_w(S_*) - \tau/t] \tag{6.77}$$

where S_* is the solution of $f'_w(S) = [f_w(S) - f_w(S_{wr})]/(S - S_{wr})$, $s(\tau/t)$ is the solution of the equation $f'_w(s) = \tau/t$, and H is the Heaviside step function, whose properties are discussed in Section 2.4.1. For any given t, the water saturation as a function of τ is given by the shock wave solution in Eq. (6.77). As illustrated in Fig. 6.1, S decreases gradually from $(1 - S_{nr})$ at $\tau = 0$ to S_* at $\tau_* = f'_w(S_*)t$ and then jumps to S_{wr}. In a randomly heterogeneous porous medium, the travel time at a particular location is a random variable whose statistical moments depend on those of the (total) velocity.

Saturation moments The random function $S(\tau, t) = S[\tau(x; x_o), t]$ may be estimated with its ensemble mean (expected value),

$$\langle S(x, t) \rangle = \int_0^\infty S(\tau, t) p(\tau; x, x_o) \, d\tau \tag{6.78}$$

where $p(\tau; x, x_o)$ is the probability density function (PDF) of the (one-particle) travel time τ. Therefore, the evaluation of the mean saturation reduces to that of travel time PDF, which only depends on the total velocity field and does not vary with the two-phase composition. The lower limit in Eq. (6.78) is zero because $S \equiv 0$ for $\tau < 0$; and the upper limit can be reduced by recognizing that $S \equiv 0$ when $\tau > f'_w(S_*)t$ as given by Eq. (6.77). The uncertainty associated with the estimation can be evaluated with the aid of variance and/or covariance. The variance at (x, t) is given by

$$\sigma_S^2(x, t) = \langle S'^2(x, t) \rangle = \langle S^2(x, t) \rangle - \langle S(x, t) \rangle^2$$

$$= \int_0^\infty S^2(\tau, t) p(\tau, x; x_o) \, d\tau - \langle S(x, t) \rangle^2 \tag{6.79}$$

The covariance reads as

$$C_S(x, t; x_1, t_1) = \langle S'(x, t) S'(x_1, t_1) \rangle$$

$$= \int_0^\infty \int_0^\infty S(\tau, t) S(\tau_1, t_1) p(\tau, x; \tau_1, x_1; x_o) \, d\tau \, d\tau_1$$

$$- \langle S(x, t) \rangle \langle S(x_1, t_1) \rangle \quad (6.80)$$

where $\tau_1(x_1, x_o)$ is the travel time of a particle from x_o to x_1, and $p(\tau, x; \tau_1, x_1; x_o)$ is the joint PDF of travel times of two particles, one from x_o to x and the other to x_1.

Another quantity of interest is $q_n(x, t; x_o) = \{1 - f_w[S(x, t)]\}q$, which is the flux of the oil phase. If the production well is at $x = L$, $q_n(L, t; x_o)$ is the rate of oil production. Since both q and f_w are random variables, so is q_n. It can be estimated by its expected value,

$$\langle q_n(x, t; x_o) \rangle = \langle q \rangle - \langle q(x) f_w[S(x, t)] \rangle$$

$$= \langle q \rangle - \int_{-\infty}^\infty \int_0^\infty q(x) f_w[S(\tau, t)] p(\tau, q; x, x_0) \, d\tau \, dq \quad (6.81)$$

where $p(\tau, q; x, x_0)$ is the joint PDF of q at x and the travel time τ of a particle from x_o to x. In general, $p(\tau, q; x, x_0) = p(\tau; x, x_0|q) p(q)$ where $p(\tau; x, x_0|q)$ is the travel time PDF conditioned on a fixed q and $p(q)$ is the PDF of flux $q(x)$. The mean cumulative production is given as

$$\langle Q_n(x, t; x_o) \rangle = \int_0^t \langle q_n(x, t'; x_0) \rangle \, dt' \quad (6.82)$$

Travel time moments and PDF In a one-dimensional domain, for steady-state flow the divergence-free condition requires the total flux q to be constant spatially. However, it can be either a deterministic or random constant. The latter represents the ensemble of 1-D domains each having a different q or one domain with a random q as a boundary condition. On the other hand, the spatial variation in porosity $\phi(x)$ will render the (total) velocity $u(x)$ random whether q is random or not. In this section, we consider the case that q is a specified constant (injection rate q_o) so that the spatial variability in porosity is the only source of randomness in the velocity $u(x) = q_o/\phi(x)$. The porosity is assumed to be stationary such that its mean $\langle \phi \rangle$ is constant and its covariance $C_\phi(x, x')$ depends only on the relative distance $r = x - x'$ between two points. The covariance function C_ϕ is taken to be exponential,

$$C_\phi(r) = \sigma_\phi^2 \exp\left(-\frac{|r|}{\lambda_\phi}\right) \quad (6.83)$$

where σ_ϕ^2 and λ_ϕ are the variance and correlation scale of porosity, respectively.

In this case, the travel time can be rewritten as

$$\tau(x; x_o) = \int_{x_o}^{x} \frac{\phi(\chi)}{q_o} d\chi \tag{6.84}$$

Hence, its moments are given exactly as

$$\langle \tau(x; x_o) \rangle = \frac{(x - x_o)\langle \phi \rangle}{q_o} \tag{6.85}$$

$$\sigma_{\tau\tau_1}(x, x_1; x_o) = \frac{1}{q_o^2} \int_{x_o}^{x} \int_{x_o}^{x_1} C_\phi(\chi - \chi') d\chi d\chi' \tag{6.86}$$

$$
\sigma_\tau^2(x; x_o) = \frac{1}{q_o^2} \int_{x_o}^{x} \int_{x_o}^{x} C_\phi(\chi - \chi') d\chi d\chi'
$$
$$
= \frac{2}{q_o^2} \int_{x_o}^{x} (x - \chi) C_\phi(\chi - x_o) d\chi
$$
$$
= \frac{2\sigma_\phi^2}{q_o^2} \left\{ \lambda_\phi(x - x_o) - \lambda_\phi^2 \left[1 - \exp\left(-\frac{x - x_o}{\lambda_\phi} \right) \right] \right\} \tag{6.87}
$$

It is seen that the mean travel time increases linearly with $(x - x_o)$ and the variance of travel time increases almost linearly with $(x - x_o)$ when λ_ϕ is small and sublinearly when λ_ϕ is large. If the travel time obeys a lognormal distribution, its PDF can be described with the first two moments,

$$p(\tau; x, x_o) = \frac{1}{(2\pi)^{1/2} \tau \sigma_{\ln(\tau)}} \exp\left\{ -\frac{\{\ln(\tau) - \langle \ln[\tau(x, x_o)] \rangle\}^2}{2\sigma_{\ln(\tau)}^2} \right\} \tag{6.88}$$

where $\langle \ln(\tau) \rangle = -0.5 \ln[\langle \tau \rangle^2 + \sigma_\tau^2] + 2 \ln(\langle \tau \rangle)$ and $\sigma_{\ln(\tau)}^2 = \ln[\langle \tau \rangle^2 + \sigma_\tau^2] - 2 \ln(\langle \tau \rangle)$. The joint two-particle travel time PDF can be written similarly (see Eq. (2.106) in Section 2.2.8). There are two important limits for the two-particle PDF [Cvetkovic *et al.*, 1992],

$$\lim_{|x - x_1| \to 0} p(\tau, x; \tau_1, x_1; x_o) = p(\tau; x, x_o)\delta(\tau - \tau_1) \tag{6.89}$$

$$\lim_{|x - x_1| \to \infty} p(\tau, x; \tau_1, x_1; x_o) = p(\tau; x, x_o)p(\tau_1; x_1, x_o) \tag{6.90}$$

With the first limit, Eq. (6.80) reduces to Eq. (6.79) when $x = x_1$ and $t = t_1$. The second limit confirms the intuitive expectation that the travel times at two locations separated by a large distance are independent. Other forms of travel time PDFs are possible. Below, we will briefly investigate the impact of distributional forms on the mean saturation profile and the mean oil production rate.

In this case, the injection rate is a specified constant so $\sigma_q^2 = 0$, and $p(\tau, q; x, x_o) = p(\tau; x, x_o)\delta(q - q_o)$. Hence, the mean production rate is given as

$$\langle q_n(x, t; x_o) \rangle = q_o - q_o \langle f_w[S(x, t)] \rangle$$

$$= q_o - q_o \int_0^\infty f_w[S(\tau, t)] p(\tau; x, x_0) \, d\tau \qquad (6.91)$$

Illustrative examples and results   Recently, Dagan and Cvetkovic [1996] and Cvetkovic and Dagan [1996] evaluated the mean production rate $\langle q_n \rangle$, mean cumulative production $\langle Q_n \rangle$ and other temporal and spatial moments, although they did not study the mean and variance of saturation $S(x, t)$. More recently, Zhang and Tchelepi [1999] evaluated the expected values and variances of saturation besides other flow quantities. Below we illustrate the Lagrangian approach and some key findings with the cases considered by the latter two authors. For simplicity, the following Corey-type relative permeability functions with $S_{nr} = S_{wr} = 0$ are used:

$$k_{rw} = S^2, \qquad k_{rn} = (1 - S)^2 \qquad (6.92)$$

Hence

$$f_w(S) = \frac{m S^2}{m S^2 + (1 - S)^2} \qquad (6.93)$$

$$f_w'(S) = \frac{2m S(1 - S)}{[m S^2 + (1 - S)^2]^2} \qquad (6.94)$$

where $m = \mu_n/\mu_w$. Based on Eq. (6.77), for this case the saturation decreases gradually from $S = 1$ at $\tau = 0$ to $S_* = 1/(1 + m)^{1/2}$ at $\tau_* = f_w'(S_*)t$ and then jumps to $S = 0$.

Figure 6.2 shows the expected values (ensemble means) of water saturation and the associated standard deviations at two different times ($t = 2$ and 4) and for two different viscosity ratios ($m = 2$ and 0.5). Here, $q_o = 0.3$ and $\langle \phi \rangle = 0.3$. For the case of random media, the standard deviation of porosity is $\sigma_\phi = 0.2$ and its correlation scale $\lambda_\phi = 1.0$. Note that q_o [L/T], λ_ϕ [L], t [T], and x [L] are given in a set of consistent units. In homogeneous media ($\sigma_\phi = 0$), the saturation profiles exhibit a sharp *discontinuity* at the front (Figs. 6.2a and b, curves without any symbol), which is well understood in the literature [e.g., Marle, 1981]. In the presence of medium heterogeneity, the discontinuities disappear in the mean saturation $\langle S \rangle$ profiles (Figs. 6.2a and b) and the standard deviations σ_S (Figs. 6.2c and d) are the largest near where the discontinuities are supposed to develop in the absence of heterogeneity. The dispersive behavior, which may be termed macrodispersion or *heterogeneity-induced dispersion*, can be explained by realizing that the expected saturation profile is an average of many possible shock waves traveling in a random velocity field. This dispersive, or spreading, behavior is observable for both $m = 2.0$ where oil is more viscous than water and $m = 0.5$ where oil is less viscous. In a homogeneous medium, the unfavorable viscosity ratio (i.e. $m = 2$) case has a smaller saturation jump at the front and exhibits more spreading (larger transition zone) than the $m = 0.5$ case.

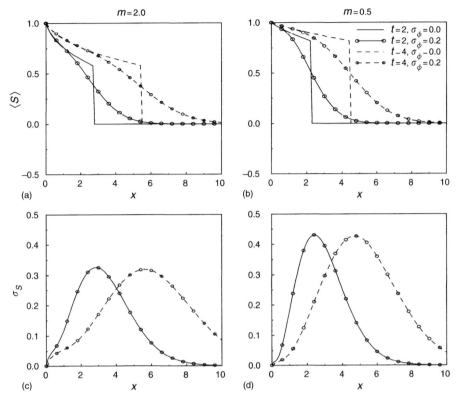

Figure 6.2. The expected values $\langle S \rangle$ and standard deviations σ_S of water saturation at two different times ($t = 2$ and 4): (a) and (c) for a more viscous oil; (b) and (d) for a less viscous oil. (Adapted from Zhang and Tchelepi [1999]. Copyright 1999 by the Society of Petroleum Engineers.)

In the presence of random heterogeneity, the spreading due to the combined effects of heterogeneity and viscosity differences is expected to be larger when the oil is more viscous than the water (i.e. $m = 2.0$). It follows that at a particular time, the σ_S profile for $m = 0.5$ has a higher peak and is more compact than the corresponding profile of the $m = 2.0$ displacement. However, a better quantitative measure of uncertainty is probably the coefficient of variation, $C_v = \sigma_S/\langle S \rangle$. The C_v at where the peak of σ_S is located is smaller in the case of a less viscous oil than in the case of a more viscous oil. In both cases, heterogeneity-induced dispersion increases with time. The expected saturation profile and the associated standard deviation may be used to construct confidence intervals for the saturation field. In constructing the confidence intervals, such as by adding and subtracting some multiple of σ_S to and from $\langle S \rangle$, one must take physical constraints into consideration. For example, the saturation must be between 0 and 1. It should be pointed out that these confidence intervals are given in an approximate sense when the saturation is not normally distributed. When the saturation distribution is far from symmetric, higher moments are needed for constructing these confidence intervals.

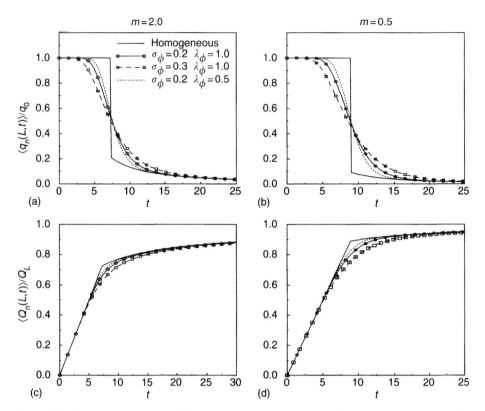

Figure 6.3. The expected values of oil production rate q_n and cumulative oil production Q_n for various statistical parameters: (a) and (c) for a more viscous oil; (b) and (d) for a less viscous oil. (Adapted from Zhang and Tchelepi [1999]. Copyright 1999 by the Society of Petroleum Engineers.)

Figure 6.3 shows the expected oil production rate and cumulative oil production as a function of time t for different statistical parameters. The former is normalized by the (total) flux q_o and the latter by Q_L (the total volume of extractable oil from the injection well at $x = 0$ to the producing well at $x = L$). Here, $L = 10$. One may define a dimensionless time, $t_D = \langle u \rangle t / L$. In this example, $t_D = t/10$. In the homogeneous case, the oil production rate is one (only oil is produced) until water breakthrough occurs, then the oil production rate drops to a much lower value and decreases slowly toward zero. In the presence of medium heterogeneity (e.g., solid curves with circles for $\sigma_\phi = 0.2$ and $\lambda_\phi = 1.0$), water breakthrough occurs much earlier and the oil production rate does not jump but decreases gradually with time (Figs. 6.3a and b). The early breakthrough time and higher water–oil ratio thereafter are, again, due to the so-called heterogeneity induced dispersion. When the variance of porosity increases, the dispersive, or spreading, behaviors become even more apparent for both more viscous and less viscous oil. Also, dispersion increases with the correlation scale in porosity. It is clear from Figs. 6.3c and d that oil recovery takes longer in heterogeneous media due to tailing caused by dispersion (heterogeneity fingering) than

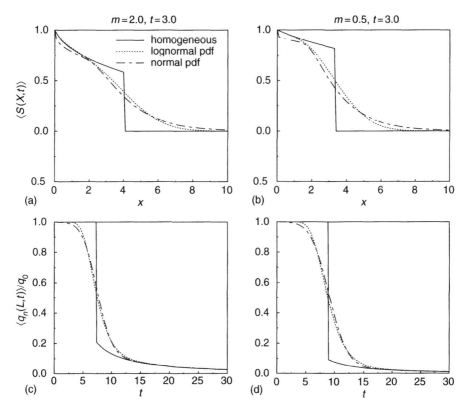

Figure 6.4. Comparisons of expected values of saturation and oil production rate based on lognormal and normal travel time PDFs: (a) and (c) for a more viscous oil; (b) and (d) for a less viscous oil. (Adapted from Zhang and Tchelepi [1999]. Copyright 1999 by the Society of Petroleum Engineers.)

in homogeneous media. Note that for both homogeneous and heterogeneous cases, the recovery takes longer in the case of a more viscous oil (Fig. 6.3c) than in the case of a less viscous oil (Fig. 6.3d).

In Figs 6.2 and 6.3, the travel time PDF is assumed to follow a lognormal distribution. As mentioned earlier, other forms of travel time PDFs are possible. Figure 6.4 compares the resultant expected saturation profiles and normalized oil production rates for two different travel time PDFs. The results for the corresponding homogeneous case are also presented for reference. The general dispersive behaviors of earlier water breakthrough and later oil arrival are the same for these two distributional forms, although the details are different. A large difference exists at early time or at locations close to the injection well. This is due to the fact that with the lognormal PDF the travel time is bounded to be positive, while with the normal PDF it can take any value. In the case where flow is from the injection well to the producing well, a normal PDF for the travel time is probably not reasonable, especially at locations near the pumping well. Nevertheless, the differences that result from these two distributional forms may be neglected for all practical purposes. Between the lognormal PDF and another reasonable PDF such as inverse-Gaussian [Simmons, 1982], the discrepancy should be even smaller.

6.3.2 Two-Dimensional Flow

In this section, we consider the two-dimensional (2-D) version of Eq. (6.72),

$$\frac{\partial S(\mathbf{x}, t)}{\partial t} + f'_w(S) \left[u_1(\mathbf{x}) \frac{\partial S(\mathbf{x}, t)}{\partial x_1} + u_2(\mathbf{x}) \frac{\partial S(\mathbf{x}, t)}{\partial x_2} \right] = 0 \tag{6.95}$$

With the transformation

$$\frac{d\tau}{dx_1} = \frac{1}{u_1(x_1, \eta)}, \qquad \frac{d\eta}{dx_1} = \frac{u_2(x_1, \eta)}{u_1(x_1, \eta)} \tag{6.96}$$

we can rewrite Eq. (6.95) in the coordinates $(\mathbf{x}, t)$ as

$$\frac{\partial S(\boldsymbol{\chi}, t)}{\partial t} + f'_w(S) \left\{ v \frac{\partial S(\boldsymbol{\chi}, t)}{\partial \chi_1} + [u_2(\mathbf{x}) - v u_2(x_1, \eta)] \frac{\partial S(\boldsymbol{\chi}, t)}{\partial \chi_2} \right\} = 0 \tag{6.97}$$

in terms of the coordinates $(\boldsymbol{\chi}, t)$ attached to the streamlines $\chi_1 = \tau(x_1; \mathbf{x}_o)$ and $\chi_2 = x_2 - \eta(x_1; \mathbf{x}_o)$. In the above, $v = u_1(\mathbf{x})/u_1(x_1, \eta)$. At $\tau = 0$, $\eta = x_{2o}$ for $x_1 = x_{1o}$; and $x_2 = \eta(x_1; \mathbf{x}_o)$ is simply the equation of the streamlines of the steady-state velocity field. Along the streamline $\chi_2 = 0$, i.e., $x_2 = \eta$, one obtains $v = 1$ and the following equation:

$$\frac{\partial S(\boldsymbol{\chi}, t)}{\partial t} + f'_w(S) \frac{\partial S(\boldsymbol{\chi}, t)}{\partial \tau} = 0 \tag{6.98}$$

This equation is the same as the one-dimensional equation (6.76) except for the additional dependency on $\chi_2 = x_2 - \eta(x_1; x_o)$. It is also exactly the same as the starting equation for the streamline approaches by the Stanford research group, among others [see, e.g., Blunt *et al.*, 1996; Thiele *et al.*, 1996]. However, in the latter approach the detailed variability of reservoir properties are assumed to be known. Hence, the velocity field can be solved for and the streamlines can be mapped out explicitly. In the present study, either permeability or porosity, or both, is assumed to be a random process with only some known statistical moments. In turn, the velocity and the streamlines are random processes. This is exactly where our approach differs from the deterministic streamline approach. Our objective is to estimate the saturation profiles and the associated uncertainty based on statistical moments of permeability and/or porosity by the method of moment equations. On the other hand, one may generate many equally likely realizations of a reservoir based on the known statistics of permeability and/or porosity (see Section 3.11 for details), and then efficiently solve each realization by the deterministic streamline approach. Our results should be comparable to those obtained by averaging the many sets of numerical simulations. This is the so-called Monte Carlo simulation approach, which has been discussed in Section 3.11.

Saturation moments As in the 1-D case, we assume that the streamlines change little with time so that the total velocity components u_1 and u_2 are functions of space only. The domain is assumed to be initially occupied by oil such that the water saturation $S = 0$, and a continuous injection of water of $S = 1$ at rate q is placed along a line of length $2L_o$ at $x_1 = x_{1o}$ and centered about x_{2o}. For a streamline starting from $\mathbf{x}_o$,

$$S(\mathbf{x}, t) = S[\tau(x_1; \mathbf{x}_o), t], \quad \text{for } x_2 = \eta(x_1; \mathbf{x}_o) \tag{6.99}$$

where $S(\tau, t)$ is given by Eq. (6.77). The contribution of an infinitesimal stream tube of size da_2 starting from $\mathbf{x}_o(x_{1o}, a_2)$ of a source area is

$$dS(\mathbf{x}, t) = S[\tau(x_1; x_{1o}, a_2), t]\delta[x_2 - \eta(x_1; x_{1o}, a_2)]\, da_2 \tag{6.100}$$

Hence, the solution for the saturation field associated with the line injection at $a_2 \in [-L_o + x_{2o}, L_o + x_{2o}]$ can be written as

$$S(\mathbf{x}, t) = \int_{-L_o+x_{2o}}^{L_o+x_{2o}} S[\tau(x_1; x_{1o}, a_2), t]\delta[x_2 - \eta(x_1; x_{1o}, a_2)]\, da_2 \tag{6.101}$$

where both τ and η are random variables when the velocity is. The expected value of $S(\mathbf{x}, t)$ is given as

$$\langle S(\mathbf{x}, t) \rangle = \int_{-L_o+x_{2o}}^{L_o+x_{2o}} \int_0^\infty \int_{-\infty}^\infty S(\tau, t)\delta(x_2 - \eta) \cdot p[\tau, \eta; x_1, \mathbf{x}_o(x_{1o}, a_2)]\, d\eta\, d\tau\, da_2$$

$$= \int_{-L_o+x_{2o}}^{L_o+x_{2o}} \int_0^\infty S(\tau, t)p[\tau(x_1; x_{1o}, a_2), \eta(x_1; x_{1o}, a_2) = x_2]\, d\tau\, da_2 \tag{6.102}$$

where $p(\tau, \eta; x_1, \mathbf{x}_o)$ is the joint PDF of $\tau(x_1; x_{1o}, a_2)$ and $\eta(x_1; x_{1o}, a_2)$.

By virtue of

$$S^2(\mathbf{x}, t) = \int_{-L_o+x_{2o}}^{L_o+x_{2o}} \int_{-L_o+x_{2o}}^{L_o+x_{2o}} S[\tau_1(x_1; x_{1o}, a_2), t]S[\tau_2(x_1; x_{1o}, a_2'), t]$$

$$\cdot \delta[x_2 - \eta_1(x_1; x_{1o}, a_2)]\delta[x_2 - \eta_2(x_1; x_{1o}, a_2')]\, da_2\, da_2' \tag{6.103}$$

we obtain the variance of saturation as

$$\sigma_S^2(\mathbf{x}, t) = \int_{-L_o+x_{2o}}^{L_o+x_{2o}} \int_{-L_o+x_{2o}}^{L_o+x_{2o}} \int_0^\infty \int_0^\infty S[\tau_1(x_1; x_{1o}, a_2), t]$$

$$\cdot S[\tau_2(x_1; x_{1o}, a_2'), t]p[\tau_1(x_1; x_{1o}, a_2), \eta_1(x_1; x_{1o}, a_2)$$

$$= x_2; \tau_2(x_1; x_{1o}, a_2'), \eta_2(x_1; x_{1o}, a_2') = x_2]$$

$$\cdot d\tau_1\, d\tau_2\, da_2\, da_2' - \langle S(\mathbf{x}, t)\rangle^2 \tag{6.104}$$

where $p[\tau_1(x_1; x_{1o}, a_2), \eta_1(x_1; x_{1o}, a_2); \tau_2(x_1; x_{1o}, a_2'), \eta_2(x_1; x_{1o}, a_2')]$ is the joint PDF of two particles, one started from (x_{1o}, a_2) at time $t = 0$, and the other from (x_{1o}, a_2').

Physically speaking, $\tau(x_1; \mathbf{x}_o)$ is the travel time of a particle from $\mathbf{x}_o = (x_{1o}, a_2)$ to x_1 and $\eta(x_1; \mathbf{x}_o)$ is the transverse displacement of the particle when it reaches x_1. Since $S(\tau, t)$ is known and is the same as in 1-D, the evaluation of the saturation moments in Eqs. (6.102) and (6.104) becomes that of the travel time and transverse PDFs, which are functions of the (total) velocity field and do not vary with the two-phase composition under some conditions. Therefore, the problem of nonlinear displacement in multidimensional random media reduces to two subproblems: one being nonlinear displacement in 1-D homogeneous media, and the other being linear advection in random media.

In general, $p(\tau, \eta; x_1, \mathbf{x}_o) = p(\tau; x_1, \mathbf{x}_o)p(\eta|\tau; x_1, \mathbf{x}_o)$, where $p(\eta|\tau)$ is the PDF of η for a given τ. However, it was shown by Dagan *et al.* [1992] for uniform mean flow that τ and η are separable in that η can be determined based on the velocity field independent of τ, and that their PDFs are thus also separable. That is to say, $p(\tau, \eta; x_1, \mathbf{x}_o) = p(\tau; x_1, \mathbf{x}_o)p(\eta; x_1, \mathbf{x}_o)$. Therefore, for uniform mean flow

$$\langle S(\mathbf{x}, t) \rangle = \int_{-L_o+x_{2o}}^{L_o+x_{2o}} \left\{ \int_0^\infty \int_{-\infty}^\infty S(\tau, t)p[\tau; x_1, \mathbf{x}_o(x_{1o}, a_2)]\delta(x_2 - \eta) \right.$$

$$\left. \cdot p[\eta; x_1, \mathbf{x}_o(x_{1o}, a_2)] \, d\eta \, d\tau \right\} da_2$$

$$= \int_{-L_o+x_{2o}}^{L_o+x_{2o}} p[\eta = x_2; x_1, \mathbf{x}_o(x_{1o}, a_2)]$$

$$\cdot \left\{ \int_0^\infty S(\tau, t)p[\tau; x_1, \mathbf{x}_o(x_{1o}, a_2)] \, d\tau \right\} da_2 \qquad (6.105)$$

The generalization of the independence of τ and η to two particles is generally not true even for uniform mean flows [Zhang *et al.*, 2000b]. However, it has been shown [Zhang *et al.*, 2000b, p. 2114] that for a spatially stationary flow the correlation between τ and η can be neglected when the two particles started from a line either parallel or normal to the mean flow. Hence, under the condition of uniform mean flow, the joint PDF of two particles started from (x_{1o}, a_2) and (x_{1o}, a_2') can be simplified as

$$p[\tau_1(x_1; x_{1o}, a_2), \eta_1(x_1; x_{1o}, a_2); \tau_2(x_1; x_{1o}, a_2'), \eta_2(x_1; x_{1o}, a_2')]$$

$$= p[\tau_1(x_1; x_{1o}, a_2), \tau_2(x_1; x_{1o}, a_2')]p[\eta_1(x_1; x_{1o}, a_2), \eta_2(x_1; x_{1o}, a_2')] \qquad (6.106)$$

For a spatially nonstationary flow, the full PDF needs to be evaluated [Zhang *et al.*, 2000a].

Travel time and displacement moments Unlike in the 1-D case, in general the statistical moments of τ and η can only be given approximately in terms of those of the medium properties. The travel time PDFs may be evaluated with the aid of particle displacement PDF which is, in turn, related to the random velocity field, as done by Dagan and Nguyen [1989] and Cvetkovic and Dagan [1996]. Here we directly relate the statistical moments of travel time to those of velocity, in a manner similar to Cvetkovic *et al.* [1992]. The travel time of a particle from x_{1o} at $t = 0$ to x_1 can be written, via Eq. (6.96), as

$$\tau(x_1; \mathbf{x}_o) = \int_{x_{1o}}^{x_1} \frac{1}{u_1(\xi_1, \eta)} \, d\xi_1 \qquad (6.107)$$

where η in $u_1(\xi_1, \eta)$ is a random variable. We may expand u about $\langle \eta(\xi_1; \mathbf{x}_o) \rangle$ by Taylor series,

$$u_1(\xi_1, \eta) = \langle u_1 \rangle + u_1'(\xi_1, \langle \eta \rangle) + \eta' \frac{\partial u_1'(\xi_1, \langle \eta \rangle)}{\partial \langle \eta \rangle} + \cdots \qquad (6.108)$$

where the mean velocity $\langle u_1 \rangle$ is assumed to be constant, $u_1' = u_1 - \langle u_1 \rangle$ with $\langle u_1' \rangle \equiv 0$, and $\eta' = \eta - \langle \eta \rangle$ with $\langle \eta' \rangle \equiv 0$. When $\langle u_2 \rangle = 0$, $\langle \eta(\xi_1; \mathbf{x}_o) \rangle = x_{2o}$. By keeping terms up to first order in the above series, Eq. (6.107) can be rewritten as

$$\tau(x_1; \mathbf{x}_o) \approx \int_{x_{1o}}^{x_1} \frac{1}{\langle u_1 \rangle + u_1'(\xi_1, x_{2o})} \, d\xi_1$$

$$= \frac{1}{\langle u_1 \rangle} \int_{x_{1o}}^{x_1} \left[1 - \frac{u_1'(\xi_1, x_{2o})}{\langle u_1 \rangle} + \cdots \right] d\xi_1 \qquad (6.109)$$

Hence we have, to first order,

$$\langle \tau(x_1; \mathbf{x}_o) \rangle = \frac{x_1 - x_{1o}}{\langle u_1 \rangle} \qquad (6.110)$$

$$\tau'(x_1; \mathbf{x}_o) = -\frac{1}{\langle u_1 \rangle^2} \int_{x_{1o}}^{x_1} u_1'(\xi_1, x_{2o}) \, d\xi_1 \qquad (6.111)$$

To derive these expressions, the velocity coefficient of variation (i.e., velocity standard deviation normalized by mean velocity) must be (much) smaller than one. This condition may be satisfied for many practical situations where the standard deviation of log-transformed permeability is moderately large. This is so because the coefficient of variation of velocity is usually smaller than the standard deviation of log permeability. With Eq. (6.111), we obtain the variance for the travel time as

$$\sigma_\tau^2(x_1; \mathbf{x}_o) = \frac{1}{\langle u_1 \rangle^4} \int_{x_{1o}}^{x_1} \int_{x_{1o}}^{x_1} C_{u_1}(\xi_1 - \xi_1', 0) \, d\xi_1 \, d\xi_1'$$

$$= \frac{2}{\langle u_1 \rangle^4} \int_{x_{1o}}^{x_1} (x_1 - \xi_1) C_{u_1}(\xi_1 - x_{1o}, 0) \, d\xi_1 \qquad (6.112)$$

where $C_{u_1}(\xi_1 - \xi_1', 0) = \langle u_1'(\xi_1, x_{2o}) u_1'(\xi_1', x_{2o}) \rangle$ is the covariance function of the total velocity. In the above, the velocity fluctuation is assumed to be stationary such that the covariance depends on the relative distance of two points rather than their actual locations. The (total) velocity covariance is derived by linearizing the stochastic flow equations, as discussed in Chapter 3. With the covariance, the travel time moments can be evaluated either analytically or numerically. Then the travel time PDF can be obtained via Eq. (6.88).

Now let us look at how to derive the statistical moments for η, which can be rewritten as

$$\eta(x_1; \mathbf{x}_o) = \int_{x_{1o}}^{x_1} \frac{u_2(\xi_1, \eta)}{u_1(\xi_1, \eta)} \, d\xi_1 \tag{6.113}$$

Similar to Eq. (6.108), we have

$$u_2(\xi_1, \eta) = u_2'(\xi_1, \langle\eta\rangle) + \eta' \frac{\partial u_2'(\xi_1, \langle\eta\rangle)}{\partial \langle\eta\rangle} + \cdots \tag{6.114}$$

Note again that we have assumed that $\langle u_2 \rangle = 0$. Therefore, on the basis of Eqs. (6.108) and (6.114), we have, to first order,

$$\eta'(x_1; \mathbf{x}_o) = \frac{1}{\langle u_1 \rangle} \int_{x_{1o}}^{x_1} u_2'(\xi_1, x_{2o}) \, d\xi_1 \tag{6.115}$$

$$\sigma_\eta^2(x_1; \mathbf{x}_o) = \frac{2}{\langle u_1 \rangle^2} \int_{x_{1o}}^{x_1} (x_1 - \xi_1) C_{u_2}(\xi_1 - x_{1o}, 0) \, d\xi_1 \tag{6.116}$$

Since $\langle \eta \rangle - x_{2o} = 0$, there is no preferential direction for the transverse displacement. Its PDF should be symmetric around $\eta = x_{2o}$ and may be well approximated by a Gaussian (normal) distribution,

$$p(\eta; x_1, \mathbf{x}_o) = \frac{1}{(2\pi)^{1/2} \sigma_\eta(x_1; \mathbf{x}_o)} \exp\left[-\frac{(\eta - x_{2o})^2}{2\sigma_\eta^2(x_1; \mathbf{x}_o)} \right] \tag{6.117}$$

With Eqs. (6.111) and (6.115), one may obtain other covariances such as $\langle \tau_1' \tau_2' \rangle$ and $\langle \eta_1' \eta_2' \rangle$, which are required for evaluating the two-particle joint PDFs $p(\tau_1, \tau_2)$ and $p(\eta_1, \eta_2)$ in Eq. (6.106). Some explicit expressions for these PDFs can be found in Section 2.2.8 when τ and η are assumed lognormal and normal, respectively. Note that in this section, the covariances are given under the condition of spatially stationary flow. For nonstationary flow, these and other covariances required are much more complicated, and they have recently been derived and evaluated by Zhang *et al.* [2000a].

Solution Together with the observation from Eqs. (6.110), (6.112), and (6.116) that the moments of travel time and transverse displacement do not depend on x_{2o} or a_2, substitution of Eq. (6.117) into Eq. (6.105) yields

$$\langle S(\mathbf{x}, t)\rangle = \int_{-L+x_{2o}}^{L_o+x_{2o}} \frac{1}{(2\pi)^{1/2}\sigma_\eta(x_1; x_{1o}, x_{2o})} \exp\left[-\frac{(x_2-a_2)^2}{2\sigma_\eta^2(x_1; x_{1o}, x_{2o})}\right] da_2$$

$$\cdot \int_0^\infty S(\tau, t)p(\tau; x_1, x_{1o})\, d\tau$$

$$= \frac{1}{2}\left\{\mathrm{erf}\left[\frac{x_2+L_o-x_{2o}}{\sqrt{2}\sigma_\eta(x_1; x_{1o}, x_{2o})}\right] - \mathrm{erf}\left[\frac{x_2-L_o-x_{2o}}{\sqrt{2}\sigma_\eta(x_1; x_{1o}, x_{2o})}\right]\right\}$$

$$\cdot \int_0^\infty S(\tau, t)p(\tau; x_1, x_{1o})\, d\tau \qquad (6.118)$$

For any finite $x_2 - x_{2o}$,

$$\lim_{L_o \to \infty} \mathrm{erf}\left[\frac{x_2 - x_{2o} + L_o}{\sqrt{2}\sigma_\eta(x_1; x_{1o}, x_{2o})}\right] = 1$$

$$\lim_{L_o \to \infty} \mathrm{erf}\left[\frac{x_2 - x_{2o} - L_o}{\sqrt{2}\sigma_\eta(x_1; x_{1o}, x_{2o})}\right] = -1$$

$$(6.119)$$

Hence, for the case of a large line injection normal to the mean flow the 2-D expression of $\langle S(\mathbf{x}, t)\rangle$ reduces to

$$\langle S(\mathbf{x}, t)\rangle = \int_0^\infty S(\tau, t)p(\tau; x_1, x_{1o})\, d\tau \qquad (6.120)$$

which is identical to Eq. (6.78) in the one-dimensional case. Although the statistical moments of flow and hence those of travel time may be different in one- and two-dimensional cases, we expect the mean saturation profiles to behave similarly as in 1-D. For the case of a finite line/area injection, there is additional dispersion (spreading) in the transverse direction due to medium heterogeneity. This is reflected in that the argument before the integral on the right-hand side of Eq. (6.118) is less than or equal to 1. The expression for saturation variance, which is more complex in 2-D and involves two-particle joint PDFs, has recently been evaluated by Zhang *et al.* [2000a]. It is expected for the case of uniform mean (total) flow the variance generally decreases as the transverse length L_o increases, and when L_o is sufficiently larger than the correlation scale of heterogeneity the ergodic condition is approached, i.e., $S(\mathbf{x}, t) \approx \langle S(\mathbf{x}, t)\rangle$.

Illustrative examples and results The general methodology developed in this section applies to any type of (steady or quasi-steady state) flow field (with some modifications) once the required moments of velocity are available. In the following examples, we consider uniform mean (horizontal) flow in two dimensions. This corresponds to displacement between two large arrays of injection and pumping wells where a constant mean gradient is maintained. In this case, the mean (total) velocity $\langle u_i\rangle = \delta_{i1}\mu$ (μ being the magnitude of the mean velocity) and the velocity covariance can be given as a function of the statistical moments of permeability. The permeability field is assumed to be second-order stationary with the following exponential covariance:

$$C_f(\mathbf{r}) = \sigma_f^2 \exp\left[-\left(\frac{r_1^2}{\lambda_1^2} + \frac{r_2^2}{\lambda_2^2}\right)^{1/2} \right]$$ (6.121)

where $\mathbf{r}$ is the separation vector between two points, σ_f^2 is the variance of log permeability, and λ_i is the correlation (integral) scale of log permeability in the r_i direction. For the isotropic form of this covariance ($\lambda_1 = \lambda_2 = \lambda_f$), analytical expressions of the velocity covariance have been obtained for unbounded domains with first-order approximations in both two and three dimensions (see Section 3.5.1). With these, the statistical moments of travel time and transverse displacement can be evaluated numerically, say, by Simpson's rule. The results for unbounded domains may be good approximations for bounded domains when the points of interest are few integral scales away from the boundaries. On the other hand, the velocity moments in bounded domains can be obtained numerically as discussed in Section 3.4.

Figure 6.5 shows the expected (mean) saturation profiles along the horizontal direction at $x_2 = 0$ and the transverse direction at $x_1 = 3$ [L] for different cases. Figures 6.5a and b illustrate the impact of the transverse length L_o of an injection zone. Here, $\lambda_f = 1$ [L], $\sigma_f = 0.5$, $\langle u_1 \rangle = 1$ [L/T], and $t = 3$ [T]. The producing wells are located 10 [L] from the injection wells. When $L_o = 0.1$ [L], the mean water saturation is much lower in both the horizontal and transverse profiles compared to the homogeneous case due to dispersion (heterogeneity fingering) in both the horizontal and transverse directions. The relative effect of dispersion on the whole saturation profile decreases as the transverse length L_o increases. At $L_o = 2$ [L], the mean saturation profiles near the center of the injection area can be well approximated by the 1-D analogy, Eq. (6.120). That is to say, transverse dispersion is insignificant for an injection of large size normal to the mean flow. Figures 6.5c and d show the effect of medium heterogeneity on the mean saturation profiles based on the case of $L_o = 2$ [L]. It is seen that the dispersive behavior becomes more apparent as the variance or the correlation scale of log permeability f increases. In Fig. 6.5, the oil is more viscous than water ($m = 2$). Though not shown, the general behavior is the same for a less viscous oil.

Figure 6.6 shows the distribution of mean saturation obtained from the moment-based stochastic approach and Monte Carlo simulations. Two hundred high-resolution realizations of the permeability field are used in the Monte Carlo results. For each realization, the governing equations of two-phase flow are solved on a fine grid with a total variation diminishing (TVD) finite-difference method [Chen *et al.*, 1993]. In the Monte Carlo simulations, the total velocity field is fixed at steady state in order to conform to the assumption made early in this section. In this example, we specify the following parameters for both approaches: $\lambda_f = 1$ [L], $\sigma_f = 0.5$, $\langle u_1 \rangle = 1$ [L/T], $t = 5$ [T], $m = 2$, and transverse length $2L_o = 5$ [L]. It is seen that heterogeneity-induced dispersion is also present in the mean saturation contours obtained from Monte Carlo simulations. Excellent agreement is found for these two approaches based on this example. We have also made other comparisons and found good agreements with σ_f being as large as one.

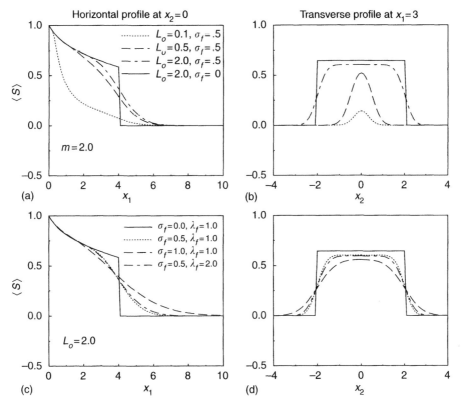

Figure 6.5. Horizontal and transverse mean saturation profiles in a two-dimensional domain: (a) and (b) for different injection size L_o; (c) and (d) for different statistical parameters. (Adapted from Zhang and Tchelepi [1999]. Copyright 1999 by the Society of Petroleum Engineers.)

6.3.3 Summary and Discussion

In this section, we developed a Lagrangian, stochastic model of immiscible nonlinear two-phase displacement in heterogeneous porous media in the absence of capillary pressure and gravity effects. The model results are the expected value of saturation and the associated uncertainty as well as the expected value of oil production rate. The expected value and the standard deviation may be used to construct confidence intervals for saturation and to evaluate the risk associated with performance predictions due to incomplete knowledge of the reservoir properties. These confidence intervals are given in an approximate sense when the saturation is not normally distributed. When the saturation distribution is far from symmetric, higher moments are needed for constructing these confidence intervals.

Through transforming to coordinates attached to streamlines, the evaluation of saturation statistical moments is simplified to that of travel time and transverse displacement PDFs. We assume that the latter do not vary with the two-phase composition but depend entirely on the total velocity, which we assume is a weak function of time. Hence, the problem of

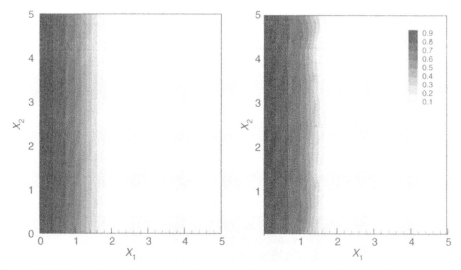

Figure 6.6. Comparison of mean saturation contours between moment-based stochastic approach (left plot) and Monte Carlo simulations (right plot). (Adapted from Zhang and Tchelepi [1999]. Copyright 1999 by the Society of Petroleum Engineers.)

nonlinear two-phase displacement in heterogeneous (random) reservoirs has been reduced to two subproblems: one being a series of nonlinear displacements, each along a streamline, and the other a multidimensional description of the total velocity in the heterogeneous medium.

Through some one- and two-dimensional examples, we found that the well-known discontinuities existing in saturation profiles and oil production curves for homogeneous media disappear in the case of heterogeneous media. This is due to heterogeneity fingering, or the so-called "heterogeneity-induced dispersion". This finding is confirmed with Monte Carlo simulations. The effect of heterogeneity on oil production is earlier water breakthrough and later oil arrival. The oil production is expected to take longer in heterogeneous media than in homogeneous media.

There are a number of simplifying assumptions and approximations made in this section: (1) the effects of capillarity and gravitation are negligible; (2) the total flow is at steady state; (3) the statistical moments of travel time and velocity are derived as first-order approximations; and (4) the mean flow is unidirectional and constant in space, and the flow domain is unbounded. First, the stochastic (Lagrangian) approach is limited to advection-dominated problems; and to formally account for capillarity a different approach such as the Eulerian one developed in Section 6.2 may be used. However, the effect of capillarity had been found to be local and only significant in zones of abrupt changes in saturation for homogeneous media; and this assumption is expected to be even more justified in heterogeneous media. The absence of gravity may not be a good assumption in many situations. To include gravity is a challenging problem and shall be a topic of future study. Second, the assumption of steady-state total flow implies neglecting saturation dependence in the total flow. This has been examined in the petroleum literature where it has been found that it does not have a large impact on heterogeneous fingering for modest mobility ratio. When it is important, the Lagrangian approach still applies if the total velocity moments are updated with time. Third, theoretically speaking, the first-order approximations require the variance of log permeability to be (much) smaller than one. However, numerical and field experimental evidence indicated that these low-order results may be applicable even for large variance systems. In addition, since the variability of velocity is usually much smaller than that of log permeability, the assumption of small variability in velocity may be valid for many practical situations when flow is not extremely rapid. Finally, the assumption of uniform mean flow in an unbounded domain is only made to take advantage of available analytical results for velocity moments. The approach developed is applicable to more general flow situations such as convergent flows in bounded domains. With the numerical moment equation approaches discussed in Section 3.4 for single-phase, steady-state flow, the velocity moments can be obtained for cases involving injection/producing wells of arbitrary configuration in bounded domains. With these moments, more realistic cases of immiscible two-phase flows have recently been investigated by Zhang *et al.* [2000a].

6.4 EULERIAN APPROACH

As mentioned in Section 6.1, for hydrological applications the governing equations for two-phase flow are usually formulated as two-pressure equations (6.4) and (6.5) and the capillary mechanism is often of great importance. The two-pressure equations are essentially the generalized version of the Richards equation (5.93). Hence, the moment equation methods introduced in Chapter 5 for unsaturated flow are applicable to the two-pressure equations. These methods are Eulerian, relative to the Lagrangian approach discussed in Section 6.3. In this section, we first utilize an Eulerian perturbative approach to derive moment equations for steady-state two-phase flow and then discuss the special case of unidirectional, uniform mean flow.

6.4.1 Moment Partial Differential Equations

At steady state, the governing flow equations are written from Eqs. (6.4) and (6.5) as

$$\nabla \cdot \{\lambda_w(\mathbf{x})[\nabla P_w(\mathbf{x}) + \rho_w \mathbf{g}]\} = 0 \tag{6.122}$$

$$\nabla \cdot \{\lambda_n(\mathbf{x})[\nabla P_n(\mathbf{x}) + \rho_n \mathbf{g}]\} = 0 \tag{6.123}$$

subject to appropriate boundary conditions. By letting $Y_w(\mathbf{x}) = \ln \lambda_w(\mathbf{x})$ and $Y_n(\mathbf{x}) = \ln \lambda_n(\mathbf{x})$, Eqs. (6.122) and (6.123) can be rewritten as

$$\frac{\partial^2 P_w(\mathbf{x})}{\partial x_i^2} + \frac{\partial Y_w(\mathbf{x})}{\partial x_i}\left[\frac{\partial P_w(\mathbf{x})}{\partial x_i} + \rho_w g \delta_{i1}\right] = 0 \tag{6.124}$$

$$\frac{\partial^2 P_n(\mathbf{x})}{\partial x_i^2} + \frac{\partial Y_n(\mathbf{x})}{\partial x_i}\left[\frac{\partial P_n(\mathbf{x})}{\partial x_i} + \rho_n g \delta_{i1}\right] = 0 \tag{6.125}$$

where summation for repeated indices is implied, g is the gravitational factor, and x_1 is directed upward. As for unsaturated flow, we may express P_i and Y_i in formal series: $P_l(\mathbf{x}) = P_i^{(0)}(\mathbf{x}) + P_i^{(1)}(\mathbf{x}) + P_i^{(2)}(\mathbf{x}) + \cdots$, and $Y_i(\mathbf{x}) = Y_i^{(0)}(\mathbf{x}) + Y_i^{(1)}(\mathbf{x}) + Y_i^{(2)}(\mathbf{x}) + \cdots$. The $Y_i^{(n)}(\mathbf{x})$ term depends on the variabilities of absolute permeability and other parameters characterizing the relative permeability functions, and its explicit forms are given later. It is assumed that the order of $P_i^{(n)}$ is in terms of $Y_i^{(n)}$. Substituting these series into Eqs. (6.124) and (6.125) and collecting terms at the same order, we have

$$\frac{\partial^2 P_w^{(0)}(\mathbf{x})}{\partial x_i^2} + \frac{\partial Y_w^{(0)}(\mathbf{x})}{\partial x_i}\left[\frac{\partial P_w^{(0)}(\mathbf{x})}{\partial x_i} + \rho_w g \delta_{i1}\right] = 0 \tag{6.126}$$

$$\frac{\partial^2 P_n^{(0)}(\mathbf{x})}{\partial x_i^2} + \frac{\partial Y_n^{(0)}(\mathbf{x})}{\partial x_i}\left[\frac{\partial P_n^{(0)}(\mathbf{x})}{\partial x_i} + \rho_n g \delta_{i1}\right] = 0 \tag{6.127}$$

$$\frac{\partial^2 P_w^{(1)}(\mathbf{x})}{\partial x_i^2} + \frac{\partial Y_w^{(1)}(\mathbf{x})}{\partial x_i}\left[\frac{\partial P_w^{(0)}(\mathbf{x})}{\partial x_i} + \rho_w g \delta_{i1}\right] + \frac{\partial Y_w^{(0)}(\mathbf{x})}{\partial x_i}\frac{\partial P_w^{(1)}(\mathbf{x})}{\partial x_i} = 0 \tag{6.128}$$

$$\frac{\partial^2 P_n^{(1)}(\mathbf{x})}{\partial x_i^2} + \frac{\partial Y_n^{(1)}(\mathbf{x})}{\partial x_i}\left[\frac{\partial P_n^{(0)}(\mathbf{x})}{\partial x_i} + \rho_n g \delta_{i1}\right] + \frac{\partial Y_n^{(0)}(\mathbf{x})}{\partial x_i}\frac{\partial P_n^{(1)}(\mathbf{x})}{\partial x_i} = 0 \tag{6.129}$$

Equations governing higher-order terms can be written similarly.

Equations (6.124) and (6.125) or (6.126)–(6.129) are incomplete without specifying the constitutive relations of relative permeabilities k_{ri} versus saturation S or versus capillary pressure P_c. There are no universal theoretical relationships. Instead, empirical relationships are usually used. For simplicity, Chang *et al.* [1995] used the following exponential model:

$$k_{rw} = \exp(-\alpha P_c), \qquad k_{rn} = 1 - \exp(-\alpha P_c) \tag{6.130}$$

where α is a parameter that depends on the fluid and the rock type. The applicability of this simple model is not known. However, we adopt it for simplicity and for ease to discuss the results of Chang *et al.* [1995] and Abdin *et al.* [1995]. In this section, both the absolute permeability $k(\mathbf{x})$ and the two-phase parameter $\alpha(\mathbf{x})$ are treated as random space functions with known statistical properties. With Eq. (6.130), we have

$$Y_w(\mathbf{x}) = \ln \mu_w + f(\mathbf{x}) - \alpha(\mathbf{x})P_c(\mathbf{x}) \tag{6.131}$$

$$Y_n(\mathbf{x}) = \ln \mu_n + f(\mathbf{x}) + \ln\{1 - \exp[\alpha(\mathbf{x})P_c(\mathbf{x})]\} \tag{6.132}$$

where $f = \ln k(\mathbf{x})$. Writing $P_c(\mathbf{x}) = P_c^{(0)}(\mathbf{x}) + P_c^{(1)}(\mathbf{x}) + P_c^{(2)}(\mathbf{x}) + \cdots$, $f(\mathbf{x}) = \langle f \rangle + f'(\mathbf{x})$, and $\alpha(\mathbf{x}) = \langle \alpha \rangle + \alpha'(\mathbf{x})$ (where f and α are treated as second-order stationary random space functions), expanding the last term of Y_n by Taylor expansions around $\langle \alpha \rangle$ and $P_c^{(0)}$, and collecting terms at separate orders, leads to

$$Y_w^{(0)} = \ln \mu_w + \langle f \rangle - \langle \alpha \rangle P_c^{(0)} \tag{6.133}$$

$$Y_n^{(0)} = \ln \mu_n + \langle f \rangle + \ln \left[1 - \exp \left(- \langle \alpha \rangle P_c^{(0)} \right) \right] \tag{6.134}$$

$$Y_w^{(1)} = f' - \langle \alpha \rangle P_c^{(1)} - \alpha' P_c^{(0)} \tag{6.135}$$

$$Y_n^{(1)} = f' + a \left(\langle \alpha \rangle P_c^{(1)} + \alpha' P_c^{(0)} \right) \tag{6.136}$$

where $a(\mathbf{x}) = \exp[-\langle \alpha \rangle P_c^{(0)}(\mathbf{x})]/\{1 - \exp[-\langle \alpha \rangle P_c^{(0)}(\mathbf{x})]\}$.

Substituting Eqs. (6.133) and (6.134) into Eqs. (6.126) and (6.127) and using $P_n^{(0)} = P_w^{(0)} + P_c^{(0)}$ yields

$$\frac{\partial^2 P_w^{(0)}(\mathbf{x})}{\partial x_i^2} - \langle \alpha \rangle \frac{\partial P_c^{(0)}(\mathbf{x})}{\partial x_i} \left[\frac{\partial P_w^{(0)}(\mathbf{x})}{\partial x_i} + \rho_w g \delta_{i1} \right] = 0 \tag{6.137}$$

$$\frac{\partial^2 P_n^{(0)}(\mathbf{x})}{\partial x_i^2} + \langle \alpha \rangle a(\mathbf{x}) \frac{\partial P_c^{(0)}(\mathbf{x})}{\partial x_i} \left[\frac{\partial P_n^{(0)}(\mathbf{x})}{\partial x_i} + \rho_n g \delta_{i1} \right] = 0 \tag{6.138}$$

These two equations are coupled with $P_c^{(0)} = P_n^{(0)} - P_w^{(0)}$. They can be rewritten as

$$\frac{\partial^2 P_w^{(0)}(\mathbf{x})}{\partial x_i^2} - \langle \alpha \rangle [J_{ni}(\mathbf{x}) - J_{wi}(\mathbf{x}) - b_i] \left[\frac{\partial P_w^{(0)}(\mathbf{x})}{\partial x_i} + \rho_w g \delta_{i1} \right] = 0 \tag{6.139}$$

$$\frac{\partial^2 P_n^{(0)}(\mathbf{x})}{\partial x_i^2} + \langle \alpha \rangle a(\mathbf{x})[J_{ni}(\mathbf{x}) - J_{wi}(\mathbf{x}) - b_i] \left[\frac{\partial P_n^{(0)}(\mathbf{x})}{\partial x_i} + \rho_n g \delta_{i1} \right] = 0 \tag{6.140}$$

where $J_{wi}(\mathbf{x}) = (\partial P_w^{(0)}(\mathbf{x})/\partial x_i) + \rho_w g \delta_{i1}$, $J_{ni}(\mathbf{x}) = (\partial P_n^{(0)}(\mathbf{x})/\partial x_i) + \rho_n g \delta_{i1}$, and $b_i = (\rho_n - \rho_w) g \delta_{i1}$.

Substituting Eqs. (6.133)–(6.136) into Eqs. (6.128) and (6.129) leads to

$$\frac{\partial^2 P_w^{(1)}(\mathbf{x})}{\partial x_i^2} + \langle \alpha \rangle [J_{wi}(\mathbf{x}) - J_{ci}(\mathbf{x})] \frac{\partial P_w^{(1)}(\mathbf{x})}{\partial x_i} - \langle \alpha \rangle J_{wi}(\mathbf{x}) \frac{\partial P_n^{(1)}(\mathbf{x})}{\partial x_i}$$

$$= -J_{wi}(\mathbf{x}) \left[\frac{\partial f'(\mathbf{x})}{\partial x_i} - P_c^{(0)}(\mathbf{x}) \frac{\partial \alpha'(\mathbf{x})}{\partial x_i} - J_{ci}(\mathbf{x}) \alpha'(\mathbf{x}) \right] \tag{6.141}$$

$$\frac{\partial^2 P_n^{(1)}(\mathbf{x})}{\partial x_i^2} + \langle \alpha \rangle a(\mathbf{x})[J_{ni}(\mathbf{x}) + J_{ci}(\mathbf{x})] \frac{\partial P_n^{(1)}(\mathbf{x})}{\partial x_i} + \langle \alpha \rangle \frac{\partial a(\mathbf{x})}{\partial x_i} J_{ni}(\mathbf{x}) P_n^{(1)}(\mathbf{x})$$

$$- \langle \alpha \rangle a(\mathbf{x}) J_{ni}(\mathbf{x}) \frac{\partial P_w^{(1)}(\mathbf{x})}{\partial x_i} - \langle \alpha \rangle \frac{\partial a(\mathbf{x})}{\partial x_i} J_{ni}(\mathbf{x}) P_w^{(1)}(\mathbf{x})$$

$$= -J_{ni}(\mathbf{x}) \left\{ \frac{\partial f'(\mathbf{x})}{\partial x_i} + a(\mathbf{x}) P_c^{(0)}(\mathbf{x}) \frac{\partial \alpha'(\mathbf{x})}{\partial x_i} \right.$$

$$\left. + \left[\frac{\partial a(\mathbf{x})}{\partial x_i} P_c^{(0)}(\mathbf{x}) + a(\mathbf{x}) J_{ci}(\mathbf{x}) \right] \alpha'(\mathbf{x}) \right\} \tag{6.142}$$

where $J_{ci} = J_{ni} - J_{wi} + (\rho_w - \rho_n) g \delta_{i1}$.

It can be shown that $\langle P_i^{(0)} \rangle = P_i^{(0)}$ and $\langle P_i^{(1)} \rangle = 0$ where $i = w$ or n. Hence, the mean pressure is $\langle P_i \rangle = P_i^{(0)}$ to zeroth or first order. The pressure fluctuation is $P_i' = P_i^{(1)}$ to first order. Equations for the (first-order) covariances $C_{P_w}(\mathbf{x}, \boldsymbol{\chi}) = \langle P_w^{(1)}(\mathbf{x}) P_w^{(1)}(\boldsymbol{\chi}) \rangle$ and $C_{P_n}(\mathbf{x}, \boldsymbol{\chi}) = \langle P_n^{(1)}(\mathbf{x}) P_n^{(1)}(\boldsymbol{\chi}) \rangle$ are given on the basis of Eqs. (6.141) and (6.142),

$$
\frac{\partial^2 C_{P_w}(\mathbf{x}, \boldsymbol{\chi})}{\partial x_i^2} + \langle \alpha \rangle [J_{wi}(\mathbf{x}) - J_{ci}(\mathbf{x})] \frac{\partial C_{P_w}(\mathbf{x}, \boldsymbol{\chi})}{\partial x_i} - \langle \alpha \rangle J_{wi}(\mathbf{x}) \frac{\partial C_{P_n P_w}(\mathbf{x}, \boldsymbol{\chi})}{\partial x_i}
$$
$$
= -J_{wi}(\mathbf{x}) \left[\frac{\partial C_{f P_w}(\mathbf{x}, \boldsymbol{\chi})}{\partial x_i} - P_c^{(0)}(\mathbf{x}) \frac{\partial C_{\alpha P_w}(\mathbf{x}, \boldsymbol{\chi})}{\partial x_i} - J_{ci}(\mathbf{x}) C_{\alpha P_w}(\mathbf{x}, \boldsymbol{\chi}) \right] \quad (6.143)
$$

$$
\frac{\partial^2 C_{P_n}(\mathbf{x}, \boldsymbol{\chi})}{\partial x_i^2} + \langle \alpha \rangle a(\mathbf{x}) [J_{ni}(\mathbf{x}) + J_{ci}(\mathbf{x})] \frac{\partial C_{P_n}(\mathbf{x}, \boldsymbol{\chi})}{\partial x_i}
$$
$$
+ \langle \alpha \rangle \frac{\partial a(\mathbf{x})}{\partial x_i} J_{ni}(\mathbf{x}) C_{P_n}(\mathbf{x}) - \langle \alpha \rangle a(\mathbf{x}) J_{ni}(\mathbf{x}) \frac{\partial C_{P_w P_n}(\mathbf{x}, \boldsymbol{\chi})}{\partial x_i} - \langle \alpha \rangle \frac{\partial a(\mathbf{x})}{\partial x_i} J_{ni}(\mathbf{x}) C_{P_w P_n}(\mathbf{x}, \boldsymbol{\chi})
$$
$$
= -J_{ni}(\mathbf{x}) \left\{ \frac{\partial C_{f P_n}(\mathbf{x}, \boldsymbol{\chi})}{\partial x_i} + a(\mathbf{x}) P_c^{(0)}(\mathbf{x}) \frac{\partial C_{\alpha P_n}(\mathbf{x}, \boldsymbol{\chi})}{\partial x_i} \right.
$$
$$
\left. + \left[\frac{\partial a(\mathbf{x})}{\partial x_i} P_c^{(0)}(\mathbf{x}) + a(\mathbf{x}) J_{ci}(\mathbf{x}) \right] C_{\alpha P_n}(\mathbf{x}, \boldsymbol{\chi}) \right\} \quad (6.144)
$$

where the cross-covariances $C_{f P_w}(\mathbf{x}, \boldsymbol{\chi})$, $C_{f P_n}(\mathbf{x}, \boldsymbol{\chi})$, $C_{\alpha P_w}(\mathbf{x}, \boldsymbol{\chi})$, and $C_{f P_n}(\mathbf{x}, \boldsymbol{\chi})$ are governed by the following equations:

$$
\frac{\partial^2 C_{f P_w}(\mathbf{x}, \boldsymbol{\chi})}{\partial \chi_i^2} + \langle \alpha \rangle [J_{wi}(\boldsymbol{\chi}) - J_{ci}(\boldsymbol{\chi})] \frac{\partial C_{f P_w}(\mathbf{x}, \boldsymbol{\chi})}{\partial \chi_i} - \langle \alpha \rangle J_{wi}(\boldsymbol{\chi}) \frac{\partial C_{f P_n}(\mathbf{x}, \boldsymbol{\chi})}{\partial \chi_i}
$$
$$
= -J_{wi}(\boldsymbol{\chi}) \left[\frac{\partial C_f(\mathbf{x}, \boldsymbol{\chi})}{\partial \chi_i} - P_c^{(0)}(\boldsymbol{\chi}) \frac{\partial C_{f\alpha}(\mathbf{x}, \boldsymbol{\chi})}{\partial \chi_i} - J_{ci}(\boldsymbol{\chi}) C_{f\alpha}(\mathbf{x}, \boldsymbol{\chi}) \right] \quad (6.145)
$$

$$
\frac{\partial^2 C_{f P_n}(\mathbf{x}, \boldsymbol{\chi})}{\partial \chi_i^2} + \langle \alpha \rangle a(\boldsymbol{\chi}) [J_{ci}(\boldsymbol{\chi}) + J_{ni}(\boldsymbol{\chi})] \frac{\partial C_{f P_n}(\mathbf{x}, \boldsymbol{\chi})}{\partial \chi_i}
$$
$$
+ \langle \alpha \rangle \frac{\partial a(\boldsymbol{\chi})}{\partial \chi_i} J_{ni}(\boldsymbol{\chi}) C_{f P_n}(\mathbf{x}, \boldsymbol{\chi}) - \langle \alpha \rangle a(\boldsymbol{\chi}) J_{ni}(\boldsymbol{\chi}) \frac{\partial C_{f P_w}(\mathbf{x}, \boldsymbol{\chi})}{\partial \chi_i}
$$
$$
- \langle \alpha \rangle \frac{\partial a(\boldsymbol{\chi})}{\partial \chi_i} J_{ni}(\boldsymbol{\chi}) C_{f P_w}(\mathbf{x}, \boldsymbol{\chi})
$$
$$
= -J_{ni}(\boldsymbol{\chi}) \left\{ \frac{\partial C_f(\mathbf{x}, \boldsymbol{\chi})}{\partial \chi_i} + a(\boldsymbol{\chi}) P_c^{(0)}(\boldsymbol{\chi}) \frac{\partial C_{f\alpha}(\mathbf{x}, \boldsymbol{\chi})}{\partial \chi_i} \right.
$$
$$
\left. + \left[\frac{\partial a(\boldsymbol{\chi})}{\partial \chi_i} P_c^{(0)}(\boldsymbol{\chi}) + a(\boldsymbol{\chi}) J_{ci}(\boldsymbol{\chi}) \right] C_{f\alpha}(\mathbf{x}, \boldsymbol{\chi}) \right\} \quad (6.146)
$$

$$\frac{\partial^2 C_{\alpha P_w}(\mathbf{x}, \boldsymbol{\chi})}{\partial \chi_i^2} + \langle\alpha\rangle[J_{wi}(\boldsymbol{\chi}) - J_{ci}(\boldsymbol{\chi})]\frac{\partial C_{\alpha P_w}(\mathbf{x}, \boldsymbol{\chi})}{\partial \chi_i} - \langle\alpha\rangle J_{wi}(\boldsymbol{\chi})\frac{\partial C_{\alpha P_n}(\mathbf{x}, \boldsymbol{\chi})}{\partial \chi_i}$$

$$= -J_{wi}(\boldsymbol{\chi})\left[\frac{\partial C_{\alpha f}(\mathbf{x}, \boldsymbol{\chi})}{\partial \chi_i} - P_c^{(0)}(\boldsymbol{\chi})\frac{\partial C_{\alpha}(\mathbf{x}, \boldsymbol{\chi})}{\partial \chi_i} - J_{ci}(\boldsymbol{\chi})C_{\alpha}(\mathbf{x}, \boldsymbol{\chi})\right] \qquad (6.147)$$

$$\frac{\partial^2 C_{\alpha P_n}(\mathbf{x}, \boldsymbol{\chi})}{\partial \chi_i^2} + \langle\alpha\rangle a(\boldsymbol{\chi})[J_{ci}(\boldsymbol{\chi}) + J_{ni}(\boldsymbol{\chi})]\frac{\partial C_{\alpha P_n}(\mathbf{x}, \boldsymbol{\chi})}{\partial \chi_i}$$

$$+ \langle\alpha\rangle\frac{\partial a(\boldsymbol{\chi})}{\partial \chi_i}J_{ni}(\boldsymbol{\chi})C_{\alpha P_n}(\mathbf{x}, \boldsymbol{\chi}) - \langle\alpha\rangle a(\boldsymbol{\chi})J_{ni}(\boldsymbol{\chi})\frac{\partial C_{\alpha P_w}(\mathbf{x}, \boldsymbol{\chi})}{\partial \chi_i}$$

$$- \langle\alpha\rangle\frac{\partial a(\boldsymbol{\chi})}{\partial \chi_i}J_{ni}(\boldsymbol{\chi})C_{\alpha P_w}(\mathbf{x}, \boldsymbol{\chi})$$

$$= -J_{ni}(\boldsymbol{\chi})\left\{\frac{\partial C_{\alpha f}(\mathbf{x}, \boldsymbol{\chi})}{\partial \chi_i} + a(\boldsymbol{\chi})P_c^{(0)}(\boldsymbol{\chi})\frac{\partial C_{\alpha}(\mathbf{x}, \boldsymbol{\chi})}{\partial \chi_i}\right.$$

$$\left. + \left[\frac{\partial a(\boldsymbol{\chi})}{\partial \chi_i}P_c^{(0)}(\boldsymbol{\chi}) + a(\boldsymbol{\chi})J_{ci}(\boldsymbol{\chi})\right]C_{\alpha}(\mathbf{x}, \boldsymbol{\chi})\right\} \qquad (6.148)$$

Note that $C_f(\mathbf{x}, \boldsymbol{\chi}) = C_f(\mathbf{x}-\boldsymbol{\chi}), C_{\alpha}(\mathbf{x}, \boldsymbol{\chi}) = C_{\alpha}(\mathbf{x}-\boldsymbol{\chi}), C_{f\alpha}(\mathbf{x}, \boldsymbol{\chi}) = C_{f\alpha}(\mathbf{x}-\boldsymbol{\chi})$ are given as input and that for any random fields p and q, $C_{pq}(\mathbf{x}, \boldsymbol{\chi}) = C_{qp}(\boldsymbol{\chi}, \mathbf{x})$. In Eqs. (6.143) and (6.144), the cross-covariance $C_{P_n P_w}(\mathbf{x}, \boldsymbol{\chi})$ can be solved from

$$\frac{\partial^2 C_{P_n P_w}(\mathbf{x}, \boldsymbol{\chi})}{\partial \chi_i^2} + \langle\alpha\rangle[J_{wi}(\boldsymbol{\chi}) - J_{ci}(\boldsymbol{\chi})]\frac{\partial C_{P_n P_w}(\mathbf{x}, \boldsymbol{\chi})}{\partial \chi_i} - \langle\alpha\rangle J_{wi}(\boldsymbol{\chi})\frac{\partial C_{P_n}(\mathbf{x}, \boldsymbol{\chi})}{\partial \chi_i}$$

$$= -J_{wi}(\boldsymbol{\chi})\left[\frac{\partial C_{f P_n}(\boldsymbol{\chi}, \mathbf{x})}{\partial \chi_i} - P_c^{(0)}(\boldsymbol{\chi})\frac{\partial C_{\alpha P_n}(\boldsymbol{\chi}, \mathbf{x})}{\partial \chi_i} - J_{ci}(\boldsymbol{\chi})C_{\alpha P_n}(\boldsymbol{\chi}, \mathbf{x})\right] \qquad (6.149)$$

These moment equations need to be solved numerically in some iterative manner. After the mean pressures $P_w^{(0)}$ and $P_n^{(0)}$ are solved from the coupled equations (6.139) and (6.140), the cross-covariances $C_{f P_w}$ and $C_{f P_n}$ can be solved from the coupled equations (6.145) and (6.146) as $C_{\alpha P_w}$ and $C_{\alpha P_n}$ are solved from Eqs. (6.147) and (6.148). Then, the pressure covariances C_{P_w}, C_{P_n}, and $C_{P_n P_w}$ can be solved from the three coupled equations, Eqs. (6.143), (6.144), and (6.149). It is thus seen that these moment equations are less strongly coupled than Eqs. (6.50)–(6.58) resulting from the Eulerian approach on the basis of the fractional flow representation. To the writer's knowledge, the two-pressure-based moment equations have not been solved except for the case of uniform mean flow in unbounded domains. In the next section, we discuss such a special case.

6.4.2 Unidirectional, Uniform Mean Flow

Chang *et al.* [1995] considered the case of unidirectional, uniform mean flow with zero mean capillary pressure gradient: $\partial P_c^{(0)}/\partial x_i = 0$. For such a flow, $J_{wi} = \partial P_w^{(0)}/\partial x_i + \rho_w g \delta_{i1} = J_w \delta_{i1}$ and $J_{ni} = \partial P_n^{(0)}/\partial x_i + \rho_w g \delta_{i1} = J_n \delta_{i1}$, where J_w and J_n are constants.

Thus, $P_c^{(0)} = P_n^{(0)}(\mathbf{x}) - P_w^{(0)}(\mathbf{x})$ are uniform throughout the domain, while $P_n^{(0)}(\mathbf{x})$ and $P_w^{(0)}(\mathbf{x})$ may vary linearly in the vertical (x_1) direction. It is clear that under these conditions, $\partial P_w^{(0)}/\partial x_1 = \partial P_n^{(0)}/\partial x_1$. In turn, the (first-order) pressure fluctuation equations (6.141) and (6.142) can be simplified as

$$\frac{\partial^2 P_w^{(1)}(\mathbf{x})}{\partial x_i^2} + \langle\alpha\rangle J_w \frac{\partial P_w^{(1)}(\mathbf{x})}{\partial x_1} - \langle\alpha\rangle J_w \frac{\partial P_n^{(1)}(\mathbf{x})}{\partial x_1} = -J_w \left[\frac{\partial f'(\mathbf{x})}{\partial x_1} - P_c^{(0)} \frac{\partial \alpha'(\mathbf{x})}{\partial x_1} \right]$$
(6.150)

$$\frac{\partial^2 P_n^{(1)}(\mathbf{x})}{\partial x_i^2} + \langle\alpha\rangle a J_n \frac{\partial P_n^{(1)}(\mathbf{x})}{\partial x_1} - a\langle\alpha\rangle J_n \frac{\partial P_w^{(1)}(\mathbf{x})}{\partial x_1} = -J_n \left[\frac{\partial f'(\mathbf{x})}{\partial x_1} + a P_c^{(0)} \frac{\partial \alpha'(\mathbf{x})}{\partial x_1} \right]$$
(6.151)

Note that Eq. (6.150) is exactly the same as Eq. (8a) of Chang *et al.* [1995] under these conditions. Equation (6.151) would be the same as Eq. (8b) of Chang *et al.* [1995] if the term $a = \exp(-\langle\alpha\rangle P_c^{(0)})/[1 - \exp(-\langle\alpha\rangle P_c^{(0)})]$ is expanded by Taylor series. The latter expansion requires the product of the two mean quantities $\langle\alpha\rangle$ and $P_c^{(0)}$ to be sufficiently small. However, this expansion is avoided when the procedure outlined in Section 6.4.1 is utilized. With the further simplification of unbounded domains, the pressures become spatially stationary with their means being constant and their covariances dependent on the separation distance. The covariance equations of Eqs. (6.143)–(6.149) can also be simplified significantly. The resulting stationary covariance equations may be solved analytically.

As discussed in Section 3.6, the stationary spectral method provides a convenient way to formulate and solve moment equations in the case of stationary fluctuations. Chang *et al.* [1995] took advantage of this method in their work. In particular, the fluctuations f', α', $P_w' = P_w^{(1)}$, and $P_n' = P_n^{(1)}$ are expressed by the following stochastic Fourier–Stieltjes integral representations:

$$f'(\mathbf{x}) = \int \exp(\iota\mathbf{k} \cdot \mathbf{x}) \, dZ_f(\mathbf{k})$$
(6.152)

$$\alpha'(\mathbf{x}) = \int \exp(\iota\mathbf{k} \cdot \mathbf{x}) \, dZ_\alpha(\mathbf{k})$$
(6.153)

$$P_w'(\mathbf{x}) = \int \exp(\iota\mathbf{k} \cdot \mathbf{x}) \, dZ_{P_w}(\mathbf{k})$$
(6.154)

$$P_n'(\mathbf{x}) = \int \exp(\iota\mathbf{k} \cdot \mathbf{x}) \, dZ_{P_n}(\mathbf{k})$$
(6.155)

where $\mathbf{k} = (k_1, \ldots, k_d)^T$ is the wave number space vector (where d is the number of space dimensions), $\iota \equiv \sqrt{-1}$, and $dZ_f(\mathbf{k})$, $dZ_\alpha(\mathbf{k})$, $dZ_{P_w}(\mathbf{k})$, and $dZ_{P_c}(\mathbf{k})$ are the complex Fourier increments of the fluctuations at $\mathbf{k}$. The integration is d-fold from $-\infty$ to ∞. The properties of the stochastic Fourier–Stieltjes integral are discussed in Section 2.4.4. Of immediate use is the orthogonality property,

$$\langle dZ_p(\mathbf{k}) dZ_q^*(\mathbf{k}') \rangle = \delta(\mathbf{k} - \mathbf{k}') S_{pq}(\mathbf{k}) \, d\mathbf{k} \, d\mathbf{k}' \tag{6.156}$$

where $p, q = f, \alpha, P_w$ or P_n, dZ_q^* is the complex conjugate of dZ_q, and S_{pq} is the (auto- or cross-) spectrum or spectral density of p and q.

Substituting Eqs. (6.152)–(6.155) into Eqs. (6.150) and (6.151) results in

$$dZ_{P_w}(\mathbf{k}) = \frac{J_w k_1 \left\{ [\iota k^2 + (a+1)\langle\alpha\rangle J_n k_1] dZ_f(\mathbf{k}) - \iota P_c k^2 dZ_\alpha(\mathbf{k}) \right\}}{k^4 - i\langle\alpha\rangle(a J_n + J_w) k_1 k^2} \tag{6.157}$$

$$dZ_{P_n}(\mathbf{k}) = \frac{J_n k_1 \left\{ [\iota k^2 + (a+1)\langle\alpha\rangle J_w k_1] dZ_f(\mathbf{k}) + \iota a P_c k^2 dZ_\alpha(\mathbf{k}) \right\}}{k^4 - i\langle\alpha\rangle(a J_n + J_w) k_1 k^2} \tag{6.158}$$

The spectral density functions $S_{P_w}(\mathbf{k})$, $S_{P_n}(\mathbf{k})$, and $S_{P_n P_w}(\mathbf{k})$ can be obtained from Eqs. (6.157) and (6.158) with the orthogonality property of Eq. (6.156). These spectral expressions are in terms of the spectral density functions $S_f(\mathbf{k})$, $S_\alpha(\mathbf{k})$, and $S_{f\alpha}(\mathbf{k})$, which are the Fourier transforms of the input covariances $C_f(\mathbf{r})$, $C_\alpha(\mathbf{r})$, and $C_{f\alpha}(\mathbf{r})$ (where $\mathbf{r} = \chi - \chi$). The covariances $C_{P_w}(\mathbf{r})$, $C_{P_n}(\mathbf{r})$, and $C_{P_n P_w}(\mathbf{r})$ are obtained by inverse Fourier transforms.

Chang et al. [1995] evaluated the pressure variances, $\sigma_{P_w}^2$ and $\sigma_{P_n}^2$, with the spectral method for uniform mean flow in one dimension with hole covariances for f and α and in three dimensions with exponential covariances. They considered the following three cases for cross-correlation of f and α: (a) α is a deterministic constant, (b) f and α are linearly correlated, and (c) they are uncorrelated. Abdin et al. [1995] performed Monte Carlo simulations of two-phase flow under similar conditions albeit in bounded domains. It was found that under their specific boundary conditions, the mean capillary pressure is constant except for the regions near the boundaries. The agreement between the Monte Carlo simulations and the spectral results of Chang et al. [1995] is good when the variabilities of f and α are small (e.g., $\sigma_f = 0.1$ and 0.5), and the agreement deteriorates as the variabilities increase.

6.5 EXERCISES

1. Show that in the presence of capillarity and gravity, the wetting phase fluid flux $\mathbf{q}_w$ is related to the total flux $\mathbf{q}$ and the fractional flow function f_w via Eq. (6.13).

2. For the linear relative permeability model of $k_{rw} = S$ and $k_{rn} = 1 - S$, the fractional flow function is

$$f_w = \frac{mS}{mS + (1 - S)} \tag{6.159}$$

where $m = \mu_n/\mu_w$ is the viscosity ratio. In addition, for the case of unit viscosity ratio (i.e., $m = 1$), $f_w = S$. Thus the first derivative of f_w is one and the second derivative is zero. Show that under these conditions, the mean saturation equation (6.60) is linear and reduces to that for solute transport in a random velocity field.

3. For the case of uniform mean flow, derive the auto- and cross-spectral density functions $S_{P_w}(\mathbf{k})$, $S_{P_n}(\mathbf{k})$, $S_{P_n P_w}(\mathbf{k})$, $S_{f P_w}(\mathbf{k})$, and $S_{\alpha P_w}(\mathbf{k})$ with Eqs. (6.157) and (6.158). Invert these spectral expressions analytically or with fast Fourier transform to obtain the pressure covariances $C_{P_w}(\mathbf{r})$, $C_{P_n}(\mathbf{r})$, and $C_{P_n P_w}(\mathbf{r})$.

4. For steady-state two-phase flow, the phase fluxes $\mathbf{q}_w$ and $\mathbf{q}_n$ can be expanded in formal series as $\mathbf{q}_i = \mathbf{q}_i^{(0)} + \mathbf{q}_i^{(1)} + \cdots$. Find the explicit expressions for $\mathbf{q}_w^{(0)}$, $\mathbf{q}_n^{(0)}$, $\mathbf{q}_w^{(1)}$, and $\mathbf{q}_n^{(1)}$, and apply the stochastic Fourier–Stieltjes integral representations of Eqs. (6.152)–(6.155) to derive spectral expressions for the flux moments.

5. For transient two-phase flow, the two-pressure representation of flow is Eqs. (6.4) and (6.5). Try to formulate moment equations for P_w and P_n by the Eulerian approach discussed in Section 6.4. Compare them with their steady-state counterparts and outline a procedure to solve them.

7

FLOW IN FRACTURED
POROUS MEDIA

7.1 INTRODUCTION

Discrete features such as fractures and macropores may greatly impact subsurface flow and transport in porous media. In such media, flow and transport usually take place preferentially through the discrete features. Therefore, these features are crucial for understanding and modeling subsurface flow and transport, although they may only occupy a small fraction of the media. In fractured porous media, flow may occur predominantly in an interconnected network of fractures while most of the fluid storage takes place in the relatively low permeability matrix blocks. Fluid flow and solute transport in fractured porous media may be described by either explicitly or implicitly accounting for the discrete fractures. In Section 7.1.1, we discuss some discrete fracture network models, in which fracture networks are considered explicitly; in Section 7.1.2, we introduce dual-porosity representations of the fractured porous media, in which the fractures are accounted for implicitly.

7.1.1 Discrete Network Models

Discrete fracture network models are largely motivated by the fact that some fractures can be identified and mapped in underground openings, boreholes, and surface outcrops. Such models attempt to delineate the geometry of the subsurface flow system explicitly. They usually consist of discrete polygonal or oval-shaped planes of finite sizes embedded in a rock matrix.

Because it is impossible to map out the detailed geometry and measure other characteristics of fractures at a field scale in the subsurface, fracture networks are usually generated on the basis of observations and measurements in outcrops, boreholes, and subsurface openings. Besides the geometry of the fracture planes, fracture hydraulic properties must be assigned stochastically or determined based on other information such as aperture distribution data. There are a number of techniques for generating fracture networks

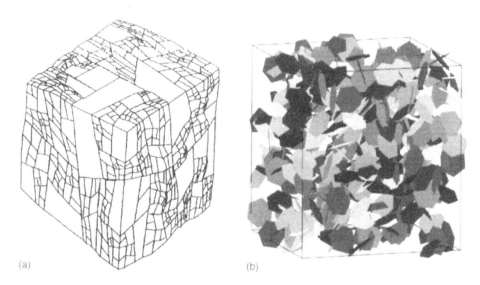

(a)

(b)

Figure 7.1. Generated fracture networks. (a) Fractal fracture network. (Adapted from Acuna and Yortsos [1995]. Copyright 1995 by the American Geophysical Union.) (b) Random plane polygonal fractures. (Adapted from Koudina *et al.* [1998]. Copyright 1998 by the American Physical Society.)

including stochastic pattern generation [Long *et al.*, 1982; Dershowitz, 1984; Billaux *et al.*, 1989], simulated annealing [Long *et al.*, 1991], fractal fracture network models [Barton and Hsieh, 1989; Acuna and Yortsos, 1995], and mechanical fracture models. Reviews of such techniques have recently been given by Chiles and de Marsily [1993] and Sahimi [1995]. Figure 7.1 shows two synthetic fracture networks.

To numerically model fluid flow and solute transport in a discrete fracture network, one needs to account for each fracture and each matrix block in the computational mesh. In practice, this is difficult, if not impossible, for fairly densely fractured media due to the large number of mesh grids required. For example, Zimmerman *et al.* [1996] estimated the number of fractures in the welded tuffs to be on the order of 10^9 in the vicinity of the potential radioactive waste repository site at Yucca Mountain, Nevada. It is because of their high computational requirements that discrete fracture network models were usually used on a scale of a few tens of meters in fractured rock [e.g., Dverstorp and Andersson, 1989; Cacas *et al.*, 1990; Dverstorp *et al.*, 1992].

7.1.2 Dual-Porosity/Double-Permeability Models

The conceptual framework behind the discrete fracture modeling approach as well as its practicality for modeling a field scale site has long been debated in the literature [e.g., Neuman, 1987, 1988; Tsang and Neuman, 1997]. Besides being a computational issue, the practicality of the fracture network models is often questioned on the grounds that existing field techniques are inadequate for measuring with any reasonable degree of fidelity either the geometry of the subsurface fracture system or the hydraulic properties of its individual components. As such, many approaches have been developed to account for the fractures implicitly. The most popular is the "dual-porosity" model [Barenblatt *et al.*, 1960; Warren and Root, 1963]. In this approach, the fractured media are represented by two completely overlapping continua, one representing the fractures and the other representing the porous matrix. The two continua exchange fluids and solutes through a coupling term due to pressure head and concentration gradients. Therefore, at any point there are two pressure heads, two flow velocities, two water contents, and two solute concentrations, one each for these two overlapping continua.

In typical fractured media, the matrix blocks have high porosity for fluid storage, while the fractures have high permeabilities. Hence, in some "dual-porosity" models the matrix material is constrained to communicate only with the neighboring fractures [Bibby, 1981; Moench, 1984; Zimmerman *et al.*, 1993]. In some other "dual-porosity" models, flow is allowed not only in the fractures but also within the matrix blocks [Duguid and Lee, 1977; Gerke and van Genuchten, 1993a,b]. In this book, we make a distinction between these two types of models, as also done by others [e.g., Zyvoloski *et al.*, 1995; Ho *et al.*, 1995]. The first type is termed the *dual-porosity approach*, and the second is called the double-porosity/double-permeability approach, or, the *double-permeability approach* for brevity. In both the dual-porosity and the double-permeability models, different pressures can exist in the fractures and matrix, which allow for flow to occur between the two continua. The difference is that in the dual-porosity model, flow is not allowed within the matrix. Thus, the dual-porosity model can be considered as a special case of the double-permeability model. Another special case of the latter is the so-called *equivalent continuum model*, which is obtained when the pressures in fractures and matrix are at equilibrium [Dykhuizen, 1987; Peters and Klavetter, 1988; Pruess *et al.*, 1990]. The assumption of pressure equilibrium implies that the resistance to flow between fractures and matrix is negligible. In the equivalent continuum model, the flow through a fracture–matrix system is simplified as flow through a *composite* porous medium whose hydraulic conductivity and storage coefficient (or moisture capacity) are represented by the weighted sum of the contributions from the fractures and matrix blocks.

In the rest of this chapter, we discuss double-permeability models for both saturated and unsaturated flows and develop stochastic moment equations on the basis of them. Some numerical examples are used to illustrate the developed stochastic continuum models.

7.2 SATURATED FLOW

In this section, a double-permeability model is adopted to describe single-phase flow in a fractured medium. Assuming that Darcy's law is valid in both the low-permeability pore system (matrix) and the high-permeability pore system (fracture), flow in the double-permeability medium satisfies the following equations [e.g., Gerke and van Genuchten, 1993a]:

$$w_f(\mathbf{x})S_f \frac{\partial h_f(\mathbf{x}, t)}{\partial t} = \nabla \cdot [w_f(\mathbf{x})K_{Sf}(\mathbf{x})\nabla h_f(\mathbf{x}, t)] - Q_w(\mathbf{x}, t) \tag{7.1}$$

$$w_m(\mathbf{x})S_m \frac{\partial h_m(\mathbf{x}, t)}{\partial t} = \nabla \cdot [w_m(\mathbf{x})K_{Sm}(\mathbf{x})\nabla h_m(\mathbf{x}, t)] + Q_w(\mathbf{x}, t) \tag{7.2}$$

where S_p ($p = f$ or m with f standing for fracture and m for matrix) represents the specific storage coefficient (L^{-1}) and is assumed to be constant in the current problem, $h_p(\mathbf{x}, t)$ is the total head (L), $K_{Sp}(\mathbf{x})$ is the hydraulic conductivity, $w_p(\mathbf{x})$ is the relative volume fraction of each pore system and $Q_w(\mathbf{x}, t)$ is a coupling term for describing fluid flow between the fracture and the matrix. There are two general ways to describe the coupling (exchange) term: the quasi-steady state formulation [e.g., Barenblatt *et al.*, 1960; Warren and Root, 1963; Gerke and van Genuchten, 1993a,b; Zyvoloski *et al.*, 1995] and the transient formulation [e.g., Kazemi, 1969; Moench, 1984; Dykhuizen, 1990; Zimmerman *et al.*, 1993]. In this work, we will adopt the former, in which the coupling term is assumed to be proportional to the difference in the pressure heads between the fracture and the matrix:

$$Q_w(\mathbf{x}, t) = b(\mathbf{x})[h_f(\mathbf{x}, t) - h_m(\mathbf{x}, t)] \tag{7.3}$$

where $b(\mathbf{x})$ is a first-order *fluid transfer coefficient* [Gerke and van Genuchten, 1996a,b] defined as

$$b(\mathbf{x}) = BK_{Sm}(\mathbf{x}), \quad B = \frac{\iota}{a^2}\gamma \tag{7.4}$$

In Eq. (7.4), ι is related to the geometry of matrix block, a is the characteristic half-length of the matrix block, and γ is an empirical scaling factor. We will regard B as deterministic in this work. Taking the logarithm of both sides of the first equation in Eq. (7.4), we have

$$\beta(\mathbf{x}) = \ln b(\mathbf{x}) = \ln B + \ln K_{Sm}(\mathbf{x}) \tag{7.5}$$

Thus, Eq. (7.3) can also be written as

$$Q_w(\mathbf{x}, t) = \exp[\beta(\mathbf{x})][h_f(\mathbf{x}, t) - h_m(\mathbf{x}, t)] \tag{7.6}$$

The water fluxes corresponding to Eqs. (7.1) and (7.2) are

$$\mathbf{q}_f(\mathbf{x}, t) = -K_{Sf}(\mathbf{x})\nabla h_f(\mathbf{x}, t) \tag{7.7}$$

$$\mathbf{q}_m(\mathbf{x}, t) = -K_{Sm}(\mathbf{x})\nabla h_m(\mathbf{x}, t) \tag{7.8}$$

The *composite flux* is given as $\mathbf{q}(\mathbf{x}, t) = w_f\mathbf{q}_f(\mathbf{x}, t) + w_m\mathbf{q}_m(\mathbf{x}, t)$.

In this section, we treat the saturated hydraulic conductivity $K_{Sp}(\mathbf{x})$, the relative volume fraction $w_p(\mathbf{x})$, and the fluid transfer coefficient $b(\mathbf{x})$ as random space functions. Hence, Eqs. (7.1) and (7.2) become stochastic partial differential equations. In the next subsection, we derive equations governing the statistical moments of flow quantities in the fracture and matrix systems on the basis of these stochastic equations.

7.2.1 Moment Partial Differential Equations

We can rewrite Eqs. (7.1) and (7.2) as

$$\exp[-Y_p]\left[S_p^*\frac{\partial h_p}{\partial t} + Q_p(\mathbf{x}, t)\right] = \frac{\partial^2 h_p}{\partial x_i^2} + \frac{\partial Y_p}{\partial x_i}\frac{\partial h_p}{\partial x_i} \tag{7.9}$$

subject to the following initial and boundary conditions:

$$h_p(\mathbf{x}, 0) = H_p^o(\mathbf{x}), \qquad \mathbf{x} \in \Omega \tag{7.10}$$

$$h_p(\mathbf{x}, t) = H_p(\mathbf{x}, t), \quad \mathbf{x} \in \Gamma_D \tag{7.11}$$

$$-n_i(\mathbf{x})K_{Sp}(\mathbf{x})\left[\frac{\partial h_p(\mathbf{x}, t)}{\partial x_i}\right] = N_p(\mathbf{x}, t), \quad \mathbf{x} \in \Gamma_N \tag{7.12}$$

where $Q_p = Q_w$ when $p = f$, $Q_p = -Q_w$ when $p = m$, $S_p^* = w_p S_p$ denotes the specific storage coefficient which is scaled by the relative volume fraction, and Y_p is defined as

$$Y_p(\mathbf{x}) = \ln[w_p(\mathbf{x})K_{Sp}(\mathbf{x})] = g_p(\mathbf{x}) + f_p(\mathbf{x}) \tag{7.13}$$

where $g_p(\mathbf{x}) = \ln w_p(\mathbf{x})$ and $f_p(\mathbf{x}) = \ln K_{Sp}(\mathbf{x})$. In Eqs. (7.10)–(7.12), $H_p^o(\mathbf{x})$ is the initial head distribution for medium p in the domain Ω, $H_p(\mathbf{x}, t)$ is the prescribed head distribution on the Dirichlet boundary Γ_D, $n_i(\mathbf{x})$ is the i-component of an outward unit vector normal to the Neumann boundary Γ_N, and $N_p(\mathbf{x}, t)$ is the prescribed flux across Γ_N. In the current problem, $N_p(\mathbf{x}, t)$ is positive when it is recharge and negative otherwise. In addition, we assume that the boundary conditions are deterministic. In Eqs. (7.9)–(7.12), Einstein summation is implied for i but not for p.

For the following analysis, we write $Y_p(\mathbf{x})$ and $S_p^*(\mathbf{x})$ as the sums of their means and their fluctuations,

$$Y_p = \langle Y_p \rangle + Y_p',$$
$$S_p^* = \langle S_p^* \rangle + S_p^{*'} \tag{7.14}$$

where $\langle\ \rangle$ represents ensemble mean and the terms with a prime represent zero-mean fluctuations. On the basis of the definition of Y_p and S_p^* we have

$$\langle Y_p \rangle = \langle g_p \rangle + \langle f_p \rangle$$

$$Y_p' = g_p' + f_p'$$

$$\langle S_p^* \rangle = S_p \exp[\langle g_p \rangle] = S_p G_p \qquad (7.15)$$

$$S_p^{*'} = S_p G_p g_p' + \cdots$$

where $G_p = \exp[\langle g_p \rangle]$ is the geometric mean of the volume fraction w_p and $K_{Gb} = \exp[\langle \beta \rangle] = B \exp[\langle f_m \rangle]$ is the geometric mean of the fluid transfer coefficient b. It can be seen from Eq. (7.9) that the variability of h_p depends on those of Y_f, S_f^*, Y_m, and S_m^*. We may express $h_p(\mathbf{x}, t)$ in an infinite series in the following form:

$$h_p = h_p^{(0)} + h_p^{(1)} + h_p^{(2)} + \cdots \qquad (7.16)$$

The terms of the right-hand side of Eq. (7.16) are in terms of the variability of fracture and matrix properties. With this, we have

$$Q_p = Q_p^{(0)} + Q_p^{(1)} + Q_p^{(2)} + \cdots$$

$$Q_p^{(0)} = \exp[\langle \beta \rangle][h_p^{(0)} - h_q^{(0)}] = K_{Gb}[h_p^{(0)} - h_q^{(0)}] \qquad (7.17)$$

$$Q_p^{(1)} = K_{Gb}\{[h_p^{(1)} - h_q^{(1)}] + [h_p^{(0)} - h_q^{(0)}]\beta'\}$$

where $p = f$ or m, $q = m$ or f, but p and q are not equal. Here and subsequently, we assume, for simplicity, that the random variables $g_p(\mathbf{x})$, $f_p(\mathbf{x})$, and $\beta(\mathbf{x})$ are second-order stationary. Substituting Eqs. (7.14)–(7.17) into Eqs. (7.9)–(7.12) and collecting terms at different orders yields

$$\frac{\partial^2 h_p^{(0)}(\mathbf{x}, t)}{\partial x_i^2} - K_{GY_p}^{-1} K_{Gb} h_p^{(0)}(\mathbf{x}, t) - K_{GY_p}^{-1}\langle S_p^* \rangle \frac{\partial h_p^{(0)}(\mathbf{x}, t)}{\partial t} = -K_{GY_p}^{-1} K_{Gb} h_q^{(0)}(\mathbf{x}, t)$$

$$h_p^{(0)}(\mathbf{x}, 0) = H_p^o(\mathbf{x}), \quad \mathbf{x} \in \Omega$$

$$h_p^{(0)}(\mathbf{x}, t) = H_p(\mathbf{x}, t), \quad \mathbf{x} \in \Gamma_D \qquad (7.18)$$

$$-n_i(\mathbf{x}) K_{Gf_p}\left[\frac{\partial h_p^{(0)}(\mathbf{x}, t)}{\partial x_i}\right] = N_p(\mathbf{x}, t), \quad \mathbf{x} \in \Gamma_N$$

and

$$
\frac{\partial^2 h_p^{(1)}}{\partial x_i^2} - J_{pi} \frac{\partial}{\partial x_i} [g_p' + f_p']
$$

$$
= K_{GY_p}^{-1} \left\{ \langle S_p^* \rangle \frac{\partial h_p^{(1)}}{\partial t} - \langle S_p^* \rangle J_{pt} f_p' - K_{Gb} [h_p^{(0)} - h_q^{(0)}](g_p' + f_p' - \beta') \right.
$$

$$
\left. + K_{Gb} [h_p^{(1)} - h_q^{(1)}] \right\}
$$

$$
h_p^{(1)}(\mathbf{x}, 0) = 0, \quad \mathbf{x} \in \Omega
$$

$$
h_p^{(1)}(\mathbf{x}, t) = 0, \quad \mathbf{x} \in \Gamma_D
$$

$$
n_i \left[\frac{\partial h_p^{(1)}(\mathbf{x}, t)}{\partial x_i} - J_{pi}(\mathbf{x}, t) f_p'(\mathbf{x}) \right] = 0, \quad \mathbf{x} \in \Gamma_N
$$

(7.19)

where $K_{GY_p} = \exp[\langle Y_p \rangle] = G_p K_{Gf_p}$ is the geometric mean of the volume fraction weighted hydraulic conductivity of medium p, $K_{Gf_p} = \exp[\langle f_p \rangle]$ is the geometric mean of the hydraulic conductivity of medium p, $J_{pi} = -\partial h_p^{(0)}/\partial x_i$ is the negative of the mean spatial gradient of pressure head in medium p, and $J_{pt} = \partial h_p^{(0)}/\partial t$ is the mean temporal head gradient. The higher-order terms $h^{(n)}$ ($n \geq 2$) can be written similarly. In writing the initial and boundary conditions for the head perturbation equation, we have made use of the assumption that $h_p(\mathbf{x}, t)$ is known initially and is deterministic on the boundary. Since there is no random quantity in Eq. (7.18), we have $\langle h_p^{(0)} \rangle = h_p^{(0)}$. It is also obvious from Eq. (7.19) that $\langle h_p^{(1)} \rangle = 0$. Hence, the mean head is $\langle h_p \rangle = h_p^{(0)}$ to zeroth or first order in the variability of medium properties and $\langle h_p \rangle = h_p^{(0)} + \langle h_p^{(2)} \rangle$ to second order. The head fluctuation is $h_p' = h_p^{(1)}$ to first order. Therefore, to first order we have the head covariance $C_{h_p}(\mathbf{x}, t; \boldsymbol{\chi}, \tau) = \langle h_p^{(1)}(\mathbf{x}, t) h_p^{(1)}(\boldsymbol{\chi}, \tau) \rangle$.

The equation for the first-order head covariance $C_{h_p}(\mathbf{x}, t; \boldsymbol{\chi}, \tau)$ is obtained by multiplying Eq. (7.19) by $h_p^{(1)}(\boldsymbol{\chi}, \tau)$ and taking ensemble expectation,

$$
\frac{\partial^2 C_{h_p}(\mathbf{x}, t; \boldsymbol{\chi}, \tau)}{\partial x_i^2} - K_{GY_p}^{-1} K_{Gb} C_{h_p}(\mathbf{x}, t; \boldsymbol{\chi}, \tau) - K_{GY_p}^{-1} \langle S_p^* \rangle \frac{\partial C_{h_p}(\mathbf{x}, t; \boldsymbol{\chi}, \tau)}{\partial t}
$$

$$
= J_{pi}(\mathbf{x}, t) \frac{\partial}{\partial x_i} [C_{g_p h_p}(\mathbf{x}; \boldsymbol{\chi}, \tau) + C_{f_p h_p}(\mathbf{x}; \boldsymbol{\chi}, \tau)]
$$

$$
- K_{GY_p}^{-1} \{ \langle S_p^* \rangle J_{pt}(\mathbf{x}, t) C_{f_p h_p}(\mathbf{x}; \boldsymbol{\chi}, \tau) + K_{Gb} [\langle h_p(\mathbf{x}, t) \rangle
$$

$$
- \langle h_q(\mathbf{x}, t) \rangle][C_{g_p h_p}(\mathbf{x}; \boldsymbol{\chi}, \tau) + C_{f_p h_p}(\mathbf{x}; \boldsymbol{\chi}, \tau) - C_{\beta h_p}(\mathbf{x}; \boldsymbol{\chi}, \tau)]
$$

(7.20)

$$
+ K_{Gb} C_{h_p h_q}(\mathbf{x}, t; \boldsymbol{\chi}, \tau) \}
$$

$$
C_{h_p}(\mathbf{x}, 0; \boldsymbol{\chi}, \tau) = 0, \quad \mathbf{x} \in \Omega
$$

$$
C_{h_p}(\mathbf{x}, t; \boldsymbol{\chi}, \tau) = 0, \quad \mathbf{x} \in \Gamma_D
$$

$$
n_i \left[\frac{\partial C_{h_p}(\mathbf{x}, t; \boldsymbol{\chi}, \tau)}{\partial x_i} - J_{pi}(\mathbf{x}, t) C_{f_p h_p}(\mathbf{x}; \boldsymbol{\chi}, \tau) \right] = 0, \quad \mathbf{x} \in \Gamma_N
$$

The unknown cross-covariances $C_{g_p h_p}(\mathbf{x}; \boldsymbol{\chi}, \tau)$, $C_{f_p h_p}(\mathbf{x}; \boldsymbol{\chi}, \tau)$, and $C_{\beta h_p}(\mathbf{x}; \boldsymbol{\chi}, \tau)$ are in turn solvable from the following moment equations:

$$\frac{\partial^2 C_{g_p h_p}(\mathbf{x}; \boldsymbol{\chi}, \tau)}{\partial \chi_i^2} - K_{GY_p}^{-1} K_{Gb} \, C_{g_p h_p}(\mathbf{x}; \boldsymbol{\chi}, \tau)$$

$$- K_{GY_p}^{-1} \langle S_p^* \rangle \frac{\partial C_{g_p h_p}(\mathbf{x}; \boldsymbol{\chi}, \tau)}{\partial \tau}$$

$$= J_{pi}(\boldsymbol{\chi}, \tau) \frac{\partial C_{g_p}(\mathbf{x}; \boldsymbol{\chi})}{\partial \chi_i} - K_{GY_p}^{-1} \{ K_{Gb}[\langle h_p(\boldsymbol{\chi}, \tau) \rangle$$

$$- \langle h_q(\boldsymbol{\chi}, \tau) \rangle] C_{g_p}(\mathbf{x}; \boldsymbol{\chi}) + K_{Gb} C_{g_p h_q}(\mathbf{x}; \boldsymbol{\chi}, \tau) \} \qquad (7.21)$$

$$\frac{\partial^2 C_{f_p h_p}(\mathbf{x}; \boldsymbol{\chi}, \tau)}{\partial \chi_i^2} - K_{GY_p}^{-1} K_{Gb} \, C_{f_p h_p}(\mathbf{x}; \boldsymbol{\chi}, \tau)$$

$$- K_{GY_p}^{-1} \langle S_p^* \rangle \frac{\partial C_{f_p h_p}(\mathbf{x}; \boldsymbol{\chi}, \tau)}{\partial \tau}$$

$$= J_{pi}(\boldsymbol{\chi}, \tau) \frac{\partial C_{f_p}(\mathbf{x}; \boldsymbol{\chi})}{\partial \chi_i} - K_{GY_p}^{-1} \{ \langle S_p^* \rangle J_{pt}(\boldsymbol{\chi}, \tau) C_{f_p}(\mathbf{x}; \boldsymbol{\chi})$$

$$+ K_{Gb}[\langle h_p(\boldsymbol{\chi}, \tau) \rangle - \langle h_q(\boldsymbol{\chi}, \tau) \rangle][C_{f_p}(\mathbf{x}; \boldsymbol{\chi}) - C_{f_p \beta}(\mathbf{x}; \boldsymbol{\chi})]$$

$$+ K_{Gb} C_{f_p h_q}(\mathbf{x}; \boldsymbol{\chi}, \tau) \} \qquad (7.22)$$

$$\frac{\partial^2 C_{\beta h_p}(\mathbf{x}; \boldsymbol{\chi}, \tau)}{\partial \chi_i^2} - K_{GY_p}^{-1} K_{Gb} \, C_{\beta h_p}(\mathbf{x}; \boldsymbol{\chi}, \tau)$$

$$- K_{GY_p}^{-1} \langle S_p^* \rangle \frac{\partial C_{\beta h_p}(\mathbf{x}; \boldsymbol{\chi}, \tau)}{\partial \tau}$$

$$= J_{pi}(\boldsymbol{\chi}, \tau) \frac{\partial C_{\beta f_p}(\mathbf{x}; \boldsymbol{\chi})}{\partial \chi_i} - K_{GY_p}^{-1} \{ \langle S_p^* \rangle J_{pt}(\boldsymbol{\chi}, \tau) C_{\beta f_p}(\mathbf{x}; \boldsymbol{\chi})$$

$$+ K_{Gb}[\langle h_p(\boldsymbol{\chi}, \tau) \rangle - \langle h_q(\boldsymbol{\chi}, \tau) \rangle][C_{\beta f_p}(\mathbf{x}; \boldsymbol{\chi}) - C_\beta(\mathbf{x}; \boldsymbol{\chi})]$$

$$+ K_{Gb} C_{\beta h_q}(\mathbf{x}; \boldsymbol{\chi}, \tau) \} \qquad (7.23)$$

subject to the initial and boundary conditions

$$C_{l h_p}(\mathbf{x}; \boldsymbol{\chi}, 0) = 0, \quad \boldsymbol{\chi} \in \Omega$$

$$C_{l h_p}(\mathbf{x}; \boldsymbol{\chi}, \tau) = 0, \quad \boldsymbol{\chi} \in \Gamma_D \qquad (7.24)$$

$$n_i \left[\frac{\partial C_{l h_p}(\mathbf{x}; \boldsymbol{\chi}, \tau)}{\partial \chi_i} - J_{pi}(\boldsymbol{\chi}, \tau) C_{l f_p}(\mathbf{x}; \boldsymbol{\chi}, \tau) \right] = 0, \quad \boldsymbol{\chi} \in \Gamma_N$$

where l can be either g_p, f_p, or β. The remaining coupling terms in Eq. (7.20), $C_{h_p h_q}(\mathbf{x}, t; \boldsymbol{\chi}, \tau)$, can be obtained from the moment equation below:

$$
\frac{\partial^2 C_{h_p h_q}(\mathbf{x}, t; \boldsymbol{\chi}, \tau)}{\partial \chi_i^2} - K_{GY_q}^{-1} K_{Gb}\, C_{h_p h_q}(\mathbf{x}, t; \boldsymbol{\chi}, \tau)
$$

$$
- K_{GY_q}^{-1} \langle S_q^* \rangle \frac{\partial C_{h_p h_q}(\mathbf{x}, t; \boldsymbol{\chi}, \tau)}{\partial \tau}
$$

$$
= J_{qi}(\boldsymbol{\chi}, \tau) \frac{\partial}{\partial \chi_i} [C_{h_p g_q}(\mathbf{x}, t; \boldsymbol{\chi}) + C_{h_p f_q}(\mathbf{x}, t; \boldsymbol{\chi})]
$$

$$
- K_{GY_q}^{-1} \{ \langle S_p^* \rangle J_{qt}(\boldsymbol{\chi}, \tau) C_{h_p f_q}(\mathbf{x}, t; \boldsymbol{\chi}) + K_{Gb}[\langle h_q(\boldsymbol{\chi}, \tau) \rangle
$$

$$
- \langle h_p(\boldsymbol{\chi}, \tau) \rangle][C_{h_p g_q}(\mathbf{x}, t; \boldsymbol{\chi}) + C_{h_p f_q}(\mathbf{x}, t; \boldsymbol{\chi}) \tag{7.25}
$$

$$
- C_{h_p \beta}(\mathbf{x}, t; \boldsymbol{\chi})] + K_{Gb} C_{h_p}(\mathbf{x}, t; \boldsymbol{\chi}, \tau) \}
$$

$$
C_{h_p h_q}(\mathbf{x}, t; \boldsymbol{\chi}, 0) = 0, \quad \boldsymbol{\chi} \in \Omega
$$

$$
C_{h_p h_q}(\mathbf{x}, t; \boldsymbol{\chi}, \tau) = 0, \quad \boldsymbol{\chi} \in \Gamma_D
$$

$$
n_i \left[\frac{\partial C_{h_p h_q}(\mathbf{x}, t; \boldsymbol{\chi}, \tau)}{\partial \chi_i} - J_{qi}(\boldsymbol{\chi}, \tau) C_{h_p f_q}(\mathbf{x}; \boldsymbol{\chi}, \tau) \right] = 0, \quad \boldsymbol{\chi} \in \Gamma_N
$$

The moment equations (7.21)–(7.24) are obtained by rewriting Eq. (7.19) in terms of $(\boldsymbol{\chi}, \tau)$, premultiplying it with $g_p(\mathbf{x})$, $f_p(\mathbf{x})$, or $\beta(\mathbf{x})$, and then taking ensemble expectation; Eq. (7.25) is obtained similarly by writing an equation for $h_q^{(1)}(\boldsymbol{\chi}, \tau)$ and premultiplying it with $h_p^{(1)}(\mathbf{x}, t)$. In these equations, four more unknown cross-covariances arise, i.e., $C_{g_p h_q}(\mathbf{x}; \boldsymbol{\chi}, \tau)$, $C_{f_p h_q}(\mathbf{x}; \boldsymbol{\chi}, \tau)$, $C_{\beta h_q}(\mathbf{x}; \boldsymbol{\chi}, \tau)$, and $C_{h_q}(\mathbf{x}, t; \boldsymbol{\chi}, \tau)$. It can be seen that $C_{\beta h_q}(\mathbf{x}; \boldsymbol{\chi}, \tau)$ and $C_{h_q}(\mathbf{x}, t; \boldsymbol{\chi}, \tau)$ can be solved from Eqs. (7.20)–(7.23) by merely interchanging the order of p and q in the respective equation, while the other cross-covariances have to be solved from the following moment equations:

$$
\frac{\partial^2 C_{g_p h_q}(\mathbf{x}; \boldsymbol{\chi}, \tau)}{\partial \chi_i^2} - K_{GY_q}^{-1} K_{Gb} C_{g_p h_q}(\mathbf{x}; \boldsymbol{\chi}, \tau) - K_{GY_q}^{-1} \langle S_q^* \rangle \frac{\partial C_{g_p h_q}(\mathbf{x}; \boldsymbol{\chi}, \tau)}{\partial \tau}
$$

$$
= J_{qi}(\boldsymbol{\chi}, \tau) \frac{\partial C_{g_p g_q}(\mathbf{x}; \boldsymbol{\chi})}{\partial \chi_i} - K_{GY_q}^{-1} \{ K_{Gb}[\langle h_q(\boldsymbol{\chi}, \tau) \rangle
$$

$$
- \langle h_p(\boldsymbol{\chi}, \tau) \rangle] C_{g_p g_q}(\mathbf{x}; \boldsymbol{\chi}) + K_{Gb} C_{g_p h_p}(\mathbf{x}; \boldsymbol{\chi}, \tau) \} \tag{7.26}
$$

$$
\frac{\partial^2 C_{f_p h_q}(\mathbf{x}; \boldsymbol{\chi}, \tau)}{\partial \chi_i^2} - K_{GY_q}^{-1} K_{Gb} C_{f_p h_q}(\mathbf{x}; \boldsymbol{\chi}, \tau) - K_{GY_q}^{-1} \langle S_q^* \rangle \frac{\partial C_{f_p h_q}(\mathbf{x}; \boldsymbol{\chi}, \tau)}{\partial \tau}
$$

$$
= K_{GY_q}^{-1} \{ K_{Gb}[\langle h_q(\boldsymbol{\chi}, \tau) \rangle - \langle h_p(\boldsymbol{\chi}, \tau) \rangle] C_{f_p \beta}(\mathbf{x}; \boldsymbol{\chi}) - K_{Gb} C_{f_p h_p}(\mathbf{x}; \boldsymbol{\chi}, \tau) \} \tag{7.27}
$$

both subject to

$$C_{lh_q}(\mathbf{x}; \chi, 0) = 0, \quad \chi \in \Omega$$

$$C_{lh_q}(\mathbf{x}; \chi, \tau) = 0, \quad \chi \in \Gamma_D \tag{7.28}$$

$$n_i \left[\frac{\partial C_{lh_q}(\mathbf{x}; \chi, \tau)}{\partial \chi_i} - J_{qi}(\chi, \tau) C_{lf_q}(\mathbf{x}; \chi, \tau) \right] = 0, \quad \chi \in \Gamma_N$$

where l represents g_q for Eq. (7.26) and f_q for Eq. (7.27). In deriving the equations governing the auto- and cross-covariances of head, Eqs. (7.20)–(7.28), we have assumed $C_{f_p g_q}$, $C_{f_p f_q}$, $C_{g_p \beta}$, and some of the cross-covariances between the input variables g_p, f_p, and β, to be zero (see Zhang and Sun [2000, Appendix] for justification). The other auto- and cross-covariances such as $C_{g_p g_q}$, C_β, and $C_{f_p \beta}$ are given as functions of the input statistics of f_m, f_f, and w_f [Zhang and Sun, 2000, Appendix].

7.2.2 Numerical Solutions

Moment equations (7.20) through (7.28) are coupled equations and thus have to be solved either simultaneously or sequentially. In the current case, the number (13) of coupled equations makes simultaneously solving these equations very difficult. As a result, Zhang and Sun [2000] pursued solving the system of equations sequentially in an iterative manner. On the other hand, it is worthwhile noting that like flow in nonfractured porous media discussed in Chapters 3 and 4, the first two moments of pressure heads for flow in fractured porous media are governed by the same type of equations but with different forcing terms. The solution of these equations is facilitated with this recognition. At each iteration, with coupling terms given based either on initial guesses or on the results of the previous iteration, these equations can be solved by using the strategies outlined in Chapters 3 and 4 for flow in nonfractured porous media. The moment equations may be solved by the numerical technique of finite differences with the spatial derivatives being approximated by the central-differences scheme and the temporal derivatives by the implicit (backward) method. At each iteration, the resulting linear algebraic equations (LAEs) may be solved by lower–upper (LU) decomposition with forward and back substitutions. Since the left-hand side is the same for all the moment equations, the coefficient matrix of these LAEs only has to be decomposed once for each time step and does not change with time as long as the time step is kept the same.

Zhang and Sun [2000] implemented the moment equations in two dimensions by finite differences. They first solved Eq. (7.18) for the mean pressure heads h_p ($p = f$ and m) by iterations. The heads h_f and h_m at each node on the grid and at every time step from 0 to t were used to compute the spatial and temporal head gradients J_{pi} and J_{pt}, which are the required information for solving the auto- and cross-covariances of head in Eqs. (7.20)–(7.28). Then with the input statistics of f_m, f_f, and w_f, at each time step the covariance equations were solved sequentially by iteration. At each time step, convergence was achieved after several iterations.

As for nonfractured porous media discussed in Chapters 3 and 4, the numerical moment equation approach has flexibility in handling complex flow configurations, (moderately) irregular geometry, medium nonstationarity caused by the presence of geological layers, zones, and facies, and different input covariance functions, all of which are important factors in real-world applications. Another approach with similar flexibilities is Monte Carlo simulation, which is based on the idea of approximating stochastic processes by a large number of equally likely realizations. As discussed in Section 3.11, the moment equation and the Monte Carlo approaches are two complementary stochastic methods. Each has its own advantages and disadvantages. For the situation studied in this section, the Monte Carlo approach needs to solve only two coupled equations albeit many (sometimes, up to several thousand) times, while the moment equation approach must solve 13 coupled equations, although only once. It is not a simple matter to evaluate the relative computational efficiency of these two approaches, which depends on the size of the problem, the particular algorithms and solvers for each approach, statistical sampling for the Monte Carlo approach, and other factors. It is of interest to note that both approaches may benefit from recent improvements in computer memory and speed, and with the availability of well-integrated massively parallel machines, because both approaches are computationally demanding, especially for large size problems, and have inherent parallel structures.

7.2.3 Illustrative Examples

We illustrate the moment equation approach through some two-dimensional examples under various transient conditions. In the following examples, the flow domain is of size 100 [L] by 100 [L] (where L is an arbitrary length unit). Unless otherwise stated, the left and right sides are specified as constant head boundaries; the two lateral sides are no-flow boundaries. The specified heads H_p may be different for the fracture and the matrix at each of the two constant head boundaries. In all examples, the autocovariance structures of $f_p(\mathbf{x})$ and $w_p(\mathbf{x})$ are assumed to follow the exponential model,

$$C_s(\mathbf{x} - \boldsymbol{\chi}) = \sigma_s^2 \exp\left\{-\left[\frac{(x_1 - \chi_1)^2}{\lambda_s^2} + \frac{(x_2 - \chi_2)^2}{\lambda_s^2}\right]^{1/2}\right\} \tag{7.29}$$

where σ_s^2 is the variance and λ_s is the correlation (integral) scale of $s = f_m, f_f, w_m$, or w_f. In addition, the covariance functions are assumed to be isotropic. It is, however, worthwhile to note that the numerical moment equation model is able to easily handle other covariance structures and statistical anisotropy.

Baseline case In the baseline case (Case 1), the required parameters are specified as follows. For log-transformed hydraulic conductivity in the fracture, $\langle f_f \rangle = -1$ (i.e., with a geometric mean of 0.368 [L/T]), $\sigma_{f_f}^2 = 0.5$, and $\lambda_{f_f} = 10$ [L]; for log-transformed hydraulic conductivity in the matrix, $\langle f_m \rangle = -3$ (i.e., with a geometric mean of 0.050 [L/T]), $\sigma_{f_m}^2 = 0.25$, and $\lambda_{f_m} = 10$ [L]; for the relative volume fraction of the fracture,

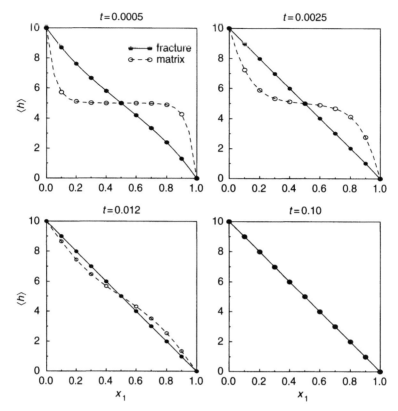

Figure 7.2. Expected values of pressure heads in the fracture and the matrix along the horizontal centerline of a square domain at different times (Case 1). (Adapted from Zhang and Sun [2000]. Copyright 2000 by the American Geophysical Union.)

$\langle w_f \rangle = 0.01$, $\sigma^2_{w_f} = 0.0001$, and $\lambda_{w_f} = 10$ [L]; the specific storage coefficients $S_f = 10^{-7}$ and $S_m = 10^{-6}$ [L^{-1}]; and the geometric factor B in Eq. (7.4) is equal to 10^{-6}.

We first look at the dynamic behavior of the pressure head field in both the fracture and the matrix systems for the baseline case. In this example, the flow domain is initially at a static state with pressure head $H_p^o = 5$ [L]. At time $t \geq 0$, the boundary condition $H_f = H_m = 10$ [L] is imposed at $x_1 = 0$, and $H_f = H_m = 0$ at $x_1 = 100$ [L]. Figure 7.2 shows the mean pressure head profiles in the fracture (solid curves) and the matrix (dashed curves) at four different times $t = 0.0005, 0.0025, 0.012$, and 0.10 [T]. The profiles are depicted along the horizontal centerline $x_2 = 50$ [L]. The horizontal axis x_1 in these plots is normalized with respect to the horizontal length (100 [L]) of the domain. It is seen that at early times the head profiles in the matrix block differ greatly from those in the fracture. Because of the high permeability and the low specific storage coefficient in the fracture, its head field approaches the steady state rapidly compared to its counterpart in the matrix. In the matrix block, at early times the flow is only significant at the two boundary sides, in that

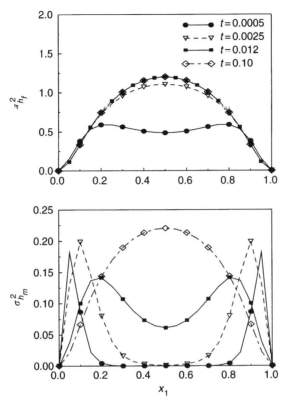

Figure 7.3. Variances of pressure heads in the fracture and the matrix along the horizontal center-line of a square domain at different times (Case 1). (Adapted from Zhang and Sun [2000]. Copyright 2000 by the American Geophysical Union.)

the pressure head changes little in the center of the domain; with time the pressure head field propagates through the whole domain. In each system, the dynamic behavior of the mean pressure head is analogous to that for flow in nonfractured porous media as discussed in Chapter 4. At late times, the mean pressure profiles reach their steady state and become identical in the two systems. Then the mean fluid transfer between the two systems vanishes. However, the actual (random) h_f and h_m at each point can be different, resulting in a random zero-mean fluid transfer between the fracture and the matrix. Figure 7.3 shows the dynamic behavior of the pressure variances in the two systems along the horizontal centerline of the domain. In the fracture system (upper plot of Fig. 7.3), the pressure variance is bimodal at early times and unimodal at late times. This transition is more apparent for the variance in the matrix (the lower plot). At time $t = 0.0005$ [T], the pressure variance is zero except near the two boundary sides. With time, the two variance peaks travel away from the boundaries and towards the domain center; at late times, the two peaks merge into a single one, which increases with time until reaching its steady state. The pressure variances are measures of the uncertainty associated with pressure predictions in the fracture and the matrix systems. The dynamic uncertainty behavior may be explained by a simple physical argument. At early time, with the specific initial and boundary conditions, the center of the domain is largely unaffected by flow, resulting in the lowest uncertainty there; the fact that the flow is significant near the two horizontal boundaries leads to the largest uncertainties there. At late times when the flow has propagated through the whole domain, the domain center has the largest prediction uncertainties because along the horizontal direction the domain center is the farthest from the two boundaries with known pressures. Like the mean pressures, the pressure variance reaches its steady state much faster in the fracture than in the matrix. However, the steady-state pressure variance is much smaller in the matrix than in the fracture. This difference in the pressure variances is partially due to that in the variances of fracture and matrix permeability. Figure 7.4 shows the cross-covariance $\sigma_{h_f h_m}(\mathbf{x}, t) = C_{h_f h_m}(\mathbf{x}, t; \mathbf{x}, t)$ between the fracture and matrix pressures at the same point along the horizontal centerlines. It is seen that the correlation between the two pressures is a strong function of time and space. At early times (e.g., $t = 0.0005$ and 0.0025), the covariances are all positive with their peaks near the boundaries. At late times (e.g., $t = 0.10$), the covariances are positive at the domain center and negative near the boundaries. That is, the pressures in the two systems are negatively correlated near the boundaries and positively correlated at the domain center at late times. Hence, we find that near the boundaries the fracture and matrix pressures change with time from having a positive correlation to having a negative one.

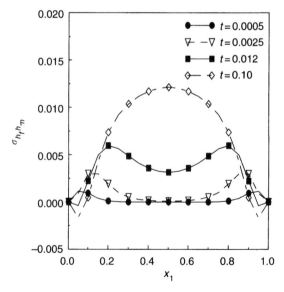

Figure 7.4. Cross-covariances of pressure heads in the fracture and the matrix along the horizontal centerline of a square domain at different times (Case 1). (Adapted from Zhang and Sun [2000]. Copyright 2000 by the American Geophysical Union.)

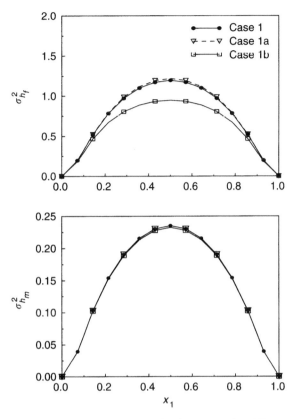

Figure 7.5. Comparisons of steady-state pressure head variances for different values of the mean log matrix hydraulic conductivity $\langle f_f \rangle$ and the geometric factor B (Case 1). (Adapted from Zhang and Sun [2000]. Copyright 2000 by the American Geophysical Union.)

Figure 7.5 shows the steady-state pressure variances (reported at $t = 1.0$ [T]) in both the fracture and the matrix systems as functions of the mean log hydraulic conductivity of matrix $\langle f_m \rangle$ and the geometric factor B based on two cases. In Case 1a, the mean log hydraulic conductivity is $\langle f_m \rangle = -4$, while other parameters are kept the same as those for Case 1; in Case 1b, the geometric factor is $B = 10^{-5}$, while others remain to be the same as in Case 1. Comparing Case 1 and Case 1a reveals that the steady-state pressure variance in the fracture is slightly affected by $\langle f_m \rangle$, while that of the matrix is not a function of $\langle f_m \rangle$. This can be explained by looking at the pressure variance equation (7.20) and the transfer coefficient of Eq. (7.4). At steady state, all time-dependent terms vanish in Eq. (7.20). As a result, all terms containing $K_{GY_p}^{-1}$ are also multiples of K_{Gb}. Because of the particular choice $K_{Gb} = B \exp[\langle f_m \rangle]$, the $\langle f_m \rangle$ term cancels out in the product $K_{GY_m}^{-1} K_{Gb}$ for the matrix system and thus has no impact on the matrix pressure variance at steady state. Before reaching steady state (though not shown here), the pressure variances in both systems are affected by $\langle f_m \rangle$. It is seen from a comparison between Case 1b and Case 1 that both pressure variances are affected by the choice of B. However, the pressure variance in the fracture system is much more sensitive to the geometric factor B than that of the matrix.

Case with distinct boundary conditions In the previous cases, the boundary conditions were the same for both the fracture and the matrix systems. In the following example (Case 2), we look at the effect of different boundary conditions in the fracture and the matrix pressure heads. Here, the required parameters are the same as in Case 1 except that $\langle f_f \rangle = -2$, $\langle f_m \rangle = -4$ and $B = 10^{-5}$. The domain is initially in a static state with $H_f^o = H_m^o = 3$ [L]. At time $t \geq 0$, the boundary conditions are specified as follows: at the left-hand side ($x_1 = 0$), $H_f = 10$ [L] and $H_m = 5$ [L]; at the right-hand side ($x_1 = 100$ [L]), $H_f = H_m = 0$ [L]; and the two lateral sides are no-flow boundaries. Figure 7.6 shows

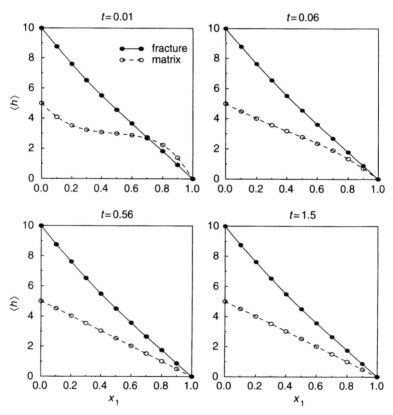

Figure 7.6. Expected values of pressure heads in the fracture and the matrix along the horizontal centerline of a square domain at different times (Case 2). (Adapted from Zhang and Sun [2000]. Copyright 2000 by the American Geophysical Union.)

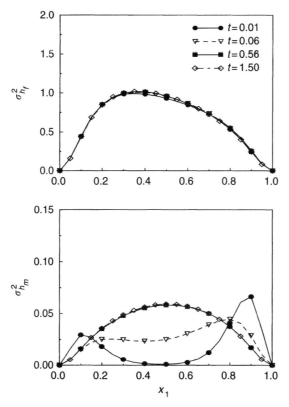

Figure 7.7. Variances of pressure heads in the fracture and the matrix along the horizontal center-line of a square domain at different times (Case 2). (Adapted from Zhang and Sun [2000]. Copyright 2000 by the American Geophysical Union.)

the mean pressure profiles along the horizontal centerline at different times $t = 0.01, 0.06,$ $0.56,$ and 1.50; Figure 7.7 shows the corresponding pressure variances. Like in Case 1, flow propagates through the fracture system rapidly, in that both the mean fracture pressure and the pressure variance have almost reached their respective steady-state profiles as early as $t = 0.06$ [T]. The flow in the matrix system is, however, much slower. At $t = 0.01$, the matrix flow is only significant near the two ends where the mean pressure has the highest gradient and the pressure variance has its peaks. Like Case 1, the pressure variance peaks travel with time away from the boundaries as the flow propagates through the matrix system. However, unlike Case 1, the pressure variance profiles are no longer symmetric about the domain center in both systems. At $t = 1.50$, it seems that flows in both systems have reached their respective steady states. However, unlike Case 1, the mean pressure profiles are not identical, resulting in continuing fluid transfer between the two systems even under steady state. The mean steady-state pressure profiles are no longer straight but curved to accommodate this fluid transfer.

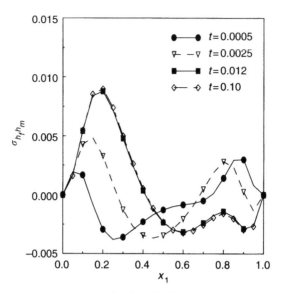

Figure 7.8. Cross-covariances of pressure heads in the fracture and the matrix along the horizontal centerline of a square domain at different times (Case 2). (Adapted from Zhang and Sun [2000]. Copyright 2000 by the American Geophysical Union.)

Figure 7.8 shows the cross-covariance between h_f and h_m at the same points along the horizontal centerline. It is seen that at early time (e.g., $t = 0.01$), the pressures in the two systems are negatively correlated in the upstream portion of the domain away from the boundary and positively correlated in the downstream. However, with time the sign of correlation is reversed. The steady-state profile of the cross-covariance is significantly different in this case than that shown in Fig. 7.4. Thus we have seen that the boundary conditions have a great impact on the dynamic behavior of the prediction uncertainties. The existence of different boundary conditions in the fracture and the matrix is plausible because the fracture and the matrix systems may be connected to different hydrological zones. For example, in a situation where the fractures penetrate through a few alternating high and lower permeability layers, the flow in the matrix of a high-permeability layer could be largely isolated, while the flow in the fracture might be greatly influenced by the conditions in the other layers.

Case involving pumping A pumping well of fixed pressure ($h_f = h_m = 1.0$ [L]) is placed in the center of the domain at time $t \geq 0$. In this case, all sides are specified as no-flow boundaries (i.e., closed boundaries); the domain is initially at uniform pressure heads $H_f^o = H_m^o = 10$ [L]; and the parameters are the same as in Case 2 except for $B = 10^{-6}$. Figure 7.9 shows the mean pressures in the fracture and the matrix systems along the horizontal centerline of the domain at different times $t = 0.005, 0.01, 0.02,$ and 0.40 [T]. It is seen that the mean pressure drop zones in the two systems propagate

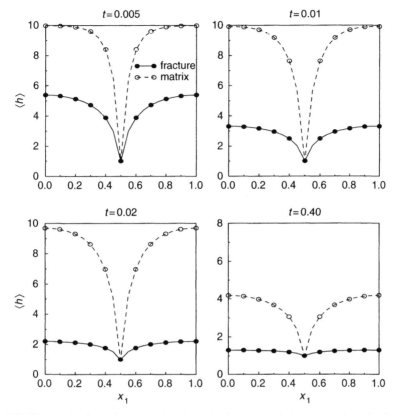

Figure 7.9. Expected values of pressure heads in the fracture and the matrix along the horizontal centerline of a square domain at different times (Case 3). (Adapted from Zhang and Sun [2000]. Copyright 2000 by the American Geophysical Union.)

with time. Fluid is depleted much faster in the fracture than in the matrix, as reflected in the mean pressure profiles. It is seen that at $t = 0.40$, the fracture medium is almost pressure-depleted. At this time the fluid extracted from the fracture mainly comes from the matrix as a result of the fluid transfer mechanism. Figure 7.10 shows the variance profiles corresponding to the mean pressure heads just shown. The pressure variance is zero at the pumping well of fixed pressure and is bimodal at early times in both systems with the peaks in the vicinity of the well. The variance peaks travel away from the well with time. The variance is zero at the outer boundaries at early times but increases with time. At later times (e.g., for the fracture system at $t = 0.01$), the pressure variance attains its maximum at the boundaries, and its magnitude decreases with time as the domain is being depleted. These mean and variance profiles are along the horizontal centerline. In this case, the behaviors of these quantities are exactly the same along the transverse centerline.

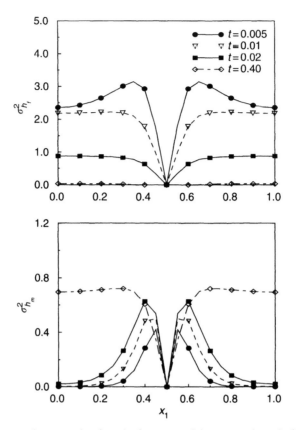

Figure 7.10. Variances of pressure heads in the fracture and the matrix along the horizontal center-line of a square domain at different times (Case 3). (Adapted from Zhang and Sun [2000]. Copyright 2000 by the American Geophysical Union.)

7.3 UNSATURATED FLOW

As for saturated flow, unsaturated flow in the fracture and matrix pore system of a double-permeability medium is described by the following equations:

$$w_f(\mathbf{x})C_f(\mathbf{x}, t)\frac{\partial \psi_f(\mathbf{x}, t)}{\partial t} = \nabla \cdot \{w_f(\mathbf{x})K_f[\psi_f(\mathbf{x}, t), \cdot]\nabla[\psi_f(\mathbf{x}, t) + x_1]\}$$

$$- Q_w(\mathbf{x}, t) \tag{7.30}$$

$$w_m(\mathbf{x})C_m(\mathbf{x}, t)\frac{\partial \psi_m(\mathbf{x}, t)}{\partial t} = \nabla \cdot \{w_m(\mathbf{x})K_m[\psi_m(\mathbf{x}, t), \cdot]\nabla[\psi_m(\mathbf{x}, t) + x_1]\}$$

$$+ Q_w(\mathbf{x}, t) \tag{7.31}$$

where $C_p(\mathbf{x}, t)$ is the water (moisture) capacity ($p = f$ or m), $K_p[\psi_p(\mathbf{x}, t), \cdot]$ is the unsaturated hydraulic conductivity for the fracture or matrix continuum and also depends on the properties of the continuum, $\psi_p(\mathbf{x}, t)$ is the pressure head in fractures or matrix, x_1 is the vertical axis and is directed upward, and $Q_w(\mathbf{x}, t)$ is the exchange term describing the transfer of water between the fracture and matrix systems, defined as

$$Q_w(\mathbf{x}, t) = b(\mathbf{x})[\psi_f(\mathbf{x}, t) - \psi_m(\mathbf{x}, t)] \tag{7.32}$$

where b is a mass transfer coefficient for water which depends on geometric factors and hydraulic conductivity at the fracture and matrix interface. Other symbols in Eqs. (7.30) and (7.31) have already been defined in Eqs. (7.1) and (7.2).

For a dual-porosity representation, Eq. (7.31) disappears because $K_m = 0$. When the pressures at the fracture and matrix systems are at equilibrium [Peters and Klavetter, 1988], i.e., $\psi_f(\mathbf{x}, t) = \psi_m(\mathbf{x}, t) = \psi(\mathbf{x}, t)$, adding Eqs. (7.30) and (7.31) yields

$$C(\mathbf{x}, t)\frac{\partial \psi(\mathbf{x}, t)}{\partial t} = \nabla \cdot \{K[\psi(\mathbf{x}, t), \cdot]\nabla[\psi(\mathbf{x}, t) + x_1]\} \tag{7.33}$$

where $K(\mathbf{x}, t)$ and $C(\mathbf{x}, t)$ are the composite unsaturated hydraulic conductivity and water capacity for the equivalent continuum, defined as

$$K[\psi(\mathbf{x}, t), \cdot] = w_f(\mathbf{x})K_f[\psi(\mathbf{x}, t), \cdot] + [1 - w_f(\mathbf{x})]K_m[\psi(\mathbf{x}, t), \cdot] \tag{7.34}$$

$$C(\mathbf{x}, t) = w_f(\mathbf{x})C_f(\mathbf{x}, t) + [1 - w_f(\mathbf{x})]C_m(\mathbf{x}, t) \tag{7.35}$$

This implies that the composite hydraulic conductivity (or water capacity) is the weighted (arithmetic) mean of the two individual continua. Figure 7.11 illustrates the composite hydraulic conductivity and water capacity as functions of capillary pressure ($= -\psi$) for the equivalent continuum model of fractured porous media. It is seen that flow is predominant in the fracture when the capillary pressure is low (i.e., the system is relatively wet) and flow mainly occurs in the matrix when the system is dry.

No matter which approach is used, the composite porosity and water content of the fractured porous media can be related to their counterparts for the fracture and matrix continuum as

$$\phi = w_f\phi_f + (1 - w_f)\phi_m \tag{7.36}$$

$$\theta = w_f\theta_f + (1 - w_f)\theta_m \tag{7.37}$$

7.3.1 Moment Partial Differential Equations

With the transformations

$$Y_f(\cdot) = \ln\{w_f(\mathbf{x})K_f[\psi_f(\mathbf{x}, t), \cdot]\} \tag{7.38}$$

$$Y_m(\cdot) = \ln\{w_m(\mathbf{x})K_m[\psi_m(\mathbf{x}, t), \cdot]\} \tag{7.39}$$

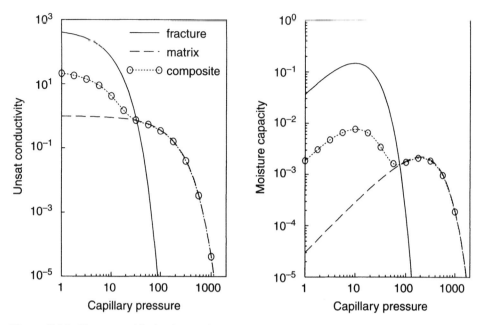

Figure 7.11. Unsaturated hydraulic conductivity and moisture capacity as functions of capillary pressure for fracture, matrix, and their composite.

we may rewrite Eqs. (7.30) and (7.31) as

$$\frac{\partial^2 \psi_p}{\partial x_i^2} + \frac{\partial Y_p}{\partial x_i}\left[\frac{\partial \psi_p}{\partial x_i} + \delta_{i1}\right] = \exp[-Y_p]\left[C_p^* \frac{\partial \psi_p}{\partial t} + Q_p\right] \tag{7.40}$$

where $p = f$ or m, $Q_f = Q_w$, $Q_m = -Q_w$, $C_p^* = w_p C_p$, and summation for repeated indices is implied. As for saturated flow in Section 7.2.1, writing $\psi_p = \psi_p^{(0)} + \psi_p^{(1)} + \psi_p^{(2)} + \cdots$, $Y_p(\cdot) = Y_p^{(0)} + Y_p^{(1)} + Y_p^{(2)} + \cdots$, $C_p^* = C_p^{*(0)} + C_p^{*(1)} + C_p^{*(2)} + \cdots$, and $Q_p = Q_p^{(0)} + Q_p^{(1)} + Q_p^{(2)} + \cdots$, and collecting terms at separate orders, leads to

$$\frac{\partial^2 \psi_p^{(0)}}{\partial x_i^2} + \frac{\partial Y_p^{(0)}}{\partial x_i}\frac{\partial \psi_p^{(0)}}{\partial x_i} = K_{po}^{-1}\left[C_p^{*(0)}\frac{\partial \langle \psi_p \rangle}{\partial t} + Q_p^{(0)}\right] \tag{7.41}$$

$$\frac{\partial^2 \psi_p^{(1)}}{\partial x_i^2} + \frac{\partial Y_p^{(0)}}{\partial x_i}\frac{\partial \psi_p^{(1)}}{\partial x_i} = K_{po}^{-1}C_p^{*(0)}\frac{\partial \psi_p^{(1)}}{\partial t} - J_{pi}\frac{\partial Y_p^{(1)}}{\partial x_i}$$

$$+ K_{po}^{-1}\{[C_p^{*(0)}J_{pt} + Q_p^{(0)}]Y_p^{(1)} + J_{pt}C_p^{*(1)} + Q_p^{(1)}\} \tag{7.42}$$

where $J_{pi} = \partial \psi_p^{(0)}/\partial x_i + \delta_{i1}$, $J_{pt} = \partial \psi_p^{(0)}/\partial t$, and $K_{po} = \exp[Y_p^{(0)}]$. With $\beta = \ln b$ and $\beta = \langle \beta \rangle + \beta'$, on the basis of Eq. (7.32) we have

$$Q_w^{(0)} = K_{Gb}[\psi_f^{(0)} - \psi_m^{(0)}] \tag{7.43}$$

$$Q_w^{(1)} = K_{Gb}\{[\psi_f^{(1)} - \psi_m^{(1)}] + \beta'[\psi_f^{(0)} - \psi_m^{(0)}]\} \tag{7.44}$$

where $K_{Gb} = \exp[\langle \beta \rangle]$.

It can be shown that $\langle \psi_p^{(1)} \rangle = 0$. The mean pressure head is $\langle \psi_p \rangle = \psi_p^{(0)}$ to first or zeroth order (in σ) and $\langle \psi_p \rangle = \langle \psi_p^{(0)} \rangle + \langle \psi_p^{(2)} \rangle$ to second order. The pressure head fluctuation is $\psi_p' = \psi_p^{(1)}$ to first order. Hence, the head covariance is $C_{\psi_p}(\mathbf{x}, t; \boldsymbol{\chi}, \tau) = \langle \psi_p^{(1)}(\mathbf{x}, t) \psi_p^{(1)}(\boldsymbol{\chi}, \tau) \rangle$ to first order (in σ^2). To first order, we have

$$\frac{\partial^2 \langle \psi_f \rangle}{\partial x_i^2} + \frac{\partial Y_f^{(0)}}{\partial x_i} \frac{\partial \langle \psi_f \rangle}{\partial x_i} = K_{fo}^{-1}\left\{ C_f^{*(0)} \frac{\partial \langle \psi_f \rangle}{\partial t} + K_{Gb}[\langle \psi_f \rangle - \langle \psi_m \rangle] \right\} \tag{7.45}$$

$$\frac{\partial^2 \langle \psi_m \rangle}{\partial x_i^2} + \frac{\partial Y_m^{(0)}}{\partial x_i} \frac{\partial \langle \psi_m \rangle}{\partial x_i} = K_{mo}^{-1}\left\{ C_m^{*(0)} \frac{\partial \langle \psi_m \rangle}{\partial t} - K_{Gb}[\langle \psi_f \rangle - \langle \psi_m \rangle] \right\} \tag{7.46}$$

$$\frac{\partial^2 \psi_f'}{\partial x_i^2} + J_{fi} \frac{\partial Y_f^{(1)}}{\partial x_i} + \frac{\partial Y_f^{(0)}}{\partial x_i} \frac{\partial \psi_f'}{\partial x_i}$$

$$= K_{fo}^{-1} C_f^{*(0)} \frac{\partial \psi_f'}{\partial t} + K_{fo}^{-1}\left\{ C_f^{*(0)} J_{ft} + K_{Gb}[\langle \psi_f \rangle - \langle \psi_m \rangle]Y_f^{(1)} \right.$$

$$\left. + J_{ft}C_f^{*(1)} + K_{Gb}[\psi_f' - \psi_m'] + K_{Gb}[\langle \psi_f \rangle - \langle \psi_m \rangle]\beta' \right\} \tag{7.47}$$

$$\frac{\partial^2 \psi_m'}{\partial x_i^2} + J_{mi} \frac{\partial Y_m^{(1)}}{\partial x_i} + \frac{\partial Y_m^{(0)}}{\partial x_i} \frac{\partial \psi_m'}{\partial x_i}$$

$$= K_{mo}^{-1} C_m^{*(0)} \frac{\partial \psi_m'}{\partial t} K_{mo}^{-1}\left\{ C_m^{*(0)} J_{mt} - K_{Gb}[\langle \psi_f \rangle - \langle \psi_m \rangle]Y_m^{(1)} \right.$$

$$\left. + J_{mt}C_m^{*(1)} - K_{Gb}[\psi_f' - \psi_m'] - K_{Gb}[\langle \psi_f \rangle - \langle \psi_m \rangle]\beta' \right\} \tag{7.48}$$

As for unsaturated flow in nonfractured porous media discussed in Chapter 5, Eqs. (7.30) and (7.31) or (7.45)–(7.48) are not closed without some constitutive relationships of $K_p(\psi)$ and $C_p(\psi)$ for the fracture and matrix systems. For simplicity, we assume that the fractures and the matrix can separately be described by the Gardner–Russo constitutive model [Gardner, 1958; Russo, 1988],

$$K_p(\mathbf{x}, t) = K_{Sp}(\mathbf{x}) \exp[\alpha_p(\mathbf{x})\psi_p(\mathbf{x}, t)] \tag{7.49}$$

$$C_p(\mathbf{x}, t) = -\tfrac{1}{4}(\theta_{sp} - \theta_{rp})[\alpha_p(\mathbf{x})]^2 \psi_p(\mathbf{x}, t) \exp[0.5\alpha_p(\mathbf{x})\psi_p(\mathbf{x}, t)] \tag{7.50}$$

where K_{Sp} ($p = f$ or m) is the fracture or matrix saturated hydraulic conductivity, α_p is the parameter related to pore size distribution corresponding to the fracture or matrix system, θ_{rp} is the residual (irreducible) water content, and θ_{sp} is the saturated water content. In general, the fracture system has larger values for saturated hydraulic conductivity and pore size distribution parameter than does the matrix. The typical behaviors of the unsaturated hydraulic conductivity and the water capacity in both systems are illustrated in Fig. 7.11. In this section, θ_{rp} and θ_{sp} are assumed to be constant. The unsaturated parameter $\alpha_p(\mathbf{x})$, the log-transformed saturated hydraulic conductivity $f_p(\mathbf{x}) = \ln K_{ps}(\mathbf{x})$ and the log-transformed relative volume fraction $g_p = \ln w_p$ are treated as random space functions. They are further assumed to be second-order stationary such that their expected values are constant and their covariances depend on the relative distances rather than the actual locations.

It can be verified that for the Gardner–Russo model, the first-order mean and covariance equations are given as

$$\frac{\partial^2 \langle \psi_p(\mathbf{x}, t) \rangle}{\partial x_i^2} + \langle \alpha_p \rangle J_{pi}(\mathbf{x}, t) \frac{\partial \langle \psi_p(\mathbf{x}, t) \rangle}{\partial x_i}$$

$$= K_{po}^{-1}(\mathbf{x}, t) \left\{ K_{g_p} C_p^{(0)}(\mathbf{x}, t) \frac{\partial \langle \psi_p(\mathbf{x}, t) \rangle}{\partial t} + (-1)^n K_{Gb}[\langle \psi_p(\mathbf{x}, t) \rangle - \langle \psi_q(\mathbf{x}, t) \rangle] \right\}$$

(7.51)

$$\frac{\partial^2 C_{\psi_p}(\mathbf{x}, t; \chi, \tau)}{\partial x_i^2} + \gamma_{1i}^P(\mathbf{x}, t) \frac{\partial C_{\psi_p}(\mathbf{x}, t; \chi, \tau)}{\partial x_i} [\gamma_2^P(\mathbf{x}, t) + \gamma_3^P(\mathbf{x}, t)] C_{\psi_p}(\mathbf{x}, t; \chi, \tau)$$

$$= \gamma_4^P(\mathbf{x}, t) \frac{\partial C_{\psi_p}(\mathbf{x}, t; \chi, \tau)}{\partial t} - J_{pi}(\mathbf{x}, t) \left[\frac{\partial C_{g_p \psi_p}(\mathbf{x}; \chi, \tau)}{\partial x_i} + \frac{\partial C_{f_p \psi_p}(\mathbf{x}; \chi, \tau)}{\partial x_i} \right.$$

$$\left. + \langle \psi_p(\mathbf{x}, t) \rangle \frac{\partial C_{\alpha_p \psi_p}(\mathbf{x}; \chi, \tau)}{\partial x_i} + \gamma_5^P(\mathbf{x}, t) C_{\alpha_p \psi_p}(\mathbf{x}; \chi, \tau) \right]$$

$$+ \gamma_6^P(\mathbf{x}, t) [C_{g_p \psi_p}(\mathbf{x}; \chi, \tau) + C_{f_p \psi_p}(\mathbf{x}; \chi, \tau)$$

$$+ \langle \psi_p(\mathbf{x}, t) \rangle C_{\alpha_p \psi_p}(\mathbf{x}; \chi, \tau)] + \gamma_7^P(\mathbf{x}, t) C_{\beta \psi_p}(\mathbf{x}; \chi, \tau)$$

$$+ \gamma_8^P(\mathbf{x}, t) [\gamma_9^P(\mathbf{x}, t) C_{\alpha_p \psi_p}(\mathbf{x}; \chi, \tau) + \gamma_{10}^P(\mathbf{x}, t) C_{\psi_p}(\mathbf{x}, t; \chi, \tau)$$

$$+ C_p^{(0)}(\mathbf{x}, t) C_{g_p \psi_p}(\mathbf{x}; \chi, \tau)] + \gamma_3^P(\mathbf{x}, t) C_{\psi_q \psi_p}(\mathbf{x}, t; \chi, \tau)$$

(7.52)

for $p = f$ or m with $q = m$ or f and $p \neq q$; $n = 1$ when $p = f$ and $n = 2$ when $p = m$. In the above, $K_{g_f} = \exp[\langle g_f \rangle]$, $K_{g_m} = \exp[\langle g_m \rangle]$, $K_{fo}(\mathbf{x}, t) = \exp[\langle g_f \rangle + \langle f_f \rangle + \langle \alpha_f \rangle \langle \psi_f(\mathbf{x}, t) \rangle]$, $K_{mo}(\mathbf{x}, t) = \exp[\langle g_m \rangle + \langle f_m \rangle + \langle \alpha_m \rangle \langle \psi_m(\mathbf{x}, t) \rangle]$, $J_{fi}(\mathbf{x}, t) = \partial \langle \psi_f(\mathbf{x}, t) \rangle / \partial x_i + \delta_{i1}$, $J_{mi}(\mathbf{x}, t) = \partial \langle \psi_m(\mathbf{x}, t) \rangle / \partial x_i + \delta_{i1}$, $J_{ft}(\mathbf{x}, t) = \partial \langle \psi_f(\mathbf{x}, t) \rangle / \partial t$,

$J_{mt}(\mathbf{x}, t) = \partial \langle \psi_m(\mathbf{x}, t) \rangle / \partial t$, the coefficients γ^Ps are defined as

$$\gamma_{1i}^P(\mathbf{x}, t) = \langle \alpha_p \rangle [2 J_{pi}(\mathbf{x}, t) - \delta_{i1}]$$

$$\gamma_2^P(\mathbf{x}, t) = - K_{po}^{-1}(\mathbf{x}, t) \{ J_{pt}(\mathbf{x}, t) K_{g_p} C_p^{(0)}(\mathbf{x}, t)$$
$$- (-1)^n K_{Gb} [\langle \psi_p(\mathbf{x}, t) \rangle - \langle \psi_q(\mathbf{x}, t) \rangle] \} \langle \alpha_p \rangle$$

$$\gamma_3^P(\mathbf{x}, t) = (-1)^n K_{po}^{-1}(\mathbf{x}, t) K_{Gb}$$

$$\gamma_4^P(\mathbf{x}, t) = K_{po}^{-1}(\mathbf{x}, t) K_{g_p} C_p^{(0)}(\mathbf{x}, t)$$

$$\gamma_5^P(\mathbf{x}, t) = J_{pi}(\mathbf{x}, t) - \delta_{i1}$$

$$\gamma_6^P(\mathbf{x}, t) = K_{po}^{-1}(\mathbf{x}, t) \{ K_{g_p} C_p^{(0)}(\mathbf{x}, t) J_{pt}(\mathbf{x}, t)$$
$$- (-1)^n K_{Gb} [\langle \psi_p(\mathbf{x}, t) \rangle - \langle \psi_q(\mathbf{x}, t) \rangle] \}$$

$$\gamma_7^P(\mathbf{x}, t) = - (-1)^n K_{po}^{-1}(\mathbf{x}, t) K_{Gb} [\langle \psi_p(\mathbf{x}, t) \rangle - \langle \psi_q(\mathbf{x}, t) \rangle]$$

$$\gamma_8^P(\mathbf{x}, t) = K_{po}^{-1}(\mathbf{x}, t) J_{pt}(\mathbf{x}, t) K_{g_p}$$

$$\gamma_9^P(\mathbf{x}, t) = - \tfrac{1}{4} (\theta_{sp} - \theta_{rp}) \exp[0.5 \langle \alpha_p \rangle \langle \psi_p(\mathbf{x}, t) \rangle] [2 \langle \alpha_p \rangle \langle \psi_p(\mathbf{x}, t) \rangle$$
$$+ 0.5 \langle \alpha_p \rangle^2 \langle \psi_p(\mathbf{x}, t) \rangle^2]$$

$$\gamma_{10}^P(\mathbf{x}, t) = - \tfrac{1}{4} (\theta_{sp} - \theta_{rp}) \exp[0.5 \langle \alpha_p \rangle \langle \psi_p(\mathbf{x}, t) \rangle]$$
$$\cdot [\langle \alpha_p \rangle^2 + 0.5 \langle \alpha_p \rangle^3 \langle \psi_p(\mathbf{x}, t) \rangle]$$

and $C_p^{(0)}(\mathbf{x}, t) = - \tfrac{1}{4} (\theta_{sp} - \theta_{rp}) \langle \alpha_p \rangle^2 \langle \psi_p(\mathbf{x}, t) \rangle \exp[0.5 \langle \alpha_p \rangle \langle \psi_p(\mathbf{x}, t) \rangle]$. In Eq. (7.52), the cross-covariances $C_{g_p \psi_p}(\mathbf{x}; \chi, \tau)$, $C_{f_p \psi_p}(\mathbf{x}; \chi, \tau)$, $C_{\alpha_p \psi_p}(\mathbf{x}; \chi, \tau)$, $C_{\beta \psi_p}(\mathbf{x}; \chi, \tau)$, and $C_{\psi_q \psi_p}(\mathbf{x}, t; \chi, \tau)$ are solutions of the following equations under appropriate boundary conditions:

$$\frac{\partial^2 C_{g_p \psi_p}(\mathbf{x}; \chi, \tau)}{\partial \chi_i^2} + \gamma_{1i}^P(\chi, \tau) \frac{\partial C_{g_p \psi_p}(\mathbf{x}; \chi, \tau)}{\partial \chi_i}$$

$$+ [\gamma_2^P(\chi, \tau) + \gamma_3^P(\chi, \tau)] C_{g_p \psi_p}(\mathbf{x}; \chi, \tau)$$

$$= \gamma_4^P(\chi, \tau) \frac{\partial C_{g_p \psi_p}(\mathbf{x}; \chi, \tau)}{\partial \tau} - J_{pi}(\chi, \tau) \left[\frac{\partial C_{g_p}(\mathbf{x}; \chi)}{\partial \chi_i} + \frac{\partial C_{g_p f_p}(\mathbf{x}; \chi)}{\partial \chi_i} \right.$$

$$\left. + \langle \psi_p(\chi, \tau) \rangle \frac{\partial C_{g_p \alpha_p}(\mathbf{x}; \chi)}{\partial \chi_i} + \gamma_5^P(\chi, \tau) C_{g_p \alpha_p}(\mathbf{x}; \chi) \right]$$

$$+ \gamma_6^P(\chi, \tau) [C_{g_p}(\mathbf{x}; \chi) + C_{g_p f_p}(\mathbf{x}; \chi) + \langle \psi_p(\chi, \tau) \rangle C_{g_p \alpha_p}(\mathbf{x}; \chi)]$$

$$+ \gamma_7^P(\chi, \tau) C_{g_p \beta}(\mathbf{x}; \chi) + \gamma_8^P(\chi, \tau) [\gamma_9^P(\chi, \tau) C_{g_p \alpha_p}(\mathbf{x}; \chi)$$

$$+ \gamma_{10}^P(\chi, \tau) C_{g_p \psi_p}(\mathbf{x}; \chi, \tau) + C_p^{(0)}(\chi, \tau) C_{g_p}(\mathbf{x}; \chi)] + \gamma_3^P(\chi, \tau) C_{g_p \psi_q}(\mathbf{x}; \chi, \tau)$$

$$(7.53)$$

$$\frac{\partial^2 C_{f_p \psi_p}(\mathbf{x}; \chi, \tau)}{\partial \chi_i^2} + \gamma_{1i}^p(\chi, \iota) \frac{\partial C_{f_p \psi_p}(\mathbf{x}; \chi, \tau)}{\partial \chi_i} + [\gamma_2^p(\chi, \tau) + \gamma_3^p(\chi, \tau)] C_{f_p \psi_p}(\mathbf{x}; \chi, \tau)$$

$$= \gamma_4^p(\chi, \tau) \frac{\partial C_{f_p \psi_p}(\mathbf{x}; \chi, \tau)}{\partial \tau} - J_{pi}(\chi, \tau) \left[\frac{\partial C_{f_p g_p}(\mathbf{x}; \chi)}{\partial \chi_i} \right.$$

$$+ \frac{\partial C_{f_p}(\mathbf{x}; \chi)}{\partial \chi_i} + \langle \psi_p(\chi, \tau) \rangle \frac{\partial C_{f_p \alpha_p}(\mathbf{x}; \chi)}{\partial \chi_i} + \left. \gamma_5^p(\chi, \tau) C_{f_p \alpha_p}(\mathbf{x}; \chi) \right]$$

$$+ \gamma_6^p(\chi, \tau) [C_{f_p g_p}(\mathbf{x}; \chi) + C_{f_p}(\mathbf{x}; \chi) + \langle \psi_p(\chi, \tau) \rangle C_{f_p \alpha_p}(\mathbf{x}; \chi)]$$

$$+ \gamma_7^p(\chi, \tau) C_{f_p \beta}(\mathbf{x}; \chi) + \gamma_8^p(\chi, \tau) [\gamma_9(\chi, \tau) C_{f_p \alpha_p}(\mathbf{x}; \chi)$$

$$+ \gamma_{10}^p(\chi, \tau) C_{f_p \psi_p}(\mathbf{x}; \chi, \tau) + C_p^{(0)}(\chi, \tau) \mathcal{C}_{f_p g_p}(\mathbf{x}; \chi, \tau)]$$

$$+ \gamma_3^p(\chi, \tau) C_{f_p \psi_q}(\mathbf{x}; \chi, \tau) \tag{7.54}$$

$$\frac{\partial^2 C_{\alpha_p \psi_p}(\mathbf{x}; \chi, \tau)}{\partial \chi_i^2} + \gamma_{1i}^p(\chi, \tau) \frac{\partial C_{\alpha_p \psi_p}(\mathbf{x}; \chi, \tau)}{\partial \chi_i} + [\gamma_2^p(\chi, \tau) + \gamma_3^p(\chi, \tau)] C_{\alpha_p \psi_p}(\mathbf{x}; \chi, \tau)$$

$$= \gamma_4^p(\chi, \tau) \frac{\partial C_{\alpha_p \psi_p}(\mathbf{x}; \chi, \tau)}{\partial t} - J_{pi}(\chi, \tau) \left[\frac{\partial C_{\alpha_p g_p}(\mathbf{x}; \chi)}{\partial \chi_i} \right.$$

$$+ \frac{\partial C_{\alpha_p f_p}(\mathbf{x}; \chi)}{\partial \chi_i} + \langle \psi_p(\chi, \tau) \rangle \frac{\partial C_{\alpha_p}(\mathbf{x}; \chi)}{\partial \chi_i} + \left. \gamma_5^p(\chi, \tau) C_{\alpha_p}(\mathbf{x}; \chi) \right]$$

$$+ \gamma_6^p(\chi, \tau) [C_{\alpha_p g_p}(\mathbf{x}; \chi) + C_{\alpha_p f_p}(\mathbf{x}; \chi) + \langle \psi_p(\chi, \tau) \rangle C_{\alpha_p}(\mathbf{x}; \chi)]$$

$$+ \gamma_7^p(\chi, \tau) C_{\alpha_p \beta}(\mathbf{x}; \chi) + \gamma_8^p(\chi, \tau) [\gamma_9^p(\chi, \tau) C_{\alpha_p}(\mathbf{x}; \chi)$$

$$+ \gamma_{10}^p(\chi, \tau) C_{\alpha_p \psi_p}(\mathbf{x}; \chi, \tau) + C_p^{(0)}(\chi, \tau) \mathcal{C}_{\alpha_p g_p}(\mathbf{x}; \chi, \tau)]$$

$$+ \gamma_3^p(\chi, \tau) C_{\alpha_p \psi_q}(\mathbf{x}; \chi, \tau) \tag{7.55}$$

$$\frac{\partial^2 C_{\beta \psi_p}(\mathbf{x}; \chi, \tau)}{\partial \chi_i^2} + \gamma_{1i}^p(\chi, \tau) \frac{\partial C_{\beta \psi_p}(\mathbf{x}; \chi, \tau)}{\partial \chi_i} + [\gamma_2^p(\chi, \tau) + \gamma_3^p(\chi, \tau)] C_{\beta \psi_p}(\mathbf{x}; \chi, \tau)$$

$$= \gamma_4^p(\chi, \tau) \frac{\partial C_{\beta \psi_p}(\mathbf{x}; \chi, \tau)}{\partial \tau} - J_{pi}(\chi, \tau) \left[\frac{\partial C_{\beta g_p}(\mathbf{x}; \chi)}{\partial \chi_i} + \frac{\partial C_{\beta f_p}(\mathbf{x}; \chi)}{\partial \chi_i} \right.$$

$$+ \langle \psi_p(\chi, \tau) \rangle \frac{\partial C_{\beta \alpha_p}(\mathbf{x}; \chi)}{\partial \chi_i} + \left. \gamma_5^p(\chi, \tau) C_{\beta \alpha_p}(\mathbf{x}; \chi) \right]$$

$$+ \gamma_6^p(\chi, \tau) [C_{\beta g_p}(\mathbf{x}; \chi) + C_{\beta f_p}(\mathbf{x}; \chi)$$

$$+ \langle \psi_p(\chi, \tau) \rangle C_{\beta \alpha_p}(\mathbf{x}; \chi)] + \gamma_7^p(\chi, \tau) C_\beta(\mathbf{x}; \chi)$$

$$+ \gamma_8^p(\chi, \tau) [\gamma_9^p(\chi, \tau) C_{\beta \alpha_p}(\mathbf{x}; \chi) + \gamma_{10}^p(\chi, \tau) C_{\beta \psi_p}(\mathbf{x}; \chi, \tau)$$

$$+ C_p^{(0)}(\chi, \tau) \mathcal{C}_{\beta g_p}(\mathbf{x}; \chi, \tau) \Big] + \gamma_3^p(\chi, \tau) C_{\beta \psi_q}(\mathbf{x}; \chi, \tau) \tag{7.56}$$

$$
\frac{\partial^2 C_{\psi_q \psi_p}(\mathbf{x}, t; \boldsymbol{\chi}, \tau)}{\partial \chi_i^2} + \gamma_{1i}^p(\boldsymbol{\chi}, \tau) \frac{\partial C_{\psi_q \psi_p}(\mathbf{x}, t; \boldsymbol{\chi}, \tau)}{\partial \chi_i}
$$

$$
+ [\gamma_2^p(\boldsymbol{\chi}, \tau) + \gamma_3^p(\boldsymbol{\chi}, \tau)] C_{\psi_q \psi_p}(\mathbf{x}, t; \boldsymbol{\chi}, \tau)
$$

$$
= \gamma_4^p(\boldsymbol{\chi}, \tau) \frac{\partial C_{\psi_q \psi_p}(\mathbf{x}, t; \boldsymbol{\chi}, \tau)}{\partial \tau} - J_{pi}(\boldsymbol{\chi}, \tau) \left[\frac{\partial C_{\psi_q g_p}(\mathbf{x}; \boldsymbol{\chi}, \tau)}{\partial \chi_i} + \frac{\partial C_{\psi_q f_p}(\mathbf{x}; \boldsymbol{\chi}, \tau)}{\partial \chi_i} \right.
$$

$$
+ \langle \psi_p(\boldsymbol{\chi}, \tau) \rangle \frac{\partial C_{\psi_q \alpha_p}(\mathbf{x}; \boldsymbol{\chi}, \tau)}{\partial \chi_i} + \gamma_5^p(\boldsymbol{\chi}, \tau) C_{\psi_q \alpha_p}(\mathbf{x}; \boldsymbol{\chi}, \tau) \Big]
$$

$$
+ \gamma_6^p(\boldsymbol{\chi}, \tau)[C_{\psi_q g_p}(\mathbf{x}; \boldsymbol{\chi}, \tau) + C_{\psi_q f_p}(\mathbf{x}; \boldsymbol{\chi}, \tau) + \langle \psi_p(\boldsymbol{\chi}, \tau) \rangle C_{\psi_q \alpha_p}(\mathbf{x}; \boldsymbol{\chi}, \tau)]
$$

$$
+ \gamma_7^p(\boldsymbol{\chi}, \tau) C_{\psi_q \beta}(\mathbf{x}; \boldsymbol{\chi}, \tau) + \gamma_8^p(\boldsymbol{\chi}, \tau)[\gamma_9^p(\boldsymbol{\chi}, \tau) C_{\psi_q \alpha_p}(\mathbf{x}; \boldsymbol{\chi}, \tau)
$$

$$
+ \gamma_{10}^p(\boldsymbol{\chi}, \tau) C_{\psi_q \psi_p}(\mathbf{x}, t; \boldsymbol{\chi}, \tau) + C_p^{(0)}(\boldsymbol{\chi}, \tau) \mathcal{C}_{\psi_q g_p}(\mathbf{x}; \boldsymbol{\chi}, \tau)] + \gamma_3^p(\boldsymbol{\chi}, \tau) C_{\psi_q}(\mathbf{x}, t; \boldsymbol{\chi}, \tau)
$$

$$
\tag{7.57}
$$

In Eqs. (7.52)–(7.56), there are still five unknown cross-covariances between the fracture and the matrix continuum, $C_{g_p \psi_q}(\mathbf{x}; \boldsymbol{\chi}, \tau)$, $C_{f_p \psi_q}(\mathbf{x}; \boldsymbol{\chi}, \tau)$, $C_{\alpha_p \psi_q}(\mathbf{x}; \boldsymbol{\chi}, \tau)$, $C_{\beta \psi_q}(\mathbf{x}; \boldsymbol{\chi}, \tau)$, and $C_{\psi_q}(\mathbf{x}, t; \boldsymbol{\chi}, \tau)$. The last two covariances are given by Eqs. (7.56) and (7.52), respectively, via replacing p and q by q and p. The others are given by

$$
\frac{\partial^2 C_{g_p \psi_q}(\mathbf{x}; \boldsymbol{\chi}, \tau)}{\partial \chi_i^2} + \gamma_{1i}^q(\boldsymbol{\chi}, \tau) \frac{\partial C_{g_p \psi_q}(\mathbf{x}; \boldsymbol{\chi}, \tau)}{\partial \chi_i} + [\gamma_2^q(\boldsymbol{\chi}, \tau) + \gamma_3^q(\boldsymbol{\chi}, \tau)] C_{g_p \psi_q}(\mathbf{x}; \boldsymbol{\chi}, \tau)
$$

$$
= \gamma_4^q(\boldsymbol{\chi}, \tau) \frac{\partial C_{g_p \psi_q}(\mathbf{x}; \boldsymbol{\chi}, \tau)}{\partial \tau} - J_{pi}(\boldsymbol{\chi}, \tau) \left[\frac{\partial C_{g_p g_q}(\mathbf{x}; \boldsymbol{\chi})}{\partial \chi_i} + \frac{\partial C_{g_p f_q}(\mathbf{x}; \boldsymbol{\chi})}{\partial \chi_i} \right.
$$

$$
+ \langle \psi_p(\boldsymbol{\chi}, \tau) \rangle \frac{\partial C_{g_p \alpha_q}(\mathbf{x}; \boldsymbol{\chi})}{\partial \chi_i} + \gamma_5^q(\boldsymbol{\chi}, \tau) C_{g_p \alpha_q}(\mathbf{x}; \boldsymbol{\chi}) \Big]
$$

$$
+ \gamma_6^q(\boldsymbol{\chi}, \tau)[C_{g_p g_q}(\mathbf{x}; \boldsymbol{\chi}) + C_{g_p f_q}(\mathbf{x}; \boldsymbol{\chi}) + \langle \psi_p(\boldsymbol{\chi}, \tau) \rangle C_{g_p \alpha_q}(\mathbf{x}; \boldsymbol{\chi})]
$$

$$
+ \gamma_7^q(\boldsymbol{\chi}, \tau) C_{g_p \beta}(\mathbf{x}; \boldsymbol{\chi}) + \gamma_8^q(\boldsymbol{\chi}, \tau)[\gamma_9^q(\boldsymbol{\chi}, \tau) C_{g_p \alpha_q}(\mathbf{x}; \boldsymbol{\chi})
$$

$$
+ \gamma_{10}^q(\boldsymbol{\chi}, \tau) C_{g_p \psi_q}(\mathbf{x}; \boldsymbol{\chi}, \tau) + C_p^{(0)}(\boldsymbol{\chi}, \tau) \mathcal{C}_{g_p g_q}(\mathbf{x}; \boldsymbol{\chi})] + \gamma_3^q(\boldsymbol{\chi}, \tau) C_{g_p \psi_p}(\mathbf{x}; \boldsymbol{\chi}, \tau)
$$

$$
\tag{7.58}
$$

$$
\frac{\partial^2 C_{f_p \psi_q}(\mathbf{x}; \boldsymbol{\chi}, \tau)}{\partial \chi_i^2} + \gamma_{1i}^q(\boldsymbol{\chi}, \tau) \frac{\partial C_{f_p \psi_q}(\mathbf{x}; \boldsymbol{\chi}, \tau)}{\partial \chi_i} + [\gamma_2^q(\boldsymbol{\chi}, \tau) + \gamma_3^q(\boldsymbol{\chi}, \tau)] C_{f_p \psi_q}(\mathbf{x}; \boldsymbol{\chi}, \tau)
$$

$$
= \gamma_4^q(\boldsymbol{\chi}, \tau) \frac{\partial C_{f_p \psi_q}(\mathbf{x}; \boldsymbol{\chi}, \tau)}{\partial \tau} - J_{pi}(\boldsymbol{\chi}, \tau) \left[\frac{\partial C_{f_p g_q}(\mathbf{x}; \boldsymbol{\chi})}{\partial \chi_i} + \frac{\partial C_{f_p f_q}(\mathbf{x}; \boldsymbol{\chi})}{\partial \chi_i} \right.
$$

$$
+ \langle \psi_p(\boldsymbol{\chi}, \tau) \rangle \frac{\partial C_{f_p \alpha_q}(\mathbf{x}; \boldsymbol{\chi})}{\partial \chi_i} + \gamma_5^q(\boldsymbol{\chi}, \tau) C_{f_p \alpha_q}(\mathbf{x}; \boldsymbol{\chi}) \Big]
$$

$$+ \gamma_6^q(\chi, \tau)[C_{f_p g_q}(\mathbf{x}; \chi) + C_{f_p f_q}(\mathbf{x}; \chi) + \langle \psi_p(\chi, \tau) \rangle C_{f_p \alpha_q}(\mathbf{x}; \chi)]$$

$$+ \gamma_7^q(\chi, \tau) C_{f_p \beta}(\mathbf{x}; \chi) + \gamma_8^q(\chi, \tau)[\gamma_9^q(\chi, \iota) C_{f_p \alpha_q}(\mathbf{x}; \chi)$$

$$+ \gamma_{10}^q(\chi, \tau) C_{f_p \psi_q}(\mathbf{x}; \chi, \tau) + C_p^{(0)}(\chi, \tau) \mathcal{C}_{f_p g_q}(\mathbf{x}; \chi)] + \gamma_3^q(\chi, \tau) C_{f_p \psi_p}(\mathbf{x}; \chi, \tau) \quad (7.59)$$

$$\frac{\partial^2 C_{\alpha_p \psi_q}(\mathbf{x}; \chi, \tau)}{\partial \chi_i^2} + \gamma_{1i}^q(\chi, \tau) \frac{\partial C_{\alpha_p \psi_q}(\mathbf{x}; \chi, \tau)}{\partial \chi_i} + [\gamma_2^q(\chi, \tau) + \gamma_3^q(\chi, \tau)] C_{\alpha_p \psi_q}(\mathbf{x}; \chi, \tau)$$

$$= \gamma_4^q(\chi, \tau) \frac{\partial C_{\alpha_p \psi_q}(\mathbf{x}; \chi, \tau)}{\partial \tau} - J_{pi}(\chi, \tau) \left[\frac{\partial C_{\alpha_p g_q}(\mathbf{x}; \chi)}{\partial \chi_i} + \frac{\partial C_{\alpha_p f_q}(\mathbf{x}; \chi)}{\partial \chi_i} \right.$$

$$+ \langle \psi_p(\chi, \tau) \rangle \frac{\partial C_{\alpha_p \alpha_q}(\mathbf{x}; \chi)}{\partial \chi_i} + \gamma_5^q(\chi, \tau) C_{\alpha_p \alpha_q}(\mathbf{x}; \chi) \right]$$

$$+ \gamma_6^q(\chi, \tau)[C_{\alpha_p g_q}(\mathbf{x}; \chi) + C_{\alpha_p f_q}(\mathbf{x}; \chi) + \langle \psi_p(\chi, \tau) \rangle C_{\alpha_p \alpha_q}(\mathbf{x}; \chi)]$$

$$+ \gamma_7^q(\chi, \tau) C_{\alpha_p \beta}(\mathbf{x}; \chi) + \gamma_8^q(\chi, \tau)[\gamma_9^q(\chi, \tau) C_{\alpha_p \alpha_q}(\mathbf{x}; \chi) + \gamma_{10}^q(\chi, \tau) C_{\alpha_p \psi_q}(\mathbf{x}; \chi, \tau)$$

$$+ C_p^{(0)}(\chi, \tau) \mathcal{C}_{\alpha_p g_q}(\mathbf{x}; \chi)] + \gamma_3^q(\chi, \tau) C_{\alpha_p \psi_p}(\mathbf{x}; \chi, \tau) \quad (7.60)$$

In Eqs. (7.58)–(7.60), the coefficients γ^qs are defined as γ^ps through switching p and q. If we let $p = f$ and $q = m$, $C_{\psi_f}(\mathbf{x}, t; \chi, \tau)$ is obtained by solving Eqs. (7.52)–(7.60) where $C_{\psi_m}(\mathbf{x}, t; \chi, \tau)$ is present and unknown. The latter is given by a similar set of equations with $C_{\psi_f}(\mathbf{x}, t; \chi, \tau)$ unknown. Therefore, like the mean heads in the fracture and the matrix continuum in Eqs. (7.45) and (7.46), the head covariances in the fracture and the matrix continuum have to be solved from two sets of coupled equations.

These coupled equations for the pressure head in an unsaturated system are analogous to those for saturated flow given in Section 7.2 except that there are more equations in the case of unsaturated flow. Although these covariance equations have not been implemented for unsaturated flow in fractured porous media, the solution strategy discussed in Section 7.2 for saturated flow is directly applicable to them. In addition, these equations are of the same type but with different forcing terms.

7.4 EXERCISES

1. The water fluxes in the fracture and the matrix systems are given by Eqs. (7.7) and (7.8), and the composite flux is $\mathbf{q}(\chi, \tau) = w_f \mathbf{q}_f(\chi, \tau) + (1 - w_f) \mathbf{q}_m(\chi, \tau)$ (where w_f is the relative volume fraction of fracture). Derive expressions for the first two moments of $\mathbf{q}$ in the terms of the log hydraulic conductivity and head moments for the following two cases: (a) w_f is a known constant; (b) it is a random space function with given statistical moments.

2. Consider steady-state saturated flow in fractured porous media with the double-permeability representation. Flow in such a situation is governed by the steady-state version of Eqs. (7.1)–(7.4). Derive moment equations for h_f and h_m. Solve for the head covariances analytically (with Green's function method or spectral method) under the following simplified conditions: The domain is unbounded; the mean flow is uniform in both the fracture and matrix systems with $\langle h_f \rangle = \langle h_m \rangle = $ const; and the fracture volume fraction w_f is a known constant.

3. Derive moment equations for steady-state unsaturated flow in fractured porous media on the basis of the double-permeability representation. For the special case that flow is mean gravity-dominated such that $\langle \psi_f \rangle = \langle \psi_m \rangle = x_1$, solve these moment equations analytically in unbounded domains.

REFERENCES

Ababou, R., and L.W. Gelhar, Self-similar randomness and spectral conditioning: Analysis of scale effects in subsurface hydrology, in *Dynamics of Fluids in Hierarchical Porous Media*, edited by J.H. Cushman, pp. 393–428, Academic Press, San Diego, CA., 1990.

Ababou, R., D. McLaughlin, L.W. Gelhar, and A.F.B. Tompson, Numerical simulation of three-dimensional saturated flow in randomly heterogeneous porous media, *Transp. Porous Media*, **4**, 549–565, 1989.

Abdin, A., J.J. Kaluarachchi, C.-M. Chang, and M.W. Kemblowski, Stochastic analysis of two-phase flow in porous media: II. Comparison between perturbation and Monte-Carlo results, *Transp. Porous Media*, **19**, 261–280, 1995.

Abramowitz, M., and I.A. Stegun, *Handbook of Mathematical Functions*, Dover., New York, 1970.

Acuna, J.A., and Y.C. Yortsos, Application of fractal geometry to the study of networks of fractures and their pressure transient, *Water Resour. Res.*, **31**(3), 527–540, 1995.

Adomian, G., *Stochastic Systems*, Academic Press, New York, 1983.

Arfken, G., *Mathematical Methods for Physicists*, third edition, Academic Press, San Diego, CA., 1985.

Aziz, K., and A. Settari, *Petroleum Reservoir Simulation*, Applied Science Publishers, London, 1979.

Bakr, A.A., *Stochastic Analysis of the Effects of Spatial Variations of Hydraulic Conductivity on Groundwater Flow*, Ph.D. dissertation, New Mexico Institute of Mining and Technology, Socorro, New Mexico, 1976.

Bakr, A.A., L.W. Gelhar, A.L. Gutjahr, and J.R. MacMillan, Stochastic analysis of spatial variability in subsurface flows, 1. Comparison of one- and three-dimensional flows, *Water Resour. Res.*, **14**(2), 263–272, 1978.

Barenblatt, G.I., Y.P. Zheltov, and I.N. Kochina, Basic concepts in the theory of seepage of homogeneous liquids in fissured rocks, *J. Appl. Math. Mech., Engl. Transl.*, **24**, 1286–1303, 1960.

Barton, C.C., and P.A. Hsieh, *Physical and Hydrologic-Flow Properties of Fractures, Field Trip Guide Book*, Vol. T 385, AGU, Washington, D.C., 1989.

Bear, J., *Dynamics of Fluids in Porous Media*, Dover., New York, 1972.

Bellin, A., P. Salandin, and A. Rinaldo, Simulation of dispersion in heterogeneous porous formations: Statistics, first-order theories, convergence of computations, *Water Resour. Res.*, **28**(9), 2211–2227, 1992.

Beran, M.J, *Statistical Continuum Theories*, Wiley Interscience, New York, 1968.

Bharucha-Reid, A.T., *Random Integral Equations*, Mathematics in Science and Engineering, Vol. 96, Academic Press, 1972.

Bibby, R., Mass transport of solutes in dual-porosity media, *Water Resour. Res.*, **17**, 1075–1081, 1981.

Billaux, D., J.P. Chiles, K. Hestir, and J.C.S. Long, Three-dimensional statistical modelling of a fractured rock mass: An example from the Fanay-Augeres mine, *Int. J. Rock Mech. Min. Sci. Geomech. Abstr.*, **26**(3–4), 281–299, 1989.

Binning, P., and M.A. Celia, Practical implementation of the fractional flow approach to multi-phase flow simulation, *Adv. Water Resour.*, **22**(5), 461–478, 1999.

Blunt, M.J., K. Liu, and M.R. Thiele, A generalized streamline method to predict reservoir flow, *Petrol. Geosci.*, **2**, 259–269, 1996.

Bonilla, F.A., and J.H. Cushman, Role of boundary conditions in convergence and nonlocality of solutions to stochastic flow problems in bounded domains, *Water Resour. Res.*, **36**(4), 981–997, 2000.

Brooks, R.H., and A.T. Corey, Hydraulic properties of porous media, *Hydrol. Pap.* 3, Colo. State Univ., Fort Collins, 1964.

Cacas, M.C., E. Ledoux, G. de Marsily, B. Tillie, A Barbreau, E. Durant, B. Feuga, and P. Peaudecerf, Modeling fracture flow with a stochastic discrete fracture network: Calibration and validation, 1. The flow model, *Water Resour. Res.*, **26**(3), 479–489, 1990.

Carle, S.F., and G.E. Fogg, Transition probability-based indicator geostatistics, *Math. Geol.*, **28**(4), 453–477, 1996.

Carle, S.F., and G.E. Fogg, Modeling spatial variability with one- and multi-dimensional Markov chains, *Math. Geol.*, **28**(7), 891–917, 1997.

Carslaw, H.S., and J.C. Jaeger, *Conduction of Heat in Solids*, Oxford University Press, New York, 1959.

Chang, C.-M., M.W. Kemblowski, J.J. Kaluarachchi, and A. Abdin, Stochastic analysis of two-phase flow in porous media: I. Spectral/perturbation approach, *Transp. Porous Media*, **19**, 223–259, 1995.

Chen, H., S. Chen, and R.H. Kraichnan, Probability distribution of a stochastically advected scalar field, *Phys. Rev. Lett.*, **63**(24), 2657–2660, 1989.

Chen, W.H., L.J. Durlofsky, B. Engquist, and S. Osheer, Minimization of grid orientation effects through use of higher order finite difference methods, *SPE Adv. Technol. Ser.*, **1**(2), 43–52, 1993.

Cheng, A.H-D., and D.E. Lafe, Boundary element solution for stochastic groundwater flow: Random boundary condition and recharge, *Water Resour. Res.*, **27**(2), 231–242, 1991.

Chiles, J.P., and G. de Marsily, Stochastic models of fracture systems and their use in flow and transport modeling, in *Flow and Contaminant Transport in Fractured Rock*, edited by J. Bear, C.-F. Tsang, and G. de Marsily, pp. 169–236, Academic Press, San Diego, CA., 1993.

Chin, D.A., An assessment of first-order stochastic dispersion theories in porous media, *J. Hydrol.*, **199**, 53–73, 1997.

Chin, D.A., and T. Wang, An investigation of the validity of first-order stochastic dispersion theories in isotropic porous media, *Water Resour. Res.*, **28**(6), 1531–1542, 1992.

Christakos, G., *Random Field Models in Earth Sciences*, Academic Press, San Diego, CA., 1992.

Christakos, G., D.T. Hristopulos, and C.T. Miller, Stochastic diagrammatic analysis of groundwater flow in heterogeneous porous media, *Water Resour. Res.*, **31**(7), 1687–1703, 1995.

Clifton, P.M., and S.P. Neuman, Effects of kriging and inverse modeling on conditional simulation of the Avra Valley aquifer in southern Arizona, *Water Resour. Res.*, **18**(4), 1215–1234, 1982.

Cushman, J.H., On diffusion in fractal porous media, *Water Resour. Res.*, **27**, 643–644, 1991.

Cushman, J.H., *The Physics of Fluids in Hierarchical Porous Media: Angstroms to Miles*, Kluwer Academic, Norwell, MA., 1997.

Cvetkovic, V., and G. Dagan, Reactive transport and immiscible flow in geologic media: II. Applications, *Proc. R. Soc. Lond.* A, **452**, 303–328, 1996.

Cvetkovic, V., A.M. Shapiro, and G. Dagan, A solute flux approach to transport in heterogeneous formation: 2. Uncertainty analysis, *Water Resour. Res.*, **28**, 1377–1388, 1992.

Dagan, G., Stochastic modeling of groundwater flow by unconditional and conditional probabilities: 1. Conditional simulation and the direct problem, *Water Resour. Res.*, **18**(4), 813–833, 1982a.

Dagan, G., Stochastic modeling of groundwater flow by unconditional and conditional probabilities: 2. The solute transport, *Water Resour. Res.*, **18**(4), 835–848, 1982b.

Dagan, G., Analysis of flow through heterogeneous random aquifer: 2. Unsteady flow in confined formations, *Water Resour. Res.*, **18**(5), 1571–1585, 1982c.

Dagan, G., Solute transport in heterogeneous porous formations, *J. Fluid Mech.*, **145**, 151–177, 1984.

Dagan, G., Stochastic modeling of groundwater flow by unconditional and conditional probabilities: The inverse problem, *Water Resour. Res.*, **21**(1), 65–72, 1985.

Dagan, G., *Flow and Transport in Porous Formations*, Springer-Verlag, New York, 1989.

Dagan, G., Higher-order correction of effective conductivity of heterogeneous formations of lognormal conductivity distribution, *Transp. Porous Media*, **12**, 279–290, 1993.

Dagan, G., and V. Cvetkovic, Reactive transport and immiscible flow in geologic media: I. General theory, *Proc. R. Soc. Lond.* A, **452**, 285–301, 1996.

Dagan, G., and V. Nguyen, A comparison of travel time and concentration approaches to modeling transport by groundwater, *J. Contam. Hydrol.*, **4**, 79–91, 1989.

Dagan, G., V. Cvetkovic, and A.M. Shapiro, A solute flux approach to transport in heterogeneous formation: 1. The general framework, *Water Resour. Res.*, **28**, 1369–1376, 1992.

Dagan, D., A. Bellin, and Y. Rubin, Lagrangian analysis of transport in heterogeneous formations under transient flow conditions, *Water Resour. Res.*, **32**(4), 891–899, 1996.

de Marsily, G., *Quantitative Hydrogeology*, Academic Press, San Diego, CA., 1986.

Deng, F.-W., and J.H. Cushman, On higher-order corrections to the flow velocity covariance tensor, *Water Resour. Res.*, **31**(7), 1659–1672, 1995.

Deng, F.-W., and J.H. Cushman, Higher-order corrections to the flow velocity covariance tensor, revisited, *Water Resour. Res.*, **34**(1), 103–106, 1998.

Dershowitz, W.S., *Rock Joint Systems*, Ph.D. thesis, Massachusetts Institute of Technology, 1984.

Desbarats, A.J., Numerical estimation of effective permeability in sand–shale formations, *Water Resour. Res.*, **23**(2), 273–286, 1987.

Desbarats, A.J., Spatial averaging of transmissivity in heterogeneous fields with flow toward a well, *Water Resour. Res.*, **28**(3), 757–767, 1992.

Dettinger, M.D., and J.L. Wilson, First-order analysis of uncertainty in numerical models of groundwater: Part 1. Mathematical development, *Water Resour. Res.*, **17**(1), 149–161, 1981.

Deutsch, C.V., and A.G. Journel, *GSLIB: Geostatistical Software Library and User's Guide*, second edition, Oxford University Press, New York, 1998.

De Wit, A., Correlation structure dependence of the effective permeability of heterogeneous porous media, *Phys. Fluids*, **7**(11), 2553–2562, 1995.

Di Federico, V., and S.P. Neuman, Scaling of random fields by means of truncated power variograms and associated spectra, *Water Resour. Res.*, **33**(5), 1075–1085, 1997.

Di Federico, V., and S.P. Neuman, Flow in multiscale log conductivity fields with truncated power variograms, *Water Resour. Res.*, **34**(5), 975–987, 1998.

Duguid, J.O., and P.C.Y. Lee, Flow in fractured porous media, *Water Resour. Res.*, **13**, 558–566, 1977.

Durlofsky, L.J., Accuracy of mixed and control volume finite element approximations to Darcy velocity and related quantities, *Water Resour. Res.*, **30**(4), 965–973, 1994.

Dverstorp, B., and J. Andersson, Application of the discrete fracture network concept with field data: Possibilities of model calibration and validation, *Water Resour. Res.*, **25**(3), 540–550, 1989.

Dverstorp, B., J. Andersson, and W. Nordqvist, Discrete fracture network interpretation of field tracer migration in sparsely fractured rock, *Water Resour. Res.*, **28**(9), 2327–2343, 1992.

Dykaar, B.B., and P.K. Kitanidis, Determination of the effective hydraulic conductivity for heterogeneous porous media using a numerical spectral approach, 2. Results, *Water Resour. Res.*, **28**(4), 1155–1166, 1992.

Dykhuizen, R.C., Transport of solutes through unsaturated fractured media, *Water Res.*, **21**, 1531–1539, 1987.

Dykhuizen, R.C., A new coupling term for dual-porosity models, *Water Resour. Res.*, **26**, 351–356, 1990.

Eckhardt, R., Stan Ulam, John von Neumann, and the Monte Carlo method, *Los Alamos Sci.*, Los Alamos National Laboratory, **15**, 131–137, 1987.

Elishakoff, L., *Probability Methods in the Theory of Structures*, John Wiley & Sons, New York, 1983.

Ewing, R.E., Aspects of upscaling in simulation of flow in porous media, *Adv. Water Resour.*, **20**(5–6), 349–358, 1997.

Feynman, R.P., Space–time approach to nonrelativistic quantum mechanics, *Rev. Mod. Phys.*, **20**, 367–387, 1948.

Foussereau, X., W.D. Graham, and P.S.C. Rao, Stochastic analysis of transient flow in unsaturated heterogeneous soils, *Water Resour. Res.*, **36**(4), 891–910, 2000.

Freeze, R.A., A stochastic-conceptual analysis of one-dimensional groundwater flow in nonuniform homogeneous media, *Water Resour. Res.*, **11**(5), 725–741, 1975.

Freeze, R.A., and J.A. Cherry, *Groundwater*, Prentice-Hall, Englewood Cliffs, NJ, 1979.

Frisch, U., Wave propagation in random media, in *Probabilistic Methods in Applied Mathematics*, edited by A.T. Bharucha-Reid, Vol. 1, pp. 75–198, Academic Press, New York, 1968.

Gardiner, G.W., *Handbook of Stochastic Methods for Physics, Chemistry and the Natural Sciences*, second edition, Springer-Verlag, New York, 1985.

Gardner, W.R., Some steady state solutions of unsaturated moisture flow equations with application to evaporation from a water table, *Soil Sci.*, **85**, 228–232, 1958.

Gelhar, L.W., *Stochastic Subsurface Hydrology*, Prentice-Hall, Englewood Cliffs, NJ, 1993.

Gelhar, L.W., and C.L. Axness, Three-dimensional stochastic analysis of macrodispersion in aquifers, *Water Resour. Res.*, **19**(1), 161–180, 1983.

Gerke, H.H., and M.T. van Genuchten, A dual-porosity model for simulating the preferential movement of water and solutes in structured porous media, *Water Resour. Res.*, **29**, 305–319, 1993a.

Gerke, H.H., and M.T. van Genuchten, Evaluation of a first-order water transfer term for variably saturated dual-porosity models, *Water Resour. Res.*, **29**, 1225–1238, 1993b.

Ghanem, R.G., and P.D. Spanos, *Stochastic Finite Elements: A Spectral Approach*, Springer-Verlag, New York, 1991.

Gomez-Hernandez, J.J., *A Stochastic Approach to the Simulation of Block Conductivity Fields Conditioned upon Data Measured at a Smaller Scale*, Ph.D. dissertation, Stanford University, 1991.

Gradshteyn, I.S., and I.M. Ryzhik, *Table of Integrals, Series, and Products*, Academic Press, San Diego, CA., 1980.

Graham, W.D., and D. McLaughlin, Stochastic analysis of nonstationary subsurface solute transport: 1. Unconditional moments, *Water Resour. Res.*, **25**(2), 215–232, 1989a.

Graham, W.D., and D. McLaughlin, Stochastic analysis of nonstationary subsurface solute transport: 2. Conditional moments, *Water Resour. Res.*, **25**(11), 2331–2355, 1989b.

Granger, W.J., and M. Hatanaka, *Spectral Analysis of Economic Time Series*, Princeton University Press, Princeton, NJ, 1964.

Guadagnini, A., and S.P. Neuman, Nonlocal and localized analyses of conditional mean steady state flow in bounded, randomly nonuniform domains: 1. Theory and computational approach, *Water Resour. Res.*, **35**(10), 2999–3018, 1999a.

Guadagnini, A., and S.P. Neuman, Nonlocal and localized analyses of conditional mean steady state flow in bounded, randomly nonuniform domains: 2. Computational examples, *Water Resour. Res.*, **35**(10), 3019–3039, 1999b.

Gutjahr, A.L., Fast Fourier transforms for random field generation, *Project Report for Los Alamos Grant to New Mexico Tech, contract number 4-R58-2690R*, Department of Mathematics, New Mexico Institute of Mining and Technology, Socorro, New Mexico, 1989.

Gutjahr, A.L., and L.W. Gelhar, Stochastic models of subsurface flows: Infinite versus finite domains and stationarity, *Water Resour. Res.*, **17**(2), 337–350, 1981.

Gutjahr, A.L., L.W. Gelhar, A.A.M. Bakr, and J.R. MacMillan, Stochastic analysis of spatial variability in subsurface flows: 2. Evaluation and application, *Water Resour. Res.*, **14**(5), 953–960, 1978.

Harter, Th., *Unconditional and Conditional Simulation of Flow and Transport in Heterogeneous, Variably Saturated Porous Media*, Ph.D. dissertation, University of Arizona, 1994.

Harter, Th., and T.-C.J. Yeh, Stochastic analysis of solute transport in heterogeneous, variably saturated soils, *Water Resour. Res.*, **32**(6), 1585–1595, 1996a.

Harter, Th., and T.-C.J. Yeh, Conditional stochastic analysis of solute transport in heterogeneous, variably saturated soils, *Water Resour. Res.*, **32**(6), 1597–1609, 1996b.

Harter, Th., and D. Zhang, Water flow and solute spreading in heterogenous soils with spatially variable water content, *Water Resour. Res.*, **35**(2), 415–426, 1999.

Hassan, A.E., J.H. Cushman, and J.W. Delleur, Monte Carlo studies of flow and transport in fractal conductivity fields: Comparison with stochastic perturbation theory, *Water Resour. Res.*, **33**(11), 2519–2534, 1997.

Hewett, T.A., Fractal distribution of reservoir heterogeneity and their influence in fluid transport, paper presented at *Annual Technical Conference and Exhibition*, Soc. of Pet. Eng., New Orleans, LA., 1986.

Ho, C.K., S.J. Altman, and B.W. Arnold, Alternative conceptual models and codes for unsaturated flow in fractured tuff: Preliminary assessment for GWTT-95, Sandia National Laboratory Technical Report SAND95-1546, 1995.

Hoeksema, R.J., and P.K. Kitanidis, An application of the geostatistical approach to the inverse problem in two-dimensional groundwater modeling, *Water Resour. Res.*, **20**(7), 1003–1020, 1984.

Hoeksema, R.J., and P.K. Kitanidis, Analysis of the spatial structure of properties of selected aquifers, *Water Resour. Res.*, **21**(4), 563–572, 1985a.

Hoeksema, R.J., and P.K. Kitanidis, Comparison of Gaussian conditional mean and kriging estimation in geostatistical solution of the inverse problem, *Water Resour. Res.*, **21**(6), 825–836, 1985b.

Hsu, K.-C., A general method for obtaining analytical expressions for the first-order velocity covariance in heterogeneous porous media, *Water Resour. Res.*, **35**(7), 2273–2277, 1999.

Hsu, K.-C., and G.L. Lamb Jr., On the second-order correction to velocity covariance for two-dimensional statistically isotropic porous media, *Water Resour. Res.*, **36**(1), 349–353, 2000.

Hsu, K.-C., and S.P. Neuman, Second-order expressions for velocity moments in two- and three-dimensional statistically anisotropic media, *Water Resour. Res.*, **33**(4), 625–637, 1997.

Hsu, K.-C., D. Zhang, and S.P. Neuman, Higher-order effects on flow and transport in randomly heterogeneous porous media, *Water Resour. Res.*, **32**(3), 571–582, 1996.

Hughson, D.L., and T.-C.J. Yeh, An inverse model for three-dimensional flow in variably saturated porous media, *Water Resour. Res.*, **36**(4), 829–840, 2000.

Indelman, P., Averaging of unsteady flows in heterogeneous media of stationary conductivity, *J. Fluid Mech.*, **310**, 39–60, 1996.

Indelman, P., and B. Abramovich, A higher-order approximation to effective conductivity in media of anisotropic random structure, *Water Resour. Res.*, **30**(6), 1857–1864, 1994a.

Indelman, P., and B. Abramovich, Nonlocal properties of nonuniform averaged flows in heterogeneous media, *Water Resour. Res.*, **30**(12), 3385–3393, 1994b.

Indelman, P., and Y. Rubin, Flow in heterogeneous media displaying a linear trend in log conductivity, *Water Resour. Res.*, **31**(5), 1257–1265, 1995.

Indelman, P., D. Or, and Y. Rubin, Stochastic analysis of unsaturated steady state flow through bounded heterogeneous formations, *Water Resour. Res.*, **29**, 1141–1148, 1993.

Indelman, P., A. Fiori, and G. Dagan, Steady flow toward wells in heterogeneous formations: Mean head and equivalent conductivity, *Water Resour. Res.*, **32**(7), 1975–1983, 1996.

Isaaks, E.H., and R.M. Srivastava, *Applied Geostatistics*, Oxford University Press, New York, 1989.

Jaekel, U., and H. Vereecken, Renormalization group analysis of macrodispersion in a directed random flow, *Water Resour. Res.*, **33**(10), 2287–2299, 1997.

James, A.I., and W.D. Graham, Numerical approximation of head and flux covariances in three dimensions using mixed finite elements, *Adv. Water Resour.*, **22**(7), 729–740, 1999.

Journel, A.G., and C.J. Huijbregts, *Mining Geostatistics*, Academic Press, London, 1978.

Kazemi, H., Pressure transient analysis of naturally fractured reservoirs with uniform fracture distribution, *Trans. Soc. Pet. Eng. AIME*, **246**, 451–462, 1969.

Kemblowski, M.W., and J.C. Wen, Contaminant spreading in stratified soils wth fractal permeability distribution, *Water Resour. Res.*, **29**(2), 419–425,1993.

King, P.R., The use of field theoretic methods for the study of flow in a heterogeneous porous medium, *J. Phys. A: Math. Gen.*, **20**, 3935–3947, 1987.

King, P.R., The use of renormalization for calculating effective permeability, *Transp. Porous Media*, **4**, 37–58, 1989.

King, P.R., Upscaling permeability: Error analysis for renormalization, *Transp. Porous Media*, **23**, 337–354, 1996.

Kitanidis, P.K., Prediction by the method of moments of transport in a heterogeneous formation, *J. Hydrol.*, **102**, 453–473, 1988.

Klimenko, A.Y., and R.W. Bilger, Conditional moment closure for turbulent combustion, *Prog. Energy Combust. Sci.*, **25**, 595–687, 1999.

Koudina, N., R.G. Garcia, J.-F. Thovert, and P.M. Adler, Permeability of three-dimensional fracture networks, *Phys. Rev. E*, **57**(4), 4466–4479, 1998.

Kraichnan, R.H., Dynamics of nonlinear stochastic systems, *J. Math. Phys.*, **2**, 124–148, 1961.

Langlo, P., and M.S. Espedal, Macrodispersion for two-phase, immiscible flow in porous media, *Adv. Water Resour.*, **17**(5), 297–316, 1994.

Leslie, D.C., *Developments in the Theory of Turbulence*, Oxford University Press, London, 1973.

Li, B., and T.-C.J. Yeh, Sensitivity and moment analyses of head in variably saturated regimes, *Adv. Water Resour.*, **21**(6), 477–485, 1998.

Li, L., and W.D. Graham, Stochastic analysis of solute transport in heterogeneous aquifers subject to spatially random recharge, *J. Hydrol.*, **206**, 16–38, 1998.

Li, S.-G., and D. McLaughlin, A nonstationary spectral method for solving stochastic groundwater problems: Unconditional analysis, *Water Resour. Res.*, **27**(7), 1589–1605, 1991.

Li, S.-G., and D. McLaughlin, Using the nonstationary spectral method to analyze flow through heterogeneous trending media, *Water Resour. Res.*, **31**(3), 541–551, 1995.

Liedl, R., A conceptual perturbation model of water movement in stochastically heterogeneous soils, *Adv. Water Resour.*, **17**, 171–179, 1994.

Loaiciga, H.A., R.B. Leipnik, P.F. Hubak, and M.A. Marino, Stochastic groundwater flow analysis in heterogeneous hydraulic conductivity fields, *Math. Geol.*, **25**(2), 161–176, 1993.

Loaiciga, H.A., R.B. Leipnik, P.F. Hubak, and M.A. Marino, Effective hydraulic conductivity of nonstationary aquifers, *Stochastic Hydrol. Hydraul.*, **8**(1), 1–18, 1994.

Long, J.C.S., J.S. Remer, C.R. Wilson, and P.A. Witherspoon, Porous media equivalents for networks of discontinuous fractures, *Water Resour. Res.*, **18**(3), 645–658, 1982.

Long, J.C.S., K. Karasaki, A. Davey, J. Peterson, M. Landsfeld, J. Kemeny, and S. Martel, Inverse approach to the construction of fracture hydrology models conditioned by geophysical data: An example from the validation exercises at the Stripa mine, *Int. J. Rock Mech. Min. Sci. Geomech. Abstr.*, **28**(2–3), 121–142, 1991.

Lu, Z., *Nonlocal Finite Element Solutions for Steady State Unsaturated Flow in Bounded Randomly Heterogeneous Porous Media Using the Kirchhoff Transformation*, Ph.D. dissertation, University of Arizona, 2000.

Lumley, J.L., and H.A. Panofsky, *The Structure of Atmospheric Turbulence*, John Wiley & Sons, New York, 1964.

Mantoglou, A., Digital simulation of multivariate two- and three-dimensional stochastic processes with a spectral turning bands method, *Math. Geol.*, **19**, 129–149, 1987.

Mantoglou, A., A theoretical approach for modeling unsaturated flow in spatially variable soils: Effective flow models in finite domains and nonstationarity, *Water Resour. Res.*, **28**(1), 251–267, 1992.

Mantoglou, A., and L.W. Gelhar, Stochastic modeling of large-scale transient unsaturated flow systems, *Water Resour. Res.*, **23**(1), 37–46, 1987.

Mantoglou, A., and J.L. Wilson, The turning bands method for simulation of random fields using line generation by a spectral method, *Water Resour. Res.*, **18**, 1379–1394, 1982.

Markov, K.Z., Application of Volterra–Wiener series for bounding the overall conductivity of heterogeneous media, I. General procedure, *SIAM J. Appl. Math.*, **4**, 831–849, 1987.

Marle, C.M., *Multiphase Flow in Porous Media*, Gulf Publishing Co., Houston, 1981.

Matheron, G., *Elements Pour une Theorie des Milieux Poreux*, Masson, Paris, 1967.

Matheron, G., The intrinsic random functions and their application, *Adv. Appl. Prob.*, **5**, 438–468, 1973.

McLaughlin, D., and E.F. Wood, A distributed parameter approach for evaluating the accuracy of groundwater model predictions: 1. Theory, *Water Resour. Res.*, **24**(7), 1037–1047, 1988a.

McLaughlin, D., and E.F. Wood, A distributed parameter approach for evaluating the accuracy of groundwater model predictions: 2. Application to groundwater flow, *Water Resour. Res.*, **24**(7), 1048–1060, 1988b.

Metropolis, N.C., The beginning of the Monte Carlo method, *Los Alamos Sci.*, Los Alamos National Laboratory, **15**, 125–130, 1987.

Mizell, S.A., A.L. Gutjahr, and L.W. Gelhar, Stochastic analysis of spatial variability in two-dimensional steady groundwater flow assuming stationary and nonstationary heads, *Water Resour. Res.*, **18**(4), 1053–1067, 1982.

Moench, A.F., Double-porosity models for a fissured groundwater reservoir with fracture skin, *Water Resour. Res.*, **20**, 831–846, 1984.

Molz, F.J., H.H. Liu, and J. Szulga, Fractional Brownian motion and fractional Gaussian noise in subsurface hydrology: A review, presentation of fundamental properties, and extensions, *Water Resour. Res.*, **33**(10), 2273–2286, 1997.

Mose, R., P. Siegel, P. Ackerer, and G. Chavent, Application of the mixed hybrid finite element approximation in a groundwater flow model: Luxury or necessity? *Water Resour. Res.*, **30**(11), 3001–3012, 1994.

Naff, R.L., Radial flow in heterogeneous porous media: An analysis of specific discharge, *Water Resour. Res.*, **27**(3), 307–316, 1991.

Naff, R.L., and A.V. Vecchia, Stochastic analysis of three-dimensional flow in a bounded domain, *Water Resour. Res.*, **22**(5), 695–704, 1986.

Neuman, S.P., A statistical approach to the inverse problem of aquifer hydrology: 3. Improved solution method and added perspective, *Water Resour. Res.*, **16**(2), 331–346, 1980.

Neuman, S.P., Stochastic continuum representation of fractured rock permeability as an alternative to the REV and fracture network concepts, in *28th US Symposium on Rock Mechanics*, edited by I.W. Farmer, J.J.K. Daemen, C.S. Desai, C.E. Glass, and S.P. Neuman, Balkema, Rotterdam, The Netherlands, pp. 533–561, 1987.

Neuman, S.P., *A Proposed Conceptual Framework and Methodology for Investigating Flow and Transport in Swedish Crystalline Rocks*, Arbetsapport 88–37, SKB Swedish Nuclear Fuel and Waste Management Co., Stockholm, September, 1988.

Neuman, S.P., Universal scaling of hydraulic conductivities and dispersivities in geologic media, *Water Resour. Res.*, **26**(8), 1749–1758, 1990.

Neuman, S.P., Reply to Comments by M. Anderson on "Universal scaling of hydraulic conductivities and dispersivities in geologic media", *Water Resour. Res.*, **27**(6), 1383–1384, 1991.

Neuman, S.P., Eulerian–Lagrangian theory of transport in space–time nonstationary velocity fields: Exact nonlocal formalism by conditional moments and weak approximations, *Water Resour. Res.*, **29**(3), 633–645, 1993.

Neuman, S.P., Generalized scaling of permeabilities: Validation and effect of support scale, *Geophys. Res. Lett.*, **21**, 349–352, 1994.

Neuman, S.P., Stochastic approach to subsurface flow and transport: A view of the future, in *Subsurface Flow and Transport: A Stochastic Approach*, edited by G. Dagan and S.P. Neuman, pp. 231–241, Cambridge University Press, New York, 1997.

Neuman, S.P., and J.S. Depner, Use of variable-scale pressure test data to estimate the log hydraulic conductivity covariance and dispersivity of fractured granites near Oracle, Arizona, *J. Hydrology*, **102**, 475–501, 1988.

Neuman, S.P., and E.A. Jacobson, Analysis of nonintrinsic spatial variability by residual kriging with application to regional groundwater levels, *Math. Geol.*, **16**(5), 499–521, 1984.

Neuman, S.P., and S. Orr, Prediction of steady state flow in nonuniform geologic media by conditional moments: Exact nonlocal formalism, effective conductivities, and weak approximation, *Water Resour. Res.*, **29**(2), 341–364, 1993.

Neuman, S.P., C.L. Winter, and C.M. Newman, Stochastic theory of field-scale Fickian dispersion in anisotropic porous media, *Water Resour. Res.*, **23**(3), 453–466, 1987.

Neuman, S.P., Tartakovsky, T.C. Wallstrom, and C.L. Winter, Correction to "Prediction of steady state flow in nonuniform geologic media by conditional moments: Exact nonlocal formalism, effective conductivities, and weak approximation", *Water Resour. Res.*, **32**(5), 1479–1480, 1996.

Noetinger, B., The effective permeability of a heterogeneous porous medium, *Transp. Porous Media*, **15**, 99–127, 1994.

Noetinger, B., and Y. Gautier, Use of the Fourier–Laplace transform and of diagrammatical methods to inperpret pumping tests in heterogeneous reservoirs, *Adv. Water Resour.*, **21**, 581–590, 1998.

Oliver, L.D., and G. Christakos, Boundary condition sensitivity analysis of the stochastic flow equation, *Adv. Water Resour.*, **19**(2), 109–120, 1996.

Orr, S., *Stochastic Approach to Steady State Flow in Nonuniform Geologic Media*, Ph.D. dissertation, University of Arizona, 1993.

Orr, S., and S.P. Neuman, Operator and integro-differential representations of conditional and unconditional stochastic subsurface flow, *Stochastic Hydro. Hydraul.*, **8**, 157–172, 1994.

Osnes, H., Stochastic analysis of head spatial variability in bounded rectangular heterogeneous aquifers, *Water Resour. Res.*, **31**(12), 2981–2990, 1995.

Osnes, H., Stochastic analysis of velocity spatial variability in bounded rectangular heterogeneous aquifers, *Water Resour. Res.*, **32**, 203–215, 1996.

Paleologos, E.K., S.P. Neuman, and D. Tartakovsky, Effective hydraulic conductivity of bounded, strongly heterogeneous porous media, *Water Resour. Res.*, **32**(5), 1333–1341, 1996.

Papoulis, A., *Probability, Random Variables and Stochastic Processes*, third edition, McGraw-Hill, New York, 1991.

Peters, R.R., and E.A. Klavetter, A continuum model for water movement in an unsaturated fractured rock mass, *Water Resour. Res.*, **24**, 416-430, 1988.

Pope, S.B., PDF methods for turbulent reactive flows, *Prog. Energy Combust. Sci.*, **11**, 119–192, 1985.

Pope, S.B., Lagrangian PDF methods for turbulent flows, *Annu. Rev. Fluid Mech.*, **26**, 23–63, 1994.

Press, W.H., B.P. Flannery, S.A. Teukolsky, and W.T. Vetterling, *Numerical Recipes: The Art of Scientific Computing*, second edition, Cambridge University Press, New York, 1992.

Priestley, M.B., *Multivariate Series, Prediction and Control, Part B, Vol. 2, Spectral Analysis and Time Series*, Academic Press, San Diego, CA., 1981.

Protopapas, A.L., and R.L. Bras, Uncertainty propagation with numerical models for flow and transport in the unsaturated zone, *Water Resour. Res.*, **26**(10), 2463–2474, 1990.

Pruess, K., J.S.Y. Wang, and Y.W. Tsang, On thermohydrologic conditions near high-level nuclear wastes emplaced in partially saturated fractured tuff: 2. Effective continuum approximation, *Water Resour. Res.*, **26**, 1249–1261, 1990.

Ragab, R., and J.D. Cooper, Variability of unsaturated zone water transport parameters: Implications for hydrological modelling, 1. In-situ measurements, *J. Hydrol.*, **148**, 109–131, 1993a.

Ragab, R., and J.D. Cooper, Variability of unsaturated zone water transport parameters: Implications for hydrological modelling, 2. Predicted vs. in-situ measurements and evaluation methods, *J. Hydrol.*, **148**, 133–147, 1993b.

Rajaram, H., and D. McLaughlin, Identification of large-scale spatial trends in hydraulic data, *Water Resour. Res.*, **26**(10), 2411–2423, 1990.

Rehfeldt, K.R., J.M. Boggs, and L.W. Gelhar, Field study in a heterogeneous aquifer, 3. Geostatistical analysis of hydraulic conductivity, *Water Resour. Res.*, **28**(12), 3309–3324, 1992.

Renard, Ph., and G. de Marsily, Calculating equivalent permeability: A review, *Adv. Water Resour.*, **20**(5–6), 253–278, 1997.

Robin, M.J.L., A.L. Gutjahr, E.A. Sudicky, and J.L. Wilson, Cross-correlated random field generation with the direct Fourier transform method, *Water Resour. Res.*, **29**, 2385–2397, 1993.

Ross, S.M., *Introduction to Probability Models*, sixth edition, Academic Press, New York, 1997.

Rubin, Y., Stochastic modeling of macrodispersion in heterogeneous media, *Water Resour. Res.*, **26**(1), 133–142, 1990.

Rubin, Y., Prediction of tracer plume migration in disordered porous media by the method of conditional probabilities, *Water Resour. Res.*, **27**(6), 1291–1308, 1991.

Rubin, Y., Flow and transport in bimodal heterogeneous media, *Water Resour. Res.*, **31**(10), 2461–2468, 1995.

Rubin, Y., and A. Bellin, The effects of recharge on flow nonuniformity and macrodispersion, *Water Resour. Res.*, **30**(4), 939–948, 1994.

Rubin, Y., and G. Dagan, Stochastic identification of transmissivity and effective recharge in steady groundwater aquifers: 1. Theory, *Water Resour. Res.*, **23**(7), 1185–1192, 1987.

Rubin, Y., and G. Dagan, Stochastic analysis of boundaries effects on head spatial variability in heterogeneous aquifers: 1. Constant head boundary, *Water Resour. Res.*, **24**(10), 1689–1697, 1988.

Rubin, Y., and G. Dagan, Stochastic analysis of boundaries effects on head spatial variability in heterogeneous aquifers: 2. Impervious boundary, *Water Resour. Res.*, **25**(4), 707–712, 1989.

Rubin, Y., and G. Dagan, A note on head and velocity covariances in three-dimensional flow through heterogeneous anisotropic porous media, *Water Resour. Res.*, **28**(5), 1463–1470, 1992.

Rubin, Y., and K. Seong, Investigation of flow and transport in certain cases of nonstationary conductivity fields, *Water Resour. Res.*, **30**(11), 2901–2911, 1994.

Russo, D., Determining soil hydraulic properties by parameter estimation: On the selection of a model for the hydraulic properties, *Water Resour. Res.*, **24**, 453–459, 1988.

Russo, D., Stochastic modeling of macrodispersion for solute transport in a heterogeneous unsaturated porous formation, *Water Resour. Res.*, **29**, 383–397, 1993.

Russo, D., On the velocity covariance and transport modeling in heterogeneous anisotropic porous formations: 1. Saturated flow, *Water Resour. Res.*, **31**(1), 129–137, 1995a.

Russo, D., Stochastic analysis of the velocity covariance and the displacement covariance tensors in partially saturated heterogeneous anisotropic porous formations, *Water Resour. Res.*, **31**(7), 1647–1658, 1995b.

Russo, D., and M. Bouton, Statistical analysis of spatial variability in unsaturated flow parameters, *Water Resour. Res.*, **28**(7), 1911–1925, 1992.

Russo, D., I. Russo, and A. Laufer, On the spatial variability of parameters of the unsaturated hydraulic conductivity, *Water Resour. Res.*, **33**(5), 947–956, 1997.

Sahimi, M., *Flow and Transport in Porous Media and Fractured Rock: From Classical Methods to Modern Approaches*, VCH, New York, 1995.

Salandin, P., and V. Fiorotto, Solute transport in highly heterogeneous aquifers, *Water Resour. Res.*, **34**(5), 949–961, 1998.

Sanchez-Vila, X., Radially convergent flow in heterogeneous porous media, *Water Resour. Res.*, **33**(7), 1633–1641, 1997.

Sanchez-Vila, X., C.L. Axness, and J. Carrera, Upscaling transmissivity under radially convergent flow in heterogeneous media, *Water Resour. Res.*, **35**(3), 613–621, 1999.

Schetzen, M., *The Volterra and Wiener Theories of Nonlinear Systems*, John Wiley & Sons, New York, 1980.

Serrano, S.E., Semianalytical methods in stochastic groundwater transport, *Appl. Math. Modelling*, **16**, 1992.

Shvidler, M.I., *Filtration Flows in Heterogeneous Media*, Consultants Bureau, New York, 1964.

Simmons, C.S., A stochastic-convective transport representation of dispersion in one-dimensional porous media systems, *Water Resour. Res.*, **18**, 1193–1214, 1982.

Smith, L., and R.A. Freeze, Stochastic analysis of steady state groundwater flow in a bounded domain: 2. Two-dimensional simulations, *Water Resour. Res.*, **15**(6), 1543–1559, 1979.

Sudicky, E.A., A natural gradient experiment on solute transport in a sand aquifer: Spatial variability of hydraulic conductivity and its role in the dispersion process, *Water Resour. Res.*, **20**(19), 2069–2082, 1986.

Sun, A.Y., and D. Zhang, Prediction of solute spreading during vertical infiltration in unsaturated, bounded heterogeneous porous media, *Water Resour. Res.*, **36**(3), 715–723, 2000.

Sun, N.-Z., and W.-G. Yeh, A stochastic inverse solution for transient groundwater flow: Parameter identification and reliability analysis, *Water Resour. Res.*, **28**(12), 3269–3280, 1992.

Sykes, J.F., J.L. Wilson, and R.W. Andrews, Sensitivity analysis for steady state groundwater using adjoint operators, *Water Resour. Res.*, **21**(3), 359–371, 1985.

Tartakovsky, D.M., Prediction of steady-state flow of real gases in randomly heterogeneous porous media, *Physica D*, **133**, 463–468, 1999.

Tartakovsky, D.M., and I. Mitkov, Some aspects of head-variance evaluation, *Computat. Geosci.*, **3**, 89–92, 1999.

Tartakovsky, D.M., and S.P. Neuman, Transient flow in bounded randomly heterogeneous domains: 1. Exact conditional moment equations and recursive approximations, *Water Resour. Res.*, **34**(1), 1–12, 1998a.

Tartakovsky, D.M., and S.P. Neuman, Transient flow in bounded randomly heterogeneous domains: 2. Localization of conditional moment equations and temporal nonlocality effects, *Water Resour. Res.*, **34**(1), 13–20, 1998b.

Tartakovsky, D.M., and S.P. Neuman, Transient effective hydraulic conductivity under slowly and rapidly varying mean gradients in bounded three-dimensional random domain, *Water Resour. Res.*, **34**(1), 21–32, 1998c.

Tartakovsky, D.M., S.P. Neuman, and Z. Lu, Conditional stochastic averaging of steady state unsaturated flow by means of Kirchhoff transformation, *Water Resour. Res.*, **35**(3), 731–745, 1999.

Thiele, M.R., R.P. Batycky, M.J. Blunt, and F.M. Orr, Simulating flow in heterogeneous systems using streamtubes and streamlines, *Soc. Petrol. Eng. Res. Eng.*, **11**(1), 5–11, 1996.

Tompson, A.F.B., R. Ababou, and L.W. Gelhar, Implementation of the three-dimensional turning bands random field generator, *Water Resour. Res.*, **25**, 227–2243, 1989.

Tsang, C.-F., and S.P. Neuman, Introduction and general comments on INTRAVAL Phase 2, Working Group 2, Test Cases, *The International Intraval Project, Phase 2*, Paris, NEA/OECD, 1997.

Unlu, K., D.R. Nielsen, and J.W. Biggar, Stochastic analysis of unsaturated flow: One-dimensional Monte Carlo simulations and comparisons with spectral perturbation analysis and field observations, *Water Resour. Res.*, **26**(9), 2207–2218, 1990a.

Unlu, K., D.R. Nielsen, J.W. Biggar, and F. Morkoc, Statistical parameters characterizing the spatial variability of selected soil hydraulic properties, *Soil Sci. Soc. Am. J.*, **54**, 1537–1547, 1990b.

van Genuchten, M.Th., A closed-form equation for predicting the hydraulic conductivity of unsaturated soils, *Soil Sci. Soc. Am. J.*, **44**, 892–898, 1980.

van Kampen, N.G., *Stochastic Processes in Physics and Chemistry*, North-Holland, Amsterdam, 1981.

Van Lent, T., and P.K. Kitanidis, Effects of first-order approximations on head and specific discharge covariances in high-contrast log conductivity, *Water Resour. Res.*, **32**(5), 1197–1207, 1996.

Vanmarcke, E., *Random Fields: Analysis and Synthesis*, MIT Press, Cambridge, MA., 1983.

Warren, J.E., and H.S. Price, Flow in heterogeneous porous media, *Soc. Petrol. Eng. J.*, **1**, 153–169, 1961.

Warren, J.E., and P.J. Root, The behavior of naturally fractured reservoirs, *Soc. Petrol. Eng. J.*, **3**, 245–255, 1963.

Wels, C., and L. Smith, Retardation of sorbing solutes in fractured media, *Water Resour. Res.*, **30**(9), 2547–2563, 1994.

White, I., and M.J. Sully, Macroscopic and microscopic capillary length and time scales from field infiltration, *Water Resour. Res.*, **23**(8), 1514–1522, 1987.

White, I., and M.J. Sully, On the variability and use of the hydraulic conductivity alpha parameter in stochastic treatment of unsaturated flow, *Water Resour. Res.*, **28**(1), 209–213, 1992.

Wilson, K.G., The renormalization group: Critical phenomena and the Kondo problem, *Rev. Mod. Phys.*, **47**(4), 773–840, 1975.

Winter, C.L., and D.M. Tartakovsky, Mean flow in composite porous media, *Geophys. Res. Lett.*, **27**(12), 1759–1762, 2000.

Winter, C.L., C.M. Newman, and S.P. Neuman, A perturbation expansion for diffusion in a random velocity field, *SIAM J. Appl. Math.*, **44**(2), 411–424, 1984.

Wolfram, S., *Mathematica*, second edition, Addison-Wesley, Reading, MA., 1991.

Yaglom, A.M., *Correlation Theory of Stationary and Related Random Functions I: Basic Results*, Springer-Verlag, New York, 1987.

Yang, J., R. Zhang, and J. Wu, An analytical solution of macrodispersivity for adsorbing solute transport in unsaturated soils, *Water Resour. Res.*, **32**, 355–362, 1996.

Yeh, T.-C.J., and J., Zhang, A geostatistical inverse method for variably saturated flow in the vadose zone, *Water Resour. Res.*, **32**(9), 2757–2766, 1996.

Yeh, T.-C., L.W. Gelhar, and A.L. Gutjahr, Stochastic analysis of unsaturated flow in heterogeneous soils: 1. Statistically isotropic media, *Water Resour. Res.*, **21**(4), 447–456, 1985a.

Yeh, T.-C., L.W. Gelhar, and A.L. Gutjahr, Stochastic analysis of unsaturated flow in heterogeneous soils: 2. Statistically anisotropic media with variable α, *Water Resour. Res.*, **21**(4), 457–464, 1985b.

Zeitoun, D.G., and C. Braester, A Neumann expansion approach to flow through heterogeneous formations, *Stochastic Hydrol. Hydraul.*, **5**, 207–226, 1991.

Zhan, H., and S. Wheatcraft, Macrodispersivity tensor for nonreactive solute transport in isotropic and anisotropic porous media: Analytical solutions, *Water Resour. Res.*, **32**(12), 3461–3474, 1996.

Zhang, D., Numerical solutions to statistical moment equations of groundwater flow in nonstationary, bounded heterogeneous media, *Water Resour. Res.*, **34**, 529–538, 1998.

Zhang, D., Quantification of uncertainty for fluid flow in heterogeneous petroleum reservoirs, *Physica D*, **133**, 488–497, 1999a.

Zhang, D., Nonstationary stochastic analysis of transient unsaturated flow in randomly heterogeneous media, *Water Resour. Res.*, **35**, 1127–1141, 1999b.

Zhang, D., and S.P. Neuman, Comment on "A note on head and velocity covariances in three-dimensional flow through heterogeneous anisotropic porous media" by Y. Rubin and G. Dagan, *Water Resour. Res.*, **28**(12), 3343–3344, 1992.

Zhang, D., and S.P. Neuman, Eulerian–Lagrangian analysis of transport conditioned on hydraulic data: 1. Analytical–numerical approach, *Water Resour. Res.*, **31**(1), 39–51, 1995a.

Zhang, D., and S.P. Neuman, Eulerian–Lagrangian analysis of transport conditioned on hydraulic data: 2. Effects of log transmissivity and hydraulic head measurements, *Water Resour. Res.*, **31**(1), 53–63, 1995b.

Zhang, D., and S.P. Neuman, Head and velocity covariances under quasi-steady state flow and their effects on advective transport, *Water Resour. Res.*, **32**(1), 77–83, 1996.

Zhang, D., and A.Y. Sun, Stochastic analysis of groundwater flow in fractured porous media: A double-permeability approach, *Water Resour. Res.*, **36**(4), 865–874, 2000.

Zhang, D., and H. Tchelepi, Stochastic analysis of immiscible two-phase flow in heterogeneous media, *Soc. Petrol. Eng. J.*, **4**(4), 380–388, 1999.

Zhang, D., and C.L. Winter, Nonstationary stochastic analysis of steady-state flow through variably saturated, heterogeneous media, *Water Resour. Res.*, **34**(5), 1091–1100, 1998.

Zhang, D., and C.L. Winter, Moment equation approach to single phase fluid flow in heterogeneous reservoirs, *Soc. Petrol. Eng. J.*, **4**(2), 118–127, 1999.

Zhang, D., T.C. Wallstrom, and C.L. Winter, Stochastic analysis of steady-state unsaturated flow in heterogeneous media: Comparison of the Brooks–Corey and Gardner–Russo models, *Water Resour. Res.*, **34**(6), 1437–1449, 1998.

Zhang, D., L. Li, and H.A. Tchelepi, Stochastic formulation for uncertainty assessment of two-phase flow in heterogeneous reservoirs, *Soc. Petrol. Eng. J.*, **5**(1), 60–70, 2000a.

Zhang, D., R. Andricevic, A.Y. Sun, X. Hu, and G. He, Solute flux approach to transport through spatially nonstationary flow in porous media, *Water Resour. Res.*, **36**(8), 2107–2120, 2000b.

Zhang, J., and T.-C.J. Yeh, An iterative geostatistical inverse method for steady flow in the vadose zone, *Water Resour. Res.*, **33**(1), 63–71, 1997.

Zhang, Q., Transient behavior of mixing induced by a random velocity field, *Water Resour. Res.*, **31**(3), 577–591, 1995.

Zhang, Q., Multi-length-scale theories for scale-up problem and renormalized perturbation expansion, *Adv. Water Resour.*, **20**(5–6), 317–333, 1997.

Zhang, Y.-K., and J. Lin, Numerical solutions of transport of non-ergodic solute plumes in heterogeneous aquifers, *Stochastic Hydro. Hydraul.*, **12**, 117–140, 1998.

Zhu, J., Specific discharge covariance in a heterogeneous semiconfined aquifer, *Water Resour. Res.*, **34**(8), 1951–1957, 1998.

Zhu, J., and J.F. Sykes, Head variance and macrodispersivity tensor in a semiconfined fractal porous medium, *Water Resour. Res.*, **34**(8), 203–212, 2000.

Zimmerman, R.W., G. Chen, T. Hadgu, and G.S. Bodvarsson, A numerical dual-porosity model with semianalytical treatment of fracture/matrix flow, *Water Resour. Res.*, **29**, 2127–2137, 1993.

Zimmerman, R.W., T. Hadgu, and G.S. Bodvarsson, A new lumped-parameter model for flow in unsaturated dual-porosity media, *Adv. Water Resour.*, **19**(5), 317–327, 1996.

Zyvoloski, G.A., B.A. Robinson, Z.V. Dash, and L.L. Trease, Models and methods summary for the FEHMN applications, Los Alamos National Laboratory Report No. LA-UR-94-3787, Rev. 1, March, 1995.

INDEX

Printed and bound by CPI Group (UK) Ltd, Croydon, CR0 4YY

03/10/2024

01040321-0008